GEOMETRODYNAMICS PROCEEDINGS 1985

# GEOMETRODYNAMICS

# PROCEEDINGS 1985

Edited by Agostino Pràstaro

**World Scientific**

*Published by*

**World Scientific Publishing Co Pte Ltd.**
**P. O. Box 128, Farrer Road, Singapore 9128**
242, Cherry Street, Philadelphia PA 19106-1906, USA

**Library of Congress Cataloging in Publication Data**

Main entry under title:

Geometrodynamics proceedings, 1985.

   1. Geometrodynamics — Congresses.  I. Pràstaro, Agostino.
QC173.59.G44G46     1985     530.1'4     85-14605
ISBN 9971-978-63-6



Printed in Singapore by Singapore National Printers (Pte) Ltd.

GEOMETRODYNAMICS PROCEEDINGS 1985

Βουληθεὶς γάρ ὁ θεὸς ἀγαθὰ μὲν
πάντα, φλαῦρον δὲ μηδὲν εἶναι κατὰ δύναμιν, οὕτω δὴ πᾶν
ὅσον ἦν ὁρατὸν παραλαβὼν οὐχ ἡσυχίαν ἄγον ἀλλὰ
κινούμενον πλημμελῶς καὶ ἀτάκτως, εἰς τάξιν αὐτὸ
ἤγαγεν ἐκ τῆς ἀταξίας, ἡγησάμενος ἐκεῖνο τούτου πάντως
ἄμεινον. — Platone, Timeo, 30.[·]

## INTRODUCTION

This book combines the communications of an international meeting on the "Geometric Methods in Non-Linear Field Theory" held in Camigliatello Silano (Cosenza) from 15 to 19 January 1985. Really this was the fifth edition of a series of workshops in Mathematical Physics patronized by the Mathematics Department of the University of Calabria, principally interesting the geometrization of Physics.

The conference consisted of invited lectures and communications that was given in general sessions all about the following main topics: fibered spaces, jet-derivative spaces and connections, symplectic structures and hamiltonian formalism, calculus of variations and lagrangian formalism, Lie pseudogroups and partial differential equations, complex manifolds and algebraic topology, functorial language and theory of geometric objects, gauge theories, super-gravity, graded Lie algebras and super-manifolds, geometric point of views for the quantization of non-linear field theories.

The geometrization of Physics is a field of research where the modern techniques of the mathematics concur to give powerful tools able to clarify some aspects of the physical world. In particular the non-linear field theory is the natural area of application of modern differential geometry. Rather, we can say that these two disciplines interact between them in such a strict way that are mutually stimulated for more general and profound developments.

*Consenza, March 1985*                                   Agostino Pràstaro

THANKS. We would like to thank World Scientific Publishing Co. of Singapore which offered us the opportunity to publish this second edition of the Geometrodynamics Proceedings.[†]

[·]                                    Le Dieu a
voulu que toutes choses fussent bonnes: il a exclu, autant
qu'il était en son pouvoir, toute imperfection, et ainsi, toute
cette masse visible, il l'a prise, dépourvue de tout repos, chan-
geant sans mesure et sans ordre et il l'a amenée du désordre
à l'ordre, car il avait estimé que l'ordre vaut infiniment
mieux que le désordre. Et oncques ne fut permis, oncques
n'est permis au meilleur de rien faire, sinon le plus beau.

[†] The first book was: Geometrodynamics Proceedings (1983), Pitagora Ed., Bologna 1984.

GEOMETRODYNAMICS PROCEEDINGS (1985), pp. 3-24
edited by A. Pràstaro


# A GEOMETRICAL INTERPRETATION OF THE 1-COCYCLES OF A LIE GROUP

S.Benenti and W.M.Tulczyjew

Istituto di Fisica Matematica J.-L.Lagrange
Università di Torino, Torino, Italy.

## 0.- Introduction.

In this paper we give a geometrical interpretation of the 1-cocycles of a Lie group and their relation with the 2-cocycles on the correponding Lie algebra. 1-cocycles are interpreted as classes of 1-forms on the Lie group G satisfying a suitable differential equation. The images of 1-forms of a class give rise to a foliation of the cotangent bundle $T^*G$ of the group. This foliation can be also constructed by means of an invariant closed 2-form on the group, representing a 2-cocycle. This form is used to correct the canonical symplectic structure of $T^*G$ and also to correct the canonical lift of vector fields of G to $T^*G$. In this approach the structure of the cotangent bundle of G plays an essential role. We do not use the natural trivialization of $T^*G$. Instead, we use both Lie algebras of left and right infinitesimal generators and the corresponding dual spaces of right and left-invariant 1-forms. This paper is a continuation of the analysis of the cotangent bundle of a Lie group initiated in [3]. The geometrical interpretation of 1-cocycles and 2-cocycles given here has been suggested by results of an analysis concerning the geometrical interpretations of Hamiltonian actions in terms of coisotropic submanifolds [4][2].

4

## 1.- Notation.

Let G be a Lie group.  We consider a vector field on the manifold G
as a section $X:G \to TG$ of the tangent bundle $\tau_G:TG \to G$ and a 1-form on G as
a section of the cotangent bundle $\pi_G:T^*G \to G$.  The spaces  of smooth vec-
torfields and $k$-forms on  G are denoted by $\mathcal{X}(G)$  and  $\Phi_k(G)$  respectively.
Symbols $[X,Y]$,  $d_X$ and $i_X$ denote the Lie brackets of vector fields X and Y,
the Lie derivative and the interior product with  respect to a vector field
X. We consider the left and the right translations on G:

$$\lambda:G \times G \to G:(g,g') \mapsto \lambda_g(g') = gg',$$
$$\rho:G \times G \to G:(g,g') \mapsto \rho_g(g') = g'g^{-1}.$$

They are smooth left actions of G onto itself.  We denote  by $\ell_G$  and $\tau_G$
the Lie  algebras of  the infinitesimal generators  (vector fields)  of the
actions $\lambda$ and $\rho$ respectively.   The linear dual spaces   $\ell_G^*$  and  $\tau_G^*$  are
identified with the spaces  of right-ivariant and left-invariant 1-forms on
G:

$$\mu \in \tau_G^* \iff \lambda_g^* \mu = \mu \ , \ \forall g \in G \iff d_X \mu = 0, \ \forall X \in \ell_G,$$
$$\nu \in \ell_G^* \iff \rho_g^* \nu = \nu \ , \ \forall g \in G \iff d_Y \nu = 0, \ \forall Y \in \tau_G.$$

Since  the two  actions  commute,  we  have $[X,Y] = 0$ for each $X \in \ell_G$  and
$Y \in \tau_G$.  Consequently: $d_X d_Y - d_Y d_X = d_{[X,Y]} = 0$.

We denote by $\theta_G$  the Liouville 1-form on $T^*G$.  The 2-form $d\theta_G$  is
the canonical symplectic form on $T^*G$.

The  canonical lift  of a diffeomorphism  $\psi:G \to G$ is the unique diffeomorphism $\hat{\psi}:T^*G \to T^*G$ such that

$$(1.1) \quad \hat{\psi}^* \theta_G = \theta_G \ , \quad \pi_G \circ \hat{\psi} = \psi \circ \pi_G.$$

It can be also defined by equation

$$(1.2) \quad \langle v, \hat{\psi}(k)\rangle = \langle T\psi^{-1}(v),k\rangle \qquad (k \in T^*_g G, \ v \in T_{\psi(g)}G),$$

where T denotes the tangent functor. We use the following formula relating $\hat{\psi}$ with the pull-back $\psi^*: \Phi_1(G) \to \Phi_1(G)$:

$$(1.3) \quad (\psi^*\mu)(g) = \hat{\psi}^{-1}(\mu(\psi(g))) \qquad (g \in G, \ \mu \in \Phi_1(G)).$$

The canonical lift of a vector field X on G is the unique vector field $\hat{X}$ on $T^*G$ such that

$$(1.4) \quad d_{\hat{X}}\theta_G = 0 \quad , \quad T\pi_G \circ \hat{X} = X \circ \pi_G.$$

It can be also defined by equation

$$(1.5) \quad i_X d\theta_G = - dE_X,$$

where

$$(1.6) \quad E_X : T^*G \to R : k \mapsto \langle X \circ \pi_G(k),k\rangle.$$

The lifted actions

$$\hat{\lambda} : G \times T^*G \to T^*G : (g,k) \mapsto \hat{\lambda}_g(k),$$
$$\hat{\rho} : G \times T^*G \to T^*G : (g,k) \mapsto \hat{\rho}_g(k),$$

are generated by the canonical lifts $\hat{X}$ and $\hat{Y}$ of the vector fields $X \in \ell_G$ an $Y \in \imath_G$ respectively. The images of the left-invariant (resp. right invariant) 1-forms are the orbits of $\hat{\lambda}$ (resp. of $\hat{\rho}$).

We have two representations of G on the vector space $\Phi_1(G)$:

$$\lambda^*:G \times \Phi_1(G) \rightarrow \Phi_1(G):(g,\nu) \mapsto \lambda_g^{*-1}\nu,$$

$$\rho^*:G \times \Phi_1(G) \rightarrow \Phi_1(G):(g,\mu) \mapsto \rho_g^{*-1}\mu.$$

When restricted to $\ell_G^*$ and to $\mathfrak{r}_G^*$ we have the <u>left</u> and the <u>right</u> <u>coadjoint</u> <u>representations</u> of G respectively.

## 2.- The 1-cocycle differential equation.

We consider the differential equation

$$(2.1) \quad d_X d_Y \gamma = 0 \quad , \quad \forall \ X \in \ell_G, \ Y \in \tau_G,$$

where $\gamma : U \to T^*G$ is a 1-form on an open set $U \subset G$.  This equation is linear in $\gamma$ and symmetric with respect to X and Y.  Sums of left and right-invariant 1-forms are <u>trivial</u> <u>solutions</u> of (2.1).  We call <u>normal</u> <u>solution</u> a solution $\gamma$ in the neighborhood of the identity e of the group G such that $\gamma(e) = 0$.  A trivial normal solution is of the kind $\gamma = \mu - \nu$ with $\mu \in \tau_G^*$, $\nu \in \ell_G^*$, and $\mu(e) = \nu(e)$.  The space of solutions of (2.1) is invariant under addition of left and right-invariant 1-forms and the pull-back actions $\lambda^*$ and $\rho^*$.

By extending local solutions we can obtain global solutions of (2.1).  A <u>global</u> <u>solution</u> is a maximal connected submanifold of $T^*G$ obtained as union of images of local solutions.  The existence and the geometrical meaning of global solutions will be discussed in the next sections.  In the present section we assume that all global solutions are 1-forms defined on all of G.  They form a subspace $S^1(G)$ of $\mathcal{\Phi}_1(G)$.  We denote by $T^1(G)$ the subspace of trivial solutions, i.e. $T^1(G) = \ell_G^* + \tau_G^*$.  The quotient space $H^1(G) = S^1(G)/T^1(G)$ is the space of the equivalence classes of the following equivalence relation in $S^1(G)$:

$$\gamma \sim \gamma' \iff \gamma' = \gamma + \mu + \nu \ , \quad \mu \in \tau_G^*, \ \nu \in \ell_G^*.$$

We denote by $[\gamma]$ the class represented by the solution $\gamma$.  For each $g \in G$ let us consider the subspaces $S_g^1(G) = \{\gamma \in S^1(G); \ \gamma(g) = 0\}$, $T_g^1(G) = \{\gamma \in T^1(G); \ \gamma(g) = 0\} \subset S_g^1(G)$.  In particular, $S_e^1(G)$ is the space of the normal solutions and $T_e^1(G)$ is the space of the trivial normal solutions.  We have natural isomorphisms

$$S_g^1(G) \simeq S_{g'}^1(G) \quad , \quad T_g^1(G) \simeq T_{g'}^1(G),$$

8

defined by means of the representations $\lambda^*$ and $\rho^*$, and also natural isomorphisms

$$S^1(G)/T^1(G) \simeq S^1_g(G)/T^1_g(G).$$

In particular:

$$S^1(G)/T^1(G) \simeq S^1_e(G)/T^1_e(G).$$

PROPOSITION 2.1.- A 1-form $\gamma \in \Phi_1(G)$ is a solution of equation (2.1) if and only if for each $g \in G$ the 1-form $\gamma - \lambda^*_g \gamma$ is right-invariant (resp. $\gamma - \lambda^*_g \gamma$ is left-invariant).

Proof.- Equation (2.1) means that $d_Y \gamma$ is a left-invariant 1-form for each $Y \in \tau_G$. Thus $\lambda^*_g d_Y \gamma = d_Y \gamma$ for each $g \in G$. Since $Y$ is left-invariant, we have $\lambda^*_g d_Y \gamma = d_Y \lambda^*_g \gamma$. It follows that $d_Y(\gamma - \lambda^*_g \gamma) = 0$, for each $Y \in \tau_G$, hence that $\gamma - \lambda^*_g \gamma$ is right-invariant. The reasoning is reversible. (Q.E.D.)

PROPOSITION 2.2.- Let $\gamma$ be a solution of equation (2.1). The mapping

$$(2.2) \quad \theta : G \to \ell^*_G : g \mapsto \gamma - \lambda^*_{g^{-1}} \gamma$$

(resp. $\theta : G \to \tau^*_G : g \mapsto \gamma - \rho^*_{g^{-1}} \gamma$) satisfies equation

$$(2.3) \quad \theta(gg') = \lambda^*_{g^{-1}} \theta(g') + \theta(g)$$

(resp. $\theta(gg') = \rho^*_{g^{-1}} \theta(g') + \theta(g)$) for each $g, g' \in G$, i.e. it is a 1-cocycle of $G$ with respect to the left coadjoint representation on $\ell^*_G$ (resp. right coadjoint representation on $\tau^*_G$). If $\gamma$ is a trivial solution, then $\theta$ is a 1-coboundary.

Proof.- Because of Prop. 2.1 the mapping $\theta$ is well defined. Moreover, $\theta(gg') = \gamma - (\lambda^*_{gg'})^{-1}\gamma = \gamma - \lambda^*_{g^{-1}}(\lambda^*_{g'^{-1}}\gamma) = \gamma - \lambda^*_{g^{-1}}\theta(g') - \lambda^*_{g^{-1}}\gamma = \theta(g) - \lambda^*_{g^{-1}}\theta(g')$. This shows that $\theta$ is a cocycle. If $\gamma = \mu + \nu$, where $\mu \in \tau^*_G$ and $\nu \in \ell^*_G$, then $\theta(g) = \gamma - \lambda^*_{g^{-1}}\gamma = \mu + \nu - \lambda^*_{g^{-1}}(\mu + \nu) = \mu + \nu - \lambda^*_{g^{-1}}\nu - \mu = \nu - \lambda^*_{g^{-1}}\nu$. This shows that $\theta$ is a coboundary. (Q.E.D.)

REMARK 2.3.- If a mapping $\theta : G \to \Phi_1(G)$ is defined as in (2.2), then for each $g \in G$:

$$(2.4) \quad \theta(g)(g) = \gamma(g) - \hat{\lambda}_g(\gamma(e))$$

(we use identity (1.3)).

PROPOSITION 2.4.- Equation

$$(2.5) \quad \gamma(g) = \theta(g)(g) \quad , \quad \forall\, g \in G,$$

defines an isomorphism of the space of 1-forms $\gamma \in \Phi_1(G)$ and the space of mappings $\theta : G \to \ell^*_G$ (or $\theta : G \to \tau^*_G$). The 1-form $\gamma$ is a normal (resp. a trivial normal) solution of equation (2.1) if and only if $\theta$ is a 1-cocycle (resp. a 1-coboundary).

Proof.- If $\gamma \in \Phi_1(G)$ is given, then for each $g \in G$ there exists a unique 1-form $\theta(g) \in \ell^*_G$ whose value at $g$ itself is equal to $\gamma(g)$, because the images of the right-invaraiant 1-forms span a foliation of T*G. Thus mapping $\theta : G \to \ell^*_G$ is well defined by (2.5). Conversely, if a mapping $\theta : G \to \ell^*_G$ is given, then (2.5) defines a unique section $\gamma : G \to $ T*G of $\pi_G$, i.e. a 1-form on G. The correspondence so defined is linear, hence it is an isomorphism. Let $\gamma$ be a normal (resp. trivial normal) solution of (2.1). Because of Prop. 2.2 the mapping defined by $\theta(g) = \gamma - \lambda^{*-1}_g\gamma$ is a 1-cocylce (resp. a coboundary). Because of Remark 2.3 this mapping coincides with that one defined in 2.5 (since $\gamma(e) = 0$). Conversely, let

$\theta : G \to \ell_G^*$ be a 1-cocylce and $\gamma : G \to T^*G$ be the 1-form defined by (2.5). We remark first of all that, because of (2.3), $\theta(e) = 0$; hence $\gamma(e) = 0$. In the following calculation we use identity (1.3) and also identity $\lambda^*_{g^{-1}} \theta(g^{-1}g') = \theta(g') - \theta(g)$ which comes from (2.3) by substituting g' with $g^{-1}g'$:

$$
\begin{aligned}
(\gamma - \lambda^*_{g^{-1}} \gamma)(g') &= \theta(g')(g') - \hat{\lambda}_g(\gamma(g^{-1}g')) \\
&= \theta(g')(g') - \hat{\lambda}_g(\theta(g^{-1}g')(g^{-1}g')) \\
&= \theta(g')(g') - (\lambda^*_{g^{-1}} \theta(g^{-1}g'))(g') \\
&= \theta(g')(g') - (\theta(g') - \theta(g))(g') \\
&= \theta(g)(g').
\end{aligned}
$$

This shows that $\gamma - \lambda^*_{g^{-1}} \gamma = \theta(g) \in \ell_G^*$, hence that $\gamma - \lambda^*_{g^{-1}} \gamma$ is a right-invariant 1-form for each $g \in G$. From Prop. 2.1 it follows that $\gamma$ is a solution of equation (2.1). If $\theta$ is a coboundary (for instance in the coadjoint representation in $\ell_G^*$), then $\theta(g) = \nu - \lambda^*_{g^{-1}} \nu$ where $\nu \in \ell_G^*$. It follows that $\gamma - \nu = \lambda^*_{g^{-1}}(\gamma - \nu)$, i.e. that $\gamma - \nu \in z_G^*$, which means that $\gamma$ is a trivial solution. (Q.E.D.)

CONCLUSION.- The quotient space $S_e^1/T_e^1$, which is isomorphic to the quotient space $H^1(G) = S^1(G)/T^1(G)$, is isomorphic to the first cohomology group of G with respect to the left or right coadjoint representation.

An equivalent interpretation of the mapping $\gamma : G \to T^*G$ representing a 1-cocycle as a Lie group homomorphism is due to Marle (private communication).

## 3.- The lift of a vector field by means of a 2-form.

Let B a 2-form on G. With each vector field X on G we associate a vector field $\bar{X}$ on T*G defined by equation

$$(3.1) \quad i_{\bar{X}} \bar{\omega} = - dE_X,$$

where $E_X$ is defined in (1.6), and

$$\bar{\omega} = d \theta_G + \pi_G^* B,$$

and a vector field $\tilde{X}$ on T*G defined by equation

$$(3.2) \quad i_{\tilde{X}} d \theta_G = - \pi_G^* i_X B.$$

PROPOSITION 3.1.- The vector field $\tilde{X}$ is vertical with respect the projection $\pi_G$, i.e. $T\pi_G \circ \tilde{X} = 0$.

Proof.- If $v \in T_k T^*G$ and $T\pi_G(v) = 0$, then $i_v i_{\tilde{X}} d \theta_G = \langle T\pi_G(v) \wedge X, B \rangle = 0$. Since v is tangent to the fibre of $\pi_G$ at $k \in T^*G$ and the fibre is a Lagrangian submanifold, it follows that also X(k) is tangent to the fibre, hence $T\pi_G(\tilde{X}(k)) = 0$. (Q.E.D.)

PROPOSITION 3.2.- For each vector field X on G we have $\bar{X} = \hat{X} + \tilde{X}$ where $\hat{X}$ is the canonical lift of X. The vector field $\bar{X}$ is $\pi_G$-projectable onto X, i.e. $T\pi_G \circ \bar{X} = X \circ \pi_G$.

Proof.- Since $\tilde{X}$ is vertical, $i_X \pi_G^* B = 0$. Hence, $i_{\hat{X} + \tilde{X}}(d \theta_G + \pi_G^* B) = i_{\hat{X}} d \theta_G + i_{\hat{X}} \pi_G^* B + i_{\tilde{X}} d \theta_G = - dE_X + i_X \pi_G^* B + i_{\tilde{X}} d \theta_G = - dE_X.$ (Q.E.D.)

PROPOSITION 3.3.- For each pair $(X_1, X_2)$ of vector fields on $G$ the following identity holds:

$$(3.3) \quad i_{[\bar{X}_1, \bar{X}_2]} \bar{\omega} = - dE_{[X_1, X_2]} + \pi_G^*(di_{X_2} i_{X_1} B + i_{X_1} i_{X_2} dB).$$

Proof.- We use identities $i_{\hat{X}_1} i_{\hat{X}_2} d\theta_G = - i_{\hat{X}_1} dE_{X_2} = - E_{[X_1, X_2]}$.

$$
\begin{aligned}
i_{\bar{X}_1} i_{\bar{X}_2} \bar{\omega} &= i_{\hat{X}_1 + \tilde{X}_1} i_{\bar{X}_2} \bar{\omega} \\
&= - i_{\hat{X}_1} dE_{X_2} + i_{\tilde{X}_1} i_{\hat{X}_2} d\theta_G \\
&= - E_{[X_1, X_2]} + i_{\hat{X}_2} \pi_G^* i_{X_1} B \\
&= - E_{[X_1, X_2]} + \pi_G^* i_{X_2} i_{X_1} B.
\end{aligned}
$$

$$
\begin{aligned}
i_{[\bar{X}_1, \bar{X}_2]} \bar{\omega} &= (d_{\bar{X}_1} i_{\bar{X}_2} - i_{\bar{X}_2} d_{\bar{X}_1}) \bar{\omega} \\
&= d i_{\bar{X}_1} i_{\bar{X}_2} \bar{\omega} - i_{\bar{X}_2} i_{\bar{X}_1} \pi_G^* dB \\
&= - dE_{[X_1, X_2]} + \pi_G^*(di_{X_2} i_{X_1} B + i_{X_1} i_{X_2} dB)
\end{aligned}
$$

(Q.E.D.)

REMARK 3.4.- Identity (3.4) ca also be written as follows:

$$(3.4) \quad i_{[\bar{X}_1, \bar{X}_2] - \overline{[X_1, X_2]}} \bar{\omega} = \pi_G^*(di_{X_2} i_{X_1} B + i_{X_1} i_{X_2} dB).$$

In the discussion above we use only the differential manifold structure of G. In the next discussion we use the Lie group structure of G. Analogous results hold by exchanging $\lambda$ with $\rho$ and $\ell_G$ with $r_G$.

REMARK 3.5.- If B is closed ($dB = 0$) then the 2-form $\bar{\omega}$ is a symplectic form on $T^*G$ and (3.1) shows that the vector field $\bar{X}$ is the Hamiltonian vector field generated by the function $E_X$ with reference to this symplectic form. In particular from (3.1) it follows that

$$(3.5) \quad d_{\bar{X}} \bar{\omega} = 0.$$

PROPOSITION 3.6.- Let $X \in \ell_G$. The vector field $\bar{X}$ defined by (3.1) has the following characteristic properties: it is $\pi_G$-projectable onto $X$ and

$$(3.6) \quad \langle \bar{X} \wedge \hat{Y}, \bar{\omega} \rangle = 0 \quad , \quad \forall \, Y \in \tau_G.$$

Proof.- If $\bar{X}$ is defined by (3.1), then because of Props. 3.1 and 3.2 we have: $\langle \bar{X} \wedge \hat{Y}, \bar{\omega} \rangle = \langle \hat{X} \wedge \hat{Y}, \pi_G^* B \rangle + \langle \tilde{X} \wedge \hat{Y}, d\theta_G \rangle = i_{\hat{Y}}(\pi_G^* i_X B + i_{\tilde{X}} d\theta_G)$. Thus (3.6) follows from (3.2). Conversely, if $\bar{X}$ is $\pi_G$-projectable onto $X$ and we use the decomposition $\bar{X} = \hat{X} + \tilde{X}$ where $\tilde{X}$ is a vertical vector field, then the above calculation shows that from (3.6) it follows that $i_{\hat{Y}} \zeta_X = 0$ where $\zeta_X \doteq \pi_G^* i_X B + i_{\tilde{X}} d\theta_G$. The 1-form $\zeta_X$ is such that $\langle v, \zeta_X \rangle = 0$ when $T\pi_G(v) = 0$. Since $\tau_G$ is transitive, it follows that $\zeta_X = 0$. (Q.E.D.)

REMARK 3.7.- If B is closed and left-invariant ($d_X B = 0$, for each $X \in \ell_G$), then the canonical lift $\hat{X}$ of $X \in \ell_G$ is a symplectic vector field with respect to the symplectic form $\bar{\omega}$, i.e.:

$$(3.7) \quad d_{\hat{X}} \bar{\omega} = 0.$$

This follows from equation (3.1), which can be written $i_{\hat{X}} \bar{\omega} = - dE_X + \pi_G^* i_X B$, and the fact that $i_X B$ is closed: $d i_X B = d_X B - i_X dB = 0$.

REMARK 3.8.- If B is closed and right-invariant then

$$(3.8) \quad [\bar{X}, \hat{Y}] = 0,$$

for each $X \in \ell_G$ and $Y \in \tau_G$. Both vector fields X and Y are indeed Hamiltonian vector fields with respect to the symplectic structure $\bar{\omega}$. Hence the Lie bracket [X,Y] is the globally Hamiltonian vector field generated by the function $\langle X \wedge Y, \bar{\omega} \rangle$, which is zero because of Prop. 3.6.

PROPOSITION 3.9.- The space of vector fields $\overline{\ell}_G = \{\overline{X}; X \in \ell_G\}$ is a Lie sub-algebra of $\mathcal{X}(T^*G)$ if and only if

(3.9) $\quad di_Y B = d_Y B - i_Y dB = 0, \quad$ for each $Y \in \tau_G$.

Proof.- Let $X_1, X_2 \in \ell_G$ and $Y \in \tau_G$. Because of (3.4) and Prop.3.5, we have:

$$\langle [\overline{X}_1, \overline{X}_2] \wedge \hat{Y}, \overline{\omega} \rangle = i_{\hat{Y}} i_{[\overline{X}_1, \overline{X}_2]} \overline{\omega}$$
$$= i_{\hat{Y}} i_{\overline{[X_1, X_2]}} \overline{\omega} + i_{\hat{Y}} \pi_G^* (di_{X_2} i_{X_1} B + i_{X_1} i_{X_2} dB)$$
$$= 0 + \pi_G^* (i_Y di_{X_2} i_{X_1} B + i_Y i_{X_1} i_{X_2} dB)$$
$$= \pi_G^* i_{X_2} i_{X_1} (d_Y B - i_Y dB).$$

If (3.9) holds, then $[\overline{X}_1, \overline{X}_2]$ is $\pi_G$-projectable onto $[X_1, X_2]$ and (3.6) holds. Thus $[\overline{X}_1, \overline{X}_2] = \overline{[X_1, X_2]}$. Conversely, if $\overline{\ell}_G$ is a sub-algebra, then $[\overline{X}_1, \overline{X}_2]$ is a linear combination of elements of $\overline{\ell}_G$. Because of (3.6), we have $i_{X_2} i_{X_1} (d_Y B - i_Y dB) = 0$, for each $X_1, X_2 \in \ell_G$ and $Y \in \tau_G$. Since $\ell_G$ is transitive, (3.9) follows. (Q.E.D.)

CONCLUSION.- With each 2-form B on G we associate a distribution $\overline{L}_G$ on $T^*G$ of rank equal to the dimension of G. This distribution is determined by the vector fields $\overline{X} \in \overline{\ell}_G$ defined as in (3.1), where $X \in \ell_G$. The distribution is completely integrable if and only if (3.9) holds. In this case the mapping

$$\ell_G \rightarrow \mathcal{X}(T^*G) : X \mapsto \overline{X}$$

is a Lie algebra homomorphism.

## 4.- Global solutions of the cocycle equation.

PROPOSITION 4.1.- The distribution $\overline{L}_G$ corresponding to a 2-form B on G is completely integrable if and only if the equation

$$(4.1) \quad i_X B + d_X \gamma = 0 \quad , \quad \forall \ X \in \ell_G,$$

where $\gamma$ is a 1-form on G, is locally integrable, i.e. for each $g \in G$ there exists a local section $\gamma : U \to T^*G$ of $\pi_G$ satisfying (4.1) such that $g \in U$. The image $\gamma(U)$ of a local solution of (4.1) is an integral manifold of $L_G$. Equation (4.1) is locally integrable and the local solutions satisfy the 1-cocycle equation (2.1) if and only if B is closed and right-invariant, i.e. $dB = 0$ and $d_Y B = 0$ for each $Y \in \mathcal{L}_G$.

LEMMA 4.2.- Let B a 2-form on G and X a vector field on G. The lift $\overline{X}$ of X (definition (3.1)) is tangent to the image $\gamma(G)$ of a 1-form $\gamma$ on G if and only if $i_X B + d_X \gamma = 0$.

Proof.- From $\gamma^*(\theta_G - \pi_G^* \gamma) = \gamma - (\pi_G \circ \gamma)^* \gamma = 0$ it follows that a vector field $\overline{X}$ on $T^*G$ is tangent to the image of the 1-form $\gamma$ if and only if

$$\gamma^* d_{\overline{X}}(\theta_G - \pi_G^* \gamma) = 0.$$

If $\overline{X} = \hat{X} + \tilde{X}$, where X is any vector field on G, then:

$$d_{\hat{X}} \theta_G = 0;$$
$$d_{\tilde{X}} \theta_G = i_{\tilde{X}} d\theta_G + d i_{\tilde{X}} \theta_G = i_{\tilde{X}} d\theta_G = -\pi_G^* i_X B;$$
$$d_{\hat{X}} \pi_G^* \gamma = \pi_G^* d_X \gamma;$$
$$d_{\tilde{X}} \pi_G^* \gamma = 0.$$

Hence, $\gamma^* d_{\overline{X}}(\theta_G - \pi_G^* \gamma) = -\gamma^* \circ \pi_G^*(i_X B + d_X \gamma) = -(i_X B + d_X \gamma)$. (Q.E.D.)

LEMMA 4.3.- Equation (4.1) is equivalent to equation

$$(4.2) \quad i_Y B - d i_Y \gamma = 0, \quad \forall Y \in \tau_G.$$

Proof.- Since X and Y commute, $i_X(d i_Y \gamma - i_Y B) = d_X i_Y \gamma - i_X i_Y B = i_Y(d_X \gamma + i_X B)$.  (Q.E.D.)

Proof of Prop. 4.1.- Let $\bar{L}_G$ be completely integrable.  Since the generating vector fields $\bar{X}$ are transverse to the fibres, integral manifolds of $L_G$ are images of local sections $\gamma : U \to T^*G$ of $\pi_G$.  They satisfy equation (4.1) because of Lemma 4.2.  Conversely, if (4.1) is locally integrable, then from (4.2) it follows that $d i_Y B = 0$, $\forall Y \in \tau_G$, i.e. the condition of complete integrability of $\bar{L}_G$ (Prop. 3.9).  By applying the Lie derivative $d_Y$ to equation (4.1) with $Y \in \tau_G$, since $[X,Y] = 0$ we find:

$$(4.3) \quad i_X d_Y B + d_X d_Y \gamma = 0.$$

If $dB = 0$ and $d_Y B = 0$ for each $Y \in \tau_G$, then $d i_Y B = 0$ and $\bar{L}_G$ is completely integrable.  From (4.3) it follows that $d_X d_Y \gamma = 0$.  Conversely, if $\bar{L}_G$ is completely integrable and $d_X d_Y \gamma = 0$, from (4.3) we derive $i_X d_Y B = 0$, hence: $d_Y B = 0$.  From integrability condition (3.9) we derive also $dB = 0$. (Q.E.D.)

PROPOSITION 4.4.- Let $\gamma$ be a 1-form satisfying the 1-cocycle equation (2.1).  Then equation

$$(4.4) \quad \langle X_1 \wedge X_2, B \rangle = i_{X_1} d_{X_2} \gamma , \quad \forall X_1, X_2 \in \ell_G,$$

defines a closed and right-invariant 2-form B satisfying equation (4.1).

Proof.- The right hand side of (4.4) is bi-linear in $X_1$ and $X_2$.  Let us apply the Lie derivative $d_Y$, with $Y \in \tau_G$, to both sides of (4.1).  We obtain: $\quad d_Y i_{X_1} d_{X_2} \gamma = i_{X_1} d_{X_2} d_Y \gamma = 0$, because of (2.1). Hence: $\langle X_1 \wedge X_2, B \rangle = - i_{X_1} i_{X_2} B = $ const., and B is right-invariant 2-form on G satisfying equation $i_{X_1}(i_{X_2} B + d_{X_2} \gamma) = 0$. It follows that B satisfies also equation (4.3). B is closed because of the last part of Prop. 4.1. (Q.E.D.)

REMARKS.

(a) A leaf (maximal connected integral manifold) of the integrable distribution $\bar{L}_G$ associated with a 2-form B represents a <u>global</u> <u>solution</u> of equation (4.1). In general a leaf $\Gamma$ form a covering of G with respect to the projection $\pi_G$ restricted to $\Gamma$. It follows that if G is connected and simply connected then global solutions are 1-forms on G. The set of global solutions is invariant under addition of left-invariant 1-forms and the action $\hat{\varrho}$. This means that if $\Gamma$ is a leaf, then the sets

(4.5) $\quad \Gamma + \mu = \{ k \in T^*G; \ k = k' + \mu \circ \pi_G(k'), \ k' \in \Gamma \}$

and $\hat{\varrho}_g(\Gamma)$ are also leaves for each $\mu \in \tau_G^*$ and $g \in G$. It follows that there is a unique global solution $\Gamma$ which contains the zero covector $0 \in T_e^*G$ at the identity e of the group, i.e. a unique <u>normal</u> <u>solution</u> (see Section 1).

(b) Closed and right-invariant 2-forms on G are 2-cocycles on $\ell_G$ with respect to the trivial representation of G on $\ell_G$. Let us denote by $[B]_\ell$ the cohomology class determined by B: it is the space of 2-forms $B' = B + dA$ where $A \in \ell_G^*$. If $\Gamma$ is a global solution of (4.1) where B is a 2-cocycle, then $\Gamma$ <u>is a global solution of the 1-cocycle equation</u> (2.1). A global solution corresponding to $B' = B + dA$ is given by $\Gamma - A$ (see definition (4.5)). The set of <u>all</u> global solutions corresponding to the class $[B]_\ell$ is then given by the class $[\Gamma]$ of the global solutions of the type $\Gamma - A + \mu$, where $\mu \in \tau_G^*$ and $A \in \ell_G^*$. Conversely, if $\Gamma$ is a

18

global solution of the 1-cocylce equation (2.1), then we can define a 2-cocycle B through formula (4.4), by any local solution $\gamma$ representing $\Gamma$, for instance in the neighborhood of the identity e of G.

(c) From the discussion above it follows that if (2.1) has only global solutions which are 1-forms, then there is an isomorphism between cohomology classes $[B]_\ell$ and $[\gamma]$.

(d) If G is semi-simple, then the 1-cocycle equation (2.1) has only trivial solutions. Indeed, for each 2-cocycle B we have B = dA with $A \in \ell_G^*$ (Whitehead Lemma); as a consequence, $\gamma' = - A$ is a global solution of (4.1). If $\gamma : U \to T^*G$ is a local solution of (2.1) and B is the 2-cocycle constructed by means of (4.4), then $\gamma - \gamma' | U = \gamma + A | U$ must be a left-invariant form $\mu$ restricted to U. Hence $\gamma = \mu | U - A | U$ and $\gamma$ is trivial.

(e) From equation (4.2) it follows that if $\gamma$ is a local solution corresponding to B, then the function

$$\bar{E}_Y = E_Y - \pi_G^* i_Y \gamma$$

is a local Hamiltonian of Y with respect to the symplectic form $\bar{\omega}$. Hence the vector fields $\hat{Y}$ are globally Hamiltonian if equations (4.1) or (4.2) have global solutions which are 1-forms.

**5.- The lift of actions by means of a 2-form.**

If a foliation of T*G is assigned, whose leaves are images of sections, then any action of G on itself can be lifted to an action on T*G in a natural way. Each infinitesimal generator X of the action on G can be lifted to an infinitesimal generator $\overline{X}$ of the lifted action. The vector field $\overline{X}$ is uniquely defined by the following conditions: (i) it is tangent to the leaves of the foliation, (ii) it is $\pi_G$-projectable onto the vector field X. For example, the left and right translations $\lambda$ and $\varrho$ are canonically lifted to the actions $\hat{\lambda}$ and $\hat{\varrho}$ by means of the foliations determined by the left-invariant and right-invariant forms respectively.

Let us consider the case of the foliation on T*G generated by a 2-cocycle B.

PROPOSITION 5.1.- Let B a 2-cocycle (closed and right-invariant 2-form on G). Let us assume that all global solutions of equation (4.1) are 1-forms and let $\theta:G \to \ell_G^*$ be the mapping defined by (2.2) where $\gamma$ is a solution of (4.1). Then the vector fields $\{\overline{X}; X \in \ell_G\}$ defined by (3.1) are the infinitesimal generators of the action $\overline{\lambda}:G \times T^*G \to T^*G$ defined by

$$(5.1) \quad \overline{\lambda}_g(k') = \hat{\lambda}_g(k') + \theta(g)(gg') \qquad (g' = \pi_G(k'))$$

for each $g \in G$ and $k' \in T^*G$.

Proof.- Let $\overline{\lambda}$ be the lifted action on T*G of the left translation $\lambda$ by means of the foliation $\overline{L}_G$ spanned by the vector fields $\overline{X}$. According to the remarks above, the vector fields $\overline{X}$ are in fact infinitesimal generators of $\overline{\lambda}$. Because of Prop.4.1 and by definition of $\overline{\lambda}$ we have:

$$\overline{\lambda}_g(k') = \gamma(gg'),$$

where $\gamma$ is the solution of equation (4.1) such that $k' = \gamma(g')$. On the other hand (see identity (1.3) and definition (2.2)):

$$\hat{\lambda}_g(k') + \theta(g)(gg') = \hat{\lambda}_g(\gamma(g')) + \theta(g)(gg')$$
$$= (\lambda_g^{*-1}\gamma + \theta(g))(gg')$$
$$= \gamma(gg'),$$

and (5.1) follows. (Q.E.D.)

REMARK 5.2.- From Remarks 3.5, 3.7 and 3.8 it follows that the actions $\bar{\lambda}$ and $\hat{\rho}$ are symplectic on $(T^*G, \bar{\omega})$ and commute.

For the pair of actions $(\bar{\lambda}, \hat{\rho})$ properties analogous to those considered in [3] for $(\hat{\lambda}, \hat{\rho})$ hold, with respect to the symplectic structure $\bar{\omega} = d\theta_G + \pi_G^*B$ on $T^*G$. In particular, the orbits of the composed action $(G \times G) \times T^*G \to T^*G : (g_1, g_2, k) \mapsto \bar{\lambda}_{g_1}\hat{\rho}_{g_2}(k)$ form a (generalized) coisotropic foliations of $T^*G$ and the corresponding reduced symplectic manifolds can be identified with the orbits of the affine action on $\ell_G^*$ corresponding to the 1-cocycle $\theta$ [7]. We do not deal with this topic here for the sake of brevity. We mention that, in a different approach, actions $\bar{\lambda}$ and $\bar{\rho}$ have been already considered in [4,5].

## 6.- Central extensions.

Let us consider a central extension  of the Lie group G by the group R, i.e.  an exact sequence of homomorphisms of Lie groups:

$$1 \to R \xrightarrow{\varepsilon} F \xrightarrow{\eta} G \to 1.$$

The homomorphism $\eta : F \to G$ is a principal fibre bundle with structural group R  whose action  on  F is  defined by  $R \times F \to F : (r,f) \mapsto \varepsilon(r)f$.  Let V be corresponding infinitesimal generator (the fundamental vector field).  This vector  field belongs  to the center of  both  Lie  algebras $\ell_F$  and  $\tau_F$: $[V,Z] = 0$, for each $Z \in \ell_F \cup \tau_F$.

Each infinitesimal generator $Z \in \ell_F$  (resp. $Z \in \tau_F$)  is $\eta$-projectable onto an infinitesimal generator of  $\ell_G$ (resp. of $\tau_G$).

A <u>connection</u> of $\eta$  is a 1-form  $\alpha \in \Phi_1(F)$  such that $i_V \alpha = 1$ and $d_V \alpha = 0$.  Since $i_V d\alpha = 0$ and  $d_V d\alpha = 0$,  there  exists  a  unique  2-form $B \in \bar{\Phi}_2(G)$  such that $d\alpha = \eta^*B$.  The closed 2-form B is the <u>curvature</u> of $\alpha$.

With  a subspace  $\delta$  of  $\ell_F$  complementar  to  V  we  associate  a $\tau_F$-invariant connection $\alpha$ defined by  equations $i_V \alpha = 1$ and $i_Z \alpha = 0$ for each $Z \in \delta$ .  ($\tau_F$-invariant means  right-invariant with  reference to the group F,  etc..) The corresponding curvature B is $\tau_G$-invariant.  If $\alpha'$ is another $\tau_F$-invariant connection and  B'  is  the corresponding  curvature, then $\alpha' - \alpha = \eta^*A$,  where  A is  a $\tau_G$-invariant 1-form  and  $B'- B = dA$. Hence,  with  a central extension  of G by R  we  associate a distinguished cohomology class [B].

The submanifold C of $T^*F$ defined by

$$C = \{h \in T^*F; \; \langle V,h \rangle = 1\}$$

is coisotropic (because it  is of codimension 1).   A **surjective submersion**

$\varkappa : C \to T^{*}G$ is defined by

(6.1) $\quad \langle v, \varkappa(h) \rangle = \langle w, h \rangle$,

where $h \in C$, $v \in T_g G$, $g = \eta(f)$, $f = \pi_F(h)$, and w is the horizontal lift of v, i.e. the vector defined by equations: $\langle w, \alpha \rangle = 0$, $T\eta(w) = v$. The fibres of $\varkappa$ coincide with the orbits of the canonical lift $\hat{\eta}$ of the action $\eta$ [6]. These orbits are characteristics of C (i.e. maximal connected integral submanifolds of the characteristic distribution). Moreover, $\varkappa^{*}(d\theta_G + \pi_G^{*}B)$ is equal to the pull-back $d\theta_F|C$ of the symplectic form $d\theta_F$ to the submanifold C. This means that $\varkappa$ defines a symplectic reduction from $(T^{*}F, d\theta_F)$ to the symplectic manifold $(T^{*}G, \overline{\omega})$ where $\overline{\omega} = d\theta_G + \pi_G^{*}B$.

PROPOSITION 6.1.- Let $\overline{\gamma}$ be a $\tau_F$-invariant or a $\ell_F$-invariant 1-form such that $i_V \overline{\gamma} = 1$. There is a unique 1-form $\gamma$ on G such that

(6.2) $\overline{\gamma} - \alpha = \eta^{*}\gamma$,

(6.3) $\varkappa \circ \overline{\gamma}(F) = \gamma(G)$.

If $\overline{\gamma}$ is $\tau_F$-invariant, then $\gamma$ is a $\tau_G$-invariant 1-form. If $\overline{\gamma}$ is $\ell_F$-invariant, then $\gamma$ is a 1-form representing the 1- cocycle associated with the curvature B, i.e. satisfying equation (4.1).

Proof.- We note that for a $\tau_F$-invariant 1-form $\overline{\gamma}$ we have $i_Z \overline{\gamma} = $ const. for each $Z \in \ell_F$. In particular $i_V \overline{\gamma} = $ const., so that hypothesis $i_V \overline{\gamma} = 1$ is formulated correctly. Since $\overline{\gamma}$ id $\tau_F$-invariant we have $d_Z \overline{\gamma} = 0$ for each $Z \in \tau_F$. In particular $d_V \overline{\gamma} = 0$ ($\overline{\gamma}$ is a connection of the principal fiber bundle $\eta$ ). It follows that $i_V(\overline{\gamma} - \alpha) = 0$ and $d_V(\overline{\gamma} - \alpha) = 0$, hence that there exists a 1-form $\gamma$ on G such that $\overline{\gamma} - \alpha = \eta^{*}\gamma$. The same reasoning holds when $\overline{\gamma}$ is $\ell_F$-invariant. From the definition (6.1) of $\varkappa$ it follows that $\langle v, \varkappa(\overline{\gamma}(f))\rangle = \langle w, \overline{\gamma}(f)\rangle = \langle w, \alpha(f) + \eta^* \gamma(f)\rangle = \langle v, \gamma\rangle$, since $\langle w, \alpha\rangle = 0$. This proves (6.3). If $\overline{\gamma}$ is $\tau$-invariant, then by applying to (6.2) the Lie derivative $d_Z$ with respect to a vector field $Z \in \tau_F$ we obtain $0 = d_Z \eta^* \gamma = \eta^* d_X \gamma$, where $X$ is the element of $\tau_G$ onto which $Z$ projects. It follows that $\gamma$ is $\tau_G$-invariant. If $\overline{\gamma}$ is $\ell_F$-invariant, then the same operation with $Z \in \ell_F$ yields the equation $- d_Z \alpha = \eta^* d_X \gamma$, where $X \in \ell_G$. Since $i_Z \alpha = 1$, $d_Z \alpha = i_Z d\alpha = i_Z \eta^* B = \eta^* i_X B$. Hence, we find $- i_X B = d_X \gamma$ for each $X \in \ell_G$. (Q.E.D.)


CONCLUSION.- A $\tau_F$-invariant connection $\alpha$ associated with the central extension defines a symplectic reduction from $(T^*F, d\theta_F)$ to $(T^*G, \overline{\omega})$, where $\overline{\omega}$ is the canonical symplectic form varied by the curvature B of $\alpha$. We can <u>reduce</u> any connection 1-form $\overline{\gamma}$ on F to a 1-form $\gamma$ on G. The reduction of the space $\ell_F^*$ of the $\tau_F$-invariant 1-forms is just the space $\ell_G^*$ of the $\tau_G$-invariant 1-forms, while the reduction of the space $\tau_F^*$ of the $\ell_F$-invariant 1-forms is the space of solutions of equation (4.1), i.e. the space of 1-forms representing the 1-cocycles associated with the 2-cocycle B.

24

**Bibliography.**

[1] - S.BENENTI,
Homogeneous Formulation of Hamiltonian Group Actions,
Proccedings of "4th Meeting on Mathematical Physics ,
Coimbra, October 15-16, 1984.

[2] - S.BENENTI, W.M.TULCZYJEW,
Momentum Relations for Hamiltonian Actions,
Proccedings of "Geometrie Symplectique et Mécanique , Colloque de la
Region Mediterranée de la S.M.F., Montpellier, 14-15 Mai 1984,
(forthcoming).

[3] - S.BENENTI, W.M.TULCZYJEW,
Sur un feuilletage coisotrope du fibré cotangent d'un groupe de Lie,
C.R.A.S. Paris, 300 (1985), 119-122.

[4] - P.LIBERMAN, C.-M.MARLE,
Géométrie symplectique base théorique de la mécanique,
Publications Mathématiques de l'Université de Paris VII,
(forthcoming).

[5] - C.-M.MARLE,
Moment de l'action hamiltonienne d'un groupe de Lie; quelques
propriétés,
Proceedings of the Meeting "Geometry and Physics",
Firenze 1982, Pitagora (Bologna, 1984), 117-133.

[6] - M.R.MENZIO, W.M.TULCZYJEW,
Infinitesimal symplectic relations and generalized Hamiltonian
dynamics,
Ann. Inst. H.Poincaré 28 (1978) 349-367.

[7] - J.M.SOURIAU,
Structures des systèmes dynamiques,
Dunod (Paris, 1970).

* * *

This research is supported by Consiglio Nazionale delle Ricerche.

---

Received May 9, 1985.

GEOMETRODYNAMICS PROCEEDINGS (1985), pp. 25-42
edited by A. Pràstaro


# SUPERMANIFOLDS AND SUPERGRAVITIES *(o)*

Yvonne Choquet-Bruhat
*Université de Paris VI*

Supersymmetry is an essential step towards the unification of the fundamental laws of physics; its aim is to represent in one multiplet bosonic and fermionic fields. The theories of supergravities which have been proposed in recent years offer dynamical models for classical (that is non quantum) such unified fields. Their mathematical coherence, equivalent to the existence of an infinitesimal "supersymmetry" of the lagrangian, relies on an identity which holds true only if the Fermi fields take anticommuting values, while the Bose fields take commuting values.

Fundamental definitions, with different point of views, for the mathematical structures relevant to such graded valued fields have been given by B. Kostant [11] and B. DeWitt [1] [8]. We shall in this article, inspired by these works give a set of definitions, which meets the two following essential requirements :

1) give a meaning to the lagrangians of supergravities as $C^1$ mappings from a locally convex topological vector space (space of $C^1$ sections of a graded bundle over ordinary d-dimensional space time) into another locally convex topological space (the fundamental graded algebra).

2) write the dynamical equations, expressing that the fields are a critical point of the lagrangian, by annuling sections of graded vector bundles, and deducing from them a hierarchy of ordinary partial differential equations.

Finally we give some comments and references on applications to supergravity.

---

*(o)* Text of a lecture given at the colloquium "Geometry and Physics" organized in Cosenza, 15-19 January 1985 by A. Prastaro.

[1] Work in the same spirit than DeWitt, but with variants which we shall use in parts, has been done by A. Rogers [12]. Further references are given in DeWitt excellent book [8].

We do not use in this article the "superspace" formalism : the lagrangian is obtained by integration over space time, an ordinary manifold modelled over $\mathbb{R}^d$. Integration over "anticommuting variables" would appear in quantization, with the Feynman path integral method.


## 1 - Graded bundles

A commutative <u>graded algebra</u> $\mathcal{U}$ is a vector space over the field R of real numbers which is the direct sum of two subspaces $\mathcal{V}_+$ (called even) and $\mathcal{V}_-$ (called odd)

$$1\text{-}1 \qquad\qquad \mathcal{U} = \mathcal{V}_+ \oplus \mathcal{V}_-$$

endowed with an associative and distributive operation, called product, such that two odd elements anticommute and even elements commute with all others [1] :

$$ab = (-1)^{d(a)d(b)}ba \qquad , \qquad a,b \in \mathcal{U}$$

where $d(a)$, degree of a, is zero if a is even (a $\in \mathcal{V}_+$) and one if a is odd (a $\in \mathcal{V}_-$) We always suppose that $\mathcal{U}$ admits a unit e; $\mathcal{U}$ can be finite or infinite dimensional. We suppose $\mathcal{U}$ endowed with a Hausdorf locally convex topology (it is uniquely determined if $\mathcal{U}$ is finite dimensional). The product $(a,b) \mapsto ab$ is assumed to be continuous from $\mathcal{U} \times \mathcal{U}$ into $\mathcal{U}$ and the sum 1-1 is assumed to be a topological sum. The fundamental example which we shall use in this article is the algebra $\mathcal{A}$ of formal series with a denumerable set of generators $\zeta^i$, i=1,2,...,N (possibly N = $\infty$) and the anticommuting property

$$1\text{-}2 \qquad\qquad \zeta^i \zeta^j = - \zeta^j \zeta^i \quad .$$

An element a of $\mathcal{A}$ is written

$$a = a_B + a_S$$

where (DeWitt)

$$a_B = a(0) = a_o e$$

---

[1] It is a special case of a $\mathbb{Z}^2$-graded algebra where $\mathcal{V}_r \mathcal{V}_s \subset \mathcal{V}_{r+s \bmod 2}$ with $\mathcal{V}_0 = \mathcal{V}_+$ and $\mathcal{V}_1 = \mathcal{V}_-$. We will in the following often omit the word "commutative", since our graded algebras will always be "graded commutative".

is called the body of a and

$$a_S = \sum_{p=1}^{N} a(p) \quad , \quad a(p) = \frac{1}{p!} a_{i_1..i_p} \zeta^{i_1}.. \zeta^{i_p}$$

is called the soul of a. The coefficients $a_o$ and $a_{i_1..i_p}$ (totally anticommuting in $i_1.. i_p$) are elements of $I\!R$ (real numbers), or eventually $C$ (complex numbers). We have

$$\mathcal{A} = \mathcal{A}_+ \oplus \mathcal{A}_-$$

where $\mathcal{A}_+$ [resp. $\mathcal{A}_-$] is constituted by formal series which contain only terms with p even [resp. p odd].

$\mathcal{A}$ is endowed with a locally convex Hausdorf topology by the family of semi-norms

$$\|a\|_{i_1..i_p} = |a_{i_1..i_p}| \quad :$$

the elements of a basis of neighborhoods of zero in $\mathcal{A}$ are given by one in equality

$$|a_{i_1..i_p}| < \varepsilon_{i_1..i_p}$$

an open set containing zero contains a set determined by a <u>finite</u> number of such inequalities (cf for instance Choquet-Bruhat and DeWitt Morette [5] p.424).
The product of formal series and the sum $\mathcal{A}_+ \oplus \mathcal{A}_-$ have the required continuity, as well as the inverse $a \mapsto a^{-1}$, when $a_B \neq 0$, which is then given by

$$a^{-1} = a_B^{-1} (e + \sum_n (-1)^n (a_S/a_o)^n) \quad .$$

A <u>graded bundle</u> $M_{\mathcal{A}}$ (denoted $\mathcal{M}$) is an $\mathcal{A}$ bundle over an ordinary (Hausdorf, para-compact) d-dimensional, $C^\infty$ manifold M : there exists a projection $p : (M,\mathcal{A}) \to M$ and a covering of M by open sets $\omega_\alpha$ such that there exist homeomorphisms $\phi_\alpha :$ $p^{-1}(\omega_\alpha) \to \omega_\alpha \times \mathcal{A}$ where : $\pi^{-1}(x) \to (x,\phi_\alpha(\pi^{-1}(x))$ and $\phi_\alpha(\pi^{-1}(x)) = \mathcal{A}$ , and the mapping $\phi_\alpha \circ \phi_{\alpha'}^{-1} : \mathcal{A} \to \mathcal{A}$ is for each $x \in \omega_\alpha \cap \omega_{\alpha'}$ deduced from a linear mapping of the generating space $(\zeta^i)$, $i = 1..N$ into itself by

$$\zeta^{i'} = \sum_j L_j^{i'}(x) \zeta^j$$

where $L^{i'}_j$ is different from zero only for a finite [1] number of indices j (when
i' is given). The bundle $(M, \mathcal{A})$ is $C^k$ if the mappings $x \mapsto L^{i'}$ are all $C^k$ mappings
from $\omega_\alpha \cap \omega_{\alpha'}$ into $\mathbb{R}$ .

Let, on the other hand, $(\omega_\beta, \varphi_\beta)$ be an atlas of coordinate charts on M. The open
sets $\omega_\alpha$ are a covering of M and the triples $(\omega_\alpha \cap \omega_\beta , \phi_\alpha, \varphi_\beta)$ are called <u>graded
charts</u> of the graded bundle $(M, \mathcal{A})$. We have the homomorphism

$$p^{-1}(\omega_\alpha \cap \omega_\beta) \to \varphi_\beta(\omega_\alpha \cap \omega_\beta) \times \mathcal{A}$$

by

$$y \mapsto (\varphi_\beta(p(x)) , \phi_\alpha(y) ) \quad .$$

<u>Note</u> A (commutative) graded sheaf in the sense of Kostant is a sheaf A of (commutative) graded algebras over a $C^\infty$ manifold M which projects onto the sheaf of $C^\infty$ functions on M.

It has been proved by M. Batchelor [2] in the case where the sheaf is locally trivial and N finite (that is there exists a covering of M by open sets $\omega_\alpha$ and isomorphisms $A|_{\omega_\alpha} \to C^\infty(\omega_\alpha) \otimes \mathcal{A}$, with a finite number N of generators in $\mathcal{A}$) that a Kostant
sheaf is isomorphic with a sheaf of sections of a graded bundle in the previous
sense.


## 2 - Graded valued fields.

A section $f : M \to \mathcal{A}$ is called a graded valued function on M. If $x \in \omega \subset M$, and
$(\omega, \phi)$ is a graded chart on $\mathcal{A}$, then $\phi \circ \varphi^{-1}$ is a map from the open set $\varphi(\omega) \subset \mathbb{R}^d$
into $\mathcal{A}$, which may be written

$$(x^\alpha)) \to \sum_p \frac{1}{p!} f_{i_1..i_p}(x^\alpha) \, \zeta^{i_1}.. \, \zeta^{i_p}$$

The function f is $C^k$ at x if the mapping $\omega \subset \mathbb{R}^d \to \mathcal{A}$ defined above is $C^k$ at x :
this will be the case, due to the topology given to $\mathcal{A}$, if each of the functions
$(x_\alpha) \to f_{i_1..i_p}(x^\alpha)$ are $C^k$ (cf a reminder on $C^k$ mappings between locally convex topological vector spaces in §2). The definition is chart independant if the manifold
M and the bundle $\mathcal{A}$ are $C^p$ with $p \geq k$.

------

[1] This condition insures that formal series in $\zeta^i$ are mapped into formal series
in $\zeta^{i'}$. A pair $(\omega, \phi)$ is called a graded chart. Definitions of equivalent atlases -
defining the same graded bundles - are the usual ones.

The simplest case of a graded bundle on M is the product M $\times$ $\mathcal{A}$ . The graded valued
functions are then the $\mathcal{A}$ valued functions on M. They generalize the real valued
functions, while graded valued functions generalize sections of vector bundles over
M with typical fiber $\mathbb{R}$.

We can  extend usual vector (or affine) bundles over M to graded bunldes. Indeed
let E $\to$ M be a vector bundle over M, with typical fiber $\mathbb{R}^n$, and group $G \subset Gl(n)$ (or
an affine group) : there exists a covering of M by open sets $\omega$ and homeomorphisms

$$e : \pi^{-1}(\omega) \to \omega \times \mathbb{R}^n \qquad by \quad y \to (x, e(y))$$

with $x = \pi(y)$, and such that then the transition functions

$$e_{\omega\omega'} = e \circ e'^{-1}$$

are smooth maps from  $\omega \cap \omega'$ into G.

We define a graded extension $\mathcal{E}$ of E to be a vector bundle over M with typical fiber
$\mathcal{A}^n$ and the same transition functions $e_{\omega\omega'}$. In particular if E $\to$ M is a vector (or
affine) bundle associated with a principal bundle of frames over M (linear frames,
orthonormal frames, spin frames, ...) by a representation r we define the graded
extension $\mathcal{E}$ $\to$ M as follows : the fiber $\pi^{-1}(x)$ is the set of equivalence classes
$(\rho,t)$ with $\rho$ a frame at x and $t \in (p^{-1}(x))^r$ (recall that $p^{-1}(x)$ is the fiber at x
in $\mathcal{M}$, isomorphis to $\mathcal{A}$). The equivalence relation is given by

$$(\rho,t) \simeq (\rho',t') \qquad if \qquad \rho' = g\rho \quad , \quad t' = r(g)t$$

<u>Remark</u> It is possible to define $\mathcal{M}_+$ [resp. $\mathcal{M}_-$], by restricting $p^{-1}(x)$ to $p_+^{-1}(x)$
[resp. $p_-^{-1}(x)$], elements whose image in a graded chart are even [resp. odd], the
definition is chart independand, due to the definition of the admissible transition
functions. It is thus possible to define even valued, or odd valued or (m,q) valued
extensions (with typical fiber $\mathcal{A}_+^m \times \mathcal{A}_-^q$)

<u>Examples</u> : we take $\mathcal{M}$ = M $\times$ $\mathcal{A}$

1. An $\mathcal{A}$ valued tangent vector at x $\in$ M is an equivalence class $(\rho,V)$ with $\rho$ a frame
in the ordinary tangent space at x, $V \in \mathcal{A}^d$ and the equivalence relation

$$V' = g^{-1} V \qquad if \qquad \rho' = g\rho \qquad , g \in Gl(d)$$

2. The differential of a $C^1$, $\mathcal{A}$ valued function at x $\in$ M is a covariant $\mathcal{A}$ valued

vector, because

$$(\partial_\alpha(f \circ \varphi^{-1})) = g(\partial_\alpha{}'(f \circ \varphi'^{-1})) \quad \text{if} \quad (\partial_\alpha) = g(\partial_\alpha{}')$$

3. Suppose that M is an ordinary pseudo riemannian manifold for instance of Lorentz signature which admits a spin structure. Then an $\mathcal{A}$ valued spinor [1] at x is an equivalence class $(\rho,\Lambda,s)$ where $\rho$ is now an orthonormal frame at x, $\Lambda$ an element of the spinor group, $s \in \mathcal{A}^{[d/2]}$ and the equivalence relation is

$$s' = \Lambda' \Lambda^{-1} s \qquad\qquad \rho' = L\rho ,$$

L the Lorentz transformation associated with $\Lambda' \Lambda^{-1}$.

$C^k$ fields of geometric objects are defined as usual by $C^k$ sections of the corresponding bundles.

<u>Proposition</u> An $\mathcal{A}$ valued $C^\infty$ tangent vector field v defines an endomorphism of the algebra $C^\infty(M,\mathcal{A})$ of $\mathcal{A}$ valued $C^\infty$ functions on M, by (the result is chart independant, due to the definitions)

$$v(f) = v^\alpha \, \partial_\alpha(f \circ \varphi^{-1})$$

where $(\omega,\varphi)$ is a chart at x and $(v^\alpha)$, $\alpha = 1,\ldots,d$ the representant of v in the chart.

This endomorphism is

(i)   additive

$$v(f + g) = v(f) + v(g)$$

(ii)   zero on constant functions $v(a) = 0$ if a is the constant functions $M \to a \in \mathcal{A}$

(iii) obeys the graded leibniz rule when v and f have a degree [2]

$$v(fg) = v(f)g + (-1)^{d(v)d(f)} f\, v(g)$$

Conversely every endomorphism of $C^\infty(M,\mathcal{A})$ which satisfies (i),(ii), and (iii) for even f is an $\mathcal{A}$ valued vector field on M.

<u>Proof</u> The properties of $v(f)$ result from a straightforward computation.

To prove the converse we use (i) and (ii) to show

$$v(f) = \Sigma \frac{1}{p!} \, v((f_{i_1..i_p}\, e)\, \zeta^{i_1}\ldots\, \zeta^{i_p}$$

---

[1] The algebra $\mathcal{A}$ is, for a general spinor, with $a_{i_1..i_p}$ complex numbers.
[2] i.e are either even or odd. A vector v is even [resp. odd] if its components $v^\alpha$ are all even [resp. odd] : the definition is chart independant.

We use then the Taylor formula for the numerical functions $f_{i_1..i_p} \circ \varphi^{-1}$ as in the classical case to obtain

$$v(f_{i_1..i_p} e) \;=\; v^\alpha \partial_\alpha (f_{i_1...i_p} \circ \varphi^{-1})$$

with, at a point $x_0 \in M$

$$v^\alpha \;=\; v((x^\alpha - x_0^\alpha)e)$$

<u>Remark</u>  It is possible to define "supertangent" vectors to M as endomorphisms of $C^\infty(M, \mathcal{A})$ satisfying (i) and (iii) but satisfying (ii) only in the case where $a = \lambda e$, $\lambda \in \mathbb{R}$. We deduce as before from (i), (iii) and the restricted (ii)' that $v(f)$ is known at $x_0$ for all $f$ when the $v((x^\alpha - x_0^\alpha)e)$ are known and also now the $v(\zeta^i)$. We denote, by analogy with classical notations, by $\partial/\partial\zeta^j$ the endomorphism of $C^\infty(M, \mathcal{A})$ defined by

$$\frac{\partial}{\partial\zeta^j}(\zeta^i) \;\overset{def}{=}\; \delta^i_j\, e \quad , \quad \frac{\partial}{\partial\zeta^j}((x^\alpha - x_0^\alpha)e) = 0$$

Then every supertangent vector can be written, at $x_0 \in M$

$$v = v^\alpha \frac{\partial}{\partial x^\alpha} + v^i \frac{\partial}{\partial\zeta^i} \quad , \text{ with } v^\alpha = v((x^\alpha - x_0^\alpha)e) \in \mathcal{A}$$
$$v^i = v(\zeta^i) \in \mathcal{A} \quad .$$


### 2 - Supersmoothness

Before giving the definition of supermanifolds we need to give the definition of supersmooth mappings from $(\mathcal{A}_+)^m \times (\mathcal{A}_-)^q$ into itself. Since $\mathcal{A}$ (and also $\mathcal{A}_+$ and $\mathcal{A}_-$) are locally convex topological vector spaces there is a standard definition of a $C^k$ mapping from an open set $U \subset \mathcal{A}$ into $\mathcal{A}$ : the mapping

$$f : \quad U \to \mathcal{A}$$

is differentiable at $a \in U \subset \mathcal{A}$ if there exists a continuous linear mapping

$$f'(a) : \mathcal{A} \to \mathcal{A}$$

such that, for all $h$ in some neighborhood $V$ of zero in $\mathcal{A}$ :

$$f(a + h) - f(a) \;=\; f'(a) \cdot h + o(h)$$

where $o(h)$ has the property that for any neighborhood of zero $u$ in $\mathcal{A}$ and any $h \in V$ there exists $\varepsilon$ such that $t^{-1} o(th) \subset u$ if $|t| < \varepsilon$. The mapping $f$ is $C^1$ on $U$ if it is differentiable at each point of U and the mapping $U \to \mathcal{L}(\mathcal{A},\mathcal{A})$ by $a \to f'(a)$ is continuous [1].

The mapping $f$ is $C^k$ if $f^{(k-1)}$ is $C^1$.

The following definition uses now the algebraic structure of $\mathcal{A}$ : note that a left [resp. right] linear mapping $\ell : \mathcal{A} \to \mathcal{A}$ is given by the product with an element of $\mathcal{A}$,

$$\ell : a \to a\,\ell(e) \qquad [\text{resp. } a \to \ell(e)a]$$

<u>Definition 1</u> The differentiable mapping $f : U \to \mathcal{A}$, $U \subset \mathcal{A}$, is said to be left [resp. right] hyperdifferentiable at $a \in U$ if its derivative $f'(a)$ is also left [resp. right] $\mathcal{A}$ linear, thus there exists $u \in \mathcal{A}$ such that :

$$2\text{-}1 \qquad f'(a).\,h = hu \qquad [\text{resp. } uh] \quad , \quad \forall h \in \mathcal{A} \ .$$

where the juxtaposition denotes the product in $\mathcal{A}$ .

$f$ is hyperdifferentiable in U if it is hyperdifferentiable at each point of U.

<u>Lemma</u> If $f$ is a $C^2$ mapping from $U \subset \mathcal{A}$ into $\mathcal{A}$ which is left [resp. right] hyperdifferentiable then the derivative $f''(a) \in \mathcal{L}(\mathcal{A} \times \mathcal{A} ,\mathcal{A})$ is also such that

$$2\text{-}2 \qquad f''(a).(h,k) = h\,k\,v$$

where $v \in \mathcal{A}$.

<u>Proof</u> (cf an analogous idea in Jadczyk and Pilch [10])

We know from the classical theory the symmetry property

$$2\text{-}3 \qquad f''(a).(h,k) = f''(a).(k,h)$$

The first derivative being given by the left (an analogous proof would hold for the right $\mathcal{A}$ linear map $h \mapsto h\,f'(a)$ the second derivative is given by

$$2\text{-}4 \qquad (h,k) \mapsto h(f''(a).k)$$

where $f''(a)$ is the derivative at $a$ of the mapping $U \to \mathcal{A}$ by $a \to f'(a)$. The mapping 2-4 is left $\mathcal{A}$ linear with respect to $h$ thus, by the symmetry property 2-3 it is

---

[1] If X and Y are Hausdorf locally convex topological vector spaces the same is true of $\mathcal{L}(X,Y)$, the semi-norms on $\mathcal{L}(X,Y)$ being $\| \ell \|_{i,j} = \sup\limits_{\|x\|_i} \| \ell(x) \|_j$

also k left linear and we have

$$2\text{-}5 \qquad f''(a).(h,k) = hk\, f''(a).(e,e) = kh\, f''(a).(e,e)$$

The equality 2-5 shows on the other hand that the definition 2-1 is too restrictive.

In fact if N is infinite we deduce from 2-5, satisfied for all $h,k \in \mathcal{A}$, $a \in U$ that

$f''(a).(e,e) = 0$, that is $f''(a) = 0$ on U, $f'(a) = $ constant if U is connected :

<u>Theorem</u> The $C^2$ and hyperdifferentiable mappings are locally affine mappings if N

is infinite.

This theorem, and the fact that even for finite N the polynomial mappings are not

hyperdifferentiable at an arbitrary point, show that this notion is too restrictive

to be useful.

The following definition uses the splitting of $\mathcal{A}$ into even and odd parts.

<u>Definition 2</u>  A $C^1$ mapping $f : U \to \mathcal{A}$ is called $G^1$, or graded $C^1$, in U if for each

$a \in U$ there exists elements u and v of $\mathcal{A}$ such that

$$2\text{-}6 \qquad f'(a).h = h_+ u + h_- v \qquad , \qquad \forall\, h = h_+ + h_- \in \mathcal{A} = \mathcal{A}_+ \oplus \mathcal{A}_-$$

We denote

$$u = \partial_+ f(a) = f'(a).e \quad , \qquad v = \partial_- f(a)$$

Remark 1 : We can also write 2-6

$$f'(a).h = u\, h_+ + (v_+ - v_-)\, h_-$$

thus a mapping is simultaneously left and right graded $C^1$, though the left and right

$\partial_-$ derivatives are not necessarily equal.

Remark 2 : If N is finite $v(a)$ is not unique : its term of order N in $\mathcal{A}$ is arbi-

trary (its product by any $h_-$ is zero).

Remark 3 : If f is $G^1$ then $f'(a)$ is $\mathcal{A}_+$ linear but not $\mathcal{A}_-$ linear (even graded),

in fact, by 2-6

$$f'(a).bh = b(f'(a).h) \qquad\qquad \text{if } b \in \mathcal{A}_+$$

while

$$f'(a).bh = -\, bh_- u + bh_+ v \qquad\qquad \text{if } b \in \mathcal{A}_-$$

<u>Definition 3</u> : A $C^2$ mapping $U \to \mathcal{A}$ is called $G^2$ at $a \in U$ if the second derivative

$f''(a)$, symmetric bilinear map $\mathcal{A} \times \mathcal{A} \to \mathcal{A}$ is such that $f''(a).(h,k)$ is a quadratic

34

form in $h_+,k_+,h_-,k_-$ with coefficients in $\mathcal{A}$ ; we write :

2-7    $f''(a).(h,k) = h_+k_+ \; \partial_{++}f(a) + h_+k_- \; \partial_{+-}f(a) + h_-k_+ \; \partial_{-+}f(a)$

It results from the known symmetry $f''(a).(h,k) = f''(a).(k,h)$ that $\partial_{+-}f = \partial_{-+}f$ and that there is no term in $h_-k_-$, that is $\partial_{--}f = 0$.

Conversely we can prove :

<u>Theorem</u> : A $C^2$ and $G^1$ mapping $U \to \mathcal{A}$ is also $G^2$ at a $\in$ U if $f''(a).(h_-,k_-) = 0$.

Proof : since f is $G^1$ we have

$$f'(a).h = h_+ \, u(a) + h_- \, v(a)$$

thus

2-8    $f''(a).(h,k) = h_+(u'(a).k) + h_-(v'(a).k)$

replacing k by $k_+ + k_-$ and using the symmetry in h and k we obtain

2-9    $$h_+(u'(a).k_+) = k_+(u'(a).h_+)$$

and

2-10    $h_+(u'(a).k_-) + h_-(v'(a).k_+) = k_+(u'(a).h_-) + k_-(v'(a).h_+)$

taking $k_+ = e$ in 2-9 gives

$$u'(a).h_+ = h_+ u'(a)$$

$k_+ = e$ and $h_+ = 0$ in 2-10 gives

$$u'(a).h_- = h_-(v'(a).e)$$

therefore if

$$f''(a).(h_-,k_-) \equiv h_-(v'(a).k_-) = 0$$

then $f''(a).(h,k)$ is given by a quadratic form with coefficients in $\mathcal{A}$ :

$$\partial_{++} f(a) = u'(a).e \quad , \quad \partial_{+-} f(a) = \partial_{-+} f(a) = v'(a).e$$

Analogous definitions and theorems are valid for $G^k$ ($k \leq \infty$) mappings.

<u>Proposition</u> : To an entire function $\mathbb{R} \to \mathbb{R}$ given by

2-11    $$f : x \to \sum_{n=0}^{\infty} c_n \, x^n$$

corresponds a $G^\infty$ mapping $\mathcal{A} \to \mathcal{A}$ by

$$f : a \to \sum_{n=0}^{\infty} c_n \, a^n$$

Proof :

1. The mapping is defined from $\mathcal{A}$ into $\mathcal{A}$ : a term $f_{i_1..i_p} \zeta^{i_1}..\zeta^{i_p}$ in $f(a)$ is well defined because it contains only a finite number of terms of the formal series which defines a, and the series 2-11 is absolutely convergent for all x.

2. f is $C^\infty$ in the locally convex topology, and $G^\infty$, indeed :

$$f(a) = \sum_{n=0}^{\infty} \left( c_n (a_+)^n + n \, c_n (a_+)^{n-1} \, a_- \right)$$

$$f'(a).h = h_+ \, \Sigma \, c_n (a_+^{n-1} + n(n-1) \, a_+^{n-2} \, a_-) + h_- \, \Sigma \, n \, c_n \, a_+^{n-1}$$

A $G^k$ mapping, with k large enough for the application at hand, is also called super-smooth.

The definition of supersmoothness extends in a standard way to mappings from $\mathcal{A}^n$ into $\mathcal{A}^p$ or, more generally from $\mathcal{A}_+^m \times \mathcal{A}_-^q$ into $\mathcal{A}_+^{m'} \times \mathcal{A}_-^{q'}$ : such a $C^1$ mapping is graded, or super $C^1$, if the derivative is given by an $(m+q) \times (m'+q')$ matrix with elements in $\mathcal{A}$, that is there exists elements of $\mathcal{A}_+^{m'} \times \mathcal{A}_-^{q'}$ denoted $\partial_{+i} f$, $\partial_{-j} f$ such that

$$f'(a).h = \sum_{i=1}^{m} \partial_{+i} f \, h_+^i + \sum_{j=1}^{q} \partial_{-j} f \, h_-^j$$

$$h_+ = (h_+^i) \in \mathcal{A}_+^m \quad , \qquad (h_-^j) = h_- \in \mathcal{A}_-^q$$

## 3 - Supermanifolds

A supermanifold is based on $(\mathcal{A}_+^m \times \mathcal{A}_-^q)$ in an analogous way as an ordinary manifold is based on $\mathbb{R}^n$

<u>Definition 1</u> : A supermanifold X of class $G^k$ is a Hausdorf topological space with a covering by open sets $U_I$, each homeomorphic to an open set of $\mathcal{A}_+^m \times \mathcal{A}_-^q$ by

$$\varphi_i : U_i \to \varphi_i(U_i) \subset \mathcal{A}_+^m \times \mathcal{A}_-^q$$

such that the transition functions

$$\varphi_i \circ \varphi_j^{-1} : \varphi_j(U_i \cap U_j) \to \varphi_i(U_i \cap U_j)$$

are $G^k$ diffeomorphisms between open sets of $\mathcal{A}_+^m \times \mathcal{A}_-^q$ (that is bijective maps which are $G^k$ as well as their inverses).

Two $G^k$ supermanifolds with the same underlying topological space are considered
as identical if the reunion of the two "atlases" $(U_i, \varphi_i)$ and $(U'_i, \varphi'_i)$ which define
them is agian a $G^k$ atlas.

<u>Definition 2</u> : A mapping $f : X \to X'$ between two $G^k$ supermanifolds is of class GP
at $x \in X$ if the mapping $\varphi' \circ f \circ \varphi^{-1}$ is a GP mapping at $\varphi(p)$ between open sets
of the modeling spaces $\mathcal{A}^m_+ \times \mathcal{A}^q_-$ and $\mathcal{A}^{m'}_+ \times \mathcal{A}^{q'}_-$. The definition is chart
independant if $p \leqq k$.

A <u>supersmooth diffeomorphism</u> $f : X \to X'$ is a bijective mapping such that $f$ and $f^{-1}$
are supersmooth (that is of some class GP with $X$ and $X'$ of some class $G^k$, $k \geqq p$.
A <u>superfibered manifold</u> $(Y, \pi, X, M)$ over the ordinary manifold $M$ is a Hausdorf topological space $Y$ with a continuous projection $\pi : Y \to M$, a covering of $M$ by open
sets $\omega_\alpha$ , and homeomorphisms :

$$\varphi_\omega : \pi^{-1}(\omega) \to \omega \times X \quad \text{by } y \to (x, \varphi_\omega(y)), \quad x = \pi(y)$$

where $X$ is a given supermanifold (the typical fiber).
The transition mappings $\varphi_{\omega\omega'} = \varphi_\omega \circ \varphi_{\omega'}^{-1}$ are smooth maps from $\omega \cap \omega' \subset M$ into the
space of supersmooth diffeomorphisms of $X$.
The graded bundles considered in §1 are superfibered manifolds.


## 4 - $C^1$ sections of a graded bundle

The differential at a point $x \in M$ of a section $f : M \to \mathcal{E}$ of a graded bundle $\mathcal{E}$ over
$M$ is a linear map $f'(x) : T_x M \to T_{f(x)} \mathcal{E}$ which can be defined as follows $(^1)$ :
Let $(\pi^{-1}(\omega), \varphi, \phi, \epsilon)$ be a chart around $\pi^{-1}(x) \subset \mathcal{E}$ admissible for its structures :
$(\omega, \varphi)$ is a chart at $x$ on $M$ and $(p^{-1}(\omega), \phi)$ is a chart for the fundamental graded
bundle $\mathcal{H}$ over $M$. In these charts the mapping $f$ is represented by a mapping from
an open set of $\mathbb{R}^d$ into $\mathbb{R}^m_c \times \mathbb{R}^q_a$ given by $f_\omega = \phi \circ \epsilon \circ f \circ \varphi^{-1}$. The mapping $f$ is
differentiable at $x$ if $f_\omega$ is differentiable at $\varphi(x)$, the definition is chart inde-
pendant due to the differentiability of the various bundle structures $(^1)$. The set
of partial derivatives $(\frac{\partial f_\omega}{\partial x^\alpha} (x^\lambda))$ gives a representant of the differential at $x$,
it is an element of $\mathbb{R}^d \otimes (\mathcal{A}^m_+ \times \mathcal{A}^q_-)$.

---

$(^1)$ We could also use the language of jets, but representants in charts are more
convenient for the writing of equations on $M$.

The set $\bigcup_{x \in M} f'(x)$ is not endowed naturally with a vector space structure by this representation [1]. Indeed by a change of coordinates in $\omega \cap \omega' \subset M$, from $(x^\alpha) = \varphi(x)$ to $(x^{\alpha'}) = \varphi'(x)$ and change of vector bundle maps, we have :

$$f_{\omega'} = \phi' \circ \varepsilon' \circ f \circ \varphi'^{-1} = (\phi' \circ \phi^{-1}) \circ (\varepsilon' \circ \varepsilon^{-1}) \circ f_\omega \circ (\varphi \circ \varphi'^{-1})$$

and the set $(\dfrac{\partial f_\omega}{\partial x^\alpha})$ is not deduced from $(\dfrac{\partial f_{\omega'}}{\partial x^{\alpha'}})$ by a linear map if $\varepsilon' \circ \varepsilon^{-1}$ or $\phi' \circ \phi^{-1}$ are not constant maps. However, we have :

<u>Lemma</u>  The $C^1$ sections of a graded bundle $\mathcal{E} \to M$ with typical fiber $\mathcal{A}_+^m \times \mathcal{A}_-^q$ are $C^0$ sections of a superfibered manifold $\mathcal{E}_1$ with base M and typical fiber $(\mathcal{A}_+^m \times \mathcal{A}_-^q)^{d+1}$ .

Proof - Let $(\pi^{-1}(\omega_\alpha), \varphi_\alpha, \phi_\alpha, \varepsilon_\alpha)$ be an admissible atlas for $\mathcal{E}$ . We define a set $F = \bigcup_\alpha F_{\omega_\alpha}$, with

$$F_\omega = \{(x,t,t_1) \ , \ x \in \omega \ , \ t \in \mathcal{A}_+^m \times \mathcal{A}_-^q \ , \ t_1 \in (\mathcal{A}_+^m \times \mathcal{A}_-^q)^d \}$$

We define the set $\mathcal{E}_1$ as the quotient of F by the equivalence relation :

4-1 $\qquad F_\omega \ni (x,t,t_1) \simeq (x,\tilde{t},\tilde{t}_1) \in F_{\tilde{\omega}}$

if

4-2 $\qquad \tilde{t} = \tilde{\phi} \circ \tilde{\varepsilon} \circ \tilde{\varepsilon}^{-1} \circ \phi^{-1} t \ , \quad \tilde{t}_1 = \tilde{\phi} \circ \tilde{\varepsilon} \circ \varepsilon^{-1} \circ \phi^{-1} t_1 + (\tilde{\phi} \circ \tilde{\varepsilon} \circ \varepsilon^{-1} \circ \phi^{-1})' t$

where the notations $\varepsilon^{-1} \circ \phi^{-1} t_1$, $t_1 = (t_1^\alpha)$, $\alpha = 1,\ldots d$, $t_1^\alpha \in \mathcal{A}_+^m \times \mathcal{A}_-^q$ denotes $(\varepsilon^{-1} \phi^{-1} t_1^\alpha)$ and the notation $(\tilde{\phi} \circ \tilde{\varepsilon} \circ \varepsilon^{-1} \circ \phi^{-1})$ denotes the derivative of the mapping $x \to (\tilde{\phi}(x) \circ \tilde{\varepsilon}(x) \circ \varepsilon^{-1}(x) \circ \phi^{-1}(x)) = (\tilde{\varepsilon} \circ \varepsilon^{-1})(x) (\tilde{\phi} \circ \phi^{-1})(x)$ .

We shall endow $\mathcal{E}_1$ with a superfibered manifold structure (in fact a graded vector bundle, though not associated with the bundle of linear frames) by the atlas whose charts have as a domain the quotient of one set $F_\omega$ by the equivalence relation $\qquad$, with corresponding homeomorphism onto $\Omega \equiv \varphi(\omega) \times (\mathcal{A}_+^m \times \mathcal{A}_-^q)^{d+1}$ the mapping which assigns to an equivalence class its representant in $F_\omega$. We verify that such an atlas endows $\mathcal{E}_1$ with a supermanifold structure by checking that the transition functions, given by the relation 4-2, are supersmooth (indeed linear on the super "coordinates"); we must also, to complete the proof, check that our definition endows $\mathcal{E}_1$ with a compatible topology, that is the sets $\Omega \cap \tilde{\Omega}$ are open in $\mathbb{R}^d \times (\mathcal{A}_+^m \times \mathcal{A}_-^q)^{d+1}$ -

---

[1] Note that the mappings $\phi$ and $\varepsilon \circ \varepsilon'^{-1}$ commute.

which is true since $\varphi(\omega) \cap \tilde{\varphi}(\tilde{\omega})$ is open in $\mathbb{R}^d$ - and that this topology is Hausdorf, which is a consequence of the fact that M is Hausdorf.

To define a topology on $C^1(M, \mathcal{E})$ we use the fact that M is paracompact, and thus admits an atlas with a countable number of charts, each of which has a relatively compact domain $\omega$. We then define semi norms on $C^1(M, \mathcal{E})$ by :

$$\| f \|_{\omega, i, j, \alpha} = \underset{x \in \omega}{\mathrm{Sup}} \{ \| f_\omega(\varphi x) \|_i + \| \frac{\partial}{\partial x^\alpha} f_\omega(\varphi x) \|_j \}$$

where $\| \ \|_i$ denotes one semi norm in $\mathcal{A}$ of one component of an element of $\mathcal{A}^m_+ \times \mathcal{A}^q_-$. The semi-norms endow $C^1(M, \mathcal{E})$ with a locally convex Hausdorf topology. It is independant of the choice of the atlas on M, with the required properties.


5 - Lagrangians.

A lagrangian $\mathcal{L}$ is a mapping from $C^1(U, \mathcal{E})$, U open set in M into the fundamental algebra $\mathcal{A}$ which is given by integration over $U \subset M$ (with respect to some volume element on M) of a mapping from U into $\mathcal{A}$ which is a convolution of maps :

$$x \mapsto f(x) \mapsto L(x, f(x), f'(x))$$

with $f \in C^1(U, \mathcal{E})$ and $L : \mathcal{E}_1 \to \mathcal{A}$ a mapping from the superfibered manifold $\mathcal{E}_1$ into $\mathcal{A}$.

Since $\mathcal{A}$ is a given locally convex space, and M an ordinary d-manifold, the integration is classically known. In particular the integral exists if the mapping $U \to \mathcal{A}$ is continuous and bounded [1] on U and U is relatively compact.

We denote by $C^1(\overline{U}, \mathcal{E})$ sections of $\mathcal{E}$ over U which are continuous and bounded [2] as well as their first derivatives.

If L is a continuous and bounded mapping from $\mathcal{E}_1$, whose restriction to each $\pi^{-1}(x)$ is $G^1$, with derivative continuous and bounded with respect to x, then

$$\mathcal{L} : C^1(\overline{U}, \mathcal{E}) \to \mathbb{R}_{ca} \qquad \text{by} \quad f \mapsto \mathcal{L}(f) \equiv \int_U (L \circ f)(x) \, d\mu(x)$$

---

[1] a mapping $f : U \to \mathcal{A}$ is bounded if for each semi norm $\| \ \|_{i_1 \ldots i_p}$ in $\mathcal{A}$ we have $\underset{x \in U}{\mathrm{Sup}} \| f(x) \|_{i_1 \ldots i_p} < + \infty$. The integral of $f = \underset{p}{\Sigma} f_{i_1 \ldots i_p} \zeta^{i_1} \ldots \zeta^{i_p}$ is

$$\int_U f \, d\mu(x) = \underset{p}{\Sigma} \left( \int_U f_{i_1 \ldots i_p} \, d\mu(x) \right) \zeta^{i_1} \ldots \zeta^{i_p}$$

[2] Definition deduced from semi-norms of representatives.

(d$\mu$(x) smooth volume element on M), is differentiable and its derivative at f a

linear map $C^1(\overline{U},\mathcal{E}) \to \mathbb{R}_{ca}$ by h $\to \mathcal{L}'(f).h$ with

$$\mathcal{L}'(f).h = \int_U L'(x,f(x);f'(x)).(h(x),h'(x))\ dx$$

where $L'(x,f(x),f'(x))$ denotes the linear mapping $\pi_1^{-1}(x) \to \mathcal{A}$ which is the deriva-

tive of L at the point $(x,f(x),f'(x))_1 \in \mathcal{E}_I$ in the direction of the fiber $\pi^{-1}(x)$

and $(h(x),h'(x))$ is a point of this fiber. It results from the hypothesis "L is

$G^1$ when restricted to a fiber" that $x \to L'(x,f(x),f'(x))$ can also be interpreted

as a section of a superfibered manifold, graded vector space dual of $\mathcal{E}_1$. Indeed

let us choose the section h $\in C^1(\overline{U},\mathcal{E})$ such that it has its support in the domain

$\omega$ of a chart of M. We denote by $(f_\omega, f'_\omega)$ the representant of $(f,f')$ in the correspon-

ding chart of $\mathcal{E}_1$, and by $L_\omega$ the representant of L. By hypothesis $L_\omega$ is a $G^1$ mapping

$(\mathcal{A}_+^m \times \mathcal{A}_-^q)^{d+1} \to \mathcal{A}$ and we have :

$$\mathcal{L}'(f).h = \int_\omega (\frac{\partial L_\omega}{\partial_+ f_\omega}\ h_{+\omega} + \frac{\partial L_\omega}{\partial_- f_\omega}\ h_{-\omega} + \frac{\partial L_\omega}{\partial_+ f'_\omega}\ h'_\omega + \frac{\partial L_\omega}{\partial_- f'_\omega}\ h'_\omega)\ d\mu_\omega(x)$$

The quantities denoted $h_+$ and $h_-$ are respectively in $C^1(\omega,\mathcal{A}_+^m)$ and $C^1(\omega,\mathcal{A}_-^q)$,

$h_{+\omega} + h_{-\omega}$ is the representant of h in the chart (recall, cf §2 , that $\frac{\partial L_\omega}{\partial_+ f_\omega}$ , or

other similar expressions, denote finite 1 X m (or 1 X m(d+1), or 1 X q, or 1 X q

(d+1)) matrices with elements in $\mathcal{A}$ acting on a vector with components in $\mathcal{A}$ by

the usual law and $\mathcal{A}$ product ; they depend continuously and boundedly on x by the

hypothesis made on L).

If {support h} $\subset \omega \cap \tilde{\omega}$ , we obtain two expressions for the integral, equal by

hypothesis.

If we suppose moreover that f is $C^2$ and h has compact support in $\omega$ we can integrate

by parts and obtain :

$$\mathcal{L}'(f).h = \int_\omega (Eul_+(f_\omega)\ h^+ + Eul_-(f_\omega)\ h_\omega^-)\ d\mu_\omega$$

with :

5-1
$$Eul_+(f_\omega) \equiv \frac{\partial L_\omega}{\partial_+ f_\omega} - div(\frac{\partial L_\omega}{\partial_+ f'_\omega}) = 0$$

5-2
$$Eul_-(f_\omega) \equiv \frac{\partial L_\omega}{\partial_- f_\omega} - div(\frac{\partial L_\omega}{\partial_- f'_\omega}) = 0$$

The expressions are respectively m X m matrices and q X q matrices, $\mathcal{A}$ valued :

detailed with indices we have, for instance (with summation on $\alpha$, det $\mu$ ordinary

40

density of the volume element)

$$\frac{\partial L_\omega}{\partial_+ f_\omega^i} - \frac{1}{(\det \mu)^{1/2}} \frac{\partial}{\partial x^\alpha} \frac{\partial (L_\omega (\det \mu)^{1/2}}{\partial_+(\partial f_\omega^i / \partial x^\alpha)} = 0 \quad , \qquad \begin{array}{l} i = 1, \ldots\ m \\[4pt] \alpha = 1, \ldots, d \end{array} \quad .$$

As in the ordinary calculus of variations we have the :

<u>Lemma</u> - Let a be a continuous mapping $\omega \to \mathcal{A}$

a) If we have :

$$\int_\omega a(x)\ b(x)\ d\mu(x) = 0$$

for every continuous mapping $b : \omega \to \mathcal{A}_+$ then $a = 0$ on $\omega$

b) If the integral vanishes for every continuous $b : \omega \to \mathcal{A}_-$ and N is infinite
we still have $a = 0$. If N is finite we may have $a_{i_1..i_N} \neq 0$.

Proof : $a(x) = \sum_p a_{i_1..i_p}(x)\ \zeta^{i_1}..\ \zeta^{i_p}$ with $a_{i_1..i_p}$ continuous functions on $\omega$.

a) Take $b(x) = b_o(x)$ e  the classical result proves that $a_{i_1..i_p} = 0$, for all p.

b) Take $b = b(x)\ \zeta^i$ with successively $i = 1,..,N$, the vanishing of the $\mathcal{A}$ valued
integral implies the vanishing of all integrals

$$\int_\omega a_{i_1..i_q}(x)\ b(x)\ d\mu(x)$$

thus $a_{i_1..i_q} = 0$ for all q if N is infinite. If N is finite the product $\zeta^{i_1}..\zeta^{i_N}\ \zeta^i$
is identically zero, thus $a_{i_1..i_N}$ is arbitrary.

We deduce from this lemma and the previous results the conclusion :

<u>Theorem</u> - A necessary condition for the mapping $f \in C^2(\overline{U}, \mathcal{E})$ to be a critical point
of the lagrangian $\mathcal{L} : C^1(\overline{U}, \mathcal{E}) \to \mathcal{A}$ by $f \to L(x,f(x),f'(x)) \mapsto \int_U L(x,f(x),f'(x))\ d\mu(x)$,
with L a continuous and $G^1$ on the fiber mapping $\mathcal{E}^1 \to \mathcal{A}$ is that f satisfies the $\mathcal{A}$
valued Euler equations, whose expression in charts is given by 5-1,5-2, when N is
infinite. When N is finite the equation 5-2 is necessary only up to terms of order
N in $\mathcal{A}$.


<u>6 - Applications to supergravities.</u>

The various lagrangians proposed for the supergravity theories since the original
one (four dimensional space time and one gravitino represented by a spinor valued
one form on space time , called d = 4, n = 1 supergravity), can all be interpreted
in the frame work given in the previous paragraphs. The fields are sections of gra-

ded vector (or affine $(^1)$) bundles, with typical fiber $\mathcal{R}_+^m \times \mathcal{R}_-^q$, on an ordinary

d dimensional manifold. In a local trivialisation of the bundle the bosonic fields

are even valued, and the fermionic fields odd valued.

The rigorous definition of the lagrangian and its Euler equations enables one to

extend to them the classical procedures known for the ordinary (numerical) situa-

tion : the identities deduced from invariances of ordinary groups (Lorentz, diffeo-

morphisms, gauge groups) are now valid as graded identities.

The consistency of a supergravity theory relies on the other hand on another identity

called "supersymmetry" which compensates for the non-well posedness of the Rarita-

Schwinger operator in curved space time : this identity can be proved to hold in

the graded context (it is not true for numerical fields). This identity expresses

the invariance of the lagrangian under a new "infinitesimal transformation" which

mixes bosons and fermions. For instance in the original (d = 4, n = 1) supergravity

we have f = (e,$\psi$) with e = ($e_a^\alpha$) a field of tetrads on M and $\psi$ = ($\psi_\alpha$) a Majorana

spinor valued 1-form and one proves (cf $[7],[9]$) that

$$\mathcal{L}'(f).h = 0$$

for all h such that ($\varepsilon$ spinor field with compact support on M)

$$h_a^\alpha = \bar{\varepsilon}\,\gamma^\alpha\,\psi_a \qquad , \qquad h_\lambda = D_\lambda\,\varepsilon\ .$$

The identity of supersymmetry can be used, together with the previous ones, to study

the well posedness of the equations of a theory of supergravity, first in an alge-

braic sense, and then in the true sense of analysis : the well posedness, and also

causality has been proved for the original supergravity, and the d = 11, n = 1

Cremmer-Julia theory (cf $[1],[3],[4]$).


Aknowledgement - I thank Bryce DeWitt for giving me a mimeographed version of his

book "Supermanifolds" prior to publication.

---

$(^1)$ In fact in standard supergravities the connexion itself is replaced as a fundamental unknown, being supposed metric, by its torsion, which is also a section of
a vector bundle.

References

[1]   D. Bao, Y. Choquet-Bruhat, J. Isenberg, P. Yasskin, to appear J. Maths. Phys.

[2]   M. Batchelor, The structure of supermanifolds, Trans. Amer. Maths. Soc. 253 (1979) 329-338.

      To approaches to supermanifolds, Trans. Amer. Maths. Soc. 258 (1980) 257- 270.

[3]   Y. Choquet-Bruhat. The Cauchy problem in extended supergravity, N=1, d=11. Communications in Maths. Phys., to appear 1985.

[4]   Y. Choquet-Bruhat, "Supergravities" in "Gravitation, Geometry and Relativistic Physics" Springer Lecture Notes in Physics, 219, P. Tourrenc ed. 1984.

[5]   Y. Choquet-Bruhat and C. DeWitt-Morette, Analysis Manifolds and Physics , revised edition, North-Holland 1982.

[6]   E. Cremmer, B. Julia, J. Scherk. Supergravity theory in 11 dimensions. Phys. Lett. 76B, 4 (1978) 409-411.

[7]   S. Deser and B. Zumino. Consistent Supergravity. Phys. Letters 62 n°3 (1976) 335-337.

[8]   B. DeWitt, Supermanifolds. Cambridge University Press, 1984.

[9]   D.Z. Freedman, P. Van Nieuwenhuisen and S. Ferrara. Progress towards a theory of Supergravity. Phys. Rev. D, 13 n°12 (1976) 3214-3218.

[10]  Jadczyk A. and Pilch K., Commun. Math. Phys. 78, 373-390 (1981).

[11]  B. Kostant, Graded manifolds in "Differential geometric methods in Mathematical physics" Springer lecture notes 570 (1977) 177-306.

[12]  A. Rogers, A global theory of supermanifolds. J. Maths Phys 21 (6) (1980) 1352-1365.

Received February 25, 1985.

GEOMETRODYNAMICS PROCEEDINGS (1985), pp. 43-57
edited by A. Pràstaro
© 1985 by World Scientific Publishing Co.

# SELF-DUAL YANG-MILLS FIELDS AND THE PENROSE TRANSFORM IN THE SPINOR CONTEXT

Albert Crumeyrolle

Laboratoire d'Analyse sur les Variétés,
Université Paul Sabatier
118, route de Narbonne,
31062 Toulouse Cedex, France

## ABSTRACT

We use particular properties of spinor geometry in dimension 4, in order to give a new approach for the introduction of self-dual connections in Yang-Mills theory (instantons), and Penrose transform. Our approach utilizes essentially pure spinors and associated totally isotropic spaces; twistors are unemployed and our methods shows that Penrose transform is nothing but a far and rather hidden resurgence of certain E. Cartan's results. We indicate finally a possible generalisation of the instantons, called by us "spintantons", connected with our definition of the enlarged spinor structures.

## I.    Introduction and recalls:

### 1.    Yang-Mills fields [8, 6, 5]

The simplest Maxwell theory makes use of a Skew-symmetric covariant tensor field $F$, in the Minkowski space $E_{1,3}$:

$$F = F_{\alpha\beta}(\theta^\alpha \wedge \theta^\beta)$$

which is assumed to satisfy the equations for no inductive medium:

$$\nabla_\beta F^{\alpha\beta} = J^\alpha \tag{1}$$

$$dF = 0 \tag{2}$$

where $J$ is the electric current, $\nabla$ the pseudo-Riemannian connection. Locally $F = dA$, $A = A_\alpha \theta^\alpha$, and we can use $A_\alpha - \partial_\alpha \psi$ instead of $A_\alpha$, where $\psi$ is a scalar function.

44

If we consider a principal bundle with U(1) as a structural group over $E_{1,3}$, and a cross-section $x \to u(x)$, then $x \to \exp(i\varphi(x) \cdot u(x))$ is also a cross section. If A defines a connection $\omega$, the components $A'_\alpha$, $A_\alpha$ are connected by:

$$A'_\alpha = A_\alpha + i\partial_\alpha \varphi = A_\alpha - \partial_\alpha \psi \quad , \qquad \text{with} \quad \psi = -i\varphi \quad .$$

The curvature form $\Omega = d\omega + \frac{1}{2}|\omega,\omega| = dA$ and (1), if $J = 0$, is translated by $d(^*\Omega) = 0$ (* is the Hodge operator, for the volume form $\eta_{\alpha\beta\gamma\delta}$). We can obtain (1) from variational principle using the Lagrangian $\mathcal{L} = \frac{1}{4} F_{\alpha\beta} F^{\alpha\beta}$, varying the $A_\alpha$.

The Yang-Mills fundamental idea was to change U(1) into SU(2).

More generally, one considers at present a principal bundle over the manifold V ($E_{1,3}$, or the Einstein space-time $V_{1,3}$ as well) with structural group G, $\omega$ is a principal connection, with curvature form $\Omega$.

If $\sigma$, $\sigma'$ are two sections such that:

$$\sigma'(x) = \sigma(x) \cdot g(x) \quad , \qquad x \in V \quad , \qquad g(x) \in G \quad ,$$

then

$$\omega' = \text{Ad}(g^{-1})\omega + g^{-1}dg$$

$$\omega = (A^I_\alpha \theta^\alpha)G_I \quad , \quad G_I \text{ is a frame in the Lie algebra } \mathcal{L}(G),$$

$$\Omega = \Omega^I_{\lambda\mu} G_I (\theta^\lambda \wedge \theta^\mu) \quad .$$

Usually G is a linear group (real or complex). In the real case one takes the Lagrangian:

$$\mathcal{L} = \frac{1}{4} \Omega^a_{b\lambda\mu} \Omega^{b\lambda\mu}_a \quad ,$$

varying the $A^a_{b\lambda}$ components of $\omega$, the Y.M. equations appear to be:

$$\nabla_\mu \Omega^{a\lambda\mu}_b + [A_\lambda, \Omega^{\lambda\mu}]^a_b = 0 \tag{3}$$

(where the covariant derivation involves only $\lambda$ and $\mu$). Often one takes into account of the vector valuedness of the curvature form and (3) is translated into

$$D(^{*}\Omega) = 0 \quad . \tag{4}$$

With a nonlinear group, we choose:

$$\mathcal{L} = \frac{1}{4} g^{\alpha\lambda} g^{\beta\mu} B(\Omega_{\alpha\beta}, \Omega_{\lambda\mu})$$

where $g$ is the fundamental metric, and $B$, the Killing form. If the group $G$ is a semi-simple group, $B$ is not degenerated and if $G$ is compact $B$ is a negative definite form.

Mathematicians prefer to write often $\mathcal{L} = -\frac{1}{4} \|\Omega\|^2$ or $-\frac{1}{4}\|\Omega\|^2 + L(\varphi, D_\mu\varphi)$ if another field $\varphi$ interacts with $\Omega$.

In the physicist's jargon: the connexion becomes the "gauge potential" and the curvature the gauge field. $G$ is the "gauge group" and an isomorphism for the principal bundle a "gauge transformation". If we choose different cross sections a change of cross sections is a "gauge change".

## 2. Self-dual and anti-self-dual fields

$V$ is a $C^\infty$ Riemannian or pseudo-Riemannian oriented manifold with $n = 2k$ dimensions and $\Lambda^p$ (resp $\Gamma(\Lambda^p)$) the bundle of the p-forms (resp. the space of the p-form cross sections). The star (or Hodge) operator $*$ is defined by:

$$\alpha \wedge (^{*}\beta) = (\alpha/\beta)\eta$$

where $(\alpha/\beta)$ is the scalar product for the differential forms, $\eta$ the n-volume form or the orientation: $\eta = \varepsilon\sqrt{|g|}\,dx^1 \wedge dx^2 \wedge \ldots \wedge dx^n$, locally, $(\varepsilon = \pm 1)$;

$$* : \Lambda^p \to \Lambda^{n-p} \quad .$$

Practically, with the local components:

$$(^{*}\alpha)^{i_{p+1},\ldots,i_n} = \eta^{i_1,\ldots,i_n} \alpha_{i_1,\ldots,i_p}$$

(modulo a normalization coefficient). We recall the classical results:

$$** (-1)^{p(n-p)} = \mathrm{Id} \quad \text{(Riemannian case)} \quad ,$$

$$**(-1)^{p(n-p)} = \varepsilon_1 \cdot \text{Id} \quad \text{(pseudo-Riemannian case)} \quad ,$$

$$\varepsilon_1 = 1, \quad \text{if} \quad \det g > 0 \quad ,$$

$$\varepsilon_1 = -1, \quad \text{if} \quad \deg g < 0 \quad .$$

If $k = 2$:

$$*^2 = \text{Id} \quad \text{(Riemannian case)} \quad ,$$

$$*^2 = -\text{Id} \quad \text{(Minkowskian case)} \quad ,$$

and $p = 2$, gives the direct sum

$$\Lambda^2 = \Lambda^2(+) \oplus \Lambda^2(-)$$

corresponding to the spectral decomposition for the star operator. The bundle $\Lambda^2$ is a direct sum of self-dual and anti-self-dual forms.

Note that $*$ is conformally invariant.

It is then natural to consider connections with curvature form $\Omega$, such that:

$$*\Omega = \pm \Omega \quad \text{(Riemannian case)} \quad ,$$
$${}^*\Omega = \pm i\Omega \quad \text{(Minkowskian case)} \quad .$$

The Y-M condition $D({}^*\Omega) = 0$ is then automatically satisfied, $D\Omega = 0$, being the Bianchi identity.

3.  Instantons $[1, 5]$

Some physicists are interested in connections giving Lagrangian with compact support so that the action integral is always defined over any domain with boundary in $\mathbf{R}^4$. Sending $\mathbf{R}^4$ onto $S^4$ by stereographic projection, one considers $C^\infty$ vector bundle (with complex rank $r$, base $S^4$) associated to a principal bundle with group $SU(r)$ and provided with a connection of which the curvature is self-dual: ${}^*\Omega = \Omega$ (self-dual connexion). This bundle is called an instanton. If ${}^*\Omega = -\Omega$ curvature is anti-self-dual. A change of orientation exchanges also the both cases.

It is difficult to discover a physical meaning for such a scheme (pseudo-particle?) except for changing  t  (the time) into  <u>it</u>  in the quadratic Minkowskian form, but unfortunately from this algebraic modification arise many topological difficulties.  Probably it would be better to consider the compactified Minkowski space isomorphic to $S_3 \underset{Z_2}{\times} S_1$ and define thus the "Minkowskian instanton" over this last base  $(^*\Omega = i\Omega)$.

A bundle with self-dual connection satisfies topological conditions, for example for an hermitian bundle the first Pontryagin number is positive |1|.

With elliptice signature self dual connections give absolute minimum for the Yang-Mills Lagrangian:  this property explains why many people are interested with instantons in spite of their artificial nature.

II.    <u>The Penrose transform as a resurgence of the Cartan's works</u>

Grosso modo the Penrose transform consists in the one to one association between an instanton and certain vector-bundle over  $P_3(\mathbb{C})$; usually the authors interested in this subject rely their approach on the rather fancy notion called "twistor" [7] and use the Pauli-Dirac matricial formalism.  This method appears rather as an intricate and heavy artifice. In point of fact in the modern frame, and according to the ideas of E. Cartan |2|, the Penrose transform is really like a far resurgence of the important notion described in |2, 3, 4a| and called "pure spinors".  In 4 and 6 dimensions any even (or odd) spinor is also a pure spinor [4b] and this remark is crucial for the following developments.

We choose the Minkowskian signature, but our results are also valuable for the elliptic signature.

1.    Pure Minkowskian and conformo-Minkowskian spinors and <u>associated spaces</u>

Our notations are like in the precedent paper [4b].  $E_{1,3}$  is the Minkowski space with orthonormal frame  $(e_1, e_2, e_3, e_4)$, $(e_1)^2 = 1$;  we define:

$$x_1 = \frac{e_1 + e_4}{2} \quad , \quad x_2 = \frac{ie_2 + e_3}{2} \quad , \quad y_1 = \frac{e_1 - e_4}{2} \quad ,$$

$$y_2 = \frac{ie_2 - e_3}{2} \qquad \text{(Witt frame)}$$

and

$$f = y_1 y_2 \quad .$$

$uf = u^+ f + u^- f$ is a Minkowskian spinor. The totally isotropic maximal subspaces (t.i.m.s.) attached to $u^+ f$ and $u^- f$ (pure spinors) respectively, are constituted by the set of points $x \in E_{1,3}$ such that $x(u^+ f) = 0$ or $x(u^- f) = 0$. Thus two families of t.i.m.s. appear (even t.i.m.s. and odd t.i.m.s.). If

$$u^- f = (ax_1 + bx_2)f \tag{5}$$

$$u^+ f = (a' + b'x_1 x_2)f \tag{6}$$

the t.i.m.s. are defined according [4b] by:

$$x = (ax_1 + bx_2)s + (ay_2 - by_1)t \qquad \text{(odd t.i.m.s.)} \tag{7}$$

or

$$x = (b'x_1 + a'y_2)s + (a'y_1 - b'x_2)t \qquad \text{(even t.i.m.s.)} \tag{8}$$

Two t.i.m.s., with different parity intersect and have an isotropic straight line in common [3].

The t.i.m.s. are bijectively attached to the [4a] pair of projective spinors represented by $(u^- f, u^+ f)$; the precedent pair defines thus biunivoquely a homogeneous straight line in $E_{1,3}^{\mathbb{C}}$ [3, 4a].

Now, we consider $E_{2,4}$ and $E_{2,4}^{\mathbb{C}}$, with orthonormal frame

$$(e_0,\ e_1,\ e_2,\ e_3,\ e_4,\ e_5) \quad \text{with} \quad (e_0)^2 = 1 \quad , \quad (e_5)^2 = -1$$

$$x_0 = \frac{e_0 + e_5}{2} \quad , \quad y_0 = \frac{e_0 - e_5}{2} \quad , \quad \hat{f} = y_0 y_1 y_2 \quad ,$$

and projective spinors, with determined parity are again in biunivoque correspondence with 3-complex t.i.m.s.

Let $\hat{u}^+ \hat{f} = (a_0 + a_1 x_0 x_1 + a_2 x_0 x_2 + a_3 x_1 x_2)\hat{f}$ be an even conformo-Minkowskian spinor, the points $\xi$ lying in the associated t.i.m.s. $(\pi)$ are defined by:

$$\xi\, \hat{u}^+ \hat{f} = 0 \quad , \quad \xi \in E_{2,4}^{\mathbb{C}} \qquad \text{according a classical result.}$$

If $\xi = \alpha x_0 + \beta x_1 + \gamma x_2 + \alpha'y_0 + \beta'y_1 + \gamma'y_2$  we obtain,  if  $a_0 \neq 0$:

$$\xi = \left(\frac{\beta'a_1 + \gamma'a_2}{a_0}\right)x_0 + \left(\frac{\gamma'a_3 - \alpha'a_1}{a_0}\right)x_1 - \left(\frac{\alpha'a_2 + \beta'a_3}{a_0}\right)x_2 + \alpha'y_0 + \beta'y_1 + \gamma'y_2$$

$$(9)$$

We introduce now, the injective image  $\mathcal{E}$  of  $E_{1,3}$  in  $E_{2,4}$  by means of
the "isotropic transform"  $v$:

$$v(x) = \frac{x^2 - 1}{2}\, e_0 + x + \frac{x^2 + 1}{2}\, e_5 = x^2 x_0 + x - y_0 \qquad . \qquad (10)$$

It's easy to verify that  $\mathcal{E}$  is contained in the isotropic cone of  $E_{2,4}$;
modulo a negligible set this image represents the compactified Minkowski
space isomorphic to  $S^1 \underset{Z_2}{\times} S^3$ .  $v$  extends naturally to the complexified
space  $E_{1,3}^{\mathbb{C}}$,  Im$(v)$  becomes  $\mathcal{E}'$.
$(\pi)$  defined by (3) is contained in  $E_{2,4}^{\mathbb{C}}$  and its intersection with it
corresponds to  $\alpha' = -1$.  We call  $(\pi_1) = \pi \cap \varepsilon'$, $(\pi_1)$  is the set of points
$\xi$  such that:

$$\xi = \frac{\beta'a_1 + \gamma'a_2}{a_0}\, x_0 + \frac{\gamma'a_3 + a_1}{a_0}\, x_1 - \frac{\beta'a_3 - a_2}{a_0}\, x_2 + \beta'y_1 + \gamma'y_2 - y_0 \qquad .$$

$$(11)$$

According the isotropic transform the points of  $(\pi_1)$  arise from the points
of a totally isotropic affine plane in  $E_{1,3}^{\mathbb{C}}$,  called  $(\pi_2)$  and defined by:

$$x = \frac{\gamma'a_3 + a_1}{a_0}\, x_1 + \frac{a_2 - \beta'a_3}{a_0}\, x_2 + \beta'y_1 + \gamma'y_2 \qquad (12)$$

and the parallel homogeneous plane  $(\pi_3)$  by:

$$x = \beta'(-a_3 x_2 + a_0 y_1) + \gamma'(a_3 x_1 + a_0 y_2) \qquad (13)$$

<u>Remark 1</u>:  The precedent calculus supposes  $a_0 \neq 0$.  If  $a_0 = 0$,  we
can suppose  $a_3 \neq 0$,  and we obtain analogous results; (13) becomes

$$x = a_3(\beta x_1 + \gamma x_2) \qquad .$$

Finally, after the choice of a frame in $E_{2,4}$, $E_{1,3}$, the set $\tilde{\Sigma}^+$ constituted by the projective even spinors for $E_{2,4}^{\mathbb{C}}$, that we can identify with the set of projective spinors for $E_{1,3}^{\mathbb{C}}$, defines the $P_3(\mathbb{C})$ space, meanwhile the points corresponding to $a_0 = a_3 = 0$ constitute the $P_1(\mathbb{C})$ space.

We write $O(P_3(\mathbb{C})) = (P_3(\mathbb{C}) - P_1(\mathbb{C}))$.

If we compare (13) and (8), we see that $(\pi_3)$ is the more general even t.i.m.s. in $E_{1,3}^{\mathbb{C}}$. We obtain:

<u>Proposition 1</u>:

After the choice of a frame in $E_{1,3}$, $E_{2,4}$, we can associate with any element of $O(P_3(\mathbb{C}))$ an even totally isotropic affine plane $(\pi_2)$ in $E_{2,4}^{\mathbb{C}}$ and an even totally isotropic affine plane $(\pi_2)$ in $E_{1,3}^{\mathbb{C}}$. Thus $O(P_3(\mathbb{C}))$ is in one to one correspondence with the set of the even totally isotropic affine planes of $E_{1,3}^{\mathbb{C}}$.

The same proposition holds for the odd planes.

<u>Remark 2</u>: We shall use below the elementary remark that an even affine plane in $E_{1,3}$ is an equivalent class of pairs

$$(x, \pi_3) \quad , \quad x \in E_{1,3}^{\mathbb{C}} \quad , \quad \pi_3 \text{ even t.i.m.s. in } E_{1,3}^{\mathbb{C}} \quad ,$$

$$(x, \pi_3) \sim (x', \pi_3') \quad \text{if and only if} \quad \pi_3 = \pi_3' \text{ and } x - x' \in \pi_3 \quad .$$

## 2. <u>Self-dual connections and vector-bundles</u>

We consider fibrations with a structural group acting effectively in the typical-fiber. $G_2^+$ denotes the Grassmannian of even t.i.m.s. in $E_{1,3}^{\mathbb{C}}$. The fundamental 4-form of $E_{1,3}$ is defined by the dual frame of $(e_1, e_2, e_3, e_4)$, written

$$\theta^1 \wedge \theta^2 \wedge \theta^3 \wedge \theta^4 \quad .$$

According to the (I, 2) we say that a plane in $E_{1,3}^{\mathbb{C}}$ defined by a decomposable 2-form $\Phi$ is self-dual (resp. anti-self-dual) if

$$^*\Phi = i\Phi \qquad (\text{resp. } ^*\Phi = -i\Phi) \quad .$$

We write down that:

$$\theta^1 \wedge \theta^3 \overset{*}{\underset{}{-}} \theta^4 \wedge \theta^1 \overset{*}{\underset{}{-}} \theta^3 \wedge \theta^2$$

$$\theta^3 \wedge \theta^4 \overset{*}{\underset{}{-}} \theta^2 \wedge \theta^1 \overset{*}{\underset{}{-}} \theta^4 \wedge \theta^3$$

$$\theta^4 \wedge \theta^2 \overset{*}{\underset{}{-}} \theta^3 \wedge \theta^1 \overset{*}{\underset{}{-}} \theta^2 \wedge \theta^4 \quad .$$

**Proposition 2:**

Every even t.i.m.s. in $E^{\mathbb{C}}_{1,3}$ is anti-self-dual.

Every odd t.i.m.s. in $E^{\mathbb{C}}_{1,3}$ is self-dual.

Indeed, from (13), $x$ is determined by the pair $-a_3 x_2 + a_0 y_1$, $a_3 x_1 + a_0 y_2$, hence, after a dual identification by the bivector:

$$\Phi = i(a_0^2 + a_3^2)(\theta^1 \wedge \theta^2) + (a_3^2 - a_0^2)(\theta^1 \wedge \theta^3) + 2a_0 a_3(\theta^1 \wedge \theta^4)$$

$$+ i(a_0^2 - a_3^2)(\theta^2 \wedge \theta^4) + 2ia_0 a_3(\theta^2 \wedge \theta^3) - (a_0^2 + a_3^2)(\theta^3 \wedge \theta^4)$$

defining the plane $(\pi_3)$.

$^*\Phi = -i\Phi$ is immediate.

The same proof works for with odd t.i.m.s.

Lemma 1: $^*\Omega = i\Omega$ is equivalent to $\Omega$ null over any even t.i.m.s. $(\pi_3)$.

We associate to $(\pi_3)$ a conjugate plane $\overline{\overline{\pi}}_3$, using a conjugation taking into account the signature. It's easy to proof that $\overline{\overline{\pi}}_3$ and $\pi_3$ have the same parity. The planes $(\pi_3)$ and $(\overline{\overline{\pi}}_3)$ carry a "real Witt frame" and the calculus of $\Phi$ above shows that this form is a linear combination of

$$\theta^1 \wedge \theta^3 + i\theta^2 \wedge \theta^4 \quad , \quad \theta^1 \wedge \theta^4 + i\theta^2 \wedge \theta^3 \quad , \quad \theta^4 \wedge \theta^3 + i\theta^1 \wedge \theta^2$$

constituting an anti-self-dual frame; supposing $\Omega$ is null over $(\pi_3)$ is equivalent to say that the three components of $\Omega$ over the anti-self-dual frame are zero, then $\Omega$ is a self-dual form.

Coming back to the fibration problems, we put in correspondence $(x, \pi_3) \in E^{\mathbb{C}}_{1,3} \times G^+_2$ and $\theta \in \varphi(P_3(\mathbb{C}))$, $\theta = \varphi(x, \pi_3)$, $\varphi$ is only a surjective map; $(\xi)$ being a vector bundle with $O(P_3(\mathbb{C}))$ as a base, $\varphi^*(\xi)$ is a bundle over $E^{\mathbb{C}}_{1,3} \times G^2_+$, with $\varphi^*(\xi)$ we can construct a bundle over $E^{\mathbb{C}}_{1,3}$

if and only if, for any points, $z \in O(P_3(\mathbb{C}))$,

$z' \in O(P_3(C))$ , $z = \varphi(x, \pi_3)$ , $z' = \varphi(x, \pi_3')$ , the fibers at $x$,

deduced from $\xi_z$ and $\xi_{z'}$ are the same. It is then easy to see that such a $z'$ describes any right line from $z$ if $z$ is fixed. $(\xi)$ is trivial over any such a line.
$\varphi^*(\xi)$ retrenches to $E_{1,3}^{\mathbb{C}}$ owns a natural parallelism between $x$ and $x'$, if $x - x' \in \pi_3$, for some even plane $(\pi_3)$. According to standard results it is as much as to say that there exists a connection with null curvature form along the even t.i.m.s. $(\pi_3)$. According to the lemma 1, the curvature is self-dual.

We can state:

<u>Proposition 3:</u>

<u>A vector bundle over $O(P_3(\mathbb{C}))$, inducing a trivial bundle over any right line defines also a vector bundle over $E_{1,3}^{\mathbb{C}}$ with the same rank, owning a self-dual connection and satisfying the Y.M. condition.</u>

Reciprocally: Let $(\eta)$ be a rank $r$ complex vector bundle with structural group, over $E_{1,3}^{\mathbb{C}}$, with a self-dual connection.
We call trivializing distinguished open set $\mathcal{U}$ of $E_{1,3}^{\mathbb{C}}$ an open set with a field of even affine t.i.m.s. such that:

$$x \longrightarrow \pi_2(x)$$

$$\eta/\mathcal{U} \simeq \mathcal{U} \times \mathbb{C}^r \quad .$$

When $x$ describes $\mathcal{U}$, we can associate to it a point $z \in O(P_3(\mathbb{C}))$, and $z$ describes then an open set $\mathcal{U}$ in $O(P_3(\mathbb{C}))$. At $z$ we construct $\xi_z$ isomorphic to the $\eta_x$ fiber. Evidently we need an identification between $\xi_z$ and $\xi_{z'}$ if

$$z = \varphi(x, \pi_2(x)) \quad \text{and} \quad z' = \varphi(x, \pi_2'(x)) \quad ,$$

and the constructed bundle is trivial over any right line.

If $(x, \pi_2(x)$ and $(x', \pi_2(x'))$ gives the same $z$, $\pi_3(x) = \pi_3(x')$ and $x' - x$ $\pi_3(x)$, but the self dual connection owns a curvature form $\Omega$, such that $\Omega$ is null over $\pi_3(x)$, then we can identify $\eta_x$ and $\eta_{x'}$, the construction of the fiber at $z$ is coherent. The differentiability comes if we consider that we have a bundle over the set of equivalence classes $(x, \pi_3)$, because $\eta$ is null over any even $(\pi_3)$ and the classes are identified with the points of $O(P_3(\mathbb{C}))$. Then:

<u>Proposition 4:</u>

<u>For any vector bundle, with a structural group, and $E^{\mathbb{C}}_{1,3}$ base, carrying a Y-M self-dual connection, we can associate a vector bundle over $O(P_3(\mathbb{C}))$ with the same rank; this bundle is trivial over any right line.</u>

Now we intend to prove that $O(P_3(\mathbb{C}))$ can be replaced by $P_3(\mathbb{C})$ in the precedent propositions (3), (4).

Indeed let $\hat{u}_0^+\hat{f} \neq 0$, be a pure spinor, it is always possible to find an open set $(\theta)$ in the space of projective spinors which contains $\hat{u}_0^+\hat{f}$ and such that $(\theta)$ is in bijective correspondence with an open set $(\theta_1)$ in $E^{\mathbb{C}}_{1,3} \times G_2^+$. For that to show we determine a new Witt frame taking:

$$\begin{cases} \sigma(y_i) = y_i \\ \sigma(x_i) = \alpha_i^j y_j + x_i \end{cases} \qquad , \qquad i = 0,1,2 \qquad , \qquad \alpha_i^j = -\alpha_j^i \qquad (14)$$

$a_0$ becomes then:

$$a_0' = a_0 + \alpha_0^1 a_1 + \alpha_0^2 a_2 + \alpha_1^3 a_3 \qquad , \qquad a_1' = a_1 \qquad ,$$

$$a_2' = a_2 \qquad , \qquad a_3' = a_3 \qquad\qquad (15)$$

and because the $a_i$, $i = 1,2,3$, are not all null we can obtain $a_0' \neq 0$.

This change of frames can be interpreted as a transformation by an element $\gamma$ belonging to the reduced special Clifford group

$$G_0'^+ = \exp(\Lambda^2 \hat{F}) \qquad , \qquad \hat{F} = (x_0, x_1, x_2)$$

$$\gamma \hat{u}^+ \hat{f} \gamma^{-1} = \gamma \hat{u}^+ \gamma^{-1} \hat{f} \qquad\qquad [4a]$$

because $G_0'^+$ is in the connected component of a spinoriality group [4a]. $\gamma$ transforms a neighborhood $(\theta)$ of $\hat{u}_0^+\hat{f}$ with $a_0 = 0$, into open set $(\theta_1)$ where the new analogous coefficient $a_0$ is different of zero.

If $(\xi)$ is a vector bundle (with a structural group) over $P_3(\mathbb{C})$, naturally associated to a principal bundle, the construction of $\varphi^*(\xi)$ above is independent of the choice of a frame, modulo the (14) change: because this change replaces $\varphi^*(\xi)$ by $\varphi^*(\gamma^*(\xi))$ where $\gamma^*$ is a principal bundle isomorphism. However, constructing a bundle over all $P_3(\mathbb{C})$ coming from the bundle $(\eta)$ above (cf. proof of proposition 4) needs the clutching of bundles obtained over the 4 open sets corresponding to $a_0 \neq 0$, or if $a_0 = 0$, $a_i \neq 0$, $i = 1,2,3$; if for example $a_2 \neq 0$ in a point of $P_3(\mathbb{C})$ reached with two frames $R_0$ $(a_0 \neq 0)$ and $R_1(a_1 \neq 0)$ we have to change simply in

$$\mathbb{C}^4 \times (G_2^+)_{\text{standard}}$$

coordinates

$$X^1 \quad \text{for} \quad \frac{a_0 + \alpha_0^1 a_1 X^1}{a_0}$$

$$X^2 \quad \text{for} \quad \frac{a_0 + \alpha_0^1 a_1 X^2}{a_0}$$

$$Y^1 \quad \text{for} \quad Y^1$$

$$Y^2 \quad \text{for} \quad Y^2$$

(in (15) above, $\alpha_0^1 \neq 0$, $\alpha_0^2 = \alpha_1^3 = 0$)
from that we obtain the clutching isomorphism. Finally, the bundles being considered through an isomorphism, we can state:

### Theorem 1

There is a natural one to one correspondence between: a) complex vector bundles with structural group, base $P_3(\mathbb{C})$, trivial over every line of $P_3(\mathbb{C})$ and b) bundles with the same rank over $E_{1,3}^{\mathbb{C}}$ having a self-dual connection.

Remarks: a) With the euclidean space $E_4$ and $E_4^{\mathbb{C}}$, nothing changes in the precedent developments except for the definition of the self duality ($^*\Omega = \Omega$ instead $^*\Omega = i\Omega$). Theorem 1 works for the euclidean case. It is the same over the $E_{2,2}$ space with neutral signature but we can define directly real spinors and consider $P^3(\mathbb{R})$ instead of $P^3(\mathbb{C})$ if you want.

b) A bundle over $E_{1,3}^{\mathbb{C}}$ is also a bundle over $E_{1,3} \times E_{1,3}$, we can then consider bundles over $E_{1,3}$ (or $E_4$) in the precedent theorem 1. We have now:

<u>Theorem 2</u>

There is a natural one to one correspondence between: a) complex vector bundles with structural group, base $P_3(\mathbb{C})$ trivial over the real lines of $P_3(\mathbb{C})$ and, b) bundles with the same rank over $S_3 \times S_1$ (or $S_4$) owning a self-dual connection.

The proof uses stereographic projection, it is almost the same for both cases: choose for example $S_4$. By means of two stereographic projections $\varphi_1$ and $\varphi_2$ with antipodic centers, one puts over $S_4$ an atlas with two charts $U_1$ and $U_2$, every $U_i$ diffeomorphic with a topological disk of $\mathbb{R}^4$. If there exists over $\mathbb{R}^4$ a Y.M. bundle with self-dual curvature, there correspond to it two bundles $\xi_1$ and $\xi_2$ with the same property over $U_1$ and $U_2$, because $\varphi_1$ and $\varphi_2$ are conformal transformations. $\xi_1$ and $\xi_2$ have isomorphic restrictions for $U_1 \cap U_2$, a clutching method permits to obtain unique bundle $\xi$ over $S_4$ with the announced property.

<u>Remark</u>: According to precedent remark (a), we have the same result with $S_2 \times S_2$, and $P_3(\mathbb{R})$.

Proposition 4, theorems 1 and 2 express the Penrose transform [5].

Often the rank of precedent bundle $(\xi)$ is 2 and the structural group SU(2, $\mathbb{C}$) [1].

Evidently any reduction of the structural group gives more particular structure for the corresponding bundle over $P_3(\mathbb{C})$: Starting from SU(2, $\mathbb{C}$) one obtains a quaternionic structure over the fibers [1, 5].

The geometry of bundles over $P_4(\mathbb{C})$ is studied intensively in [5].

3.    <u>Spintantons</u>

"Spintanton" is coined for spin-instanton. Suppose that the curved space-time oriented manifold $V_{1,3}$ owns an enlarged spinor structure [4a] [4c]. If the complexified tangent space $T_{\mathbb{C}}(V_{1,3})$, owns a field of even t.i.m.s., $V_{1,3}$ satisfies our precedent hypothesis, and reciprocally [4a, 4c].

56

Let  (U)  be an open set in  $V_{1,3}$  trivializing the tangent bundle and  $\sigma$  the spinorial projective bundle associated (fibers of the enlarged spinor bundle are minimum ideals in the tangent Clifford algebra and define projective complex space in 3 dimensions).  It is always loisible to increase the tangent bundle by means of a trivial bundle with complex rank 1, giving again in every fiber  $E_{1,3}^{\mathbb{C}}$  an immersion in  $E_{2,4}^{\mathbb{C}}$.

Suppose we have also over  $V_{1,3}$  a vector bundle  $\eta$  with structural group  G  acting effectively on it.  We recupere thus locally the situation described in propositions 3 and 4.  On account of the equivariance of the spinor group action and the special Lorentz group action which transforms even t.i.m.s. one another we can associate to the bundle  $(\eta)$  with self-dual connection a complex bundle  $(\xi)$  over the projective enlarged spinor bundle  $(\sigma)$.  Bundle  $(\xi)$  will be trivial over any subset of  $\sigma$  identifiable to some  $U \times D$,  where  D  is a right line in  $P_3(\mathbb{C})$  intrinsic property because the spinor group acts linearly in the spinor space - and where (26) is an open trivializing set, homeomorphic with a pseudo-euclidean disk.

It is clear that the results hold for any Riemannian or pseudo-Riemannian manifold in four dimensions, admitting an enlarged spinor structure.

<u>Definition: We shall call "spintanton" any G-complexe vector bundle $(\eta)$ over a manifold in four dimensions $(V_4, V_{1,3}, V_{2,2}, V_{3,1})$ owning an enlarged spinor structure and carrying a G-self-dual connection.</u>

Thus its properties are described by the geometry of certain vector bundle over an enlarged spinor bundle of typical-fiber  $P_3(\mathbb{C})$.

It is very well known that spheres have strict spinor structures, but there exists important peculiarity:  indeed  $S_4$  owns unique classes of enlarged spinor structure because it is possible to construct over  $S_4$  a global field of t.i.m.s. in  $T_{\mathbb{C}}(V_4)$  [4, d].  The enlarged spinor bundle is identifiable with  $P_3(\mathbb{C}) \times S_4$.  The same property appears for  $S_3 \times S_1$, because  $T(S_3 \times S_1)$  is trivializable and also for  $S_2 \times S_2$  because  $S_2$  is carrying a global field of isotropic directions  [4, d].

Finally, a "spintanton" over  $S_4$, $S_3 \times S_1$, $S_2 \times S_2$  reduces to an instanton and constitutes a good generalization.

The situation is particulary simple for a space-time  $V_{1,3}$  admitting a strict spinor structure because  $T(V_{1,3})$  is parallelizable and it appears again a bundle over  $P_3(\mathbb{C})$  corresponding to a self-dual connection.

Spintanton over $V_{1,3}$ and associated bundle over $\sigma$ have clearly a physical meaning.


## References


[1]  M.F. Atiyah and R.S. Ward, Communications in math. physics, no. 55 (1977), pp. 117-124.

[2]  E. Cartan, Leçons sur la théorie des spineurs, Hermann Paris (1938).

[3]  C. Chevalley, The algebraic theory of spinors, Columbia U.P., New York (1954).

[4]  A. Crumeyrolle, a) Algèbres de Clifford et spineurs, Toulouse (1974)
                      b) ... Annales I.H.P. Section A, Vol. 34 (1981),
                         pp. 351-372.
                      c) Periodice Math. Hungarica, Vol. 6(2) (1975),
                         pp. 143-171.
                      d) Kodai. Math. Journal 7 (1984).

[5]  A. Douady - J. Verdier, Les équations de Yang et Mills Astérisque 71-72, Séminaire E.N.S., 1977-78 (Ed. Soc. Math. de France).

[6]  H. Kerbrat, Ann. de l'I.H.P., Vol. 21 A, no. 4, p. 333.

[7]  R. Penrose and R.S. Ward, Twistors for flat and curved space-time, in General Relativity and Gravitation 2, pp. 283-328, Plenum Press (1980).

[8]  C.N. Yang and R.C. Mills, Phys. Rev. 96 (1954), p. 191.


Received January 19, 1985.

GEOMETRODYNAMICS PROCEEDINGS (1985), pp. 59-100
edited by A. Pràstaro


# ON SMOOTH AND ANALYTIC FUNCTIONS IN GAUGE FIELD THEORY

J. Czyz

Mathematics Institute, Polish Academy of Sciences
Warsaw 00950, P.O. Box 137, Poland

SUMMARY

This is a review of main results dated from late 70s and early 80s
which pertain to several problems in Euclidean-like gauge fields,
namely the existence of self-dual gauge fields (instantons) over
an arbitrary 4-manifold, direct constructions of such fields, the
topology of instanton moduli spaces and its role in proofs of the
existence of exotic differentiable structures on $R^4$, hypothetical
non-self-dual critical points of the Yang-Mills functional, gauge
monopoles, merons and geometry of supersymmetry and supergravity.
We pay attention to a contribution of methods of analytic functions
and methods of topology and differential (but non-analytical) geometry to the reviewed results.

## Introduction

The considered problems are related to the following scheme:  A
field over a base manifold  $M$  is given and values of the field belong to
fibres in corresponding points  $x \in M$  of a vector bundle  $V \to M$.  The field
equations  $Ax = 0$  are elliptic or quasi-elliptic (e.g. singular elliptic or
"super"-elliptic equations will be discussed).  Then it may happen that
there is such a surjective map  $p: M_c \to M$  that the induced equations
$p^*(A)x = 0$  in suitable coordinates turn out to be Cauchy-Riemann equations,
see Fig. 1 below.  Then we say that a Penrose transformation is given.
This transformation makes of the problem of an elliptic operator the problem
of analytic (in particular rational) functions.  Thus two mathematical
technics may be used for our problem:  1) topology (e.g. methods of the
topological index theorem) and differential geometry (e.g. methods of
differential operators, variational methods etc.), 2) analytic functions

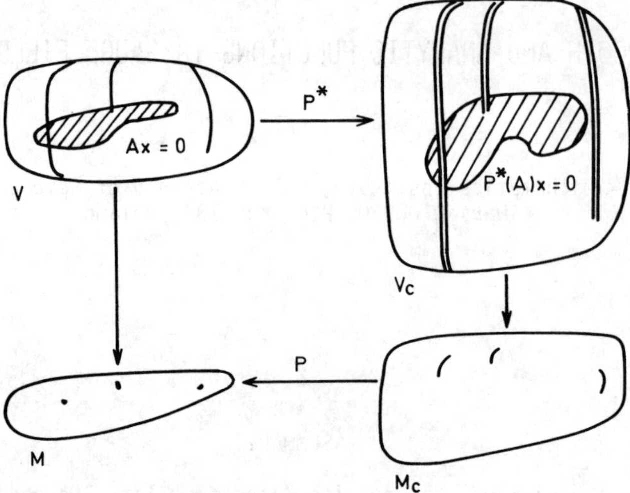



Fig. 1

(complex manifolds, analytic vector bundles) and in a particular rational
case algebraic geometry (jumping lines, spectral curves and so on).  Both
these technics are related to different systems of axioms and primeval
ideas.  Recall that each reasonable problem of analytic functions can be
expressed by means of some relations between coefficients of power series;
all quantities related to analytic functions are determined by power series
so that they may be defined via an initial condition on a zero-measure set;
well-defined analytic objects have explicit representations via coefficient
series and so on.  Instead, in topology and differential geometry we can
meet objects having no explicit representation or realization.  As an
example we will outline rudimentarily an implicit proof of the existence
of so-called exotic differentiable structures on $R^4$.  Thus methods of
topology and differential geometry seem to be nothing else but translators
of intuitions and imaginations into a language of words and formulae.

In a particular problem usually one of the above general methods
seems to be more accurate and in this paper we will try to answer which
one is it in cases of the nine discussed problems.

In the paper we will study manifolds which are defined in the
following classes of functions:  1) $C^0$ (continuous functions) then a manifold is called topological (top.), 2) $C^\infty$ (infinitely differentiable) and
then a manifold is called smooth or differentiable (diff.), $3^0$ $C^\omega$

(holomorphic), in this case a manifold is called complex (not holomorphic or analytic!). We will also say that the structure of a manifold is top., diff., or complex resp. In the first two cases we will use real coordinates, in the third one of course complex coordinates will be used.

In Appendices A and B main properties of intersection forms and K3 surfaces are recalled.

## 1. The existence of instantons over general 4-manifolds

The word "instanton" has been firstly used by t'Hooft [48] in 1976. One uses it for a self-dual connection in a principal G-bundle $E^p \to M$ up to equivalence, where the base space $M$ is a real smooth oriented Riemannian 4-manifold. The structural (gauge) group $G$ will be usually minimal (i.e. no reduction of $E^p$ to a G'-bundle, $G' \subseteq G$ is possible) and then $G$ will be compact.

As a space of field configurations one considers a vector bundle $E \to M$ which is associated to $E^p$ by a linear representation $T$ of $G$. But we will not define $T$ explicitly because in our examples $G$ is usually a classical group so that its elements may be identified with matrices. In this case it is conveninet to deal with a connection in $E$ associated to that in $E^p$.

Let $\nabla_\mu = \partial_\mu + A_\mu$ be a covariant derivative of a connection in $E$. Recall that $A_\mu$ take values in the Lie algebra $g$ of $G$. Assume that $\partial_\mu$ are orthogonal in sense of the Riemannian metric on $M$. Then the following equations define the self-duality relation

$$[\nabla_1, \nabla_2] = [\nabla_3, \nabla_4] \quad , \quad [\nabla_1, \nabla_3] = -[\nabla_2, \nabla_4] \quad ,$$

$$[\nabla_1, \nabla_4] = [\nabla_2, \nabla_3] \quad .$$

They imply simple equations for coefficients of the curvature tensor $F = \frac{1}{2} F_{\mu\nu} dx_\mu \wedge dx_\nu$, where $F_{\mu\nu} = \partial_\mu A_\nu - \partial_\mu A_\nu + [A_\mu, A_\nu]$, namely

$$F_{12} = F_{34} \quad , \quad F_{13} = -F_{24} \quad , \quad F_{14} = F_{23} \quad ,$$

which mean $F = {}^*F$, where $*$ is the Hodge star and denotes the duality in a Grassmann algebra with a scalar product.

The gauge transformation formulae for $A_\mu$ and $F_{\mu\nu}$ are

$$A_\mu \to g^{-1}A_\mu g + g^{-1}\partial_\mu g \quad , \qquad F_{\mu\nu} = g^{-1}F_{\mu\nu}g \quad . \tag{1.1}$$

A main parameter of instantons is the instanton number $k$

$$k: = \frac{1}{8\pi^2}\int_M (F_{\mu\nu}, {}^*F_{\mu\nu})dv \overset{G=SU(2)}{=} -\frac{1}{16\pi^2}\int_M \mathrm{tr}\, F_{\mu\nu}({}^*F_{\mu\nu})d^4x \quad . \tag{1.2}$$

Here $(.,.)$ denotes the normalised Ad G-invariant scalar product in $g$. The above formula is true for each connection in $E$ (not necessary self-dual). If a self-dual connection in $E$ exists then $k > 0$. If an anti-self-dual connection exists $(F = -{}^*F)$ then $k < 0$. The change of the orientation makes of anti-self-dual connections self-dual ones. Therefore we will consider only the case of a self-dual connection and $k > 0$.

The 4-form integrated in (1.2) is an element of the first Pontryagin class $p_1$ of $E$ (if $G = U(n)$, $SU(n)$ then it is the 2nd Chern class $c_2(E)$). This class may be identified with an element of $H^4(M, Z)$ since $k \in Z$. Note that the instanton number is in fact an invariant of the bundle.

Below we will note the main events in the history of discoveries of instantons:

(a) <u>1975</u>: Belavin, Polyakov, Schwarz and Tyupkin [11] derive a formula for a <u>special</u> $k = 1$, $G = SU(2)$ instanton over $\underline{S}^4$.

(b) <u>1976-1978</u>: Many people obtained the formulae for special and <u>general</u> instantons over $S^4$. Let me try to list them: Witten [87], Burlakov, Dutyshev [21], t'Hooft [48], Jackiw, Nohl, Rebbi [54], Schwarz [74], Jackiw, Rebbi [55], Bernard, Christ, Guth, Weinberg [13] and Atiyah, Hitchin, Singer [6]. Finally, Atiyah, Hitchin, Drinfeld, Manin (AHDM) [5] prove the completeness of linear algebra constructions of <u>all</u> instantons over $\underline{S}^4$ for all classical simple groups $G$ (for special Lie groups dimensions of moduli[a] spaces of k-instantons are calculated by Atiyah <u>et al</u>. [6]).

---

[a]  The name "moduli" is caused by the 1.1 relation "modulo".

(c) <u>1978</u>:  Atiyah <u>et al.</u> [6] note that $k > 0$, <u>U(1)</u>-instantons exists over 4-<u>tori</u> being Abelian varieties and that such instantons exist over each compact oriented 4-manifold having a <u>positive</u> intersection form (see Appendix A).  This fact is also noted and exploited by Donaldson in his famous paper [32]. Moreover Atiyah <u>et al.</u> prove the existence of instantons over compact 4-manifolds admitting an Einstein metric (e.g. $P^2\mathbb{C}$) in the O(3)-bundle of self-dual 2-forms.

(d) <u>1978</u>:  In the aforementioned paper [6] Atiyah <u>et al.</u> derive the formula for dimension of moduli spaces of k-instantons $\mathcal{M}(k)$ over a <u>positive</u> (scalar curvature is $> 0$) and <u>self-dual</u> (the Weyl tensor is self-dual) manifold or a positive Kähler manifold [53].

$$\dim\mathcal{M}(k) = p_1(g_E) - (\chi(M) - \tau(M))\,\frac{1}{2}\,\dim G \quad\overset{M=S^4,\ G=SU(2)}{=}\quad 8k - 3$$

$$(1.3)$$

where all quantities ($\chi$ is Euler characteristic, $\tau$ is Hirzebruch signature) are topological invariants.  Unfortunately their proof contains the <u>assumption</u> that

$$\mathcal{M}(k) \text{ is not the empty set} \qquad . \tag{1.4}$$

Note that manifolds satisfying the assumptions before (1.3) have not been classified as yet.  One knows that $S^4$, $P^2\mathbb{C}$ and $S^1 \times S^3$ belong to this class but is not the case of $S^2 \times S^2$ and K3 (see Appendix B).

(e) <u>1978</u>:  Page [68] proves that K3 is the only compact non-flat 4-manifold which admits a non-trivial self-dual Levi-Civita connection (then $E = TK3$).  The Riemannian metric which is associated to such a connection is an example of a gravitational instanton, see [41].  The proof of Page is a consequence of the Calabi conjecture which has been proved just before by Yau [88]. Page proposes also a local parametrization of the moduli space of self-dual Levi-Civita connections on K3.  The number of para-meters is 58.

(f) <u>1974-79</u>: Many people detected instantons over <u>non-compact</u> 4-manifolds. Recall that gauge field <u>monopoles</u> which were constructed in a special $k = 1$, $G = SU(2)$ case in 1974(!) by t'Hooft [47] and Polyakov [69] and for general $k > 0$ and $G$ Taubes [79] and Murray [64] can be realized as certain instantons over $\mathbf{R}^3 \times S^1$ as we note in Sec. 7. In case of $\mathbf{R}^2 \times S^2$ Taubes |78| proves the existence of $k > 0$, $G = SU(2)$ instantons solving the Ginzburg-Landau equation for <u>vortices</u>. The next result is due to Hitchin [46] who constructs a $(3n - 6)$-dimensional family of instantons over a non-compact complex 2-surface which satisfy an n-cyclic group periodicity condition in infinity.

<u>Comment</u>: Looking at the above six points one sees that until the early 80s instantons were known only in special cases and then analytic methods were mostly used in order to detect and to construct them. In particular the condition of self-duality of $M$ as it is before (1.3) is equivalent to the existence of a certain complex structure on the total space of the bundle of projective spinors over $M$.

(g) A rapid growth of a spectrum of known instantons takes place when Taubes [81] announces the following result:

each compact oriented 4-manifold $M$ having a positive intersection form $I_M$ admits instantons in almost all G-bundles $E$ such that $p_1(E) > 0$. In particular the· theorem (1.3) is true without assumption (1.4). $\qquad\qquad$ (1.5)

Recall that the notion of the intersection form at this time was very popular because a few months before Freedman [40] had classified topological compact and simply connected 4-manifolds by means of their intersection forms and a $z_2$-characteristic (this result will be described in Section 5).
The result of Taubes makes the analyticity-related assumptions as those before (1.3) unnecessary. For its proof Taubes uses as main tools the implicit function theorem in Sobolev spaces (which contain a lot of non-analytic functions) and the result of Karen Uhlenbeck [84] which enables us to prolongate considered connections on "seemingly" singular points.

(h) <u>1983-4</u>: Recall that the assumption $I_M > 0$ which is made in (1.5) excludes such popular cases as $M = S^2 \times S^2$ and K3. However recently Taubes [83] has extended his result onto much remaining cases proving that

Self-dual connections (i.e. instantons) exist over each compact oriented 4-manifolds in almost all $k > 0$, $G = SU(2)$ bundles.

The Taubes condition of the existence of instantons for every Riemannian metric is $k > \max (1, \frac{4}{3} I_M^-, -4(I_M^- - 2)^2 + 4)$, where $I_M^-$ is the number of negative eigenvalues of $I_M$. For generic such a metric it is $k > I_M^-$. In particular Donaldson proves that there is no instanton over $-P^2\mathbb{C}$ (the orientation of $-P^2$ is opposite to the holomorphic orientation of $P^2\mathbb{C}$) when $k = 1$ and the metric considered is that of Fubini-Study (this result is unpublished).

On the other hand Freed, Freedman and Uhlenbeck [38] prove that no instanton exist with respect to a generic metric if $G = SU(2)$ and $8k < 3(1 + I_M^- - b_1 (M))$. Note that in this case the "dimension" in (1.3) is negative. The simplest of such a case known is $M = K3$ and $k = 1$.

In the case of "larger" groups (than $U(3)$) Taubes [83] suggests new possibilities of splitting $\mathcal{M}(k)$ into a number of connected components, cf. Section 6. Then the non-existence of instantons in the special cases would be more difficult to be proved.

However now, after the results of Taubes, the main difficulty in a classification of all instantons over all compact 4-manifolds seems to be not the lack of results concerning the special cases of non-existence of instantons but the lack of criteria saying which intersection forms are realized by differentiable manifolds. Nevertheless studies of instantons have led to a partial result of Donaldson to this question which will be described in Sections 4 and 5.

Looking at the above outline of the history of discovering of instantons we can see that the use of non-analytical methods (like those of Sobolev spaces) has turned upside down the problem of the existence of instantons making of the question "how to prove the existence of instantons in more general cases?" the question "how to prove non-existence of instantons in the very special cases?"

## 2. Linear algebra constructions of instantons over $S^4$

Throughout this section $M = S^4$ and $G$ is a classical (matrix) Lie group.

In 1977 direct analytic formulae were known for so-called t'Hooft instantons. In this case one can derive suitable formulae for $A_\mu$ and $F_{\mu\nu}$ calculating first and second order partial derivatives of the following densities [11], [53],

$$\rho(x) = \frac{\lambda}{\lambda + (x - x_1)^2} \quad , \quad \lambda \in (0, \infty) \quad , \quad x \in R^4 \cup \{\infty\}$$

$$\text{if} \quad k > 1 \tag{2.1}$$

and

$$\rho(x) = \sum_{i=1}^{k+1} \frac{\lambda_i^2}{(x - x_i)^2} \quad , \quad \text{where} \quad \lambda_i > 0 \quad ,$$

$$\sum_{i=1}^{k+1} \lambda_i^2 = 1 \quad , \quad \text{if} \quad k > 1$$

(for $k = 2$ an additional equation must be satisfied, cf. [44]). We see that the t'Hooft instantons form a 5, 13, $(5k + 4)$-dimensional families while the moduli spaces of general $SU(2)$-instantons are $(8k - 3)$-dimensional as it is in (1.3). Jackiw and Rebbi [55] derived some perturbative formulae for instantons belonging to $(8k - 3)$-dimensional spaces, but they could prove them only locally that is close to t'Hooft instantons.

Analytic formulae for all instantons ($M = S^4$, $G$ is a simple classical Lie group so that $G \neq SO(4)$) were derived and proved by Atiyah, Hitchin, Drinfeld and Manin (AHDM) in [5] in 1978. These authors proved a 1-1 correspondence between instantons and certain families of linear operators parametrized by points of $S^4$ or $P^2\mathbb{C}$ ($P^2\mathbb{C} = S_c^4$ in sense of Fig. 1).

There are a few such linear algebra realizations of instantons. In the first one [34], [35], [3] instantons are determined by a hermitian structure in the trivial vector bundle $S^4 \times \mathbb{C}^{4k}$ and a family of projectors in the fibres. These projectors when acting on the trivial connection in

$S^4 \times \mathbb{C}^{4k}$ make of it a self-dual connection in a vector bundle $E \to S^4$. In the second realization [3] instantons are given by transformations of $\mathbb{R}^{4k}$ which preserve a given family of 1-forms. In the third (and first historically) realization of instantons a linear algebra procedure for algebraic[b] vector bundles over projective spaces due to Horrocks [51, 1965] is used. Recall that if $\nabla$ is a self-dual connection in a vector bundle $E \to S^4$ and the projection $p$ is given by

$$P^3\mathbb{C} \ni [z_1, z_2, z_3, z_4] \overset{P}{\to} [z_1 + jz_2, z_3 + jz_4] \quad , \qquad (2.2)$$

then $p^*(\nabla)$ is a Cauchy-Riemann operator which determines a <u>complex structure on the total space</u> of $p^*(E)$ [66] and <u>makes of $p^*(E) \to P^3\mathbb{C}$ a holomorphic vector bundle</u>, cf. Fig. 1 and Comment in Section 1. By virtue of the Serre theorem [76] every holomorphic vector bundle over $P^m\mathbb{C}$ is equivalent to an <u>algebraic</u> one. We will denote the resulting algebraic bundle $E_c^\nabla \to P^3\mathbb{C}$. AHDM proved that the Horrocks procedure allows us to produce every bundle $E_c^\nabla$ from a certain congruence $V_z$, $z \in P^3\mathbb{C}$, of complex k-planes in $\mathbb{C}^{2k+n}$ provided with an n-degenerate skew-symmetric form $< ., .>$ and then we have $(E_c^\nabla)_2 = V_z^\perp / V_z$.

All the quoted above linear algebra realizations of instantons may be written in a form of matrix equations. In particular $(G = SU(n))$ the equations for the first realization look as follows.

$$Q = \begin{bmatrix} I_{2n} & 0 \\ & \\ 0 & I_{2n} \end{bmatrix} \quad , \quad R_{2n} = \begin{bmatrix} R_n & 0 \\ & \\ 0 & R_n \end{bmatrix} \quad ,$$

$$R_n = \begin{bmatrix} \rho_1 & 0 \\ & \ddots & \\ 0 & & \rho_n \end{bmatrix} \quad 0 < \rho_1 < \ldots < \rho_n \quad , \quad D = \begin{bmatrix} B & C \\ & \\ -c^t & B^t \end{bmatrix}$$

$$(2.3)$$

---

[b] The bundle is called algebraic if all matrix elements of its transition functions are rational functions.

$$\text{rank } Q = 2k + n \tag{2.4}$$

$$Q \sim \tilde{Q} \leftrightarrow U^{-1}R_n U = R_n \quad , \quad U^{-1}BU = \tilde{B} \quad , \quad U^{-1}CU = \tilde{C} \quad , \quad U \in SU(n) \tag{2.5}$$

Then

$$\mathcal{M}(k) = \{Q\}/\sim \quad . \tag{2.6}$$

One recognizes in $Q$ a matrix of the hermitian structure in $S^4 \times \mathbb{C}^{4k}$, in (2.4) a gauge transformation rule and in $\mathcal{M}(k)$ the moduli space of $SU(n)$-k-instantons.

Observe that (2.5) is a serious obstruction if one tries to parametrize spaces $\mathcal{M}(k)$ or to calculate their topological invariants. Direct calculations have been made for $k = 1$ (then $\mathcal{M}(1) = B^5$, i.e. interior of $S^4$), $k = 2$ and 3. Thus the matrix representations of instantons and their direct matrix formulae turn out to be insufficient for learning something about $\mathcal{M}(k)$ (e.g. are they connected?). As we will see in Section 6 certain topological problems concerning $\mathcal{M}(k)$ have been recently solved but in the proofs variational and topological methods were used instead of those of analytic functions and linear algebra.

The proof of the self-duality of the connections generated by the AHDM matrix representations is easy and short. Much more difficult are proofs of the completeness of the AHDM realizations of instantons though one can quickly calculate that the dimensions of moduli spaces of instantons in the AHDM constructions like (2.6) are equal to dimensions of moduli spaces $\mathcal{M}(k)$ given by (1.3).

The main point of the AHDM proof [36], [33] of the completeness of their realization of instantons is the <u>vanishing theorem</u> $H^1(P^3\mathbb{C}, E_{\mathbb{C}}^{\triangledown} \otimes H^{-2}) = 0$, where $H \to P^3\mathbb{C}$ is the hyperplane bundle and $H^{-1} = H^*$. In the AHDM proof of the vanishing theorem one uses standard methods of analytic functions as divisors, Cauchy integral formulae, the Liouville theorem and so on. The remaining part of the completeness is the Riemann-Roch-Hirzebruch theorem and the standard rules of exact sequences of cohomologies $H^1(P^3\mathbb{C}, E_{\mathbb{C}}^{\triangledown} \otimes H^1)$ which take a part in the Horrocks procedure.

Note that proofs of AHDM of the completeness of other realizations of instantons also contain the vanishing theorem and the exact sequences, cf. [3].

Nevertheless one knows that instantons may be well defined in terms of differential geometry of $S^4$ so that Atiyah in 1979 proposed to look for a pure geometrical proof of the completeness of an AHDM realization without the vanishing theorem. Such a proof was elaborated in 1983 by Corrigan and Goddard (CG) in [24].

As in the AHDM proof basic tools are cohomologies like $H^1(P^3\mathbb{C}, E_\mathbb{C}^\nabla \otimes H^1)$ as in the CG proof main tools are "moments" of spinor fields in an instanton field like $\mu = \int_{S^4} \times \psi^* \psi dv$ and asymptotics of the fields $\psi$.

The CG proof is based on the standard integral calculus, formuale for the Green function and the index theorem. This proof is much longer than the AHDM proof (in spaces of differentiable functions one counts more slowly than in spaces of rational functions) but the method of CG is relevant to interesting problems, in particular

(a) The CG proof implies that the number of Dirac modes in a non-self-dual gauge field is greater than $k$. The problem is to calculate or to estimate dimensions of moduli spaces of gauge fields for which this number is $k + 21$.

(b) The CG proof suggests that moduli spaces of instantons underlying to the AHDM construction may be embedded in corresponding moduli spaces of instantons over 4-manifolds admitting a quaternionic structure and being different to $S^4$ (by virtue of (1.3) the latter ones have dimensions greater than $8k - 3$).

(c) Perhaps a linear algebra construction of all $SO(4)$ gauge fields over $S^4$ satisfying the two-fold self-duality condition (5.2) may be realized and proved in a similar way.

3.   What invariant is given by $H^1(P^3\mathbb{C}, E_\mathbb{C}^\nabla)$?

The following dimensions can be calculated by means of the exact sequences rules, the vanishing theorem and the Riemann-Roch-Hirzebruch theorem in the case $M = S^4$, $G = SU(n)$, see [3], [8], [10], [25] $(h^{(\cdot)}(\cdot) = \dim H^{(\cdot)}(\cdot))$

$$h^1(P^3\mathbb{C}, E_\mathbb{C}^\nabla \otimes H^1) = 0 \quad \text{if} \quad 1 < -2$$

$$h^1(P^3\mathbb{C}, E_\mathbb{C}^\nabla \otimes H^{-1}) = k$$

$$h^1(P^3\mathbb{C}, E_{\mathbb{C}}^{\nabla}) = 2k - n$$

$h^1(P^3\mathbb{C}, E_{\mathbb{C}}^{\nabla} \otimes H)$ is not constant on $\mathcal{M}(k)$ for $k > 2$.

The underlined dimension is clearly a topological invariant. It is equal to dimension of the degeneracy space of the hermitian form (2.2) in the AHDM construction. The question is what geometric object does correspond to $h^1(P^3\mathbb{C}, E_{\mathbb{C}}^{\nabla})$ in case of a general 4-manifold M. Such an object might help us to find more constructive realizations of instantons over 4-manifolds different to $S^4$.

## 4. Singularities of the instanton moduli spaces

Moduli spaces of instantons are integration domains in pseudo-classical non-perturbative models of quantum field theory so that their geometry and topology is interesting for physicists as well as mathematicians. The question is whether they are smooth manifolds or spaces with singularities. In theory of complex manifolds, in particular in algebraic geometry, such problems are difficult if not hopeless because there are no effective criteria selecting smooth algebraic manifolds in a class of algebraic varieties (which admit singularities). In particular the problem of smoothness of moduli spaces of <u>mathematical instantons</u> has not been solved as yet.

Recall that a mathematical instanton is said to be an algebraic vector bundle $E_{\mathbb{C}} \to P^3\mathbb{C}$ such that

1. rank $E_{\mathbb{C}} = 2$,

2. $c_1(E_{\mathbb{C}}) = 0$, $c_2(E_{\mathbb{C}}) = k > 0$,

3. $H^1(P^3\mathbb{C}, E_{\mathbb{C}} \otimes H^{-2}) = 0$ (this is the vanishing theorem).

We see that the above definition is more general than that of "instanton" algebraic bundles $E_{\mathbb{C}}^{\nabla}$ in Section 2. From a point of view of algebraic geometry mathematical instantons seem to be even more "natural" objects than the instanton bundles $E_{\mathbb{C}}^{\nabla}$. Mathematical instantons satisfy the assumptions of the Horrocks constructions and are uniquely determined by solutions of quadratic matrix equations up to the action of $Gl(k, \mathbb{C}) \times Sp(2k+2)$. These equations are given in [9].

The moduli spaces of mathematical instantons $\mathcal{M}_C(k)$ are of the complex dimension $\geq 8k - 3$ and a component of $\mathcal{M}_C(k)$ is $(8k - 3)$-dimensional. In all known cases i.e. $k = 1,2,3,4$ $\mathcal{M}_C(k)$ is a connected space.

The smoothness of $\mathcal{M}_C(k)$ is proved only if $k = 1,2,3,4$.

The moduli spaces of instantons $\mathcal{M}(k)$ may be regarded as sets of real points of $\mathcal{M}_C(k)$ and the lack of results concerning $\mathcal{M}_C(k)$ is a reason that the smoothness of $\mathcal{M}(k)$ for $k > 4$ has not been proved by means of algebraic and analytic methods.

The smoothness of moduli spaces of <u>generic</u> instantons for a general M and S has been proved recently [38] by means of functional analysis methods, especially estimations in Sobolev norms. The first step on the way to this result was made by Atiyah <u>et al</u>. [6] who proved the smoothness of moduli spaces of <u>irreducible</u> instantons (i.e. irreducible self-dual connections) if M is positive and self-dual and $\mathcal{M}(k) \neq \emptyset$, cf. 1d. Recall that the assumption about self-duality of M is related to the existence of a complex structure on the space of a bundle over M. Nevertheless Atiyah <u>et al</u>. used in their proof also Sobolev norms of considered local maps. I think that it was the first case when non-analytic self-dual gauge fields played an important part in the theory of instantons.

Observe that the assumption of irreducibility is meaningless if $H^2(M, \mathbf{R}) = 0$, e.g. $M = S^4$, and $G = SU(2), SU(3)$. Otherwise, e.g. $M = P^2\mathbb{C}$, there are usually self-dual $U(2)$-connections admitting a reduction to $U(1) \times U(1)$ which correspond to singular points of a suitable $\mathcal{M}(k)$. Such a case was noted by Atiyah <u>et al</u>. but they did not pay more attention to it. This case was examined five years later by Donaldson [32] and then became a source of a wave of surprising results in theory of 4-manifolds.

Using a similar methodology to that of Taubes [81] (implicit function theorem in Sobolev spaces, cf. 1.9) Donaldson explored instanton moduli spaces when $k = 1$ (so that $G = SU(2)$) and M is a compact connected oriented (briefly <u>c.s.o.</u>) Riemannian differential 4-manifold satisfying the following assumption: the intersection form $\underline{I_M}$ <u>is positive</u>. In this case Donaldson considered the moduli space of irreducible 1-instantons. He could not prove its smoothness but he proved that a deformation of this space is smooth (the self-duality equations were deformed). In this case the Atiyah <u>et al</u>. assumption imposing an analytic-related structure on M turned out to be unnecessary.

This result of Donaldson was improved by Freed, Freedman and Uhlenbeck [38]. They extended it onto a general case, namely $G = SU(2)$, $M$ is compact and the Atiyah et al. formula (1.3) makes sense that is

$$8k - \frac{3}{2}(\chi(M) - \tau(M)) = 8k - 3(1 + I_M^- - b_1(M)) > 0$$

($b_1$ is first Betti number). Then these authors proved that the Donaldson's deformation of the self-duality equations may be replaced by a deformation of the Riemannian metric so that the moduli spaces of suitable irreducible self-dual connections are smooth in the sense of a generic metric.

But the main point of the results of Donaldson is related to the topology of the moduli spaces $\mathcal{M}(1)$ in the considered cases. It is

> If $M$ is a c.s.o. diff. 4-manifold such that $I_M > 0$ then (in a generic case) $M$ compactifies the moduli space of 1-instantons $\mathcal{M}(1)$ to a compact manifold $\mathcal{\bar{M}}(1)$ having as the boundary and $p = b_2(M)$ singular points which correspond to the $U(1) \times U(1)$-1-instantons. Moreover each such singular point admits a neighbourhood in $\mathcal{\bar{M}}(1)$ isomorphic to the $P^2\mathbb{C}$-conus.

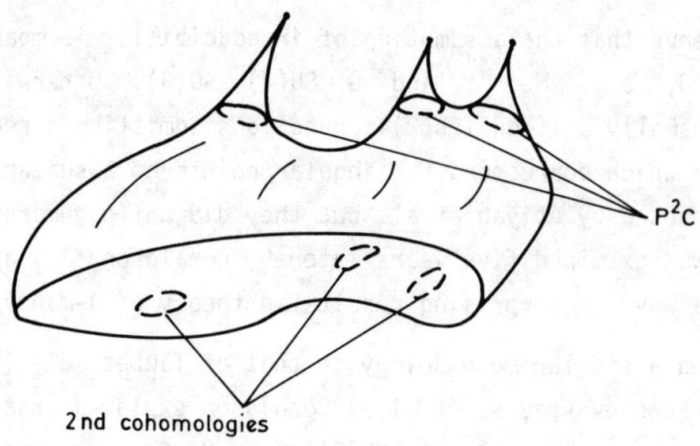


Fig. 2

Note that the M-like component of $\partial\mathcal{M}(1)$ corresponds to $\delta^4(x - x_1)$, $x_1 \in M$, densities of a topological charge, cf. (2.1) [44].

Then the theory of cobordisms implies immediately that $M$ is uniquely given in sense of a top. manifold (i.e. the corresponding manifold of $C^0$-class), namely

$$M = P^2\mathbb{C}\#\underbrace{\ldots}\#P^2\mathbb{C}$$
$$\text{p copies}$$

so that

$$I_M = I_{P^2\mathbb{C}\#\ldots\#P^2\mathbb{C}} = I_P$$

(cf. Appendix A). It means that $I_M$ may be diagonalized to the identity matrix over integers. Therefore a non z-diagonalizable positive integer unimodular bilinear form cannot be the intersection form of a c.s.o. diff. 4-manifold. But such a form is the intersection form of a c.s.o. top. 4-manifold as we see in the next section. Thus the Donaldson theorem together with the classical theory of integer unimodular forms (Appendix A) gives us the following criterion:

If a c.s.o. top. 4-manifold $M$ admits a diff. structure then $I_M$ either is z-diagonalizable or it splits in an even number of $E_8$ (see B3) and a positive number of $\begin{bmatrix} 0 & 1 \\ 1 & 0 \end{bmatrix}$.

Recall that the above criterion eliminates a lot of candidates for smooth 4-manifolds (see Appendix A).

## 5. Self-duality and exotic differential structures

In Section 2 we have seen that instantons make a lot of complex structures on the total space of a smooth vector bundle $p^*(E)$, where $E \to S^4$ and $p:P^3\mathbb{C} \to S^4$ are as in (2.1). However different complex structures being equivalent in sense of smooth manifolds are a popular phenomenon. An example is the moduli space of $K3$ surfaces in Appendix B. Recall that even $\mathbb{R}^2$ admits a continuum of different complex structures as the coordinate systems $\langle 1,i \rangle$ and $\langle a,bi \rangle\rangle$, $a, b \in \mathbb{C}-\{0\}$, $a \neq b$ determine inequivalent complex structures.

The notion of a differential structure (that is a smooth manifold) seems to be closer to usual intuitions since in dimensions 0,1,2,3 each top. manifold (i.e. a manifold of $C^0$-class) admits the only differential structure. This fact is known since 20s (dim. 0,1,2) and 50s (dim. 3), see [63]. Nevertheless in higher dimensions there are both top. manifolds admitting no differential structure and top. manifolds "carrying" different differential structures. The latter fact is known since 1956, when Milnor [60] detected non-standard differential structures of $S^7$, which are called <u>exotics</u> $S^7$. Nowadays 27 exotic $S^7$ and e.g. 16 881 176 exotic $S^{31}$ are known. In 1958 Borel and Hirzebruch [15] detected finite numbers of inequivalent differential (and complex) structures on some homogeneous spaces, e.g. $\frac{U(4)}{U(2) \times U(1) \times U(1)}$ analysing corresponding root systems. In 1968 Kirby and Siebenmann [59] pointed out a class of topological manifolds of dimension > 4 admitting no differentiable structure. In 1979 Freedmann [39] proved the existence of exotic differential structures on a 4-manifold namely $S^3 \times R$. In 1982 the same author proved a classification theorem for topological 4-manifolds (it will be quoted in this section) which implies non-existence of differentiable structures on a large class of topological 4-manifolds. This class was increased soon, when Donaldson proved the theorem about diagonalization of intersection forms for a class of differential 4-manifolds which we have seen in the previous section. But more surprising seems to be the following fact which is also a consequence of the Donaldson theorem:

There exists exotic differential structures on $\mathbf{R}^4$.

(They will be denoted $\mathcal{R}^4$). The existence of $\mathcal{R}^4$ and the gauge fields way of detecting them has changed our perception of geometry. It suggests that the nature of even such a banal abstract space as $\mathcal{R}^4$ is completely different to that of "our" 3-dimensional world and that the ideas commonly used by theoretical physicists are a source of paradoxal mathematical results.

Before we say more about $\mathcal{R}^4$ let us remind the aforementioned Freedmann classification theorem.

> Each integer unimodular symmetric form up to z-equivalence is the intersection form $I_M$ of a c.s.o. topological 4-manifold $M$. This manifold is uniquely determined by $I_M$ and a $Z_2$-index $\alpha(M)$, where $\alpha(M) = 0$ iff $M \times S^1$ admits a diff. structure. If $I_M$ is odd (see Appendix A) then $\alpha(M)$ takes both values 0 and 1. Otherwise $\alpha = \frac{1}{8}$ signature $I_M$ (mod 2)[c].

---

[c] The fact that if $M$ is smooth and $I_M$ is even then signature of $I_M$ is divisible by 16 was proved by Rochlin [71] in 1952! Moreover Milnor proved in 1958 [61] that if $M$, $M'$ are c.s.o. top. 4-manifolds and $I_M = I_{M'}$ then homotopy types of $M$ and $M'$ are the same. The index $\alpha$ was defined by Kirby and Siebenmann.

The above theorem says roughly speaking that the spectrum of c.s.o. top. 4-manifolds is in partially 2-1 and partially 1-1 correspondence with the set of integer unimodular symmetric forms and is split by $\alpha$ into two "almost symmetric" parts. The manifolds belonging to the first one do not admit any diff. structure while the manifolds which are members of the second one may admit zero, one or perhaps many diff. structures.

Before the Freedmann theorem no top. manifold having $E_8$ or $E_8 \oplus E_8$ (see B3) as the intersection form was known. So the top. manifolds $M_8$ and $M_{8 \oplus 8}$ realizing $E_8$ and $E_8 \oplus E_8$ resp. are children of this theorem. Also by virtue of the Freedmann theorem the following decomposition in the sense of top. manifolds takes place

$$K3 = M_{8 \oplus 8} \; \# \underbrace{(S^2 \times S^2 \# S^2 \times S^2 \# S^2 \times S^2)}_{CH} \quad .$$

Note that the Freedmann theorem says nothing about diff. structures on $M_{8 \oplus 8}$ in contrast to $M_8$ where it makes diff. structures prohibited. This is the Donaldson theorem (Section 4) which implies non-existence of $M_{8 \oplus 8}$ in sense of a smooth manifold. The absence of diff. structures on $M_{8 \oplus 8}$ enables us to prove shortly by means of the Freedmann theorem methods the existence of some exotic structure $\mathcal{R}^4$: it turns out to be the diff. structure of the open submanifold $B := CH - M_{8 \oplus 8}$ of $CH$ provided with the standard diff. structure ($B$ is clearly homeomorphic to $\mathbf{R}^4$). This structure is called the Freedmann-Donaldson exotic $\mathcal{R}^4$.

In order to learn why the diff. structure on $B$ is not the standard one of $\mathbf{R}^4$ observe that the boundary $\partial B$ of $B$ being a topological $S^3$ (i.e. homeomorphic to $S^3$) is not smooth or smoothable in the sense of $K3$. Otherwise the compatibility diff. structures on $B$ and $CH-B$ (which can be merged into $CH$) and of those on $K3-CH$ and $CH-B$ (which can be merged into $K3$) would imply the compatibility of the diff. structures on $B$ and $K3-CH$ so that a diff. structure on $M_{8 \oplus 8}$ would exist. The next consequence of non-existence of a smooth $M_{8 \oplus 8}$ is lack of any smooth transversal 3-sphere in the neck $N$ on Fig. 3 that is in a neighbourhood of $\partial B$ which is homeomorphic to $S^3 \times R^1$. The transversality means that such a sphere would separate the ends (i.e. the images of $S^3 \times \{\pm\infty\}$). Otherwise one could shift $\partial B$ to a $K3$-smooth 3-sphere. Thus the diff. structure of $N$ must be exotic.

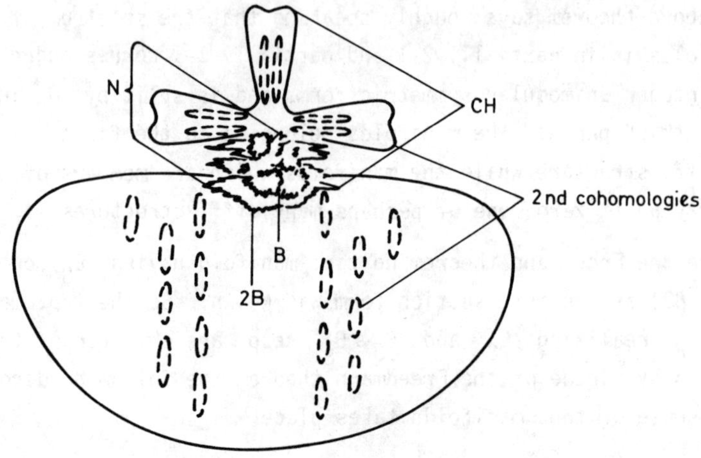



Fig. 3

On the other hand if a space homeomorphic to $S^3 \times [0, \infty]$ is glued to a top. 4-ball provided with the standard diff. structure of $\mathbf{R}^4$ then the standard diff. structure of $S^3 \times \mathbf{R}^1$ is induced on it and there is a lot of smooth 3-spheres which enclose the top. 4-ball. That is why the resulting diff. structure on B is exotic.

Note that there are local diffeomorphisms mapping cross sections of N close to $\partial$B into B. Their images as C on Fig. 4 are topological 3-spheres which are encircled by no $\mathcal{R}^4$-smooth sphere. In particular suffi-ciently large $\mathbf{R}^4$-smooth spheres are not $\mathcal{R}^4$-smooth. This paradox is related to a mysterious distribution of smooth lines and (hyper-)surfaces in $\mathcal{R}^4$. One can think that the horizon in $\mathcal{R}^4$ is extremely jagged.

Note that the exploration of the topology of such "necks" given by embeddings of manifolds homeomorphic to $S^2 \times S^2$ and their connected sums (we call them flexible Casson handles) is the most important and difficult part of the proof of the Freedmann theorem. Recall that Bowder proved in 70s that each integer unimodular form can be realised by a 1-connected top. 4-manifold and the results concerning the Casson handles helped Freedmann to realize them by means of simply connected 4-manifolds and then to prove the uniqueness of such a realization. In the Freedmann proof non-standard topological spaces as the Whitehead continuum [86] are deeply involved. One can imagine that the neck of Fig. 5 is weaved of such "threads" as the

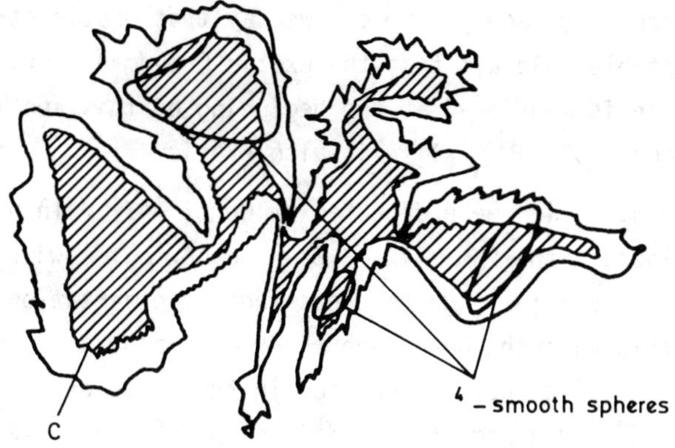



Fig. 4

famous Weierstrass teeth-function and its sections being topological non-differentiable 3-spheres look like the "horned" sphere below.  The construction of such 3-spheres in the Freedmann proof looks (methodically) like the pictorial prescription for the Peano line.

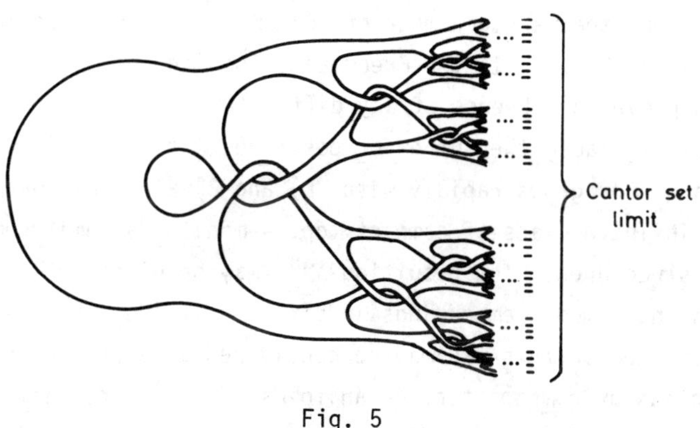



Fig. 5

However such an intuition cannot give us a proper image since $\mathcal{R}^4$ is an effect of the dimension four.

Before detecting $\mathcal{R}^4$ it was known since 70s that $\mathbf{R}^n$, $\underline{n \neq 4}$, admit the only standard diff. structures, cf. [89] (though it is unreasonable why the problem of diff. structures of $\mathbf{R}^4$ was not considered as one of more important problems of mathematics of 70s).  Thus we have $\mathcal{R}^4 \times \mathbf{R}^1 = \mathbf{R}^4 \times \mathbf{R}^1 = \mathbf{R}^5$ so that the category of diff. manifolds with the Cartesian product fails the group property $ac = bc = a = c$.  We say that exotic structures on $\mathbf{R}^4$ are $\mathbb{R}^1$-unstable.  Recall that the exotic Freedman diff. structure on $S^3 \times \mathbf{R}^1$ (which is mentioned at the beginning of this section) is $\mathbf{R}^1$-stable so that then $(S^3 \times \mathbf{R}^1)_e \times \mathbf{R}^1$ is also exotic.

Next two exotic $\mathcal{R}^4$ were detected by Gompf [43].  In his construction K3 is replaced with $P^2\mathbb{C} \# (\#_9 - P^2\mathbb{C})$ and $M_{8+8}$ — with the top. manifold realizing $- E_8 + [-1]$ up to the change of orientation.  Let us observe that in this case the Donaldson theorem is also used in order to prove that $- E_8 + [-1]$ cannot be realized in the class of diff. 4-manifolds. The exotic Gompf $\mathcal{R}^4$ admit smooth embeddings in $P^2\mathbb{C}$ and $-P^2\mathbb{C}$ resp. (but not in $S^4$!) in contrast to the exotic Freedmann-Donaldson $\mathcal{R}^4$ which cannot be embedded in $S^4$ and in any smooth e.s.o. 4-manifold $M$ such that $I_M > 0$.

Note that we do not know what we obtain if we take in the Donaldson-Freedmann construction a part realizing $E_8 + E_8 + \begin{bmatrix} 0 & 1 \\ 1 & 0 \end{bmatrix}$ or $E_8 + E_8 + \begin{bmatrix} 0 & 1 \\ 1 & 0 \end{bmatrix} + \begin{bmatrix} 0 & 1 \\ 1 & 0 \end{bmatrix}$ because we do not know whether the corresponding top. c.s.o. 4-manifolds admit diff. structures or not.

One hopes that a big number of $R^4$ may be identified in a similar way.  That is so because 1) the Freedmann, Rochlin and Donaldson theorems allow us to prove the absence of any diff. structure on top. 4-manifolds of a large class (because the number of positive, integer, unimodular, symmetric forms of rank $8n$ grows rapidly with $n$) and 2) algebraic geometry teaches us that an infinite class of compact top. 4-manifolds admit complex so that also diff. structures.  The resulting $\mathcal{R}^4$ may be different as they can realize distinct smooth embeddings in diff. manifolds.  Let us observe that all $\mathcal{R}^4$ obtained in this way may be considered as a product of a duality between a class of <u>compact</u> top. 4-manifolds admitting no diff. structure and a class of "additional" diff. structures on <u>non-compact</u> 4-manifolds. The presented method gives no result when limited to the class of non-compact manifolds because each such manifold (of any dimension $n$) has at least one diff. structure [40]).  So that if we remove one point from $M_{8+8} \subset K3$ then this proof of K3 fails.

The discoveries of $\mathcal{R}^4$ did not help us as yet to prove or to deny the "smooth Poincaré conjecture" in dimension four:  there is no exotic $S^4$. Nevertheless 13 exotic unstable diff. structures on $S^3 \times \mathbf{R}_1$ induced by $\mathcal{R}^4$ (the Freedmann-Donaldson and the Gompf one) were identified.  So that the 4-dimensional world seems to be terrible exotic.

As the milestones on the way to the place where the exotic 4-dimensional worlds becomes observable one can consider:

1. The Taubes theorems which say that SU(2)-instantons exist over each "reasonable" diff. 4-manifolds.

2. The Freedmann classification theorem saying that every integer unimodular symmetric form can be realized by top. 4-manifolds (Casson handles and Whitehead continuum).

3. The Rochlin and Donaldson theorems giving us criteria which determine a lot of compact top. 4-manifolds carrying top diff. structure.

We see that both the discussed methods of mathematics and mathematical physics, that is topology together with a "smooth" geometry and analysis functions are necessary in order to deal with the exotic 4-manifolds.

Before the outline the idea of finding exotic structures on spheres due to Milnor let us come back to the duality in Grassmann algebras. Recall that such a duality is well defined on $\Lambda^2 V_n$, where $V_n$ is an Euclidean n-dimensional vector space, iff $\underline{n = 4}$. Then we have $\Lambda^2 V_4 = \Lambda^2_+ V_4 \oplus \Lambda^2_- V_4$, where elements of $\Lambda_+ V_4 (\Lambda_- V_4)$ are self-dual (anti-self-dual). One knows that $\Lambda^2 V_4 \cong SO(4)$ and hence we have $SO(4) = SO(3) \oplus SO(3)$. This is the only such decomposition of a classical Lie algebra as $SO(4)$ is the only non-simple classical Lie algebra.

In the case of SO(4)-bundles over smooth 4-manifolds the latter decomposition when applied to coefficients $F_{\mu\nu}$ of a curvature form $F$ gives us

$$F_{\mu\nu} = F^+_{\mu\nu} + F^-_{\mu\nu} \qquad \text{so that}$$

$$F = F^+ + F^- = F^+_+ + F^+_- + F^-_+ + F^-_- \qquad ({}^*F^{(\cdot)}_+ = F^{(\cdot)}_+, \; {}^*F^{(\cdot)}_- = -F^{(\cdot)}_-) \, .$$

Thus two topological invariants of SO(4)-bundles $E$ over 4-manifolds make sense, namely

$$k = p_1(E) = \frac{1}{8\pi^2} \int_M (\| F^+_+ \|^2 + \| F^-_+ \|^2 - \| F^+_- \|^2 - \| F^-_- \|^2)dv \qquad (5.1)$$

which is the Pontryagin number, cf. (1.2), and

$$\chi = \chi(E) = \frac{1}{8\pi^2} \int_M (\| F_+^+ \|^2 + \| F_-^- \|^2 - \| F_-^+ \|^2 - \| F_+^- \|^2) dv$$

which is called the Euler characteristic (in case $E = TM$ it gives the Euler characteristic of $M$).

In case $M = S^4$ both these numbers determine uniquely an SO(4)-bundle.

If $\chi > |k|$ (e.g. $E = TS^4$) then no self-dual connection in $E$ exists but there exist two-fold self-dual connections given by

$$F_-^+ = F_+^- = 0 \ . \tag{5.2}$$

In case $M = S^4$ Bourguignon and Lawson [17] proved that moduli spaces of self-dual connections (if $k > \chi > 0$) and those of two-fold self-dual ones (if $\chi > k > 0$) are d-dimensional, where

$$d = \begin{cases} 8k - 6 & k \neq \chi \\ 8k - 3 & k = \chi \end{cases} \Big\} \ k > 0, \ \chi > 0 \\ \quad\ 0 \quad\quad k = \chi = 0 \ .$$

Observe that if $k = \chi$ then $d$ is larger than the "dimension" calculated accordingly to (1.3).

Note that neither a Penrose transformation nor AHDM-like realization of the two-fold self-dual connections are given in the literature.

The invariants $k$ and $\chi$ are used in the proof [60] of the existence of exotic 7-spheres. It turns out that the total bundle of the fibre bundle of unit balls in the fibres of $E$ is homeomorphic to $S^7$ iff $\chi = 1$ and it is not diffeomorphic to $S^7$ if $k \neq 1$ (mod 7). More profound studies allow us to detect 27 exotic 7-spheres.

Thus we see that the duality in $\Lambda^2 V_4$ helped us to point out both $\mathcal{R}^4$ and exotic $S^7$. However the ideas of the proofs were different. In the case of $S^7$ the exotic structures were consequences of certain cohomological properties of $S^7$. In the case of $R^4$ the exotic structures were obtained by means of an external construction.

Now let me try to give a summary of the present knowledge about the discussed structures of manifolds.

1.  In special cases (e.g. compact, simple connected and dimension 4) top. manifolds are known to be classified by discrete sets of

invariants.  This is not the case of dimension 3 where the Poincaré hypothesis (is top. $S^3$ determined by its homotopy?) remains unproved.

2.  Some conditions for non-existence of any diff. structure on a top. manifold were given.  The set of differentiable structures on a top. manifold  M  was proved to be discrete if  dim M $\neq$ 4. The case  dim M = 4  remains unsolved.

3.  Moduli spaces of complex structures which are equivalent in the smooth sense in known cases are usually algebraic varieties of a positive dimension (except e.g. some complex homogeneous spaces).

6.  <u>Non-self-dual critical points</u>

Self-dual connections are global minima of the Yang-Mills functional $\mathcal{L}(\nabla)$  considered in a set of all connections in a vector bundle  E → M, where

$$\mathcal{L}(\nabla) = \frac{1}{2} \int_M \| F \|^2 \, dv = \frac{1}{2} \int_M ( \| F^+ \|^2 + \| F^- \|^2 ) dv \qquad (6.1)$$

as

$$\frac{1}{2} \int_M ( \| F^+ \|^2 - \| F^- \|^2 ) dv = 4\pi^2 k$$

cf. (5.1).

However other critical points, if exist, would be also interesting. In particular contributions of such points to quantum field action functionals might be appreciable.

Atiyah and Bott [16] presented an idea of detecting sets of critical points by means of the Morse inequality and abstract topological so called classifying spaces.  In the case when  M  is a Riemmann surface of arbitrary genus  (dim$_R$ M = 2)  and  k = 1  Atiyah and Bott calculated cohomologies of moduli spaces of global minima and uniquely determined moduli spaces of the remaining critical points of the (6.1) functional (then  $F^\pm$  make no sense). Soon Sachs and Uhlenbeck [72] proved that their method fails when applied to a 4-manifold  M.

The results related to the problem of non-self-dual critical points are of two types, namely:  1) such points are absent in resulting regions of moduli spaces of connections [30], 2) "certain features of some critical points are prohibited for the non-self-dual ones".

The important result of the 2nd type is due to Bourguignon, Lawson and Simons [18], [17].  They proved that

If  M  is a homogeneous compact 4-manifold, i.e.  $M = S^4$, $S^2 \times S^2$, $S^3 \times S^1$, $S^2 \times S^1 \times S^1$, $S^1 \times S^1 \times S^1 \times S^1$, $P^2\mathbb{C}$, $G = SU(2)$, $U(2)$, $SU(3)$ and $k > 0$ then self-dual connections are the only local minima.

Thus non-self-dual critical points, if exist, are saddle points of $\mathcal{L}(\nabla)$ (6.1).

The proof of the above theorem exploits the following fact:

If a Lie algebra  $g$  satisfies the following condition:
if Lie subalgebras  $g_1$, $g_2 \subset g$  commute  $([g_1, g_2] = 0)$  then either  $g_1$  or  $g_2$  is Abelian, then  $g = SU(2)$, $U(2)$, $SU(3)$.

Burguignon _et al_. combined artfully this fact with the duality relation using a classical differential geometry calculus (in particular the Bochner-Weitzenböck formulae for laplacians).  Their ideas were continuated by Taubes [80], [82], who estimated from down the index of the hessian in the hypothetical non-self-dual critical points  ($M = S^4$, $T^4$, K3; $G = SU(2)$, $SU(3)$).  This method turned out to be insufficient in order to answer the question "do such points exist?" but it helped him much to answer "yes" another important question related to instantons, namely "are all moduli spaces of instantons  $\mathcal{M}(k)$  connected?" in case  $M = S^4$;  $G = SU(2)$, $SU(3)$  [82].  Recall that this problem has not been solved by means of the algebraic methods reviewed in Section 2.

The Taubes proof of the connectivity of  $\mathcal{M}(k)$  is purely variational and geometrical.  Apart from the index theorem it uses the Morse inequality, a "mini-max" variational technics and Sobolev spaces which are exploited probably in all papers of this author.

Observe that the last result says that the only topological invariant of instantons in considered cases is the instanton number  k.  Before proving it not all mathematicians believed that it is true.  That was so because various topological invariants are known in algebraic geometry, e.g. the Atiyah-Rees invariant  $h^1(P^3\mathbb{C}, E_c \otimes H^{-2})$  (mod 2), where  $E_c \to P^3\mathbb{C}$  is an algebraic vector bundle of rank 2, and many specialists in algebraic geometry expected that more invariants rule in the domain of instantons, cf. Section 3.

Another interesting consequence of the Taubes proof is a mysterious relation between the hypothetical non-self-dual critical points and the also hypothetical multi-component instanton moduli spaces in cases of more complicated base spaces $M$ and gauge groups $G$. In particular Taubes proved that in the case when either $M = S^4$ and rank $G > 3$ or $M$ is 1-connected, $k = 1$ and $G = SU(2)$ the existence of more than one components of $\mathcal{M}(k)$ implies the existence of non-self-dual critical points of $\mathcal{L}(\nabla)$.

On the other hand in the mostly banal case $M = S^4$, $G = SU(2)$ a main problem related to instantons seems to be the Atiyah conjecture

$$\pi_1(\mathcal{M}(k)) = z_2 \quad \text{if} \quad k > 2 \quad (\pi_1 \text{ is first homotopy group})$$

All what we know is $\pi_1(\mathcal{M}(1)) = 0$ and $\pi_1(\mathcal{M}(2)) = z_2$, cf. $[44]$.

## 7. The gauge field monopoles

Monopoles in gauge fields are a special case of instantons. That is so because if we have the Bogomolny equation $[14]$

$$\nabla \phi = {}^*F \tag{7.1}$$

(here $\phi: \mathbf{R}^3 \to g$ is such that $\|\phi\| \overset{r \to \infty}{\to} 1 + \dfrac{ck}{r} + 0(r^{-2})$, $c > 0$, $k$ is the degree of $\phi$ on a "large" 2-sphere in $\mathbf{R}^3$, $\nabla$ is a covariant derivative in $\mathbf{R}^3 \times V$ and $F$ denotes its curvature) then we can prove a 1-1 correspondence between the solutions (7.1) up to gauge transformations and these $G$-instantons on $\mathbf{R}^3 \times S^1$ which are $S^1$-invariant $(\partial_4 A_\mu = 0)$ and satisfy an asymptotic condition, see $[46]$.

Ward $[85]$ applied for monopoles a usual Penrose transformation as that in Section 2 in (2.2) and obtained in this way a formula for $k = 2$ monopoles. Hitchin $[46]$ looking for a space of data for all SU(2)-monopoles proposed another variant of the Penrose transformation which turned out to be close to the oldest known example of the Penrose transformation due to Weierstrass (1866). Recall that many historians of mathematics suppose that Weierstrass was the first one who was interested in the fact of existence and non-existence of solutions of proper analytic problems instead of possibilities of solving them explicitly.

Weierstrass used his "Penrose" transformation dealing with the problem of a minimal closed surface in $\mathbf{R}^3$ which encloses given $k$ points.

He observed that if a solution of a system of complex algebraic equations is given then one can make of it a solution of the minimal surface problem. In order to obtain such a system of equations one should identify the set of oriented lines $\ell$ in $\mathbb{R}^3$ with $TS^2$ (observe that the line is represented by a unit vector of direction and the perpendicular vector ending in the origin) and then provide $TS^2$ with the complex structure of $TP^1\mathbb{C}$.

The idea of Hitchin is to assign to each pair of solutions $(A, \phi)$ of (7.1) $(G = SU(2))$ an algebraic vector bundle $E^m_c \to TP^1\mathbb{C}$ of rank 2 such that its fibres $(E^m_c)_1$ are formed by solutions $s$ of the equation $(\nabla_1 - i\phi)s = 0$ restricted to 1 (here 1 denotes both an element of $TP^1\mathbb{C}$ and an oriented line in $\mathbb{R}^3$).

The standard rules of exact sequences enable us to prove a 1-1 correspondence between a class of algebraic vector bundles of rank 2 over $TP^1\mathbb{C}$ satisfying some conditions of "stability" and reality close to the asymptotic conditions for (7.1) and a set of so called spectral curves in $TP^1\mathbb{C}$ which are solutions of some systems of algebraic equations. Note that the spectral curves simultaneously determine solutions of the aforementioned Weierstrass problem.

Explicit computations of parameters of the spectral curves gave us the following results for monopole moduli spaces $\mathcal{M}^m(k)$ [52]

$$\dim_{\mathbb{R}}\mathcal{M}^m(k) = 4k - 1 \quad , \quad \mathcal{M}^m(1) = \mathbb{R}^3 \quad , \quad \mathcal{M}^m(2) = \mathbb{R}^3 \times T(P^2\mathbb{R})/_{Z_2}$$

By means of spectral curves Hitchin proved the completeness of the Nahm's realization of monopoles by means of a system of algebraic and 1st order ordinary differential equations [65].

Murray [64] extended the Hitchin construction onto the case of arbitrary compact simple Lie groups $G$ but his procedure is not effective if $G$ is large (recall that the system of equations in the AHDM construction is relatively simple if $k \sim n$, cf. (2.4)).

However the Hitchin method give no satisfactory (general) condition determining non-singular solutions of the Bogomolny equation. Perhaps non-analytic methods as Sobolev spaces will help more.

An interesting though pathological case is a $k = 0$ critical point of the Yang-Mills-Higgs functional which corresponds to a non-self-dual connection on $R^3 \times S^1$, see [79]. Thus one expect to detect other non-self-dual critical points by means of variational methods.

## 8.     Fractional topological charges

A topological charge meant in the sense of the normalized integral
of a Pontryagin or a Chern class of a bundle if well-defined is always an
integer, cf. (1.2).  Recall that it can be calculated by integrating a
Chern form  atrF $\wedge$ F  where  F  is the curvature form of a connection in a
given bundle as it is in the case of the instanton number.

However for a few years one can meet in literature such gauge fields
which give a fractional topological charge.  The examples are:

($\alpha$) Merons [1], [2]:  $A_\mu = \frac{1}{2} g^{-1} \partial_\mu g$,  where  $g = \dfrac{x_4 - i\vec{x}\sigma}{\sqrt{x^2}}$,  having
"instanton" number 1/2.

($\beta$) Singular (anti-) self-dual SU(2)-Yang-Mills fields on  $P^2\mathbb{C}$
defined by Charap and Duft [23], [42] whose topological charge
is 3/4 (9/4).

($\gamma$) SU(N)-twisted instantons over 4-tori whose charges are multiples
of  1/N.  Their construction is due to t'Hooft [50].

In such cases a question arises whether such a field admits a prolongation to some proper periodical gauge field over a larger base space.
It may happen that if we take  q  copies of a manifold with boundary  $M^b$
being equivalent conformally to a given base space  M  (up to a "0-measure
set") and identify their boundaries then we obtain a manifold  $M_q$  (without
boundary) over which the given field when prolongated periodically is a
well-defined connection in a vector bundle over  $M_q$.  Such a procedure might
provide us with instantons over  $S^1 \times S^3$  (in point ($\alpha$); a connected cyclic
sum of 4 copies of  $P^2\mathbb{C}$  ($\beta$) and a similar sum of  N  copies of  $T^4$  ($\gamma$).
Note that the initial fields admit singularities (possibly unremovable) in
such points of  M  which correspond to the boundary of  $M^b$.


## 9.     Supergeometry and supergravity

In the previous sections we could observe an increasing contribution
of topology and a "smooth" differential geometry in global gauge theories.
However in supergeometry in the 80s we can see an increasing interest in
analytical methods.

Before we outline some global structures relevant to geometry of

supersymmetry and supergravity recall the idea of bundles and sheaves.
A bundle (e.g. a vector bundle or a bundle of moduli of algebras) is such
a fibred space that coordinates in fibres are "stable" that is isomorphisms
between neighbouring fibres are given.  A sheaf is roughly speaking such a
fibred space that algebras of local sections are well-defined.  In analysis
on complex manifolds mostly frequent used objects are coherent sheaves
which may be embedded locally in the sheaf of holomorphic, $\mathbb{C}^r$-valued functions on their base manifolds.  Coherent sheaves correspond to vector
bundles admitting certain singularities on sets of codimension $\geq 1$, see [37].
That is why a category of coherent sheaves may be larger than a corresponding category of holomorphic vector bundles.

All objects in "super" theories are formed according to the following
prescription:  replace the field  K  ($\mathbb{R}$ or $\mathbb{C}$)  with a Grassmann algebra
everywhere where it is possible.  In this way one can obtain a precise
definition of:

1.  Supermanifolds in sense of Kostant [12]:  They are some sheaves
of Grassmann algebras.  In this case algebras of local functions on a mani-
fold  $C_{loc}(M, K) = \Gamma_{loc}(M \times K)$  are replaced with algebras  $\Gamma_{loc}(S)$  of
sections of sheaves $S$  where fibres (stalks)  $S_x$  of the sheaf  $S \to M$  have
a Grassmann algebra structure.

2.  Supermanifolds in sense of Alice Rogers [20]:  their definition
is similar to that of manifolds.  The only difference is that coordinate
functions take values in a Grassmann algebra.

3.  Graded bundles [25-29]:  they look locally like  $\overset{n}{\oplus} \Lambda E_{|U}$,  where
U  M  and  $\Lambda E \to M$  is a given Grassmann bundle, and their transition functions
are assumed to preserve the Grassmann moduli structures.  Thus the graded
bundles may be identified with Cech cohomologies  $H^1(M, GL(n, \Lambda E))$,  where
$GL(n, \Lambda E) \to M$  is the bundle of groups  $(GL(n, \Lambda E))_x = GL(n, \Lambda E_x)$,  $x \to M$.

4.  Grassmann moduli [58], [59]:  They are sheaves of Grassmann
moduli looking locally like  $\overset{n}{\oplus} S_{|U}$,  where  $S \to M$  is a supermanifold in the
sense of Kostant.

The above objects seem to be much more interesting in the analytic
case.  That is so because if we define them in classes of smooth functions
then we can prove 1-1 correspondences between them and rather well known
objects in non-super theories.  In particular we have such correspondences
between supermanifolds of Kostant and smooth Grassmann bundles and between

graded bundles, Grassmann moduli and smooth vector bundles of the type $W \otimes \Lambda E$, see [26], [29]. Only the supermanifolds of Rogers have not been classified directly by means of simple non-super objects, cf. [19].

In the analytic case we don't hope to classify easily the main structures of a global supergeometry. Let us regard the point 1 and consider a holomorphic Kostant supermanifold $S \to M$. The natural $z_2$-gradation of Grassmann algebras induces the local decomposition $S = S_0 \oplus S_1$. Let us put $G = S_0 + S_1^2$ and define $Gr_1 = G/G^2$. Assuming the quasi-coherence of $S_0$ with respect to $S_0$ and that $S = S_0 + S_1$ holds in global we can prove that $Gr_1 S$ is the sheaf of the sections of a holomorphic Grassmann bundle $\Lambda E_c \to M$. We can associate with $S$ elements of $H^1(M, TM \otimes \Lambda^2 E_c)$, $H^1(M, TM \otimes \Lambda^4 E_c), \ldots$ in such a way that if $S$ is the sheaf of sections of $\Lambda E_c$ then all these elements vanish. Nevertheless such a correspondence is known to be 1-1 only if rank $E_c < 2$.

Graded bundles which are not isomorphic globally to $W \otimes \Lambda E$ exist if $H^1(M, E_c) \neq 0$. The inverse theorem is not true as we see in the example below.

If we restrict all Grassmann coordinates in fibres of a graded bundle to $K = \Lambda^0 E_c$ then we obtain a vector bundle $W$ called the scalar part of the graded bundle. We will consider moduli spaces of graded bundles $\mathcal{M}(E_c, W)$ having a fixed bundle $E_c$ and a given scalar part $W$. Note that $\mathcal{M}(E_c, W) \neq 0$ as $W \otimes \Lambda E_c \in \mathcal{M}(E_c, W)$.

Example: Let $M = P^1\mathbb{C}$ be. Then by virtue of the Grothendieck theorem we have $E_c = \bigoplus\limits_{k=1}^{r} H^{P_k}$, $W = \bigoplus\limits_{i=1}^{n} H^{t_i}$, $P_k, t_i \in Z$. The direct calculations give us

$$2 \sum_{i,j=1}^{n} \sum_{k=1}^{r} H(t_j - t_i - P_k - 1) \leq \dim_R G(E_{\mathbb{C}} W)$$

$$\leq 2 \sum_{i,j=1}^{n} \sum_{k=1}^{r} \sum_{1 \leq k_1 < \ldots < k_1 \leq r} H(t_j - t_i - P_{k_1} - \ldots - P_{k_1} - 1) =: 2D \quad ,$$

$$H(x) = \begin{cases} x, & x > 0 \\ 0, & x < 0 \end{cases} \quad .$$

Moreover it is

$$\mathcal{M}(E_c, W) = \mathbb{C}^D \quad \text{if} \quad t_j - t_i - p_{k_1} - \dots - p_{k_1} \geq 1 \quad \text{for all} \quad i, j, 1,$$

$$k_1, \dots, k_1 \; .$$

If $E_c = H^Q + H^{-2}$, $W = H^Q$ then $\mathcal{M}(E, W)$ is singular: it consists of two planes crossing themselves in a point.

Similar formulae may be derived if $M$ is an Abelian variety and $E_c$, $W$ split in direct sums of rank 1 complex, holomorphic bundles, cf. [28], [29].

Let $M = P^m \mathbb{C}$, $m > 1$, and $E_c = \overset{n}{\underset{i=1}{\oplus}} H^{t_i}$ be. Then one can prove that $\mathcal{M}(E, W) = \{W \times \Lambda E\}$, see [27], [29]. This is a generalization of $H^1(P^m, H^i) = 0$, $m > 1$. Looking at dimension and geometry of $\mathcal{M}(E, W)$ we can say that graded bundles are capricious objects.

Cohomological conditions for Grassmann moduli may be derived by combining those for supermanifolds and those for graded bundles.

In [58], [59] Manin explored supermanifolds of flags in flat super-spaces (i.e. spaces with distinguished even and odd elements). Then Penkov and Skornyakov proved for such manifolds an analogue of the Borel-Weil-Bott theorem which says that suitably twisted sheaves on such supermanifolds admit only one non-vanishing cohomology. This theorem enables constructions of representations of some Lie superalgebras.

Complex and analytic supermanifolds were used as parameter spaces by Ogievetsky-Sokatchev [67] and Schwarz [75] in local models with a flat base space and by Manin [59] in a more general model admitting non-flat base spaces. This is not a standard case of complex twistorial coordinates because gravity is ruled by hyperbolic equations so that there is no "normal" Penrose transformation making of them Cauchy-Riemann equations. But if we look carefully at the equations of supergravity in the complex coordinates then we see that the hyperbolicity is "compensated" by the anti-commutati-vity of fermionic fields so that the total system acquires some elliptic properties. This phenomenon is relevant to the fact that in the secondly quantized supergravity models (for any N) energy is positive [39].

Note that in the global model of Manin two strong assumptions are made 1) the space-time must be a self-dual manifold (as in (1.3)) in order

the complex coordinates to be integrable, cf. $[6]$, $[66]$, 2) $N = 1$ since
in extended supergravity some contradictions appear. The problem is
whether it is possible to remove them or not.

The case of $N = 1$ supergravity is perhaps one of a lot of hyperbolic
models which admit analytic global structures in their "superextensions".

## 10. What has changed recently in our intuitions of a 4-dimensional space?

In late 70s I had several occasions to ask a few great mathematicians as M. Atiyah, N. Hitchin, Th. Friedrich and my friend A. Derdziński, what new results in gauge fields and 4-manifolds do they expect as soon as. It would be interesting to compare theirs and other opinions at this time with our present knowledge. I think that such a comparison contains the following three points:

1. The spectrum of known smooth and analytic <u>gauge field like objects</u> (as instantons, monopoles, Einstein metrics, supergauge fields) <u>has increased</u> in last few years <u>more</u> than many mathematicians expected. But, paradoxically, some gauge fields which have been detected in this time create obstructions for compact top. manifolds to be smooth so that:

2. The category of <u>compact smooth 4-manifolds</u> now seems to be smaller than it was supposed to be a few years ago. But it turns out that a lack of any diff. structure in some compact topological spaces creates diff. structures in some non-compact top. 4-manifolds. Note that the non-smooth compact 4-manifolds existing as a consequence of the Freedmann theorem after removing any non-empty discrete subset become smooth non-compact 4-manifolds by virtue of the Quinn theorem. Furthermore such Freedmann top. 4-manifolds together with a suspended Whitehead continuum (which is not any diff. manifold too) enables Freedmann and Gompf to detect new rather unexpected exotic structures on $S^3 \times \mathbf{R}^1$ and $\mathbf{R}^4$. Thus we can say that:

3. The category of <u>non-compact smooth 4-manifolds</u> has been recently <u>enriched</u> with a lot of surprising objects.

To the end of this review let us look at the following diagram

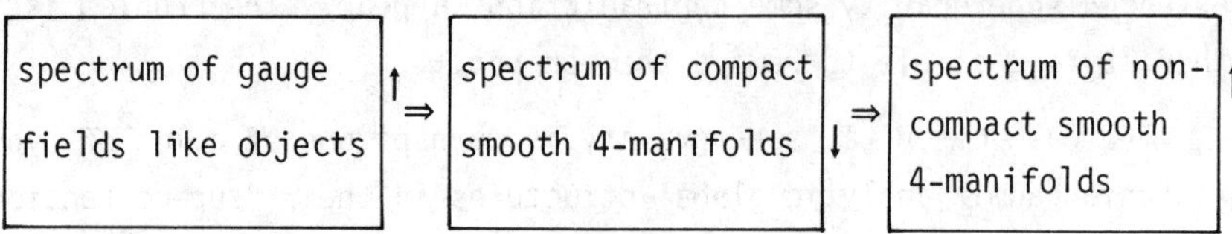



The above results imply the following conclusion for physicists writing about field in 4-dimensional spaces.  If a topology of such a space is given then one must say which differentiable structure is considered (this is usually the standard one).  Furthermore, if such a space is compactified one must check whether the compactification admits a differentiable structure.

## Appendix A:  The Intersection Form

Let  $M$  be a compact, oriented, real, smooth 2n-manifold.  Put

$$(\alpha, \beta) := \int_M \alpha \wedge \beta \quad . \tag{A.1}$$

By virtue of the Stockes theorem if  $\alpha, \beta$  are exact then it is

$$(\alpha, \beta) = (\alpha + d\gamma, \beta + d\delta)$$

so that  $(\alpha, \beta)$  is well defined on the de Rham cohomologies  $H^n(M)$, integer cohomologies  $H^n(M, Z)$  and homologies  $H_n(M, Z)$.  In the cases of  $Z$  $(\cdot,\cdot)$  is called the intersection form of  $M$  and denoted by  $I_M$.  The product  $(\alpha, \beta)$, $\alpha, \beta \in H^n(M, Z)$  may be identified with an element of  $H^{2n}(M, Z)$  so that matrix elements of  $I_M$  are <u>integers</u> (we say that  $I_M$  is an integer form).

Let us give a second definition of  $I_M$  which explains the word "intersection".  Take two cycles  $a, b$  representing (co-)homology classes  $\alpha, \beta$  resp. which intersect transversally in isolated points.  Then  $(\alpha, \beta)$  is equal to the number of points  $p$  of the intersection  $a \cap b$  in which the orientations of  $T_p M$  and  $T_p a \oplus T_p b$  are identical minus the number of these points in which the above orientations are opposite.  Note that this definition makes sense if  $M$  is a top. (not necessary smooth) oriented 2n-manifold and also some singularities of  $M$  are admissible.  Thus the intersection form is an invariant of the topological (not differential) structure.

A change of a basis in $H^n(M, Z)$ induces a Z-transformation of $I_M$.

A change of the orientation of $M$ makes the change of signs of all elements of $I_M$.

If $\alpha_1,\ldots,\alpha_k$ form a basis of $H^n(M, Z)$ and $\omega$ is an orientation then $\alpha_1' := \alpha_1\omega,\ldots,\alpha_k' := \alpha_k\omega$ form a basis of $H_n(M, Z)$ such that the matrix of the change of the basis is unimodular (determinant is $\pm1$). It follows immediately from (A.1) that $I_M(\alpha_i, \alpha_j') = \delta_{ij}$ so that $I_M$ is also unimodular.

From now on $\underline{\dim. M = 4}$. Then $I_M$ is $\underline{symmetric}$.

Integer, unimodular and symmetric forms were studied by founders of linear algebra as Gauss and Hermite.

We say that an integer bilinear form is even if its quadratic form takes only even values on integer vectors (this is the case iff all diagonal elements are even). Otherwise the form is $\underline{odd}$. One of the classical linear algebra theorems says that the rank of a definite (positive or negative) even integer unimodular symmetric form is divisible by 8. It turns out that the number of such forms of rank $8n$ grows rapidly with $n$. For $n = 1,2,3,5$ it is up to sign $1(E_8$, see B.3), 2 $(E_8 \oplus E_8, E_{16})$, 24 and many thousands respectively. Indefinite even unimodular symmetric forms are uniquely determined by the admissible rank $8n + 2p$ and signature $8n$ are given by

$$I_M = (\overset{n}{\oplus} E_8) \oplus \left(\overset{p}{\oplus} \begin{bmatrix} 0 & 1 \\ 1 & 0 \end{bmatrix}\right) \tag{A.2}$$

Furthermore, all odd integer indefinite unimodular symmetric forms are Z-diagonalizable. Here one of forms $F \oplus [\pm1]$ is diagonalizable where $F$ is integer and unimodular. Thus even definite forms dominate the class of intersection forms of 4-manifolds and the Donaldson diagonalization theorem implies that a $\underline{typical\ c.s.o.\ top.\ 4\text{-manifold admits no diff.}}$ $\underline{structure}$.

The second definition of the intersection form implies immediately

$$I_{M\#N} = I_M \oplus I_N \quad .$$

We can prove directly that

$$I_{S^4} = 0 \quad , \quad I_{S^2 \times S^2} = \begin{bmatrix} 0 & 1 \\ 1 & 0 \end{bmatrix} \quad , \quad I_{\#P^2\mathbb{C}} = I_n \quad .$$

## Appendix B:  On K3 Surfaces and the Kummer Surface

A compact Kähler manifold of the complex dimension 2 is called a  K3 surface if  $b_2(M) = 0$  (first Betti number vanishes) and  $c_1(TM) = 0$  (first Chern class also vanishes).

The name  K3  is an abbreviation of names Kummer, Kodaira and Kähler and it suggests that the class of K3 surfaces is as difficult as a Himalayan peak like  K2.  It was proposed by A. Weil who posed the conjecture that the structures of real diff. 4-manifolds corresponding to all  K3  surfaces are equivalent so that there is the only smooth manifold  K3.  His conjecture was proved by Kodaira [57].

Let us observe that both conditions in the above definition are purely topological and they are realized by flat spaces.  Thus they provide K3  with many other properties of flat spaces.  In particular the  K3  surfaces are simply connected, torsionless (i.e. there is no non-zero cohomology  $H^m(K3, Z_p)$ ) and  $\Lambda^2 TK3$  is the trivial topologically (as  $c_1(TK3) = 0$ ) and analytically (because  $b_1(K3) = 0$ ) $\mathbb{C}$ -bundle.

Nevertheless it turns out that  $b_2(K3) = 22$  so that the  K3  manifold is extremely kinked.  In order to calculate  $b_2(K3)$  observe that  $h^1(K3) = h^3(K3) = b_1(K3) = 0 = h^{1,0}(K3,\theta)$   (the last cohomology is defined in the set (sheaf) of local holomorphic functions),  $h^{2,0}(K3, V) = h^{0,2}(K3,\theta) = 1$  (this is a consequence of  $c_1(TK3) = 0$ ) and apply the Noether formula which is a case of the Riemann-Roch-Hirzebruch theorem

$$12(h^{2,0}(K3,\theta) - h^{1,0}(K3, V) + 1) = \chi(K3) + c_1^2(TK3) \tag{B.1}$$

( $\chi$  is the Euler characteristic).  Thus we have

$$\underline{b_2(K3) = \chi(K3) - 2 = \underline{22}} \quad .$$

The analytic cohomology formula for the Hirzebruch signature which is nothing else but signature of $I_{K3}$ gives us

$$\tau(I_{K3}) = 2(h^{1,0}(\cdot)+1) - h^{1,1}(\cdot) = -16 \quad . \tag{B.2}$$

It is convenient to take the anti-holomorphic orientation of K3 so that $\underline{(I_{K3}) = 16}$.

Using the Wu formula for Stiefel-Whitney classes and the definition of K3 Milnor proved that $\underline{I_{K3}}$ is even [61]. This was the last and more difficult point of the search for $\overline{I_{K3}}$ because the three underlined data uniquely determine $I_{K3}$ in the way as in Appendix A. Thus we have

$$I_{K3} = E_8 + E_8 + \begin{bmatrix} 0 & 1 \\ 1 & 0 \end{bmatrix} + \begin{bmatrix} 0 & 1 \\ 1 & 0 \end{bmatrix} + \begin{bmatrix} 0 & 1 \\ 1 & 0 \end{bmatrix}$$

where

$$E_8 = \begin{bmatrix} 21 & & & & & & & 0 \\ 121 & -1 & & & & & & \\ & 121 & & & & & & \\ & & 121 & & & & & \\ & & & 121 & & & & \\ & & & & 121 & & & \\ & & & & -1 & 121 & & \\ 0 & & & & & 12 & & \end{bmatrix} \quad \text{or; in other matrix representation,} \quad \begin{bmatrix} 2 & -1 & & & & & & 0 \\ -1 & 2 & -1 & & & & & \\ & -1 & 2 & -1 & & & & \\ & & -1 & 2 & -1 & & & \\ & & & -1 & 2 & -1 & -1 & \\ & & & & -1 & 2 & -1 & \\ & & & & & -1 & 2 & \\ 0 & & & & -1 & & & 2 \end{bmatrix} \quad .$$

$$\tag{B.3}$$

Note that in the 2nd representation $E_8$ is equal to the Cartan matrix of the exceptional Lie algebra $e_8$.

The moduli space of K3 surfaces $\mathcal{M}(K3)$ is locally isomorphic to the quadric in $P^{21}\mathbb{C}$ whose matrix form is just $I_{K3}$ so that $\dim \mathcal{M}(K3) = 20 = h^{1,1}(K3,\theta)$ (this coincidence is not accidental). Algebraic K3 surfaces, i.e. those which may be embedded in some $P^m\mathbb{C}$ (only for them the word "surface" is not ambiguous) form a countable set of 19-dimensional complex submanifolds of $\mathcal{M}(K3)$ which is in it dense everywhere. One obtains 19 as $h^{1,1}(\cdot) - h^{2,0}(\cdot)$ [73].

As a consequence of $b_2(K3) = 22$ the set of smooth tank 1 bundles over K3 is parametrized by $Z^{22}$. Furthermore, $h^{1,0}(\cdot) = 0$ and $h^{2,0}(\cdot) = 1$ imply that $Z^{21}$ of these bundles admit a holomorphic structure. In recent years we see a progress in learning moduli spaces of other simple algebraic vector bundles over K3 surfaces (the simplicity means the non-existence of any non-constant global endomorphism).

Recall that there is no compact simply connected smooth 4-manifold realizing $E_8$ (the Rochlin theorem) and $E_8 \oplus E_8$ and $E_{16}$ (the Donaldson theorem). But we do not know whether there exists a smooth 4-manifold realizing $E_8 \oplus E_8$ plus one or two copies of $\begin{bmatrix} 0 & 1 \\ 1 & 0 \end{bmatrix}$ or not. Such a manifold, if exists, does not admit any Einstein metric in contrast to $K3$ where the Einstein metrics form a 57-dimensional moduli space [19] and are very special, e.g. each such a metric is Kählerian and Ricci-flat. In this way the $K3$ manifold turns out to be the simplest known manifold being neither a homogeneous space $G/_K$ nor a connected sum of such spaces and the geometry of $K3$ seems to be particularly subtle.

A few explicit realizations of algebraic $K3$ surfaces are known. The most elementary one is the quartic surface in $P^3\mathbb{C}$ given by

$$z_1^4 + z_2^4 + z_3^4 + z_4^4 = 0 \quad .$$

However a better insight in geometry of $K3$ surfaces is given by the non-singular Kummer surfaces. We can learn it by performing the following algebraical and topological construction.

Let us take a 4-torus $T^4 = R^4/_{Z \oplus Z \oplus Z \oplus Z}$. Its (co-)homology spaces have dimensions $1, 4 = \binom{4}{1}, 6 = \binom{4}{2}, 4, 1$ since e.g. the 2nd homologies are represented by 2-subtori spanned by pairs of axial vectors. One can check that

$$I_{T4} = \begin{bmatrix} 0 & 1 \\ 1 & 0 \end{bmatrix} \oplus \begin{bmatrix} 0 & 1 \\ 1 & 0 \end{bmatrix} \oplus \begin{bmatrix} 0 & 1 \\ 1 & 0 \end{bmatrix} \quad .$$

Then let us identify each element $[x_1, x_2, x_3, x_4] \in T^4$ with $[-x_1, -x_2, -x_3, -x_4]$. The resulting space $\tilde{K}$ has 16 singular points whose coordinates $x_i$ are 0 or 1/2. They have not been identified with anyone. Observe that a "horizon" in each of turn is of one-half of the angle of a 3-sphere. Notice that the first cohomologies of $T^4$ have been killed while these identifications because both arches in the circles representing the first homologies have been identified. Thus the (co-)homology dimensions of $\tilde{K}$ are 1,0,6,0,1 and $I_{\tilde{K}} = I_M$. Note that the 2nd homologies in $\tilde{K}$ are represented by $S^2$-surfaces (not by 2-tori as they were in $T^4$).

Recall that $\tilde{K}$ may be provided with a structure of a complex singular

2-manifold if we put $\mathbb{C}^2$ instead of $\mathbb{R}^4$ in the definition of the 4-torus. Then $\tilde{K}$ is called the singular Kummer surface. Kummer defined it in 1864 |90| by means of a 4-degree homogeneous equation in $P^3\mathbb{C}$ while looking for a burning surface of a rays system. Pay attention that the complex projective space which was used by him for needs of geometrical optics has much in common with the contemporary Penrose transformation.

If we remove all the 16 singular points from $\tilde{K}$ then we obtain a non-compact smooth 4-manifold. It turns out that one can compactify it to a smooth manifold by glueing to it 16 copies of $S^2$. Such an operation cannot be carried out in 3 dimensions because in $\mathbb{R}^3$ no boundary of any connected manifold can be cancelled by joining $S^2$. Nevertheless a procedure of smoothing 2-surfaces at isolated singular points by replacing them with copies of $P^1\mathbb{C} \cong S^2$ is well defined in local complex coordinates. We call it blowing singular points up. Note that each point of the glueing $P^1\mathbb{C}$ is identified with a complex direction through the point blown up and this direction together with the direction along $P^1\mathbb{C}$ spans a complex tangent space. One can check that the space of complex directions in $\tilde{K}$ through each singular point is isomorphic to $P^1\mathbb{C}$ (these directions look locally like $\mathbb{C}-\{0\}$).

If someone blows up all the 16 singular points then we obtain the non-singular Kummer surface $K$. Direct calculation allow to check that $K$ is a four degree surface in $P^3\mathbb{C}$ and $K$ is the intersection of three quadrics in $P^5\mathbb{C}$. Observe that the blowing up has increased the number of 2nd cohomologies by 16 because each copy of $P^1\mathbb{C}$ which replaced a singular point of $\tilde{K}$ represents a 2nd cohomology class (shrinking such $P^1\mathbb{C}$ to a point makes a singularity again). Thus Betti numbers of $K$ are 1,0,22,0,1. By means of $B1$ and $B2$ one can prove that $c_1(TK) = 0$ so that non-singular Kummer surface is an algebraic Kummer surface.

Let us look at 2nd homologies of $K$. Observe that these pairs of the 2nd cohomologies which were generated by complementary 2-tori of $T^4$ (equally dashed lines on Fig. 6) preserves in $K$ a one-point intersection so that there is no obstruction for them to span in $K$ an $S^2 \times S^2$ - like submanifold. But unions of the complementary 2-tori interset in $T^4$ along some lines. This fact seems to be related to the phenomenon that all the six homology cycles generated by 2nd cohomologies of $T^4$ span a piece of $K$ having the topology of $S^2 \times S^2 \ \# \ S^2 \times S^2 \ \# \ S^2 \times S^2$.

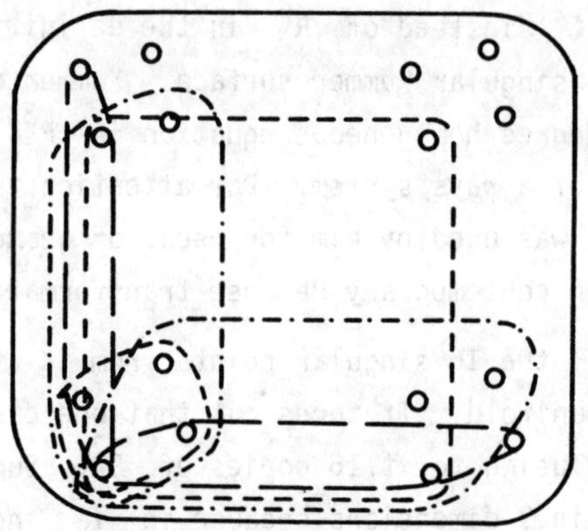

Fig. 6

We see that the part of $K$ realizing the six $T^4$-homology cycles is placed somewhere in "bowels" of $K$. Its extraction made in Section 5, which is called usually a topological surgery in this case looks like a harakiri. Such a brutal treatment may cause serious changes in the "body" of $K$ and that is why the exotic differentiable structures may appear.


## Acknowledgement

I would like to thank Professor Agostino Prastaro for the organization of this interesting conference.


## References


[1]    de Alfaro, V., S. Fubini, G. Furlan, Phys. Lett. 65B (1976) 163.

[2]    de Alfaro, V., S. Fubini, G. Furlan, Lecture Notes Math. 676 (1979) 255.

[3]    Atiyah, M., "Geometry of Yang-Mills Fields", Fermi Lectures, Pisa 1979.

[4]    Atiyah, M., in "Gauge Theories: Fundamental Interactions and Rigorous Results", Birkhäuser, 1982, p. 207.

[5]    Atiyah, M., N. Hitchin, V. Drinfeld, Yu. Manin, Phys. Lett. 65A (1978) 185.

[6]    Atiyah, M., N. Hitchin, I. Singer, Proc. Roy. Soc. London 362A (1978) 425.

[7]   Atiyah, M., E. Rees, Inv. Math. $\underline{35}$ (1976) 131.

[8]   Atiyah, M., R. Ward, Comm. Math. Phys. $\underline{55}$ (1977) 117.

[9]   Barth, W. in "Gauges Theories: Fundamental Interactions and Rigorous Results", Birkhäuser, 1982, p. 177.

[10]  Barth, W., K. Hulek, Manuscripta Math. $\underline{25}$ (1978) 323.

[11]  Belavin, A., A. Polyakov, A. Schwarz, S. Tyupkin, Phys. Lett. $\underline{59B}$ (1975) 85.

[12]  Berezin, F., "Vvedenye w algebru i analiz s antikommuturuyu-shtchimi peremennymi", MGU, 1983 (in Russian).

[13]  Bernard, G., N. Christ, A. Guth, E. Weinberg, Phys. Rev. $\underline{16D}$ (1977) 2967.

[14]  Bogomolny, E., Sov. J. Nucl. Phys. $\underline{24}$ (1976) 449.

[15]  Borel, A., F. Hirzebruch, Amer. J. Math. $\underline{80}$ (1958) 458.

[16]  Bott, R., in Proc. Differential Geometrical Methods in Math. Phys., Salamanca, 1978, p. 269.

[17]  Bourguignon, J.-M., H. Lawson, Jr., Comm. Math. Phys. $\underline{79}$ (1981) 189.

[18]  Bourguignon, J.-M., H. Lawson, Jr., J. Simons, Proc. Acad. Sci. USA $\underline{76}$ (1979) 1550.

[19]  Bourguignon, J.-M., Asterisque $\underline{80}$ (1980) 21.

[20]  Boyer, Ch., Contemporary Math. $\underline{12}$ (1982) 53.

[21]  Burlakov, D., V. Dutyshev, Zh. ETP $\underline{73}$ (1977) 377.

[22]  Casson, A., Publications Orsay 81T06 (1980).

[23]  Charap, J., M. Duff, Phys. Lett. $\underline{69B}$ (1977) 445.

[24]  Corrigan, E., P. Goddard, Ann. Phys. $\underline{154}$ (1984) 253.

[25]  Czyz, J., Lecture Notes Math. $\underline{838}$ (1981) 110.

[26]  Czyz, J. preprint Luminy-Marseille CPT-81/PE. 1328.

[27]  Czyz, J., preprint IM PAN 277 (1983).

[28]  Czyz, J., to appear in Lett. Math. Phys.

[29]  Czyz, J., to appear in Rept. Math. Phys.

[30]  Daniel, M., P. Miter, C. Vialet, Phys. Lett. $\underline{77B}$ (1978) 77.

[31]  Deser, S., C. Teitelboim, Phys. Rev. Lett. $\underline{39}$ (1977) 249.

[32] Donaldson, S., J. Diff. Geom. 18 (1983) 279.

[33] Douady, A., J.-L. Verdier, eds., Asterisque 71-72 (1978) 177.

[34] Drinfeld, V., Yu. Manin, Comm. Math. Phys. 63 (1978) 177.

[35] Drinfeld, V., Yu. Manin, Sov. J. Math. Phys. 29 (1979) 1649.

[36] Drinfeld, V., Yu. Manin, Lecture Notes Math. 732 (1979) 60.

[37] Fisher, G., Lecture Notes Math. 538 (1976).

[38] Freed, D., M. Freedmann, K. Uhlenbeck, "Gauge Theories and Four
Manifolds", preprint MSRI Berkeley, 1983.

[39] Freedmann, M., Ann. Math. 110 (1979) 177.

[40] Freedmann, M., J. Diff. Geom. 17 (1982) 357.

[41] Gibbons, G., Lecture Notes Phys. 116 (1980) 282.

[42] Gibbons, G., C. Pope, Comm. Math. Phys. 61 (1978) 239.

[43] Gompf, R., J. Diff. Geom. 18 (1983) 317.

[44] Hartshorne, R., Comm. Math. Phys. 59 (1978) 1.

[45] Hitchin, N., Math. Proc. Camb. Phil. Soc. 85 (1979) 465.

[46] Hitchin, N., Comm. Math. Phys. 89 (1983) 145.

[47] t'Hooft, G., Nucl. Phys. 79B (1974) 274.

[48] t'Hooft, G., Phys. Rev. Lett. 37 (1976) 8.

[49] t'Hooft, G., Phys. Rev. 14D (1976) 3432.

[50] t'Hooft, G., Comm. Math. Phys. 81 (1981) 267.

[51] Horrocks, G., Proc. London Math. Soc. 14 (1964) 689.

[52] Hurtubise, J. Comm. Math. Phys. 92 (1983) 195.

[53] Itoh, M., Proc. Japan Acad., 57A (1981) 176.

[54] Jackiw, R., C. Nohl, C. Rebbi, Phys. Rev. 15D (1977) 1642.

[55] Jackiw, R., C. Rebbi, Phys. Rev. 16D (1977) 1052.

[56] Kirby, R., K. Siebenmann, "Foundational Essays on Topological
Manifolds:  Smoothing and Triangulations", Princeton University
Press, 1977.

[57] Kodaira, K., Amer. J. Math. 86 (1964) 751.

[58] Manin, Yu., in:  Arithmetics and Geometry, Vol. II, Birkhäuser,
1983, p. 175.

[59]  Manin, Yu., "Geometry of Supergravity and Super-Schubert Cells" in "Differential Geometry, Lie Groups and Mechanics VI", Nauka, 1984, p. 160.

[60]  Milnor, J., Am. Math. 64 (1956) 399.

[61]  Milnor, J., in "Symposium International Topologia Algebraica", Mexico 1958, p. 122.

[62]  Milnor, J., D. Husemøller, "Symmetric Bilinear Forms", New York, Springer, 1958.

[63]  Moise, E., Ann. Math. 56 (1952) 96.

[64]  Murray, M., Comm. Math. Phys. 90 (1980) 263.

[65]  Nahm, W., in Proc. Symposium on Particle Physics, Visegrad, 1981.

[66]  Newlander, N., L. Nirenberg, Ann. Math. 65 (1957) 391.

[67]  Ogievetsky, V., E. Sokatchev, Phys. Lett. 79B (1978) 722.

[68]  Page, D., Phys. Lett. 80B (1978) 55.

[69]  Polyakov, A., Zh. ETF Lett. 20 (1974) 194.

[70]  Quinn, F., J. Diff. Geom. 17 (1982) 503.

[71]  Rochlin, V., Dokl. Akad. Nauk SSSR 84 (1952) 221.

[72]  Sachs, J., K. Uhlenbeck, Ann. Math. 113 (1981) 1.

[73]  Safarevitch, F., ed., Proc. Inst. St'eklov 75 (1965).

[74]  Schwarz, A., Comm. Math. Phys. 64 (1978) 233.

[75]  Schwarz, A., Nucl. Phys. 171B (1980) 154.

[76]  Serre, J.-P., Ann. Inst. Fourier 6 (1956) 1.

[77]  Stern, R., Math. Intelligencer 5 (1983) 39.

[78]  Taubes, C., Comm. Math. Phys. 72 (1980) 277.

[79]  Taubes, C., Comm. Math. Phys. 80 (1981) 343.

[80]  Taubes, C., Comm. Math. Phys. 91 (1983) 235.

[81]  Taubes, C., J. Diff. Geom. 17 (1982) 139.

[82]  Taubes, C., J. Diff. Geom. 19 (1984) 337.

[83]  Taubes, C., J. Diff. Geom. 19 (1984) 519.

[84]  Uhlenbeck, K., Comm. Math. Phys. 83 (1982) 11.

[85]  Ward, R., Comm. Math. Phys. 79 (1981) 317.

[86] Whitehead, J., Ann. Math. $\underline{41}$ (1940) 809.

[87] Witten, E., Phys. Rev. $\underline{38}$ (1977) 121.

[88] Yau, S.T., Comm. Pure Appl. Math. $\underline{31}$ (1978) 339.

[89] Hirsch, M., B. Mazur, "Smoothings of piecewise linear manifolds", Princeton University Press, 1974.

[90] Kummer, E., Collected papers, Springer, 1975, p. 418.

[91] Atiyah, M., J. Jones, Comm. Math. Phys. $\underline{61}$ (1978) 97.

Received April 29, 1985.

GEOMETRODYNAMICS PROCEEDINGS (1985), pp. 101-137
edited by A. Pràstaro


# AN APPLICATION OF TOPOLOGICAL METHODS TO THE STUDY OF PERIODIC SOLUTIONS OF HAMILTONIAN SYSTEMS *

G.F.Dell'Antonio
*Dipartimento di Matematica, Università di Roma, La Sapienza*

## 1. INTRODUCTION

Topological methods are often employed in Mathematical Physics to provide estimates on the number of stationary points of a functional F on a Banach manifold M, or on the zeroes of a vector field f on M.

Local information about the manifold structure of the zeroes can also be obtained.

The mathematics employed comprises several versions of M. Morse's theory, especially its equivariant form, including its generalization to flows. One can make use alternatively of category theory, along the lines of the theory of genus developed by Lÿustersnik and Schirelmann. Recently several authors [1] have applied successfully these techniques to the study of finite-action solutions of gauge theories, in particular of Yang-Mills type.

In this talk, I shall be concerned with another problem, namely the study of periodic solutions of a dynamical system, with special emphasis on the Hamiltonian case.

To understand the relation with problems in gauge theory, consider the Hamiltonian case. Here the solutions are stationary points of an action integral I, suitably defin-

---

* *Work supported in part with funds of MPI.*

ed  on a class of trajectories. In order to apply (equivariant)  Morse theory, one has  to consider a manifold M of  trajectories which has a non-trivial topological structure. In particular, some homology group should not be trivial, or alternatively some subset of M must have a non-trivial category. This can be achieved(possibly through an intermediate use of a Lyapunov-Schmidt procedure) when there is a natural action on M of a group G, which leaves the functional I invariant. Then I can be viewed as a function on the quotient space M/G, and the latter has a non-trivial topological structure.

This is in fact only "roughly" true , since M/G is not a manifold if the action of G on M is not free. When this in the case, one can instead use equivariant Morse theory.

In gauge theory, the space of connections has trivial topological structure (it is contractible). The group G is the "gauge group" which acts on connections and leaves the action functional invariant. The quotient space M/G has non-trivial homological structure [2].

Another familiar example can be constructed by taking M to be the unit sphere $S_1^{2n-1}$ in $C^n$. The group G is now the circle group $S^1$ which acts freely by multiplication on $C^n|_{\{0\}}$ and leaves $S_1^{2n-1}$ invariant. Moreover $S_1^{2n-1}/S^1 \simeq CP^n$, the complex projective space of dimension n-1.

Any function T on $S_1^{2n-1}$ which is $S^1$-invariant can be regarded as a function on $CP^n$, a manifold for which the even cohomology groups are non-trivial. Morse's theory (in particular Morse's inequalities) can be used if T is of class $C^2$. If T is of class $C^1$, one must resort to category theory.

As we shall see, in many cases the problem of finding periodic solutions of a Hamiltonian system can be reduced to variants of the finite-dimensional problem discussed above.

On periodic functions of (minimal) period T, $S^1$ acts by the shift  action $\sigma$ definedy by

$$(\sigma(\theta))f(t) = f(t + \frac{\theta T}{2\pi})$$

For Hamiltonian systems with n degrees of freedom the natural starting point is then the Hamilton action functional A on the space M  of all $2\pi$-periodic, $R^{2n}$-valued functions.

This functional is by construction invariant under the shift action of $S^1$, and defines therefore a function on $M_1 \backslash S^1$. This space is in general contractible, but one capitalizes on the fact that there exists a constant of motion, i.e. the energy. If $M_2$ is the subset of trajectories which have a prescribed energy, A is regarded as a function on $M_2 \backslash S_1$, and the latter may have non-trivial homology.

Many interesting results have been obtained in the past ten years along this line of thought, especially if the energy surface is star-shaped or convex,[ 3 ][ 4 ], or if the Hamiltonian is sub - or super - quadratic at the origin and at infinity [ 3] or for Hamiltonians of the form $H = p^2 + V(q)$, $p,q \in R^n$, if V satisfies suitable growth conditions [ 5].

In this talk, I will outline still another case, i.e. the study of families of periodic solutions near an equilibrium point for a dynamical system, in particular for a Hamiltonian one.

Part of the results presented here were obtained in collaboration with B. D'Onofrio; for further details the interested reader may consult [6].

We begin by a discussion of periodic solutions of a dynamical system described on $R^{2N}$ by the vector field $f(x) \in R^{2N}$. We assume that f is of class $C^1$ and that $f(\underline{0}) = \underline{0}$; we consider only a suitable neighborhood of the origin.

The equations are

$$\dot{\underline{y}} = f(\underline{y}) \tag{1.1}$$

In seeking periodic solutions of (1.1), one may follow two roads. One may write the integrated version of (1.1) (the Poincaré map) and look for fixed points. Or one may consider the space of $R^{2n}$-valued periodic functions and find in that space the solutions of (1.1). We shall develop both methods here.

Somewhat surprinsingly, the two approaches bring into consideration for topological analysis the same manifolds and invariance group.

Since f is of a class $C^1$, we can write

$$f = f^0 + f^1 \tag{1.2}$$

where f' is infinitesimal at the origin of order greater than one and $f^0$ is linear

$$f^0(x) = \omega_o L^o x , \qquad \omega_o > 0 \tag{1.3}$$

We make the following

ASSUMPTION 1.0   $L^o$ is antisymmetric and its eigenvalues are

$$\{\pm i\nu_k\} , \quad \nu_1 = 1, \quad \nu_k \in Z \quad k = 2\ldots s$$

We denote by $d_k$ the multeplicity of the $k^{th}$ pair of eigenvalues, so that $\sum_{k=1}^{s} d_k = N$.

Since all eigenvalues of $iL^o$ are proportional to $\omega^o$ we are in a "fully resonant" case.

Our aim is to give a lower bound on the number of families of periodic solutions of (1.1) with frequency close to $\omega^o$ in a neighborhood of the origin.

It is convenient to introduce the scaling $x \to \varepsilon x$; small values of the parameter $\varepsilon$ will then identify "small" neighborhoods of the origin.

Setting

$$f_\varepsilon (x) = \varepsilon^{-1} f(\varepsilon x) = f^o(x) + f^1(x,\varepsilon) \tag{1.4}$$

one has, for every $a > 0$

$$\lim_{\varepsilon \to 0} \sup_{|x| < a} |f^1(x,\varepsilon)| = 0$$

so that $f^1$ is a small perturbation of $f^o$ uniformly over compact sets.

Under the assumptions stated, the solution of

$$\dot{x} = \omega_o L^o x , \quad x(0) = \xi \in R^{2N} \tag{1.5}$$

is periodic with (minimal) period $\frac{2\pi}{\omega_o}$ unless $\xi$ belongs to some linear subspace of $R^{2N}$ (in which case $\frac{2\pi}{\omega_o}$ is not the minimal period). In general, "most" periodic solutions will be destroyed by the perturbation $f^1$; our purpose is to place a lower bound on the number of those which survive, more precisely on the number of families of periodic solutions of

$$\dot{x} = \omega_o L^o x + f^1(x,\varepsilon) \tag{1.6}$$

with minimal period close to $2\pi\omega_o^{-1}$.

Let $t \to x(t)$ be solution of (1.6) and let $\omega$ be its frequency.

Define $z(t)$ by

$$z(t) = x(\omega t) \tag{1.7}$$

Then $z(t)$ is $R^{2N}$-valued and $2\pi$-periodic, and moreover

$$\frac{dz}{dt} = \omega^{-1} f(z,\varepsilon) \quad , \quad f(z,\varepsilon) \doteq f_2(x) \tag{1.8}$$

To find a solution of (1.8) we can consider first the space $P$ of $R^{2N}$-valued $2\pi$-periodic functions, and find there soltuions of (1.8) when $\omega$ is close to $\omega_o$.

Let

$$P^1 \doteq P \cap C^1 \quad , \quad P^o \doteq P \cap C^o$$

and consider the bounded map (vector field)

$$P^1 \times R|_{\{0\}} \times R \rightarrow P^0$$

defined by

$$(z,\omega,\varepsilon) \rightarrow \frac{dz}{dt} - \frac{1}{\omega}f(z,\varepsilon) \doteq F(z,\omega,\varepsilon) \qquad (1.9)$$

Obviously $z(t)$ solves (1.8) iff

$$F(z,\omega,\varepsilon) = 0 \qquad (1.10)$$

so that $2\pi$-periodic solutions of (1.8) are in correspondence with the zeroes of F.

We shall consider (1.10) for small values of $\varepsilon$ and rely on the Inverse Function Theorem. A similar analisis can be done when $L^o$ depends on additional parameters $\mu_1 \ldots \mu_p$. This is necessary if one wants to study the singularity structure of the family of periodic solutions of (1.1) [7].

One has

$$F(z,\omega_o,0) = \frac{d}{dt}z - L^o z \qquad (1.11)$$

therefore the solutions of $F(z,\omega_o,0) = 0$ are

$$z_o(t) = \exp(L^o t) \cdot \xi$$

for some $\xi \in R^{2N}$.

Let $M$ be the linear space of $R^{2N}$-valued functions defined by

$$M \equiv \{t \rightarrow e^{L^o t}\xi, \ \xi \in R^{2N}\} \qquad (1.12)$$

i.e. the null-space of the vector field $F(z,\omega_o,0)$.

The Frechet differential of $F(z,\omega,0)$ at $(z_o,\omega_o;$ $z_o \in M, \omega_o \in R)$ is

$$DF(z_o,\omega_o,0)\cdot\eta = \frac{du}{dt} - L^o u(t) - \frac{\nu}{\omega_o}\frac{dz_o(t)}{dt} \in P^o \qquad (1.13)$$

when $\eta \equiv (u(t),\nu) \in P' \times R$.

Therefore

$$Ker(DF(z_o,\omega_o,0)) = M \times \{0\} \qquad (1.14)$$

and, by construction

$$Im(DF(z_o,\omega_o,0)) = \{u + \lambda\frac{dz_o}{dt}, u \in \underset{\sim}{P}, \lambda \in R\} \qquad (1.15)$$

where

$$\underset{\sim}{P} \equiv \{u \in P^o | \int_0^{2\pi} e^{(2\pi - s)L^o} u(s)ds \in Im(I - e^{2\pi L_o})\} \qquad (1.16)$$

One can verify that $DF(z_o,\omega_o 0)$ is a continuous bijection between $P^1 \times R$ and $ImDF$.

Let $M_{z_o}^\perp$ be the orthogonal complement in $M$ to $\{\lambda\frac{dz_o}{dt}, \lambda \in R\}$, (orthogonal with respect to the scalar product introduced in $M$ by the bijection $M \leftrightarrow R^{2N}$).

Then $M_{z_o}^\perp$ is complementary in $P^o$ to $ImDF(z_o,\omega_o,0)$, and, by the inverse function theorem, one can find a constant $\varepsilon_1 > 0$ such that, if $0 \leq \varepsilon \leq \varepsilon_1$, there exists a unique solution to

$$F(z_o + z, \omega_o + \delta, \varepsilon)\Big|_{\mathrm{ImDF}(z_o, \omega_o, 0)} = 0 \tag{1.17}$$

Denote this solution by

$$z = \varphi(z_o, \omega_o, \varepsilon), \qquad \delta = \psi(z_o, \omega_o, \varepsilon) \tag{1.18}$$

Then (1.10) is equivalent to

$$F(z_o + \varphi(z_o, \omega_o, \varepsilon), \omega_o + \psi(z_o, \omega_o, \varepsilon))\Big|_{M_{z_o}^{\perp}} = 0 \tag{1.19}$$

Define on $M$ the vector field

$$\underset{\sim}{F}(z_o, \omega_o, \varepsilon) \doteq F(z_o + \varphi(z_o, \omega_o, \varepsilon), \omega_o + (z_o, \omega_o, \varepsilon))\Big|_M \tag{1.20}$$

Then (1.19) reads

$$\exists \lambda \in R \text{ such that } \underset{\sim}{F}(z_o, \omega_o, \varepsilon)(t) = \lambda \frac{dz_o}{dt} \tag{1.21}$$

Using equivariance of F (a consequence of the uniqueness of the solutions (1.18)) one verifies that (1.21) is equivalent to

$$\exists \lambda \in R \text{ such that } \underset{\sim}{f}(\xi, \varepsilon) = \lambda L^o \xi, \qquad \xi \in \mathbb{R}^{2N} \tag{1.22}$$

where

$$\underset{\sim}{f}(\xi, \varepsilon) \doteq F(z_o(t), \varepsilon)(0) \tag{1.23}$$

110

The field $\underset{\sim}{f}$ is by construction invariant under the flow of the vector field $L^o\xi$.

We have proved

PROPOSITION 1.1 [7].Periodic solutions of (1.1) with period close to $\dfrac{2\pi}{\omega_o}$ are in one-to-one correspondence with the points $\xi \in R^{2N}$ where the field $f(\xi,\varepsilon)$ is parallel to $L^o\xi$. $\qquad\square$

REMARK 1.2.If natural numbers other than $\pm 1$ are in the spectrum of $iL^o$, there are subspaces of M composed of functions which are $2\pi$-periodic but a smaller *minimal* period, while for the rest of M $2\pi$ is the minimal period. In this case, some of the periodic solutions of (1.1) described in proposition 1.1 could have *minimal* period close to a submultiple of $\dfrac{2\pi}{\omega_o}$ . We shall come back later to this point. $\quad\square$

REMARK 1.3.The construction described above bears a close resemblance to the method of averaging of Bogolijubov and Mitropolski.

The vector field $\underset{\sim}{f}(z,\varepsilon)$ on $\mathbb{R}^{2n}$ is obtained through a construction in function space which involves averaging (compare (1.16) and the construction of Ker DF). The construction of $\underset{\sim}{f}$ directly in $\mathbb{R}^{2n}$ requires an averaging procedure which is defined only implicitly and may even be ill-posed for initial data which do not correspond to periodic solutions. $\qquad\square$

If $[f^{(1)}(z,\varepsilon),L^oz] = 0$, the field $\underset{\sim}{f}$ takes a particularly simple form. Indeed from the construction given above one has (see also Lemma 2.6).

PROPOSITION 1.4. If $[f^{(1)}(z,\varepsilon),L^{o}z] = 0$, then $\underset{\sim}{f}(z,\varepsilon) = f^{(1)}(z,\varepsilon)$. $\qquad\square$

DEFINITION 1.5. We shall denote by $\Gamma^{o}$ the flow of $L^{o}z$ on $\mathbb{R}^{2N}$. Topological considerations are not sufficient to provide a lower bound on orbits of $\Gamma^{o}$ on which $L^{o}z$ is parallel to $f^{(1)}(z,\varepsilon)$. This is due to the fact that $\mathbb{R}^{2N}\backslash_{\Gamma^{o}}$ is not compact. A bound can be provided if (1.6) admits a constant of motion $\mu(\varepsilon)$ which has the following property : one can find a constant $\varepsilon_1$ and a function $\delta(\varepsilon)$ such that for all $0 \leq \varepsilon \leq \varepsilon_1$, the set $D_{\varepsilon} \doteq \{z \mid \mu(z,\varepsilon) = \delta(\varepsilon)\}$ is diffeomorphic to a sphere, and moreover $\underset{0\leq \varepsilon \leq \varepsilon_1}{\cup}\ D_{\varepsilon}$ is a neighborhood of the origin.

We shall treat in detail only the Hamiltonian case, which has some special features and allows stronger results. We follow again the approach of Moser and Weinstein.

Denote by h the Hamiltonian function and assume that the symplectic form $\sigma$ is in canonical form, so that Hamilton's equations read

$$\frac{dq_i}{dt} = \frac{\partial h}{\partial p_i} \qquad \frac{dp_i}{dt} = -\frac{\partial h}{\partial q_i} \qquad i = 1\ldots N \qquad (1.24)$$

The vector field F defined in (1.9) is now

$$F(z,\omega,\varepsilon) = \{\frac{dq_i}{dt}(t) - \frac{\omega}{\omega_o}\frac{\partial h}{\partial p_i}(t), \frac{dp_i}{dt}(t) + \frac{\omega}{\omega_o}\frac{\partial h}{\partial p_i}(t)\} \qquad (1.25)$$

where $z \equiv \{q_1^{(t)} \cdot q_N^{(t)}, p_1^{(t)} \cdot p_N^{(t)}\}$ and $h(z,0) = h^{o}(z) = \frac{1}{2}\sum_{i=1}^{N} \nu_i(q_i^2 + p_i^2)$

The $\mathbb{R}^{2N}$-valued function $z(t) \in P^1$ is a solution of (1.24) (i.e. a zero of F) iff z is a critical point of the

Action functional A defined by

$$A(z(\cdot)) \doteq \int (\langle \frac{dq}{dt}, p \rangle - \frac{\omega}{\omega_o} h(q,p,\varepsilon)) dt \qquad (1.26)$$

when $\langle \xi, \eta \rangle = \sum_{i=1}^{N} \xi_i \eta_i$.

Remark parenthetically that F can be regarded as Hamiltonian vector field on $P^1$, with Hamiltonian A and symplectic structure $J_o$ given by

$$J_o(\xi,\eta) \doteq \int (\langle \dot{\xi}, \eta \rangle (t) - \langle \xi, \dot{\eta} \rangle (t)) dt \qquad (1.27)$$

(here and in what follows, for every $\xi \in P^1$, we identify $T_\xi P^1$ with $P^1$).

The reduction procedure of Moser and Weinstein consists in solving first for $z, \delta$ the equation

$$dA(z_o + z, \omega_o + \delta, \varepsilon)\Big|_{J_o \underset{\sim}{P}} = 0 \qquad (1.28)$$

By the Inverse Function Theorem, this equation has a unique solution, $z = \varphi(z_o, \varepsilon), \delta = \psi(z_o, \varepsilon)$, and one is left with

$$dA(z_o + \varphi(z_o, \varepsilon), \omega_o + \psi(z_o, \varepsilon), \varepsilon)\Big|_{M_{J_o z_o}^{\perp}} = 0 \qquad (1.29)$$

From (1.25), (1.26) one verifies that (1.29) is the condition that there exists on M a Hamiltonian vector field, equivariant under $Jdh^o$ and orthogonal at $z_o$ to $Jdh^o$.

Equivalently, using uniqueness of $z, \delta$ above and equivariance of the Moser-Weinstein construction, (1.29) is equivalent to the existence of a Hamiltonian function $\hat{h}(q, p, \varepsilon)$ such that

$$d\hat{h}\Big|_{q_o p_o} = \lambda \, dh^o\Big|_{q_o p_o} \quad , \quad (q_o, p_o) \equiv z_o \qquad (1.30)$$

for some real constant $\lambda$.


## 2. THE USE OF NORMAL FORMS

We begin by a short review of normal forms for vector fiels, with special emphasis on the Hamiltonian case. For further details we refer to [7].

LEMMA 2.1. Lef $f$ be a vector field on R, of class $C^{k+1}$, which can be written as $f(x) = L^o \cdot x + f^1(x)$, $L^o$ semi-simple, $|f^1(x)| = o(|x|)$. Assume that $[L^o x, f^{(1)}(x)] = o(|x|^k)$. One can find a vector field $\psi \in C^k$, $|\psi(x)| = o(|x|^k)$ such that

$$[L^o x, (D(id + \psi))^{-1}(x) f(x + \psi(x))] = o(|x|^{k+1})$$

where $DG(y)$ is the differential of $G$ at $y$. $\qquad \square$

PROOF. Consider the map $x \to x + \psi(x)$. Pulling back $f$ through $(id + \psi)$ one has

$$[D(id+\psi)(x)]^{-1} f(x+\psi(x)) = f(x) + L^o \cdot \psi(x) - D\psi \cdot L^o x + o(|x|^{k+1}) \qquad (2.1)$$

Consider now on vector-valued homogeneous polynomials

of order k the map $ad_{(k)}L^O$ defined by

$$ad_{(k)}L^O \cdot \psi \doteq L^O\psi(x) - D\psi \cdot L^O x \qquad (2.2)$$

Let $\{e_j\}$ be a basis of eigenvectors of $L^O$, $\{\lambda_j\}$ the cor-responding eigenvalues. Consider the monomials

$$x^{\underline{\alpha}} \cdot e_j \doteq x_1^{\alpha_1} \ldots x_m^{\alpha_m} \cdot e_j \quad , \quad \sum_1^m \alpha_j = k \quad , \quad \alpha_j \in z^+ \quad (2.3)$$

On $x^{\underline{\alpha}} e_j$ one has

$$ad_k L^O(x^{\underline{\alpha}} e_j) = (\lambda_j - \sum_{k=1}^m \alpha_k \lambda_k) x^{\underline{\alpha}} e_j$$

It follows that the eigenvalues of $ad_k L^O$ are $\lambda_j - \sum_{k=1}^m \alpha_k \lambda_k$, $j = 1 \ldots m$, and that the image of $ad_k L^O$ is trasversal to

$$\text{Ker } ad_k L^O = \text{span}\{x^{\underline{\alpha}} e_j, \lambda_j - \langle \alpha, \lambda \rangle = 0\}.$$

By the inverse function theorem one can then find a vector field $\psi \in C^k$ such that

$$[L^O, D(id + \psi)^{-1} f(x + \psi(x))] = \mathcal{O}(|x|^{k+1}) \qquad (2.4)$$

$\square$

REMARK 2.2. The requirement that $L^O$ be semi-simple is unnecessarily stringent. One can develop the same proof using only the semi-simple part of $L^O$; also, one can let $L^O$ depend on a set of parameter $\mu$, and one finds a set of constants $\mu_o$

and, if $|\mu| < |\mu_0|$, a vector field $\psi(x,\mu)$ such that

$$[L^o(\mu),D(id + \psi(\cdot,\mu))^{-1}f(x + \psi(x,\mu))] = \mathcal{O}(|x|^{k+1})$$

uniformly in $|\mu| < |\mu_0|$.

DEFINITION 2.3. We shall say that $f'$ is in normal form relative to $L^o x$ if $[f'(x),L^o x] = 0$.

With this notation, the previous lemma can be read as follows: let $f \in \mathbf{C}^{k+1}$, $f = L^o x + f_1(x) + R(x)$, where $f_1$ is in normal form relative to $L^o x$ and $|R(x)| = \mathcal{O}(|x|^k)$. One can then find a transformation of coordinates of class $\mathbf{C}^k$ such that, in the new coordinates, $f$ has the form

$$f(y) = L^o y + f_1'(y) + R'(y)$$

where $f_1'$ is in normal form relative to $L^o y$ and $|R'(y)| = \mathcal{O}(|y|^{k+1})$.

We now specialize to the case when $f$ is a Hamiltonian vector field i.e. $f = \sigma \nabla h$.

One has then

$$h = h^o + h' \qquad (2.5)$$

where $h$ is of class $\mathbf{C}^{k+2}$ and

$$h^o = \omega_o \sum_i^N \frac{1}{2} \nu_i(p_i^2 + q_i^2) \quad , \quad \nu_i = 1, \ \nu_k \in \mathbb{Z} \big|_{\{0,1\}} \qquad (2.6)$$

We shall use complex coordinates

$$\xi_i \doteq \frac{1}{\sqrt{2}} (p_i + iq_i) \qquad i = i\ldots n$$

and take explicitely into account the multeplicities of the eigenvalues $\nu_i$.

Therefore

$$h^o = \omega_o \sum_i^s \nu_k \xi^{(k)} \bar\xi^{(k)} \tag{2.7}$$

where $\underline\xi^{(k)} \equiv \{\xi_i^{(k)},\ldots,\xi_{d_k}^{(k)}\}$ and $\sum_{k=1}^{s} d_k = N$.

In complex coordinates, Hamilton's equation are

$$\dot\xi_p^{(k)} = i \frac{\partial h}{\partial \bar\xi_p^{(k)}} \qquad k = 1\ldots s, p = 1\ldots d_k. \tag{2.8}$$

Our previous study of normal forms can be specialized to the Hamiltonian case. One verifies that the map $id + \psi$ can be taken to be a canonical trasformation; in fact through the inverse function theorem one can construct its generator.

Denote by $\{A,B\}$ the Poisson bracket of $A,B$ (relative to $\sigma$). One proves

LEMMA 2.4. Let $h \in \mathbf{C}^{k+1}$, $h = h^o + h'$, $|h'(\xi)| = o(|z|^2)$, $|\{h^o,h'\}| = o(|z|^k)$, where we have denoted collectively by $z \in \mathbf{C}^N$ the variables $\xi_q^p$.

One can find a canonical transformation $\varphi$ of class $\mathbf{C}^k$, such that

$$\{h^o \circ \varphi, h' \circ \varphi\} = \mathcal{O}(|z|^{k+1})$$

Equivalently, if $h = h^o + h_1 + R$, $\{h^o, h_1\} = 0$, $R = \mathcal{O}(|z|^k)$, then in the new (canonical) variables $z'$ one has

$$h = h^o + h_1' + R \quad , \quad \{h^o, h_1'\} = 0 \quad , \quad R' = \mathcal{O}(|z'|^{k+1}).$$

$\square$

DEFINITION 2.5. If $h = h^o + h_2 + R$, $\{h^o, h_2\} = 0$, $|R(z)| = \mathcal{O}(|z|^k)$, we shall say that $h_2$ is a normal form of $h$ to order $k$.

We add here two brief remarks on normal forms; further details are given in [6],[7].

Let $\Gamma$ be the group of those symplectic (canonical linear) transformations which leave $h^o$ invariant. Let, as before, $\Gamma^o$ be the one-parameter group generated by $h^o$. The action of $\Gamma^o$ is an $S^1$-action, namely

$$\xi^{(k)} \to e^{i\nu_k \theta} \xi^{(k)} \quad , \quad \xi^{(k)} \in \mathbf{C}^{d_k} \tag{2.9}$$

One has $h^o \circ \gamma = h^o$ if $\gamma \in \Gamma$, and if $R = \mathcal{O}(|z|^k)$, also $R \circ \gamma = \mathcal{O}(|z|^k)$ since $\gamma$ acts linearly. Therefore if $h_2$ is a normal form of $h$ to order $k$, so is also $h_2 \circ \gamma$. Since in general $h_2 \circ \gamma \neq h_2$, one can see that in general a Hamiltonian has many normal forms to any given order (in fact an entire $\Gamma$-orbit); it is often convenient to choose a particular one.

For our purposes, the following relations will be important.

1) If $h_2 = P(h^o)$ for some polynomial P, then $h_2 \circ \gamma = P(h^o)$ for all $\gamma \in \Gamma$. In this case we say that h is trivial to order k. Indeed it is straight-forward to verify that one can find a polynomial Q, $Q(x) = \mathcal{O}(|x|^2)$, such that $h + Q(h) = \mathcal{O}(|x|^{k+2})$. Since h and h + Q(h) have the same periodic trajectories (their flows differ only by a time-change along each trajectory), the case when h is trivial to order k is equivalent to one in which $h = h^o + h'$, $h' = \mathcal{O}(|x|^{k+1})$.

2) Suppose $h_2$ is normal with respect to $h^o$, and that there exists a point $\underline{z}_o \in \mathbb{C}^n$ and a constant $\lambda$ such that

$$dh^o(\underline{z}_o) = \lambda dh(\underline{z}_o) \tag{2.10}$$

Then

a) (2.10) holds on the entire orbit of $\Gamma^o$ through $\underline{z}_o$.

b) For every $\gamma \in \Gamma$, $dh^o(\gamma(\underline{z}_o)) = \lambda d(h_2 \circ \gamma)(\gamma(\underline{z}_o))$. Therefore the number of orbits of $\Gamma^o$ for which (2.10) holds depends only on the equivalence class of $h_2$.

We are now ready to discuss the connection between normal forms of a vector field (or of a Hamiltonian) and the vector field $\underset{\sim}{f}$ which was constructed in Sect 1 and played a central rôle in finding periodic solutions.

We shall consider only the Hamiltonian case; the next two Lemmas generalize easily to any vector field of the class considered in Section 1. Remark that in the Hamiltonian case the condition $[\sigma \nabla h^1, \sigma \nabla h^2] = 0$ is equivalent to $\{h^1, h^2\} = $ constant, where $\{,\}$ is the Poisson bracket.

LEMMA 2.6. Let $h = h^o + h^1$, $\{h^1, h^o\} = 0$, $h'(z) = \mathcal{O}(|z|^2)$. Let $\hat{h}$ be as described in Section 1. Then $\hat{h} = h$. $\qquad \square$

PROOF. Since $\{h^o,h\} = 0$, the bijection $R^{2N} \leftrightarrow M$ is equivariant under the flow of h. Therefore M is left invariant by the flow of the vector field F, i.e. $F(z_o,\omega_o,\varepsilon)$ is tangent to M for all $z_o(t) \in M$. The unique solution described in (1.18) is therefore $\varphi = \psi = 0$. It follows that

$$F(z_o,\omega_o,\varepsilon) = F(z_o,\omega_o,\varepsilon)$$

and, by equivariance of $\sigma\nabla h$ under the flow of $h^o$, $\hat{h} = h$. $\quad\square$

COROLLARY 2.7. Let $h \in C^{k+1}$, $h^o$ as in Section 1, and let $P^{(k)}$ be a normal form for h to order k, so that

$$h = h^o + P^{(k)} + R^{(k)}, \quad R^{(k)} = \mathcal{O}(|z|^k) \qquad (2.11)$$

Then

$$\hat{h} = h^o + P^{(k)} + Q^{(k)} \qquad (2.12)$$

where $\{h^o,P^{(k)}\} = 0$ and $Q^{(k)} = \mathcal{O}(|z|^k)$. $\quad\square$

REMARK 2.8. If the frequencies of the linear part are not in full resonance, the role of M is played by

$$N \doteq \{e^{L^o t}\xi, \ \xi \in N \doteq \text{Ker}(\exp 2\pi L^o - I)\}$$

Lemma 2.6 still holds, and $\hat{h} = h|_N$, i.e. $\hat{h}$ is obtained from h by setting equal to zero all normal modes which do not resonate. $\quad\square$

From Lemma 2.6 and Proposition 1.1 one can deduce the following. If $\{\tau_\varepsilon(t), 0 \le \varepsilon \le \varepsilon_o\}$ is a family of periodic solutions of equation (2.8) with $|\tau_\varepsilon(t)| = O(\varepsilon)$, and frequency $\omega(\varepsilon) = \omega_o + O(\varepsilon)$, then there exists a family $\tau_\varepsilon^o(t)$ of periodic solutions for the Hamiltonian $h^o + P^k$ with frequency $\omega^o(\varepsilon)$ such that

$$\sup_{t} \inf_{t'} \left| \tau_\varepsilon(t) - \tau_\varepsilon^o(t') \right| = \mathcal{O}(\varepsilon), \qquad \left| \omega(\varepsilon) - \omega^o(\varepsilon) \right| = \mathcal{O}(\varepsilon) .$$

Therefore a necessary (not always sufficient) condition in order to have N families of periodic solutions of (1.6) is that the same be true for the vector field $L^o x + f_{(k)}(x)$, where $f_{(k)}(x)$ is a normal form for $f_\varepsilon(x)$ to order k.

We shall give a direct proof of this fact, working in $R^{2N}$. This will make more transparent the rôle of normal forms and the geometric meaning of the condition $\underset{\sim}{f} \parallel L^o x$.

We give the proof for a generic vector field but restrict ourselves to the case when $f_{(k)}$ is a homogeneous polynomial of order $k(k > 1)$. The general case is treated similarly, but there are additional technical difficulties which are due to the fact that the orbits $\tau_\varepsilon^o(t)$ no longer satisfy the identity $\tau_\varepsilon^o(t) = \varepsilon \tau_1^o(t)$.

We prove first.

LEMMA 2.9. Let f be a vector field on $R^m$ which can be written as $f = L^o x + f'(x)$, $k \geq 2$, where $L^o$ is such that $\exp(2\pi L^o) = I$ and $f'$ is a homogeneous polynomial of order $k_o$. Suppose moreover that $[L^o x, f'(x)] = 0$. One can than find a constant $\varepsilon_o > 0$ such that for every $0 < \varepsilon < \varepsilon_o$ a point $x_o \in R^m, |x_o| = \varepsilon$ belongs to a periodic orbit of f with period close to $2\pi$ if and only if one can find $\lambda(x_o) \in R$ such that $f'(x_o) = \lambda(x_o) L x_o$.

The periodic orbit of f passing through $x_o$ is then also a periodic orbit of $L^o x$, and the ratio of the two periods is $1 + \lambda(x_o)$; in particular $\lambda(x_o) = 0(|x_o|^{k-1})$. $\qquad \square$

REMARK 1.10. In this case, due to homogeneity of f', one as for a > 0

$$f'(ax_o) = a^k f'(x_o) \quad , \quad L^o \cdot ax_o = aL^o x_o.$$

Therefore one has $f'(x_o) = \lambda(x_o)L^o x_o$ iff

$$f'(\hat{x}_o) = |x_o|^{k-1}\lambda(\hat{x}_o)L^o\hat{x}_o, \text{ where } \hat{x}_o = x_o \cdot |x_o|^{-1}.$$

Therefore it suffices to find the points on $S_1^m$ where $f_{(k)}$ is parallel to $L^o x$. $\qquad\qquad\square$

Proof of Lemma 1.9. Scaling $x \to \varepsilon x$ one can write

$$\frac{1}{\varepsilon} f(\varepsilon x) = f_\varepsilon(x) = L^o x + \varepsilon^{k-1} f'(x)$$

Denote by $\exp(tf)$ the exponential map of the field $f$ at "time" $t$. Then, for all $0 < \varepsilon < \varepsilon_o, |z_o| = 1$, $\varepsilon z_o$ is on a periodic orbit of $f$ with period $T_\varepsilon(z_o)$ iff

$$\exp(T_\varepsilon f_\varepsilon) \cdot z_o = z_o \qquad\qquad (2.13)$$

Set $f_o(x) \equiv L^o x$. Since $[f', f_o] = 0$, for all $b \in R$ $\exp(bf_\varepsilon) = \exp(bf')\circ\exp(bf_o)$. Let $T_\varepsilon = 2\pi + c_1(z_o)\varepsilon^{k-1} + o(\varepsilon^{k-1})$ and recall that by assumption $\exp 2\pi f_o = I$. Expanding (2.13) in powers of $\varepsilon$ to order $\varepsilon^{k-1}$ one has

$$2\pi f_1(z_o) + c(z_o)f_o(z_o) = 0 \qquad\qquad (2.14)$$

Therefore (2.14) is a necessary condition for (2.13)

It is also sufficient. Indeed, if (2.12) holds, $f'$ and $f_o$ are parallel along the entire orbit of $f_o$ through $z_o$, since $[f',f_o] = 0$. Therefore, if

$$z(t) \equiv \exp(tf_o)\varepsilon z_o,$$

then

$$\frac{d}{dt} z(t) = f_o(z(t)) = f(z(t)) \cdot \frac{2\pi}{T_\varepsilon}$$

Therefore $z(t)$ is a periodic solution for $f$ with period $T(\varepsilon)$. $\qquad\Box$

LEMMA 2.11. Let $f$ be a vector field on $\mathbb{R}^m$ which can be written as $f = f_o + f' + R_k$ with $f_o$ and $f'$ as in Lemma 2.9 and $R_k(z) = o(|z|^k)$.

Assume moreover that one can find constants $\varepsilon_o > 0$, $C \in R$ and a point $x \in R^m$ such that for $0 < \varepsilon < \varepsilon_o$ there is a one-parameter family $z(t)$ of periodic solutions of $\dot{z} = f$ with period $T_\varepsilon = 2\pi + c\,\varepsilon^{k-1} + o(\varepsilon^{k-1})$ and satisfying $z^\varepsilon(0) = x\varepsilon + o(\varepsilon)$.

Then one can find a point $z_o \in \mathbb{R}^m$ such that (2.14) holds and moreover $d(z^\varepsilon(t),\varepsilon z_o(t)) = o(\varepsilon)$ where $z_o(t)$ is the periodic solution of $f_o + f'$ through $z_o$, as described in Lemma 2.9. If $T^o$ is the period of this solution, one has $T_\varepsilon - T_\varepsilon^o = o(\varepsilon^{k-1})$. $\qquad\Box$

PROOF. Remark that, for all $a \in R$,

$$\exp a(f_o + f' + R_k) = \exp a(f_o + f') \cdot (I + o(|x|^k)).$$

Proceeding as in the proof of Lemma 2.9 one concludes

$$2\pi f_1\left(\frac{z_\varepsilon(t)}{\varepsilon}\right) + \frac{T(\varepsilon) - 2}{\varepsilon^{k-1}} f_o\left(\frac{z_\varepsilon(t)}{\varepsilon}\right) = \frac{1}{\varepsilon} o(\varepsilon)$$

uniformly in $0 \leq t \leq 4\pi$. For $t = 0$, take the limit $\varepsilon \to 0$. Then

$$2\pi f_1(x) + cf_o(x) = 0.$$

The conclusions of Lemma 2.11 follow now from Lemma 2.9.

$\square$

We specialize now to the case in which $f$ is a Hamiltonian vector field, $f = \sigma \nabla h$.

Now

$$h = h^o + P^k + R^k \tag{2.15}$$

where $P^{(k)}$ is homogeneous of order $k$, $\{h^o, P^k\} = 0$, $R^k = \mathcal{O}(|z|^k)$. Condition (2.14) reads now

$$dP^{(k)}(z_o) - \lambda dh^o(z_o) = 0 \tag{2.16}$$

while condition (2.13) reads

$$dP^{(k)}(z(\varepsilon)) - \lambda dh^o(z(\varepsilon)) = \varepsilon F(\varepsilon, z(\varepsilon)) \tag{2.17}$$

where $\sup\limits_{0 \leq \varepsilon < \varepsilon_o} |F(\varepsilon, z(\varepsilon))| < \mathbf{C} < \infty$.

Let $z_o$ be a solution of (2.16); a sufficient condition that equation (2.17) have a unique solution of the form

$$z(\varepsilon) = z_o \varepsilon + o(\varepsilon)$$

would be that the Jacobian $J$ of (2.16) be non singular.

However this condition cannot be satisfied, since both $P^{(k)}$ and $h^o$ are invariant under the Hamiltonian flow of $h^o$,

124

so that the vector $\sigma\nabla h^o$ is in the kernel of the Jacobian matrix.

One then considers the restriction $J'$ of the Jacobian to a linear space complementary to $\sigma\nabla h^o$. $J'$ is the Jacobian of the Poincarè map $z(0) \xrightarrow{\pi_o} z(T_\varepsilon)$ in a neighborhood of $\varepsilon z_o$, along the flow of $h^o + P^{(k)}$. One readily verifies that the Poincarè map $\pi_\varepsilon$ along the flow of $h^o + P^{(k)} + R^k$ exists in a neighborhood of $\varepsilon z(0)$, since $R^k = o(|x|^k)$. The point $\varepsilon z_o$ is a fixed point of $\pi_o$, and if $J'$ is non-singular there is a fixed point $z(\varepsilon)$ of $\pi_\varepsilon$, for $0 \le \varepsilon \le \varepsilon_1$. This fixed point correpsonds to a periodic orbit of (1.6) with period close to $T$.

It is also straight forward to verify that $J'$ is non-singular at $\xi$ precisely if the restriction of $P^{(k)}$ to $\sum_1^o$, where $\sum_1^o \equiv \{z | h^o(z) = 1\}$, has a non-singular Hessian.

A lower bound on the number of solutions of (2.16) can now be obtained by combining topological and algebraic techniques. We shall give an example below. We remark however that the techniques are purely topological if either $h^o$ or $P^{(k)}$ satisfy suitable convexity assumptions. If these assumption are satisfied, one has then directly a lower bound on the number of periodic solutions of (1.6).

One has indeed:

THEOREM 2.12. Let $h = h^o + P^{(k)} + R^k$ where $P^{(k)}$ is a polynomial of order up to k in normal form, $P^{(k)}(z) = O(|z|^3)$, $R^k = o(|z|^k)$, $z \in R^{2N}$. Assume that $h^o$ has frequencies $\omega_o \nu_i$, $\nu_1 = 1$, $\nu_i \in z|_{\{0\}}$, $i = 1,\ldots,s$, with multeplicity $d_i$, $\sum_{i=1}^s d_i = 2N$. Assume that one can find a function $f: R \times R \to R$, $f(0,0) = 0$, with the following property. Define

$$N_\varepsilon \doteq \{z \mid f(h^o(z), P^{(k)}(z)) = \varepsilon\}$$

One can find constants $\varepsilon_o > 0$, $\delta > 0$ such that, for $0 < \varepsilon < \varepsilon_o$, $N_\varepsilon$ is diffeomorphic to a sphere and $\underset{0 < \varepsilon < \varepsilon_o}{\cup} N_\varepsilon \supset B_\delta \setminus \{0\}$, where $B_\delta \equiv \{z \mid |z| < \delta\}$. Assume moreover that either $\dfrac{\partial f}{\partial x_1} \neq 0$ on $N_\varepsilon$ or $\dfrac{\partial f}{\partial x_2} \neq 0$ on $N_\varepsilon$. Then in $B_\delta$ there are at least N families of periodic solutions of (1.6) each of which has frequency close to $\omega_o \nu_i$ for some i.

REMARK 2.13. If $h^o$ is of definite sign at the origin (i.e. if $\nu_k > 0$ $k = 1 \ldots s$), then one can choose $f(x,y) = x$. This is the result of Moser and Weinstein. Another simple case is when $f(x,y) = y$, and $\{z \mid P^{(k)}(z) = \varepsilon\}$ is diffeomorphic to a sphere. A simple example is the following: $P^{(k)}(z) = \sum\limits_{i=1}^{s} c_i |\underline{\xi}^{(i)}|^4$, $c_i > 0$. The case $f(x,y) \neq x$ was not considered before. $\qquad\square$

PROOF. For each $0 < \varepsilon < \varepsilon_o$ consider the manifold

$$N'_\varepsilon = \{z \mid f(h^o(z), \underset{\sim}{h}(z) - h^o(z)) = \varepsilon\}$$

Since $\underset{\sim}{h}(z) - h^o(z) - P^{(k)}(z)$ is infinitesimal w.r. to $P^{(k)}$, one can find $\varepsilon_1 \leq \varepsilon_o$ such that, for $0 < \varepsilon < \varepsilon_1$, $N'_\varepsilon$ is diffeomorphic to a sphere. It is also invariant under $\Gamma^o$ (the Hamiltonian flow of $h^o$), since $\{h^o, \underset{\sim}{h}\} = 0$.

If $\dfrac{\partial f}{\partial x_2} \neq 0$, consider the problem of finding the number of orbits $\Gamma^o$ which are composed of points where $dh^o$ is parallel to $dh$, i.e. of stationary points of the restriction of

$h^o$ to $N'_\varepsilon$.

Both $h^o$ and $N'_\varepsilon$ are invariant under $\Gamma^o$, and it follows from equivariant Morse theory that there are at least N such orbits.

Since $df = \dfrac{\partial f}{\partial x_1} dh^o + \dfrac{\partial f}{\partial x_2} d\underset{\sim}{h}$ and $\dfrac{\partial f}{\partial x_2} \neq 0$, one concludes that $d\underset{\sim}{h} \| dh^o$ at those points. Theorem 2.12 follows now from the analysis given in Section 1.

If $\dfrac{\partial f}{\partial x_i} \neq 0$ on $N'_\varepsilon$, the same result holds and the proof is the same as above, with $\underset{\sim}{h}$ playing the role of $h^o$. $\square$

If the conditions of Theorem 2.12 are not met, it is in general much more difficult to give estimates on the number of families of periodic solutions. In [8] we discuss some cases in which the multeplicity of each frequency is one. A general approach, with some examples is discussed in [6]. Here we shall only outline the approach and give results which complement those of [6].

We remark first that the arguments developed in the course of the proof of Theorem 2.12 are sufficient to prove.

LEMMA 2.14. Assume that $P^{(k)}(z)$ is of order $k_o$ at the origin and that for all $0 < \varepsilon < \varepsilon_o$ there is on $\{z \mid h^o(z) = \varepsilon\}$ a point $z_o(\varepsilon)$ at which $dh^o$ is parallel to $dP^{(k)}_\varepsilon$, where $P^{(k)}_\varepsilon(z) \doteq \varepsilon^{-k_o} P^{(k)}(\varepsilon z)$. Assume that $z_o(\varepsilon)$ dependes smoothly on $\varepsilon$ and that the Jacobian of $dP^{(k)}_\varepsilon$ at $z_o(\varepsilon)$ is non-singular when restricted to the orthogonal complement of the linear space spanned by $\nabla h^o$ and $\sigma \nabla h^o$.

Then there exists on $\{z \mid h(z) = \varepsilon\}$ a point $z_1(\varepsilon)$, such that $|z_1(\varepsilon) - z_o(\varepsilon)| = o(\varepsilon)$ and such that the orbit of the Hamiltonian flow of h through $z_1(\varepsilon)$ is periodic with period close to $2\pi\omega_o^{-1}$. $\square$

This lemma was proved in [8] in a slightly different form, using the fact that the energy is a constant of motion.

We shall not  d well  on this point here, and limit our-selves to a short remark which may be in order since it makes contact with the very interesting approach to "detuning" developed in [7].

REMARK 2.15. Let $\xi_i^k$, $i = 1\ldots d_k$ be the canonical vari-ables (in complex notation) associated to the frequency $\nu_k$, $k = 1\ldots s$, so that the Hamiltonian flow of $h^o$ is given by

$$t \to \exp(i\nu_k\omega_o t)\xi_i^k$$

We shall again use the notation

$$z \equiv \{\xi_i^k, \quad k = 1\ldots s, \quad i = 1\ldots d_k\}.$$

Suppose that at the point $\underline{z}$ one has $dP^{(k)} = \lambda dh^o$ for some $\lambda \in R$. Without loss of generality, we can assume that $\underline{\xi}_i^k = 0$, $i \geq 2$. Then the flow of $h^o + P^{(k)}$ through $\xi_i^k$ is given by

$$\xi_1^k(t) = \exp(i\nu_k(1 + \lambda)^{-1}\omega_o t)\xi_1^k(0), \quad \xi_i^k(t) = 0 \quad i \geq 2$$

To use the inverse function theorem one is then lead to study in neighborhood of $\underline{z}$ a linear system of equations which describes N harmonic oscillators, one of which is detuned by a factor $(1 + \lambda)$ with respect to the others (which are in resonance) but is also coupled linearly to remaining ones.

One wants to find conditions under which the system of coupled oscillators either has no zero normal mode or, if there are zero normal modes, their eigenvectors are orthogonal to $\underline{z}$. The detuning and coupling are of the same order of magnitude. It is straightforward to verify that these conditions can be expressed by stating that the Jacobian of the restriction of $P^{(k)}$ to the orthogonal complement of $\alpha \nabla h^o + \beta \sigma \nabla h^o$, $\alpha$, $\beta \in R$, either has no zero eigenvalue or, if there are zero eigenvalues, $\underline{z}$ is orthogonal to the zero subspace.

In particular, if for some $p, q$

$$\frac{\partial^2 P^{(k)}}{\partial \bar\xi_1^m \partial \xi_q^P} = \frac{\partial^2 P^{(k)}}{\partial \bar\xi_1^m \partial \bar\xi_q^P} = 0$$

$\forall\, m = 1 \ldots s$, then the normal mode $\xi_q^P$ is decoupled from the detuned one, and one can equivalently consider the problem in which $\xi_q^P \equiv 0$. If a zero mode of the coupled system is not orthogonal to $\underline{z}$, a more detailed analysis is needed, involving terms of higher order. In that case one has secular terms and the existence of periodic solutions of period close to $\frac{1 + \lambda}{\nu_1} \cdot 2\pi$ depends in general on stability arguments, and therefore on the definiteness in sign of $h^o$ or on the topological structure of $\{z \mid P^{(k)}(z) = c\}$.

## 3. A GENERAL PROCEDURE BASED ON NORMAL FORMS

We shall now sketch a general method which yields a lower bound on the number of orbits of $\Gamma^o$ along which $dP^{(k)}$ is parallel to $dh^o$. This approach can be used when neither $h^o$ nor $P^{(k)}$ satisfy a positiveness requirement; it consists in separating the problem in a topological part and an algebraic one.

We shall exemplify in a simple case; further details
and other examples can be found in [6].

Assume that $P^{(k)}$ is homogeneous of order k and denote
again by

$$\xi_i^P \quad , \quad p = 1 \ldots s \quad , \quad i = 1 \ldots dp \qquad (3.1)$$

a set of canonical variables (in complex form) such that

$$h^o = \sum_{m=1}^{s} \nu_m |\xi^m|^2 \quad , \quad \nu_1 = 1 \quad , \quad \nu_m \in Z|_{\{0\}}, \quad \xi^m \equiv \{\xi_i^m \ldots \xi_{ds}^m\} \qquad (3.2)$$

The frequency considered has been normalized to one .
We also use the notation

$$z \equiv \{\xi^P\} \quad , \quad p = 1 \ldots s \quad , \quad z \in \mathbf{C}^N \quad , \quad N = \sum_{m=1}^{s} d_m \qquad (3.3)$$

One can introduce generalized spherical coordinates

$$\xi^i \to (\rho_i, \hat{\xi}^{(i)}) \quad , \quad \rho_i = |\xi^i| \quad , \quad |\hat{\xi}^{(i)}| = 1 \qquad (3.4)$$

The points where $\rho_i = 0$ for some $i = 1 \ldots s$ must be
treated separately in an obvious way.

We shall denote by $S_1^d$ the unit sphere in $\mathbf{C}^d$.

One can now regard $P^{(k)}$ as a function F on
$(R^+)^s \times (S_1^{d_1} \times \ldots \times S_1^{d_s})$ . Both F and $(S_1^{d_1} \times \ldots \times S_1^{d_s})$ are invariant
under the flow of $h^o$.

For fixed $\rho \equiv \{\rho_1, \ldots, \rho_s\} \in (R^+)^s$, one can regard F as
a function, denoted by $F_\rho$, on $S_1^{d_1} \times \ldots \times S_1^{d_s} \backslash \Gamma^o$. The latter has

in general non-trivial homology groups, and this provides a lower bound on the number of points at which $dF_\rho = 0$. They correspond to orbits of $\Gamma^o$ at which $dF$ vanishes on vectors tangent to $S_1^{d_1} \times \ldots \times S_1^{d_s}$ (the fibration by $\Gamma^o$ is in general not of Hopf type, and the analysis described above is incorrect as it stands, since $S_1^{d_1} \times \ldots \times S_1^{d_s} \backslash_{\Gamma^o}$ is in general not a manifold. The lower bound considered is in fact obtained by the use of equivariant Morse theory).

Let $\{\hat{\xi}^{(i)}\} = \zeta_\rho \in S_1^{d_1} \times \ldots \times S_1^{d_s}$ be a point where $dF = 0$. Consider on $(R^+)^s$ the function

$$F(\rho) = F_\rho(\zeta_\rho) \tag{3.5}$$

The algebraic problem which remains to be solved is the determination of the zeroes of $D(F|_{F_\varepsilon})$, where

$$F_\varepsilon \equiv \{z \in \mathbf{C}^n, \ h^o(z) = \varepsilon\} \tag{3.6}$$

This procedure can be reformulated in the following equivalent way. Remark that $P^{(k)}$ is a homogeneous polynomial in normal form. If one takes a minimal basis (of cardinality M) of monomials in normal form, $P^{(k)}$ can be identified with a point $\sigma \in \mathbf{C}^M$. We denote by $\varphi$ the identification map, so that $P^{(k)} = \varphi(\sigma)$. The group $\Gamma$ act on canonical coordinates, and therefore on polynomials of order k in normal form (since $\Gamma$ leaves $h^o$ invariant), and by duality on $\mathbf{C}^M$. We denote by $\Gamma_M$ the corresponding representation.

Define now a subset $M_{\underline{z}}^\lambda \in M^M$, $\underline{z} \equiv \{z_i \in \mathbf{C}\}$, $\lambda \in R$, by

$$M_z^\lambda \equiv \{y \in \mathbb{R}^M; d(\varphi^{-1}(y)) = \lambda dh^o \text{ at } \xi_1^i = z_i e^\theta \text{ for some } \theta \in \mathbb{R},$$

$$\xi_m^i = 0 \quad m \geq 2 \quad i = 1 \ldots s\} \qquad (3.7)$$

Denote by $\overset{\sim}{M}_z^\lambda$ the subset of $M_z^\lambda$ composed of those points for which the Jacobian of $d\varphi^{-1}(y) - \lambda dh^o$ is non singular when restricted to a subspace in $\mathbb{C}^n$ orthogonal to $\sigma\nabla h^o$.

Note that $M_z^\lambda$ is the set of zeroes of a polynomial, so it is the union of smooth submanifolds of $\mathbb{C}^M$. One can show that also the dependence on $\lambda$ is smooth, so that also

$$M_z \equiv \underset{\lambda \in R}{\cup} M_z^\lambda$$

is the union of a finite number of smooth submanifolds of $\mathbb{C}^M$.

Let $\overset{\sim}{M}_z \equiv \underset{\lambda \in R}{\cup} \overset{\sim}{M}_z^\lambda$; then $M_z - \overset{\sim}{M}_z$ is of higher co-dimension.

With these notations, there are on $E_\varepsilon$ $\ell$ points at which $dP^{(k)} = \lambda dh^o$ for some $\lambda \in R$ iff the orbit of $\Gamma_M$ through $\varphi^{-1}(P^{(k)})$ intersects $\ell$ times $M^\varepsilon(\underset{z \in (R^+)^s}{\cup} M_z) \cap E_\varepsilon$ (remark that $P^{(k)}$ is left invariant by $\Gamma^o$ so that points in $\mathbb{C}^M$ correspond to orbits of $\Gamma^o$ in $\mathbb{C}^n$). If the intersection takes place at points in $\underset{z}{\cup} \overset{\sim}{M}_z$, then the Jacobian of $dP^{(k)} - \lambda dh^o$ is non singular and the system of equation (1.6) has correspondingly a periodic solution.

This procedure was used in [8] for the case of frequency ratioes $1:-1, 1:-2, 1:-3$ and all multiplicities equal to one.

Here we shall briefly discuss the first approach, using as an example the case when the frequency ratio is $1:\pm 2$ and each frequency has multiplicity two.

We treat first the case $\nu_1 = 1$, $\nu_2 = -2$, and set $\omega_o = 1$.

132

The problem is then the following:

Determine the points in $\mathbf{C}^4$ at which $\lambda dh^o = dh'$, where $\lambda \in R$,

$$h_o = |\underline{\xi}|^2 - 2|\underline{\eta}|^2 \quad , \quad h' = (\eta_1 P^{(1)}(\underline{\xi}) + \eta_2 P^{(2)}(\underline{\xi})) + c.c. \quad (3.8)$$

$\underline{\xi} \equiv \{\xi_1, \xi_2\}$ , $\xi_i, \eta_k \in \mathbf{C}$, $P^{(i)}$ are homogeneous polynomials in $\xi_1, \xi_2$ of order two.

Consider the sets $\sum_i \equiv \{\xi | P^{(i)}(\xi) = 0\} \cap \{\xi | |\xi| = 1\}$. Both are invariant under $\Gamma^o$ and of dimension one. Moreover either $\sum_1 \cap \sum_2 = \sum_1 = \sum_2$ or $\sum_1 \cap \sum_2 = \{\phi\}$. In the former case, one has a $P_1(\xi) = b P_2(\xi)$ for some $a, b \in R$, not both equal to zero. One can then choose a basis for $\underline{\eta}$ such that $P^{(2)} = 0$, i.e. the coordinate $\eta_2$ is decoupled at the order considered. The problem is reduced to one in which $d_2 = 1$.

We shall therefore assume that

$$\min_{\xi \in S_1^2} R(\xi) \quad c_- > 0 \equiv c_- > 0 \quad\quad\quad (3.9)$$

when $R(\xi) \doteq (|P_1(\xi)|^2 + |P_2(\xi)|^2)^{1/2}$.

We also assume that $R(\xi)$ is not constant, and denote by $c_+$ the maximum value of $R(\xi)$ on $S_1^2$.

It is straight forward to verify that, for fixed $\underline{\xi}$ and for any $b > 0$, the maximum (respectively minimum) value of $h'$ on $\{\xi, \eta | |\underline{\xi}| = 1, |\underline{\eta}| = b\}$ is attained at

$$\eta^{\pm} = \{\eta_i^{\pm}(\xi)\} \quad , \quad \eta_i^{\pm}(\xi) \doteq \frac{\pm P^{(i)}(\xi)}{R(\xi)} \cdot b \quad\quad\quad (3.10)$$

and one has there

$$h'(\underline{\xi}, \eta^{\pm}(\underline{\xi})) = R(\xi) \cdot b \quad\quad\quad (3.11)$$

Let $\underline{\xi}^{(+)}$ (resp $\underline{\xi}^{(-)}$) a point at which $R(\xi)$ attains its maximum (resp. minimum value) on $\{\xi \mid |\xi| = 1\}$.

Let $\underline{z} \equiv \{\xi, \eta\} \in \mathbf{C}^4$ and choose generalized spherical coordinates $(|\xi|, |\eta|, \hat{\xi}, \hat{\eta})$. Recall that the points we are considering satisfy $|\xi| \neq 0, |\eta| \neq 0$.

In spherical coordinates, consider the decomposition of the tangent space at $z$ given by

$$T_z \mathbf{C}^4 = T_z (R^+ \times R^+) \times T(S_{|\xi|}^{d_1} \times S_{|\eta|}^{d_2}) \qquad (3.12)$$

Denote by $z_{\pm,\pm}$ the points of coordinates $\{\xi^{(\pm)}, \eta^{(\pm)}(\xi^{(\pm)})$ By construction

$$\left. dh' \right|_{T_{z_{\pm,\pm}} (S_1^{d_1} \times S_{c_\pm}^{d_2})} = 0 \qquad (3.13)$$

Let

$$E(c) \equiv \{z \mid |\xi|^2 - 2|\eta|^2 = c^2\} \qquad (3.14)$$

Then

$$T_z E(c) \simeq TM_c \times (S_{|\xi|}^{d_1} \times S_{|\eta|}^{d_2}) \qquad (3.15)$$

where $M_c \equiv \{x, y \mid x > 0, y > 0, x^2 - 2y^2 = c^2\}$ $\qquad (3.16)$

One has

$$h'(\xi, \eta) = |\xi|^2 |\eta| h'(\hat{\xi}, \hat{\eta}) \qquad (3.17)$$

134

Remark that $dh'(\sigma \nabla h^o) = dh^o(\sigma \nabla h^o) = 0$ since $h'$ and $h^o$ are invariant under $\Gamma^o$.

Therefore $dh' = \lambda dh^o$ at $z_{\pm,\pm}$ iff

$$dh' = 2xydx + x^2dy = 0 \tag{3.18}$$

under the condition

$$dh^o = xdx - 2ydy = 0 \tag{3.19}$$

Equation (3.18), (3.19) imply

$$4y^2 + 1 = 0 \tag{3.20}$$

which has *no* solution.

We have proved

PROPOSITION 3.1. If $\nu_2 = -2\nu_1$, $d_1 = d_2 = 2$, and if the normal form $P^{(3)}$ is non-zero, there is no family of periodic solutions of (1.6) near the origin with frequency close to $\nu_1$ and the regularity properties described in Lemma 2.14. $\square$

REMARK 3.2. It is straightforward to verify that for arbitrary multeplicities $d_1, d_2$ the condition (3.18), (3.19) must be satified for every point at which $dh'$ is parallel to $dh^o$. Therefore Proposition 3.1 holds true for any multiplicity of the frequencies $\nu_1$ and $-2\nu_1$. $\square$

If $\nu_2 = 2\nu_1$, the analysis coincides with the one given above up to equation (3.13). It is sufficient to substitute everywhere $\eta$ with $\bar{\eta}$. Also (3.15) and (3.17) hold, but the energy surface is now

$$E_c \equiv \{z \mid |\xi|^2 + 2|\eta|^2 = c^2\} \qquad (3.14)'$$

and one has instead of (3.16)

$$M'_c = \{x,y \mid x > 0, y > 0, x^2 + 2y^2 = c^2\}$$

Instead of (3.20) one must now satisfy

$$4y^2 - 1 = 0 \qquad (3.20)'$$

which has two solutions, $y = \pm 1/2$.

Therefore, for every $\varepsilon > 0$, on $|z| = \varepsilon$ there are at least four orbit of $\Gamma^o$ at which $dh' = \lambda dh^o$ for some $\lambda \in R$, namely the ones which pass through the points

$$z_{\pm}^{(+)} = (\frac{2}{\sqrt{5}} \, \hat{\xi}^{(+)} \varepsilon, \frac{1}{\sqrt{5}} \, \eta^{(\pm)}(\hat{\xi}^{(+)}) \cdot \varepsilon)$$

$$z_{\pm}^{(-)} \equiv (\frac{2}{\sqrt{5}} \, \hat{\xi}^{(-)} \varepsilon, \frac{1}{\sqrt{5}} \, \eta^{\pm}(\hat{\xi}^{-}) \cdot \varepsilon)$$

Remark that the value of $h^o$ on these solutions is

$$(\frac{4}{5} + \frac{2}{5} \, c_{\pm})\varepsilon^2$$

where $c_+ = \sup_{|\xi|=1} R(\xi)$, $c_- = \inf_{|\xi|=1} R(\xi)$ .

For a generic choice of the coefficients in $h'$, the vector field $\lambda dh^o - dh'$ has non-vanishing Jacobian when restricted to the orthogonal complement of $\sigma \nabla h^o$.

One can then use the implicit function theorem and Lemma 2.14 to prove.

PROPOSITION 3.3. If $\nu_2 = 2\nu_1$, $d_1 = d_2 = 2$ for a generic choice of coefficients in the normal form $P^{(3)}$, there are in a neighborhood of the origin at least four one-parameter families of periodic solutions with minimal period close to $\frac{2\pi}{\nu_1}$. They have the regularity properties described in Lemma 2.14. $\qquad\square$

REFERENCES

[1]  C.H. TAUBES: Min-max Theory for the Yang-Mills equation Comm. Math . Phys. to appear.

[2]  M. NARASHIMAN, T. RAMADAS: On the geometry of orbits in gauge theory - Comm. Math. Phys. $\underline{67}$ 21 (1979).

[3]  I. EKELAND, J. LASRY: "On the number of periodic trajectories..." Ann. Math. $\underline{112}$ 283 (1980).

A. AMBROSETTI, G. MANCINI: "On a Theorem of Ekeland and Lasry" Journ. Diff. Eq. $\underline{43}$ 249 (1982).

H. BERESTYCKI, J. LASRY; G. MANCINI, B. RUF: "Existence and multeplicity of periodic orbits...", ISAS Preprint 12/84/M.

[4]  A. WEINSTEIN: "Normal modes for non-linear Hamiltonian systems" Inv. Math. $\underline{20}$ 47 (73).

J. MOSER: "Periodic orbits near an equilibrium ..." Ann. Math. $\underline{112}$ 283 (80).

[5]  V. BENCI, A. CAPOZZI, D. FORTUNATO: "Periodic solutions of Hamiltonian systems" MRC U. of Wisconsin Report 2508, April 83.

[6]   G.F. DELL'ANTONIO, B. D'ONOFRIO: "On the number of peri-
      odic solutions for a Hamiltonian system near equilib-
      rium II", preprint.

[7]   H. DUISTERMAAT: "Bifurcation of periodic solutions near
      equilibrium" Lecture Notes in Math N°1057, 55-105,
      1983, Springer Verlag.

      J.C. VAN DER MEER: "The Hamiltonian Hopf bifurcation",
      Thesis, Univ. of Utrecht, January 1985.

[8]   G.F. DELL'ANTONIO, B. D'ONOFRIO: "On the number of peri-
      odic solutions for a Hamiltonian system near equi-
      librium I", Boll. UMI.


Received May 24, 1985.

GEOMETRODYNAMICS PROCEEDINGS (1985), pp. 139-148
edited by A. Pràstaro


# INTRODUCING SPINORS, ISOSPINORS, ETC. IN GLOBALLY NONTRIVIAL SPACE-TIMES

L. Dabrowski[*]
International School for Advanced Studies,
Trieste, Italy

## ABSTRACT

We review the situation when the structure group $G$ is to be replaced by its covering $\tilde{G}$ in a globally nontrivial space-time $M$. This is necessary in order to introduce consistently spinors, isospinors, quarks, etc. in _M. The application to the minimal conformal compactification $\tilde{M}$ of the Minkowski space-time is presented.

## 1. The G Structure and its Existence

The standard way to deal with physical objects transforming under projective representations of the group $G$ is to pass over to a covering group $\tilde{G}$ of $G$. The aim of the present note is to review the situation when the structure group $G$ is replaced in a global way by $\tilde{G}$ over a manifold (space-time) $M$. We assume that the kernel $K$ of the homomorphism $\rho$ is a discrete subgroup of $G$, contained in the centre of $G$

$$0 \longrightarrow K \longrightarrow \tilde{G} \overset{\rho}{\longrightarrow} G \longrightarrow 0 \qquad (1.1)$$

($\tilde{G}$ in the short exact sequence (1.1) is not necessarily universal). This includes the case of spinors in $M$ equipped with pseudo-Riemannian metric of signature $p, q$ and, eventually, with space and/or time orientations

---

[*] On leave of absence from I.F.T., University of Wrocław, Poland.

$$0 \to Z_2 \to Pin(p, q) \to O(p, q) \to 0$$

$$0 \to Z_2 \to Spin(p, q) \to SO(p, q) \to 0$$

$$0 \to Z_2 \to Spin_0(p, q) \to SO_0(p, q) \to 0 \qquad (1.2)$$

the case of quarks

$$0 \to Z_3 \to SU(3) \to SU(3)/Z_3 \to 0$$

or generalized isospinors

$$0 \to Z_n \to SU(n) \to SU(n)/Z_n \to 0$$

$$0 \to Z \to R \to U(1) \to 0$$

$$0 \to Z_n \to U(1) \to U(1) \to 0 \qquad (1.3)$$

and their combinations like

$$0 \to Z_2 \to (Spin \times U(1))/Z_2 \to SO \times U(1) \to 0 \qquad (1.4)$$

for charged spinors. In order to provide a general setting for fields transforming under $\tilde{G}$ one defines the $\tilde{G}$ structure.

Given a principal G-bundle $\mathcal{F} = [F, p, M, G]$, the $\underline{\tilde{G}\text{-structure}}$ is a principal $\tilde{G}$-bundle $\tilde{\mathcal{F}} = [\tilde{F}, \tilde{p}, M, \tilde{G}]$ together with the (strong) bundle map $\eta : \tilde{F} \to F$, $p \circ \eta = \tilde{n}$ which intertwines the right actions of the relative groups

$$\eta(Ah) = \eta(A)\rho(h) \quad , \quad A \in \tilde{F} \quad , \quad h \in \tilde{G} \quad . \qquad (1.5)$$

Now, the $\tilde{G}$-field $\Phi$ of type $D$, where $D$ is a representation of $\tilde{G}$ in a linear space $V$, is the assignment of coordinates in $V$ to each "frame" $A \in \tilde{F}$ (i.e. $\Phi$ is a map $\Phi : \tilde{F} \to V$) together with the correct transformation law

$$\Phi(Ah) = D^{-1}(h)\Phi(A) \quad , \quad A \in \tilde{F} \quad , \quad h \in \tilde{G} \quad . \qquad (1.6)$$

The usual local description of a field $\Phi$ as a function $\Phi_\alpha$ on an open subset $U_\alpha \subset M$ is obtained by composing $\Phi$ with a local section (choice of "frame") $\sigma_\alpha : U_\alpha \to \tilde{F}$

$$\Phi_\alpha \equiv \Phi \circ \sigma_\alpha : U_\alpha \to V \ . \tag{1.7}$$

(The G-field, which transforms under  G,  is a particular kind of $\tilde{G}$-field, coming from a representation of  $\tilde{G}$  which is trivial on  $K \subset \tilde{G}$).

The  $\tilde{G}$  structure, is the prolongation of  G  to  $\tilde{G}$  in the terminology of Trautman [2], which should be distinguished from the G-structure defined as the reduction of structure group  H  to its subgroup  $G \subset H$.  The existence conditions are also different, however by no means trivial.  In particular, not all space-times admit spinors.  This was a surprise for physicists, who encountered various paradoxes.  For instance, the Atiyah-Singer index seems to be non-integer on CP(2) [1]

$$n_R - n_L = - \frac{1}{384\pi^2} \int_{CP(2)} R_{\mu\nu\rho\lambda} {}^*R^{\mu\nu\rho\lambda} \sqrt{g}\,d^4x = - \frac{1}{8} \tag{1.8}$$

or there are contradictory prescriptions to define the parallel transport of spinors.  The last property is related to the existence of a noncontractible loop in the fibre  $p^{-1}(x) \subset F$,  which is contractible in the total space  F.  This prevents the existence of spin structure on  CP(2n),  since noncontractible loops lift to "open" paths in  $p^{-1}(c) \approx Pin(4n)$,  which cannot be deformed to a loop in  $\tilde{F}$.  For instance, the loop 1

$$< 0,\ 2n >\ni \tau \xrightarrow{\ell} \begin{bmatrix} \exp|i\tau(2n+1)|, & 0, & \dots, & 0 \\ 0 & , & 1, & \dots, & 0 \\ . & & & & . \\ . & & & & . \\ . & & & & . \\ 0 & , & 0, & \dots, & 1 \end{bmatrix} \in U(2n) \subset O(4n) \tag{1.9}$$

is noncontractible in  O(4n).  Since 1 is (2n+1)-th power of another loop, it is contractible in  $SU(2n+1)/Z_{2n+1}$,  which is (the total space of the) sub-bundle of unitary frames of the bundle of orthonormal frames on  CP(2n) [2].  Other examples of non-spin manifolds are provided by the Genoch [6] manifolds  $M_m$,  by the twisted $S^2$-bundle over  $S^2$,  [4] and by the Witten spaces  $M^{p,q,r} = SU(3) \times SU(2) \times U(1)/SU(2) \times U(1) \times U(1)$  for  p/s  even, where s  is the highest common factor of  p  and  q,  c.f. [4].

Similar obstructions arise in gauge theories.  For instance, in the Kaluza-Klein framework,  CP(3)  has been recently proposed [5] as an "internal" 6-dimensional space with an  $(SU(3) \times U(1))/Z_3 = U(3)$  gauge field, coming from the bundle

$$U(3) \longrightarrow SU(4n)/Z_4 \longrightarrow CP(3) \quad . \tag{1.10}$$

By constructing 1 analogous to (9) it can be seen that this $U(3)$-gauge field cannot be extended to an $SU(3) \times U(1)$-gauge field. Physically this means that only the fields transforming under the representations of $SU(3) \times U(1)$, which trivially represents the centre $Z_3$, are allowed.

The condition for the existence of spin structure on $M$ is the vanishing of the second Stiefel-Whitney class $w_2(TM)$ [6]. The more general case $K \to \tilde{G} \to G$, in which we are interested, has been worked out by Greub and Petry [7]. One lifts the transition functions $\phi_{\alpha\beta} : U_\alpha \cap U_\beta \to G$ of $F$ to $\tilde{\phi}_{\alpha\beta} : U_\alpha \cap U_\beta \to \tilde{G}$ and constructs $k_{\alpha\beta\gamma} \equiv \tilde{\phi}_{\beta\gamma} \tilde{\phi}_{\alpha\gamma}^{-1} \tilde{\phi}_{\alpha\beta}$. It can be shown that $k_{\alpha\beta\gamma}(x) \in K$ and $k_{\alpha\beta\gamma}$ is closed, hence it determines the class $k(F) \in \check{H}^2(M, K)$.

<u>Theorem 1.11</u>: $F$ admits a $\tilde{G}$ structure iff $k(F) = 0$ [7].

"No-go" can be evaded by introducing the extra gauge group $H$, which contains $K$ in its centre, and by coupling the fields to this $H$-gauge with unusual charges. For instance, one defines the generalized $spin_H$ structure by extending $SO(p, q) \times H/Z_2$ to $Spin_H \equiv (Spin(p, q) \times H)/Z_2$, c.f. [8]. Since representations of $Spin_H$ come from representations of $Spin \times H$, which trivially represent $Z_2$, the relation between the internal spin and statistics arises. For $M = CP(2)$ and $H = SU(2)$ the $SU(2)$-isospin of fermions is half-integer while that of bosons is integer. This residual mixing of internal and external symmetries may play some physical role. For instance, in the (Euclidean) path integral approach to quantum gravity either one restricts the integration to the family of spin manifolds or one rules out some possible gauge groups like $SU(5)$, c.f. [9].

In general, $Spin_H$ forms multiplets of spinors, however there is a particular case when $H = U(1)$. On $CP(2)$ there exists a rather natural $Spin_c \equiv Spin_{U(1)}$ structure, c.f. [10] and the twisted Atiyah-Singer index is integer

$$n_R - n_L = -\frac{1}{384\pi^2} \int R_{\mu\nu\rho\lambda} {}^*R^{\mu\nu\rho\lambda} \sqrt{g}\, d^4x + \frac{e^2}{16\pi^2} \int F_{\mu\nu} {}^*F^{\mu\nu} \sqrt{g}\, d^4x$$

$$= -\frac{1}{8} + \frac{1}{2} \left(n + \frac{1}{2}\right)^2 = \frac{n(n+1)}{2} \quad . \tag{1.12}$$

Now, bosons are integer charged while fermions are half-integer charged
and obey the half-integer Dirac condition

$$\int_{\partial\Sigma} A = \int_{\Sigma} F = \frac{2\pi}{e}\left(n+\frac{1}{2}\right) \quad . \tag{1.13}$$

In supergravity one works from the beginning with spinors, and the
underlying space-time obviously carries some sort of spin structure.  Often
spinors are Majorana in order to balance the number of bosonic and fermionic degrees of freedom.  Then, for a simple $N=1$ supergravity this
must be just the ordinary spin structure, while for $N>2$ case spinors
can build pseudo-Majorana multiplets and the generalized spin structure
can be admitted.  Also in Kaluza-Klein theories the global existence of
spinors is a reasonable postulate.  Since starting from pure gravity and
fermions in $d+4$ dimensions it is difficult to get the observed chiral
asymmetry in 4 dimensions, extra gauge fields are often introduced in $d+4$
dimensions.  Then one needs only a generalized spin structure to exist.

## 2.    Inequivalent $\tilde{G}$ structures

Assuming that $G$ can be globally extended to $\tilde{G}$ over $M$, the
extension is in general not unique.  Two $\tilde{G}$ structures $(\tilde{F}, \eta)$ and
$(\tilde{F}', \eta')$ are <u>equivalent</u> iff there exists a (strong) bundle morphism $\beta$,
$\beta(Ah) = \beta(A)h$, $A \in \tilde{F}$, $h \in \tilde{G}$, making the diagram (2.1) commute

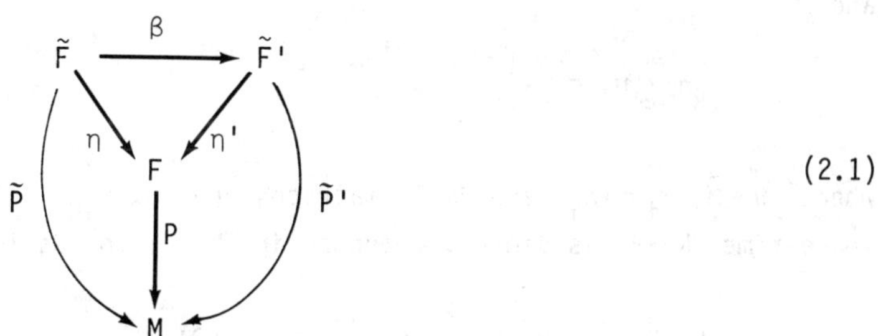

$$\tag{2.1}$$

Theorem 2.2. Inequivalent $\tilde{G}$ structures are 1-1 labelled by elements
of $H^1(M, K)$. [7]

Two $G$ structures can be inequivalent due both to the inequivalence
of $\tilde{G}$-bundles $\tilde{F}$ and $\tilde{F}'$, and to different maps $\eta$ and $\eta'$, which can be
constructed when $\pi_1(M) \neq 0$ and $\mathrm{HOM}(\pi_1(M), K) \neq 0$.

<u>Example 2.3</u>. Spinors on $M = S^1 \times R^3$, with flat metric diag(+,+,+,-).

144

Now, $\tilde{F} = \tilde{F}' = M \times Spin_0(3, 1)$ and $F = M \times SO_0(3, 1)$ is the bundle of ortho-normal frames. The two maps $\eta$ and $\eta'$ are defined by $\eta(x, h) = (x, E_x \rho(h))$ and $\eta'(x, h) = (x, E'_x \rho(h))$, where $x \in M$, $h \in Spin_0(3, 1)$ and the global frames $E_x$ and $E'_x$ differ by a nontrivial rotation $r(x)$.

$$(E'_x)_m = (E_x)_n r_{nm} \quad ,$$

$$r_{nm}(x) = \begin{bmatrix} \cos(x^1), & -\sin(x^1), & 0, & 0 \\ \sin(x^1), & \cos(x^1), & 0, & 0 \\ 0, & 0, & 1, & 0 \\ 0, & 0, & 0, & 1 \end{bmatrix}$$

$(x^1$ is the coordinate along $S^1)$ .

It is easy to see that $\beta$ does not exist.

<u>Example 2.4</u>. Spinors on the minimal conformal compactification $\bar{M}$ of Minkowski space-time $R^{1,3}$, $\bar{M} \equiv U(2) \approx \dfrac{U(1) \times SU(2)}{Z_2}$ ($\approx S^1 \times S^3$, topologi-cally). Now, $F = \bar{M} \times SO_0(3, 1)$ since $\bar{M}$ is parallelizable, and two global (orthonormal) frames ${}_L\sigma_m$ and ${}_R\sigma_m$, $m = 0,1,2,3$, are defined by

$$_L\sigma_m\varphi(u) = \left.\frac{\partial}{\partial\tau}\right|_{\tau=0} \varphi(e^{i\sigma_m}u) \tag{2.5}$$

and

$$_R\sigma_m\varphi(u) = \left.\frac{\partial}{\partial\tau}\right|_{\tau=0} \varphi(ue^{i\sigma_m}) \tag{2.6}$$

where $u \in \bar{M}$, $\sigma_i = -\underset{\sim}{\sigma}_i$ are Pauli matrices and $\sigma_0 = \underset{\sim}{\sigma}_0 = 1_2$. The Minkowski space-time $R^{1,3}$ is densely embedded in $\bar{M}$ by the Gayley map $f$

$$R^{1,3} \ni x = x^m\sigma_m \xrightarrow{f} u = \frac{i - x}{i + x} \in U(2) \quad , \tag{2.7}$$

which is conformal with conformal factor $\Omega = 4\det^{-1}(1 + x^2)$. The frames $_R\sigma_m$ and $_L\sigma_m$, which are related by the nontrivial $SO(3)$-rotation

$$_R\sigma_m(u) = {}_L(u^+\sigma_m u)(u) \tag{2.8}$$

can be used to define two inequivalent spin structures on $\bar{M}$, $\tilde{F} = \tilde{F}' =$

$M \times \mathrm{Spin}_0(3, 1)$  and

$$\eta(u, 1) = (u, {}_L\sigma(u)) \qquad , \qquad \eta'(u, 1) = (u, {}_R\sigma(u)). \qquad (2.9)$$

<u>Example 2.10.</u>  The original Kaluza-Klein internal space $S^1$, as well as $M^{p,q,r}$ for even 1, where 1 is a highest common factor of $2pr$, $rq$ and $(3p^2 + q^2)$ admit two inequivalent spin structures (provided that it exists). This follows from $\pi_1(M^{pqr}) = Z_1$, [3] and from $\mathrm{HOM}(Z_1, Z_2) = \begin{cases} 0 & \text{if 1 is odd} \\ Z_2 & \text{if 1 is even.} \end{cases}$

Physically, inequivalent  G  structures can manifest as different covariant derivatives of $\tilde{G}$-fields.  A connection on  $\tilde{F}$  obviously induces a connection on  F.  Also given a connection 1-form  $\omega$  on  F,  the connection on  $\tilde{F}$  is defined by  $\omega = \eta^*\omega$.  Given two maps  $\eta$  and  $\eta'$  the connections  $\omega$  and  $\omega'$  differ <u>locally</u> by a guage transformation.  By working in a dense submanifold of  M  one can often gauge away this difference at the price of introducing different 'twisted' boundary conditions for $\tilde{G}$-fields. This results in different spectra of various operators and therefore in different physics.

<u>Example 2.11.</u>

1. Twisted  $SU(n)/2_n$  fields in the t'Hooft treatment of confinement in a box (= torus) [12].

2. Petry's explanation of the correct quantization of the magnetic flux trapped in a superconducting ring (approximate topology $S^1 \times R^3$), due to 'exotic' spinors [13].

3. Quantum electrodynamics on  $S^1 \times R^3$, where the two kinds of fermions can 'flow' in the 1-loop corrections to the photon propagator $\multimap\!\!\bigcirc\!\!\multimap$ .  It turns out that twisted (antiperiodic) fermions on  $S^1 \times R^3$  (or in finite temperature) behave more properly:  lower  $< T_{00} >$,  no tachyonic modes etc. [14].

4. Different helicities of zero-eigenmodes of the Dirac operator coupled to nontrivial e.-m. potential  $A = \frac{1}{4} \det^{-1}(u)\, d(\det(u))$  on  $\bar{M}$  [11].

It has been shown [15], that the numbers of inequivalent spin structures and of different scalar field configurations are both labelled by $H^1(M, Z_2)$,  and twisted scalars and spinors can be meaningfully combined in supersymmetric multiplet [16].  The twists of individual fields have to sum up to zero (mod. 2) in order for the Lagrangian to be a function and have to be consistent with SUSY transformations. In general, the infinitesimal fermionic group parameter has to be twisted.

The number of local supersymmetries, which is determined by the number of local killing spinors, can be suppressed by the global periodicity conditions. The twisted spinors could in principle increase the number of globally allowed killing spinors by the possibility of antiperiodic boundary conditions. (Unfortunately this is not the case for $M^{p,p,r}$).

## 3.    Transformations of Pin Structures under Conformal Mappings

In this section we restrict our attention to the Pin structures and discuss their behaviour under mappings of manifolds [17]. The case of G structures can be treated in an analogous way. Since the metric is explicitly involved in the definition of spinors, the isometries form a natural class of transformations. Also the conformal maps $f : M \to M'$, $f^*g' = \Omega^2 g$ between the connected $(p+q)$-dimensional manifolds with metrics $g$ and $g'$ respectively (both of signature $(p, q)$), can be easily included. There is a natural lift of $f$ to a morphism $\hat{f}$ between the relative bundles of orthonormal frames over $M$ and $M'$ respectively, given by

$$\hat{f}(E_x) = \Omega_x^{-1} \circ f_x \circ E_x \quad , \tag{3.1}$$

where the o.n. frame $E_x$ at $x \in M$ is identified with the isometry $E_x : R^{p,q} \to T_x M$ of the standard pseudoeuclidean space $R^{p,q}$ onto the tangent space at $x$. This uniquely determines transformations of tensorial objects. In order to determine transformations of spinors, the analogous morphism $\tilde{f}$ between Pin structures, compatible with $\hat{f}$, has to be given.

$$
\begin{array}{ccc}
\tilde{F} & \xrightarrow{\ \ \tilde{f}\ \ } & \tilde{F}' \\
\eta \downarrow & \quad \hat{f} & \downarrow \eta' \\
F & \xrightarrow{\ \ \hat{f}\ \ } & F' \\
P \downarrow & & \downarrow P' \\
M & \xrightarrow{\ \ f\ \ } & M'
\end{array}
\tag{3.2}
$$

The situation depicted by the commutative diagram (3.2) is governed by the following proposition [17].

Proposition 3.3. A conformal map $f : M \to M'$ and the bundle morphism $\hat{f} : F \to F'$ can be lifted to a bundle morphism $\tilde{f} : \tilde{F} \to \tilde{F}'$ for exactly one (out of all inequivalent possible) pin structure $\tilde{F}$. The lifting $\tilde{f}$ is unique up to a sign.

Consider now the case $M = M'$, $F = F'$ and the group $\mathrm{Conf}(M)$ of conformal maps $f : M \to M$. In the case of its connected component $\mathrm{Conf}_0(M)$ it can be shown, that the assignment $f \to \tilde{f}$ preserves the composition rule up to a sign, and yields the representation of the double covering $\overline{\mathrm{Conf}_0(M)}$ of $\mathrm{Conf}_0(M)$ on spinor fields [17]. For discrete transformations, according to (3.3), the possibility arises, that there exists $f$, which changes the pin structure over $M$.

Example 3.4. Consider (2.4) and observe that the parity inversion

$$R^{1,3} \ni x \equiv x^m \sigma_m \;\longrightarrow\; P(x) = x^m \underset{\sim}{\sigma}_m \in R^{1,3} \tag{3.5}$$

induces the isometry of $\bar{M} = U(2)$

$$\bar{M} \ni u \;\longrightarrow\; P(u) = \epsilon u^T \epsilon^{-1} \quad , \tag{3.6}$$

where $\epsilon = i\sigma_2 = \begin{bmatrix} 0, & -1 \\ 1, & 0 \end{bmatrix}$. Now, (3.6) interchanges the global frames (2.5) and (2.6), and in fact interchanges the two Pin structures over $\bar{M}$. This proves the content of the Proposition (3.3) to be nontrivial and suggests that both Pin structures over $\bar{M}$ have to be considered when one wants to implement $P$ (and also $PT$), the fundamental geometrical transformation of the space-time.



References

[1]   S. Hawking and C. Pope, Phys. Lett. B73 (1978) 42.

[2]   A. Trautman, Lectures at SISSA, Trieste, Italy (1984).

[3]   L. Castellani, L.J. Romans and N.P. Warner, Nucl. Phys. B241 (1984) 429.

[4]   L.D. Page, Phys. Lett. B79 (1978) 235.

[5]   J. Sobczyk, Phys. Lett. B, to be published.

[6]   A. Haefliger, C.R. Acad. Sci. Paris <u>243</u> (1956) 558;
J. Milnor, L. Ens. Math. <u>9</u> (1963) 198;
A. Lichnerowicz, Bull. Soc. Math. France <u>92</u> (1964) 11;
R. Geroch, J. Math. Phys. 9 (1969) 1739 and 11 (1970) 343;
H. Baum, Teubner Texte in Math. <u>41</u>, Leipzig (1981).

[7]   W. Greub and H.R. Petry, in Lect. Notes Math. <u>676</u>, Springer-Verlag (1978).

[8]   S.J. Avis and C.J. Isham, Comm. Math. Phys. <u>72</u> (1980) 103.

[9]   P.G.O. Freund, in Proc. 1978 Tokyo Conf. on High-Energy Physics.

[10]   G. Gibbons and C.N. Pope, Comm. Math. Phys. <u>61</u> (1978) 239.

[11]   P. Budinich, L. Dabrowski and H.R. Petry, to be published.

[12]   G. t'Hooft, Comm. Math. Phys. <u>81</u> (1981) 267.

[13]   H.R. Petry, J. Math. Phys. <u>20</u> (1979) 231.

[14]   L.H. Ford, Phys. Rev. <u>D21</u> (1980) 933.

[15]   C.J. Isham, Proc. R. Soc. London, <u>A362</u> (1978) 383 and <u>A364</u> (1978) 591.

[16]   A. Chockalingham and C.J. Isham, J. Phys. <u>A13</u> (1980) 2723.

[17]   L. Dabrowski and H.R. Petry, to be published.



Received February 25, 1985.

GEOMETRODYNAMICS PROCEEDINGS (1985), pp. 149-160
edited by A. Pràstaro
© 1985 by World Scientific Publishing Co.

# THE RADON TRANSFORM ON COMPACT SYMMETRIC SPACES

Hubert Goldschmidt
*Columbia University*

1. Let $(X, g)$ be a symmetric space of compact type of dimension $n$. We shall denote by $T$ the tangent bundle of $X$, and by $T^*$ the cotangent bundle of $X$. By $\otimes^k E$, $S^k E$, $\Lambda^k E$, we shall mean the k-th tensor product, the k-th symmetric and the k-th exterior product of a vector bundle $E$, respectively; we denote by $C^\infty(E)$ the space of global sections of $E$ over $X$.

We are interested in the following problem in integral geometry on $X$. Let $\Xi$ be a family of closed totally geodesic submanifolds of dimension $\ell$, with $1 \leqslant \ell \leqslant n-1$. We consider a sub-bundle $E$ of $\otimes^q T^*$ and the space of tensors $C^\infty(E)$; assume that the integrals of an arbitrary $\omega \in C^\infty(E)$ over the elements of $\Xi$ are well-defined. The "Radon transform" of $\omega \in C^\infty(E)$ is the function $\hat{\omega}$ on $\Xi$ defined by

$$\hat{\omega}(\xi) = \int_\xi \omega,$$

for $\xi \in \Xi$. If $\omega$ satisfies the zero-energy condition, i.e. if $\hat{\omega}$ vanishes, what can one say about $\omega$? In particular, can the kernel of the Radon transform be characterized by a differential operator? In other words, can one find a vector bundle $F$ over $X$ and a linear differential operator $D: C^\infty(F) \to C^\infty(E)$ whose image is equal to the kernel of the Radon transform $\omega \mapsto \hat{\omega}$. If $D = 0$, the Radon transform would be injective.

We now consider several problems of this type.

A) Let $X$ be a projective space $\mathbb{K}\mathbb{P}^n$, which is not a sphere, with its canonical metric; here $\mathbb{K} = \mathbb{R}$, $\mathbb{C}$ or $\mathbb{H}$, with $n \geqslant 2$, or $\mathbb{K}$ is the Cayley numbers and $n = 2$. Let $\Xi$ be the set of all projective lines of $X$ and $E$ be the trivial complex line bundle. Then $C^\infty(E)$ is the space of all $C^\infty$-complex-valued functions on $X$ and according to Helgason ([8], [9]), the Radon transform is injective.

B) Let $X$ be a projective space which is not a sphere. Then all the geodesics of $X$ are closed and of length $\pi$. Let $\Xi$ be the space of these geodesics and $E$ be the cotangent bundle $T^*$ of $X$. The Radon transform of an exact 1-form is obviously equal to zero. The converse is also true and is given by

Theorem 1. Let $X$ be a projective space which is not a sphere and let $\omega$ be a 1-form on $X$. Then $\hat{\omega} = 0$ if and only if $\omega$ is exact.

This result was proved for $\mathbb{R}\mathbb{P}^n$ by Michel [13]; the case of the other projective spaces is treated in [5]. Thus the kernel of the Radon transform is equal to the image of the exterior derivative $d: C^\infty(F) \to C^\infty(T^*)$, where $F$ is the trivial line bundle.

C) Let $(X,g)$ be an arbitrary symmetric space of compact type. Let $\Xi$ be the set of all closed geodesics of $X$ and let $E = S^2 T^*$. If $h$ is a symmetric 2-form on $X$ and $\gamma$ is a closed geodesic of $X$, the integral of $h$ over $\gamma$ is

$$\int_\gamma h = \int_0^L h(\dot{\gamma}(s), \dot{\gamma}(s))\,ds,$$

where $\gamma(s)$ is the tangent vector to the geodesic $\gamma$ parametrized

by its arc-length  s  and  L  is the length of  $\gamma$.  We recall
that the Lie derivative  $L_\xi g$  of the metric  g  along a vector
field  $\xi$  on  X  always satisfies the zero-energy condition.
Let  $g_t$  be a deformation of the metric  $g_o = g$.  If the spectrum
of the lengths of the closed geodesics of  $g_t$  is independent
of the parameter  t,  then the infinitesimal deformation
$h = (d/dt)g_t|_{t = 0}$  is a symmetric 2-form satisfying the zero-
energy condition.  Determining whether the only symmetric
2-forms on  X  satisfying the zero-energy condition are the Lie
derivatives of the metric  g  is therefore an infinitesimal
rigidity problem for  g.  If  X  is a projective space, which
is not a sphere, this is the infinitesimal version of the Blaschke
conjecture: a metric on  X,  all of whose geodesics are closed
and of length  $\pi$, is isometric to the canonical metric  g  on  X .


$\underline{\text{Theorem}}$ 2.  Let  (X ,g) be equal to a product  $X_1 \times \cdots \times X_r$,
where each  $X_j$  is either a flat torus or a projective space,
which is not a sphere, with its canonical metric.  Let  h  be a
symmetric 2-form on  X.  Then the following conditions are
equivalent:

(i)  h  satisfies the zero-energy condition;

(ii) there exists a vector field  $\xi$  on  X  such that
$L_\xi g = h$.


This result was proved by Michel ([11], [12]) for  $\mathbb{R}\mathbb{P}^n$  and
for flat tori, and by Tsukamoto [14] for  $\mathbb{C}\mathbb{P}^2$.  The case of the
remaining projective spaces then follows using the work of
Michel [11]  (see [14]).  A direct proof of the above theorem
for the complex projective spaces is given in [3].
The extension to arbitrary products of these spaces is derived

152

in [6]; the special case $X = S^1 \times \mathbb{RP}^n$, with $n \geqslant 2$, is treated in [4]. In [10], Kiyohara uses Theorem 2 for the projective spaces to prove a version of the Blaschke conjecture with parameters.

For these spaces, the kernel of the Radon transform is thus equal to the image of the Killing differential operator $D_0: C^\infty(T) \to C^\infty(S^2T^*)$, sending $\xi$ into the Lie derivative $L_\xi g$.

D) Let $X$ be the complex projective space $\mathbb{CP}^n$, with $n \geqslant 2$. Let $\Xi$ be the space of all complex projective lines in $\mathbb{CP}^n$ and $E$ be the bundle $T^{1,1}$ of forms of type $(1,1)$. Let $T^*_\mathbb{C}$ be the complexification of $T^*$ and

$$D: C^\infty(T^*_\mathbb{C}) \to C^\infty(T^{1,1})$$

be the differential operator sending $\beta$ into $\pi_{1,1} d\beta$, where $\pi_{1,1}$ is the natural projection $\Lambda^2 T^*_\mathbb{C}$ onto its sub-bundle $T^{1,1}$. Clearly, the Radon transform of an element of the image of $D$ vanishes. The converse is given by the result of [7]:

Theorem 3. Let $X = \mathbb{CP}^n$, with $n \geqslant 2$, and $\alpha \in C^\infty(T^{1,1})$. Then the following statements are equivalent:

(i) the Radon transform of $\alpha$ vanishes;

(ii) there exists an element $\beta \in C^\infty(T^*_\mathbb{C})$ such that $D\beta = \alpha$.

2. We now consider the situation described above in (B) and proceed to give an outline of our proof of Theorem 1 for $X = \mathbb{CP}^n$, with $n \geqslant 2$ (see [5]). We assume the corresponding result for $\mathbb{RP}^2$. Let $x \in X$ and $\xi, \eta \in T_x - \{0\}$; if the complex lines $\mathbb{C}\xi$ and $\mathbb{C}\eta$ generated by $\xi$ and $\eta$ are orthogonal,

then $Y = \mathrm{Exp}_X(\mathbb{R}\xi \oplus \mathbb{R}\eta)$ is a real projective plane, with its metric of constant curvature 1, totally geodesic in X. If $\alpha$ is a 1-form on X with $\hat{\alpha} = 0$ and $i: Y \to X$ is the natural imbedding, then $i^*\alpha$ is a 1-form on Y with zero-energy. Hence it is exact and so $(d\alpha)(\xi,\eta) = 0$. Thus $d\alpha$ is a section of the vector bundle

$$F = \{\beta \in \Lambda^2 T^* \mid \beta(\xi,\eta) = 0 \quad \text{for all} \quad \xi,\eta \in T, \text{ with } \mathbb{C}\xi \perp \mathbb{C}\eta\}.$$

We have therefore shown that, if

$$\tilde{d}: C^\infty(T^*) \to C^\infty(\Lambda^2 T^*/F)$$

is the differential operator sending $\alpha$ into the projection of $d\alpha$ in $C^\infty(\Lambda^2 T^*/F)$, Theorem 1 holds for $X = \mathbb{C}\mathbb{P}^n$ if the complex

$$(1) \qquad\qquad C^\infty(\mathbf{1}) \xrightarrow{\ d\ } C^\infty(T^*) \xrightarrow{\ \tilde{d}\ } C^\infty(\Lambda^2 T^*/F)$$

is exact, where $\mathbf{1}$ is the trivial real line bundle over X. We denote by $F_1$, $F_2$, $F_3$ the complexifications of the vector bundles $\mathbf{1}$, $T^*$, $\Lambda^2 T^*/F$, respectively. Since X is the homogeneous space $U(n+1)/U(1) \times U(n)$ and the $F_j$ are homogeneous vector bundles endowed with Hermitian metrics, to each irreducible representation $\gamma$ of $U(n+1)$, there is associated a finite-dimensional $U(n+1)$-submodule $C^\infty_\gamma(F_j)$ of $C^\infty(F_j)$, which is isomorphic to the direct sum of a finite number (called the multiplicity of $C^\infty_\gamma(F_j)$) of copies of $V_\gamma$, an irreducible $U(n+1)$-module corresponding to $\gamma$. If $\hat{G}$ is the dual of $G = U(n+1)$, for $\gamma, \gamma' \in \hat{G}$, with $\gamma \neq \gamma'$, the subspaces $C^\infty_\gamma(F_j)$ and $C^\infty_{\gamma'}(F_j)$ of $C^\infty(F_j)$ are orthogonal with respect to the $L^2$-inner product on $C^\infty(F_j)$; moreover, the direct sum $\bigoplus_{\gamma \in \hat{G}} C^\infty_\gamma(F_j)$ is a dense subspace of $C^\infty(F_j)$. As the differential operators d and $\tilde{d}$ of the complex (1) are homogeneous, we obtain a sub-complex

$$(2) \qquad C^\infty_\gamma(F_1) \xrightarrow{\ d\ } C^\infty_\gamma(F_2) \xrightarrow{\ \tilde{d}\ } C^\infty_\gamma(F_3)$$

of the complexification of (1). Since the operator d of (1) is elliptic, the exactness of the sequence (1) is equivalent to the exactness of all the sequences (2) as $\gamma$ ranges over $\hat{G}$ (see [3]).

Let $\zeta = (\zeta_o, \zeta_1, \ldots, \zeta_n)$ be the standard coordinate system of $\mathbb{C}^{n+1}$ and $H_q$ be the space of complex bihomogeneous polynomials on $\mathbb{C}^{n+1}$, of degree q in $\zeta$ and degree q in $\bar{\zeta}$ and which are harmonic. The restriction of an element f of $H_q$ to the unit sphere $S^{2n+1}$ is $U(1)$-invariant and so induces a function $\tilde{f}$ on $\mathbb{CP}^n$. The space $\tilde{H}_q$ of functions on $\mathbb{CP}^n$ obtained from $H_q$ is an irreducible $U(n+1)$-module isomorphic to $H_q$ and is the eigenspace of the Laplacian of $(X, g)$ corresponding to the eigenvalue $4q(q+n)$. We set

$$f = \zeta_m \bar{\zeta}_o, \qquad\qquad f' = \zeta_{m-1} \bar{\zeta}_o;$$

then $f^q \in H_q$ (see [1]). The $U(n+1)$-submodule $W_q$ of $H_q \otimes \Lambda^2 H_1$ generated by $f^q \otimes f \wedge f'$ is irreducible, and therefore so is its conjugate $\bar{W}_q$ under the conjugation of $H_q \otimes \Lambda^2 H_1$.

If $T^{(p,q)}$ is the bundle of forms of type (p,q) on $\mathbb{CP}^n$, we have mappings

$$\eta' : H_q \otimes \Lambda^2 H_1 \longrightarrow C^\infty(T^{1,0}),$$

$$\eta'' : H_q \otimes \Lambda^2 H_1 \longrightarrow C^\infty(T^{0,1})$$

defined by

$$\eta'(f_o \otimes f_1 \wedge f_2) = \tilde{f}_o(\tilde{f}_1 \, \partial \tilde{f}_2 - \tilde{f}_2 \, \partial \tilde{f}_1),$$

$$\eta''(f_o \otimes f_1 \wedge f_2) = \tilde{f}_o(\tilde{f}_1 \, \bar{\partial} \tilde{f}_2 - \tilde{f}_2 \, \bar{\partial} \tilde{f}_1),$$

for $f_o \in H_q$, $f_1, f_2 \in H_1$.

The non-zero $U(n+1)$-modules $C_\gamma^\infty(F_1)$, $C_\gamma^\infty(F_2)$ are given by the following tableau:

(3)

| $\gamma \in \hat{G}$ | $C_\gamma^\infty(F_1)$ | $C_\gamma^\infty(F_2)$ |
|---|---|---|
| $H_o$ | $\tilde{H}_o$ | $0$ |
| $H_q \quad q \geqslant 1$ | $\tilde{H}_q$ | $\partial \tilde{H}_q \oplus \bar{\partial} \tilde{H}_q$ |
| $W_q \quad q \geqslant 0$ | $0$ | $\eta'(W_q)$ |
| $\bar{W}_q \quad q \geqslant 0$ | $0$ | $\eta''(\bar{W}_q)$ |

Since the kernel of $d: C^\infty(F_1) \to C^\infty(F_2)$ is equal to $\tilde{H}_o$, the space of constant functions, we see that Theorem 1 holds for $X = \mathbb{CP}^n$ if

$$\tilde{d} \, \partial H_q \neq 0, \quad \text{for } q \geqslant 1,$$

$$\tilde{d} \, \eta'(W_q) \neq 0, \quad \text{for } q \geqslant 0,$$

or if

$$\tilde{d} \partial \tilde{f}^q \neq 0, \quad \text{for } q \geqslant 1,$$

$$\tilde{d} \eta'(f^q \otimes f \wedge f') \neq 0, \quad \text{for } q \geqslant 0.$$

The two latter relations are explicitly verified in [5], concluding the proof of Theorem 1 for $X = \mathbb{CP}^n$.

3. We next describe the proof of Theorem 3.  We use the notation introduced in §2 concerning $X = \mathbb{CP}^n$, with $n \geq 2$. The canonical metric $g$ of $\mathbb{CP}^n$ is the Fubini-Study metric, whose Kähler form we denote by $\omega$. Let $\{\omega\}$ be the sub-bundle of $T^{1,1}$ generated by its section $\omega$. If

$$(4) \qquad P: C^\infty(F_1) \longrightarrow C^\infty(T^{1,1})$$

is the second-order elliptic differential operator defined by

$$Pf = \partial\bar{\partial}f + \frac{i}{2n}(\Delta f)\omega,$$

for $f \in C^\infty(F_1)$, where $\Delta$ is the Laplacian of $(X,g)$, we show that

$$(5) \qquad C^\infty(T^{1,1}) = C^\infty(\{\omega\}) \oplus PC^\infty(F_1) \oplus \partial\bar{\partial}\,{}^*C^\infty(T^{0,2})$$

$$\oplus \bar{\partial}\partial^*C^\infty(T^{2,0})$$

is an orthogonal decomposition of $C^\infty(T^{1,1})$ into closed subspaces (in the $L^2$-sense).  The four subspaces of this decomposition are always closed and orthogonal for any Kähler manifold of dimension $\geq 2$.  The vector bundles and the differential operators appearing in both sides of (5) are all homogeneous. The completeness of this decomposition is verified using the tableau (3) and the non-zero multiplicities of $C^\infty_\gamma(T^{1,1})$, for $\gamma \in \hat{G}$, which are given by

| $\gamma \in \hat{G}$ | $C_\gamma^\infty(T^{1,1})$ |
|---|---|
| $H_o$ | 1 |
| $H_q \quad q \geqslant 1$ | 2 |
| $W_q \quad q \geqslant 0$ | 1 |
| $\bar{W}_q \quad q \geqslant 0$ | 1 |

If $\gamma = H_o$, we have

$$C_\gamma^\infty(T^{1,1}) = \mathbb{C}\omega.$$

It is easily verified that the kernel of (4) is $\tilde{H}_o$ and so

$$P: C_\gamma^\infty(F_1) \longrightarrow C_\gamma^\infty(T^{1,1})$$

is injective for $\gamma = H_q$, with $q \geqslant 1$; hence

$$C_\gamma^\infty(T^{1,1}) = C_\gamma^\infty(F_1) \cdot \omega \ \oplus \ PC_\gamma^\infty(F_1).$$

For $\gamma = W_q$, with $q \geqslant 0$, one then verifies that

$$C_\gamma^\infty(T^{1,1}) = \bar{\partial}C_\gamma^\infty(T^{1,0}) = \bar{\partial}\partial*C_\gamma^\infty(T^{2,0});$$

by conjugation, we obtain

$$C_\gamma^\infty(T^{1,1}) = \partial C_\gamma^\infty(T^{0,1}) = \partial\bar{\partial}*C_\gamma^\infty(T^{0,2}),$$

for $\gamma = \bar{W}_q$, $q \geqslant 0$. In follows that $C_\gamma^\infty(T^{1,1})$ is contained in the right-hand side of (5), for all $\gamma \in \hat{G}$. Using elliptic theory, we see that the sum on the right-hand side of (5) is closed in $C^\infty(T^{1,1})$ (in the $L^2$-sense), and so is equal to

158

$C^\infty(T^{1,1})$.  If $\alpha \in C^\infty(T^{1,1})$, we therefore can write

$$\alpha = f\omega + \pi_{1,1} d\beta ,$$

for some $f \in C^\infty(F_1)$ and $\beta \in C^\infty(T_{\mathbb{C}}^*)$.  If $Y$ is a complex projective line in $\mathbb{CP}^n$, then

$$\int_Y \alpha = \int_Y f\omega = \int_Y f .$$

The equivalence of statements (i) and (ii) of Theorem 3 is now a consequence of the result of Helgason [8] mentioned above in (A).  Helgason's theorem can also be proved using the decompositio of $C^\infty(F_1)$ considered above and methods similar to ones used in the proof of Theorem 1 (see [15]).

4.  Finally, we wish to discuss our proof of Theorem 2 for $X = \mathbb{CP}^n$, with $n \geqslant 2$, given in [3].  Let $N'$ be the sub-bundle of $\Lambda^2 T^* \otimes \Lambda^2 T^*$ consisting of all $\theta \in \Lambda^2 T^* \otimes \Lambda^2 T^*$ which satisfy the following conditions:

(a)  $\theta(\xi,\eta,\xi,\eta) = 0$, for all $\xi, \eta \in T$ with $\mathbb{C}\xi \perp \mathbb{C}\eta$;

(b)  $\theta(\xi,\eta,\xi,\zeta) = 0$, for all $\xi,\eta,\zeta \in T$ whenever the subspaces $\mathbb{C}\xi$, $\mathbb{C}\eta$, $\mathbb{C}\zeta$ are mutually orthogonal;

(c)  $\theta(\eta_1,\eta_2,\eta_3,\eta_4) = 0$, for all $\eta_1, \eta_2, \eta_3, \eta_4 \in T$ whenever $(\mathbb{C}\eta_i) \subset (\mathbb{C}\eta_j)^\perp$, if $i \neq j$.

In [2], we define a differential operator

$$D_g : C^\infty(S^2 T^*) \longrightarrow C^\infty(\Lambda^2 T^* \otimes \Lambda^2 T^*)$$

such that:

        (i)   if $\xi$ is a vector field on  X,   then  $D_g L_\xi g$  is a section of  N';

        (ii) a section  $h \in C^\infty(S^2 T^*)$  satisfies the zero-energy condition if and only if  $D_g h$  is a section of  N'.

        Thus, if

$$D'_g : C^\infty(S^2 T^*) \longrightarrow C^\infty((\wedge^2 T^* \otimes \wedge^2 T^*)/N')$$

is the operator induced by  $D_g$,  we see that Theorem 1 holds for  $\mathbb{CP}^n$  if and only if the complex

$$C^\infty(T) \xrightarrow{\;D_{O'}\;} C^\infty(S^2 T^*) \xrightarrow{\;D'_g\;} C^\infty((\wedge^2 T^* \otimes \wedge^2 T^*)/N')$$

is exact.  The vector bundles and the differential operators of this complex are homogeneous; its exactness is proved by the methods used in §2 to prove that of the sequence (2).

## References


[1]    M. Berger, P. Gauduchon, E. Mazet, Le spectre d'une variété riemannienne, <u>Lecture Notes in Mathematics</u>, n° 194, Springer-Verlag, Berlin, Heidelberg, New York, 1971.

[2]    J. Gasqui, H. Goldschmidt, Déformations infinitésimales des espaces riemanniens localement symétriques. I, <u>Adv. in Math.</u>, 48(1983), 205-285.

[3]    __________________________, Déformations infinitésimales des espaces riemanniens localement symétriques. II. La conjecture infinitésimale de Blaschke pour les espaces projectifs complexes, <u>Ann. Inst. Fourier</u>, <u>Grenoble</u>, 34(1984), (2) 191-226.

[4]    __________________________, Infinitesimal rigidity of $S^1 \times \mathbb{R}\,\mathbb{P}^n$, <u>Duke Math. J.</u>, 51(1984), 675-690.

160

[5]     _________________________, Une caractérisation des formes exactes de degré 1 sur les espaces projectifs, Comment. Math. Helv. (to appear).

[6]     _________________________, Infinitesimal rigidity of products of symmetric spaces (to appear).

[7]     H. Goldschmidt, The Radon transform for forms of type (1.1) on complex projective space, Math. Ann. (to appear).

[8]     S. Helgason, The Radon transform on Euclidean spaces, compact two-point homogeneous spaces and Grassmann manifolds, Acta Math., 113(1965), 153-180.

[9]     _________, The Radon transform, Progress in Mathematics, n°5, Birkhäuser, Boston, Basel, Stuttgart, 1980.

[10]    K. Kiyohara, On deformations of the standard $C_\pi$-metrics on projective spaces (to appear).

[11]    R. Michel, Problèmes d'analyse géométrique liés à la conjecture de Blaschke, Bull. Soc. Math. France, 101(1973), 17-69.

[12]    _________, (a) Un problème d'exactitude concernant les tenseurs symétriques et les géodésiques, C.R. Acad. Sc. Paris Sér. A, 284 (1977), 183-186; (b) Tenseurs symétriques et géodésiques, C.R. Acad. Sc. Paris Sér. A, 284(1977), 1065-1068.

[13]    _________, Sur quelques problèmes de géométrie globale des géodésiques, Bol. Soc. Bras. Mat., 9(1978), 19-38.

[14]    C. Tsukamoto, Infinitesimal Blaschke conjectures on projective spaces, Ann. Sc. École Norm. Sup., (4) 14(1981), 339-356.

[15]    E.L. Grinberg, Spherical harmonics and integral geometry on projective spaces, Trans. Amer. Math. Soc., 279(1983), 187-203

Received January 18, 1985.

GEOMETRODYNAMICS PROCEEDINGS (1985), pp. 161-171
edited by A. Pràstaro


# UNCONSTRAINED DEGREES OF FREEDOM OF GRAVITATIONAL FIELD AND THE POSITIVITY OF GRAVITATIONAL ENERGY

Jerzy Kijowski

*Institute for Theoretical Physics, Polish Academy of Sciences
Aleja Lotnikow 32/46, 02-668 Warsaw, Poland.*

ABSTRACT: The space of Cauchy data for Enstein equations is effectively reduced with respect to Gauss-Codazzi constraints. The mixed initial value – boundary value problem is analysed. The role of the boundary degrees of freedom is discussed. The energy-positivity is obtained as a simple consequence of the construction used.

## 1. Symplectic structure of the space of Cauchy data

The goal of this paper is to present the construction of unconstrained degrees of freedom of gravitational field together with the analysis of the notion of gravitational energy as a hamiltonian of the system. Traditionally, the "canonical" formalisms used in General Relativity were based on: 1)integration by parts and 2) the hope, that under sufficiently strong asymptotic conditions all the "inconvenient" boundary integrals will wanish. Therefore, all the results were valid "modulo surface integrals" (e.g. we have been taught that the hamiltonian of the gravitational field is equal to zero – modulo surface integrals). Recently, the important role of surface phenomena has been stressed by many authors ( see e.g. [2] and [8] ). The coherent description of these phenomena can be given in terms of the theory of symplectic relations as proposed by W.M.Tulczyjew (see e.g. [9] and [7] ). In this theory both volume integrals and boundary integrals have equally legal status.

The analysis of the gravitational field in terms of the theory of symplectic relations leads to the so called affine formulation of General Relativity (see e.g. [5] , [7] and [4] ). In the paper [6] ,based on the affine formulation, the dynamics of the gravitational field within a finite space-time region $\Sigma$ with the boundary $\partial\Sigma$ was analysed (at the end of our considerations $\partial\Sigma$ can be shifted to space-infinity or to null-infinity and the limits of the corresponding boundary terms can be calcutated). The present paper is the straightforward continuation of [6] .

Let $\Sigma$ be a compact, smooth 3-dimensional manifold with the boundary $\partial\Sigma$ . In the present paper we limit ourselves to the simplest topological situation i.e. we assume that $\Sigma$ is diffeomorphic to the 3-disc $K(0,R) \subseteq \mathbb{R}^3$. Let $\Lambda = \partial\Sigma \times \mathbb{R}^1$ and let $\Sigma_t = \Sigma \times \{t\}$. The space $V = \Sigma \times \mathbb{R}^1$ will be the interior of our space-time tube and the boundary $\Lambda = \partial V$ will be a 1-time-like and 2-space-like surface in our space-time. Points of $\Sigma$ are observers, each of them having its own method of moving in space-time (e.g. each of them is equipped with a pre-programmed jet-engine and a clock). The change of coordinates in V ("passive gauge") is irrelevant since we can give at the very beginning the names to all the observers and to equip them with clocs. Let us therefore fix a coordinate chart $(x^\mu)$ on $V$ such that $x^0 = t$, $x^3 = \log r$, where $r$ is the radial coordinate (i.e. the value of $x^3$ lies within the half-line $]-\infty, \log R]$), and $(x^1, x^2)$ is a coordinate chart on $\partial\Sigma = S^2$ (e.g. spherical angles $\phi$ and $\theta$). We use the following convention for indices: greek indices $\mu, \nu$ run from 0 to 3; $k, l$ are coordinates on $\Sigma$ and run from 1 to 3; $i, j$ are coordinates on $\Lambda$ and run from 0 to 2; $A, B$ are coordinates on $\partial\Sigma$ and run from 1 to 2. We have $\Sigma_t = \{x^0 = t\}$; $\Lambda = \{x^3 = \log R\}$.

We will consider pseudo-riemannian geometries on $V$. Intuitively, the transformations of the geometry due to the simple "reorganization" of the observation (change of clocks and of programs of the jet-engines) should be considered as "gauge transformations". However, we are not free to decide which transformations can be called gauge. The gravitational field inside $V$ is a generalized (constrained) hamiltonian system and the gauge transformations are precisely those which correspond to the degeneracy of the symplectic form on the constraint manifold. The usual description of the symplectic structure is given by the so called A.D.M. symplectic form (see [1]):

$$\omega_{ADM} = \frac{1}{2\kappa} \int_\Sigma dP^{kl} \wedge dg_{kl} \tag{1}$$

where $\kappa = 8\pi G$ is the gravitational constant, $g_{kl}$ is a Riemannian 3-metric on $\Sigma$ and

$$P^{kl} = \sqrt{\det g}\,(\tilde{g}^{kl}\,\mathrm{Tr}\,K - K^{kl}) \tag{2}$$

By $\tilde{g}$ we denote the contravariant metric inverse to $g$ and $K$ is the extrinsic curvature of $\Sigma$ with respect to the 4-metric $g_{\mu\nu}$ which we are looking for. The phase space $P_\Sigma$ is therefore described by the two objects $g$ and $P$ (12 functions on $\Sigma$). Real physical situations correspond to fields which fulfill 4 constraint conditions induced by Gauss-Codazzi equations:

$$P^{kl}{}_{|l} = 0 \tag{3}$$

(the covariant divergence of $P$ with respect to $g$) which we call the vector-constraint, and the following "scalar constraint":

$$(\det g)\,\overset{3}{R}(g) - P^k{}_l P^l{}_k + \frac{1}{2}(P^k{}_k)^2 = 0 \tag{4}$$

The form $\omega_{ADM}$ is degenerate on the constraint space $\overline{P}_\Sigma \subset P_\Sigma$. Gauge transformations are generated by vector fields tangent to this degeneracy. There is a very disappointing result about the A.D.M. form (1) : not all the transformations of $\bar{P}_\Sigma$ which can be implemented by boundary preserving diffeomorphisms of $V$ belong to this class. The simple computation shows that the value of the form (1) is modified by a boundary term when subjected to the one-parameter group of diffeomorphisms generated by a vector field $X$ on $V$. The boundary term arises due to integration by parts and is composed of two terms: one term is proportional to the value of $X$ on $\partial V = \Lambda$ and the other term is proportional to derivatives of $X$ on $\Lambda$. For $\Lambda$ – preserving diffeomorphisms we have $X|_\Lambda = 0$ but the second term does not vanish in general (the symplectic structure is invariant with respect to gauge transformations. This is true only for transformations generated by vector fields $X$ which vanish on $\Lambda$ together with first derivatives).

Using the theory of symplectic relations it was proved in [6] that the correct sympletic structure in the space of Cauchy data differs from the A.D.M. structure by the surface term and is equal to

$$\omega = \frac{1}{2\kappa} \int_\Sigma dP^{kl} \wedge dg_{kl} + \frac{1}{\kappa} \int_{\partial\Sigma} d\lambda \wedge d\alpha \tag{5}$$

where

$$\lambda = \sqrt{\det g_{AB}} \tag{6}$$

is the 2-dimensional volume density on $\partial\Sigma$ and

$$\alpha = \text{arsh} \frac{g^{03}}{\sqrt{|g^{00}|g^{33}}} \tag{7}$$

is the "hyperbolic angle" between the hypersurfaces $\Sigma$ and $\Lambda$. Now, the phase space is the direct sum

$$P = P_\Sigma \oplus P_{\partial\Sigma} \tag{8}$$

where $P_{\partial\Sigma}$ is described by "2 functions on the boundary" (actually $\lambda$ is a scalar density on $\partial\Sigma$ and $\alpha$ is a scalar function on $\partial\Sigma$). The constraint manifold $\bar{P} \subset P$ is defined by those objects which fulfill not only (3) and (4) but also the compatibility condition between $\lambda$ and the restriction of $g$ to $\partial\Sigma$. The following theorem can be proved:

Theorem. Gauge transformations for the pair $(\bar{P}, \omega)$ are precisely the transformations which can be implemented by $\Lambda$ – preserving diffeomorphisms.

The theorem is a consequence of the fact that the modification of the "volume term" in (5) due to the derivatives of $X$ on $\Lambda$ is cancelled by the corresponding modification of the boundary term.

## 2. Gauge conditions

The reduced phase space $\tilde{P} = \bar{P} / \sim$ is the quotient space, where $\sim$ denotes the gauge-
-equivalence relation. To describe effectively the quotient we impose 4 gauge conditions
which enable us to pick up a representant within each gauge-equivalence class. The conditions are:

$$P^{kl} g_{kl} = 0 \tag{9}$$

(i.e. $\Sigma$ is a maximal surface) and

$$g_{kl} = f \, \gamma_{kl} \tag{10}$$

where $\gamma$ is the metric which satisfies 4 conditions:

$$\gamma_{33} = 1 \; ; \qquad \gamma_{3A} = 0 \; ; \qquad \sqrt{\det \gamma_{AB}} = \overset{o}{\lambda} \; . \tag{11}$$

Here $\overset{o}{\lambda}$ the standard 2-volume density on the unit sphere $S^2 = \partial K(0,1) \subset \mathbb{R}^3$. If $(x^1, x^2)$
are spherical angles $\phi$ and $\theta$ then

$$\overset{o}{\lambda} = \sin \theta \tag{12}$$

However, there is no global coordinate chart on $\partial \Sigma \cong S^2$ and therefore we keep the condition (11) in the form which does not depend of the particular choice of the coordinates
on $\Sigma$. As an example of the metric which has the representation satisfying (10)-(11) we
can take the flat metric on $\mathbb{R}^3$ which can be written in the form

$$ds^2 = r^2 [(d \log r)^2 + d\Omega^2] \tag{13}$$

where $d\Omega^2$ is the standard metric on $S^2$. Thus

$$f = r^2 = \exp(2x^3) \tag{14}$$

in this case. The possibility of finding for a given metric $g$ the representant fulfilling
(10)-(11) is equivalent to the following 2-nd order equation for the function $\rho = x^3$:

$$\left( \tilde{g}^{kl} + \frac{\nabla^k \rho}{|\nabla \rho|} \frac{\nabla^l \rho}{|\nabla \rho|} \right) \nabla_k \nabla_l \rho = 0 \tag{15}$$

The solution has to satisfy the boundary conditions: $\rho|_{\partial \Sigma} = \log R$ and $\rho \to -\infty$ at a given
point $\underset{o}{x}$ inside $\Sigma$. The choice of the point $\underset{o}{x}$ is also a gauge condition. As an example
of the function which satisfies (15) we can take $\rho = \log r$ in the flat space.

For a given solution $\rho$ of (15) we "extend" coordinates $(x^1, x^2)$ from $\partial\Sigma$ to the interior of $\Sigma$ in such a way that the vector field $\partial_3$ is orthogonal to the family of surfaces $\{x^3 = \text{const}\}$. It can be easily checked that the coordinate chart obtained this way satisfies the conditions (10)-(11). The problem wheather or not the equation (15) together with boundary conditions admits always a (unique) solution has not yet been fully solved. There are however partial results which are very favorable for the positive answer.

The equation (15) is written by help of a metric $g$. However, it is invariant with respect to conformal deformations of $g$. Therefore, only the conformal structure implied by $g$ is involved. The easiest way to verify this observation is to derive (15) from the variational principle

$$\frac{\delta L}{\delta \rho} = 0 \tag{16}$$

where the lagrangean

$$L = \sqrt{\det g}\ [\ \tilde{g}^{kl}(\partial_k \rho)(\partial_l \rho)\ ]^{3/2} \tag{17}$$

is manifestly invariant with respect to conformal modifications of the metric. The equation (15) is therefore the conformal generalization of the Laplace equation and its solutions will be called conformally-harmonic functions. Corresponding 2-surfaces $\{\rho = \text{const}\}$ will be called conformally-harmonic surfaces.

Our gauge conditions correspond thus to the 3+1 maximal slicing of $V$ and morover to the 2+1 slicing of each $\Sigma_t$ which is conformally-harmonic:

$$\Sigma - \{\underset{o}{x}\} = \chi \times\ ]-\infty, \log R\ ] \tag{18}$$

where $\chi \stackrel{\sim}{=} \partial\Sigma \stackrel{\sim}{=} S^2$.


## 3. Reduction

The following formula is easy to check

$$dP^{kl} \wedge dg_{kl} = d(f P^{kl}) \wedge d\gamma_{kl} - df \wedge d(P^{kl}\gamma_{kl}) \tag{19}$$

Therefore, due to (9)-(11) we have:

$$dP^{kl} \wedge dg_{kl} = d(f P^{AB}) \wedge d\gamma_{AB} \tag{20}$$

We define the following object:

$$p^{AB} = f P^{AB} + \frac{1}{2}\gamma^{AB} P^3_{\ 3} \tag{21}$$

where $\gamma^{AB}$ is the contravariant 2-metric on $\chi$ inverse to $\gamma_{AB}$. We have

$$d(\tfrac{1}{2}\gamma^{AB} P^3{}_3)\wedge d\gamma_{AB} = dP^3{}_3 \wedge \tfrac{1}{2}\gamma^{AB} d\gamma_{AB} + \tfrac{1}{2} P^3{}_3 \, d\gamma^{AB} \wedge d\gamma_{AB} = 0 \tag{22}$$

because

$$\tfrac{1}{2} \gamma^{AB} d\gamma_{AB} = d \log \sqrt{\det \gamma} = d \log \overset{\circ}{\lambda} = 0 \tag{23}$$

due to the condition (11). Finally our symplectic structure (5) reduces to

$$\omega = \frac{1}{2\kappa} \int_{\Sigma} dp^{AB} \wedge d\gamma_{AB} + \frac{1}{\kappa} \int_{\partial\Sigma} d\lambda \wedge d\alpha \tag{24}$$

The number of independent components of $\gamma_{AB}$ is 2 (because of the symmetry and the unimodularity condition (11)). The number of independent components of $p^{AB}$ is also 2 since the equation (9) implies: $p^{AB} \gamma_{AB} = g_{AB} P^{AB} + P^3{}_3 = 0$. The object $\gamma_{AB}$ can be called a unimodular metric on $\chi$. The reduced phase space $\tilde{P}$ is thus the collection of "2 degrees of freedom per point in $\Sigma$ " (unimodular metric and its momentum) and "1 degree of freedom per point in $\partial\Sigma$ " (the hyperbolic angle $\alpha$ and its momentum $\lambda$ ). For each collection $(p^{AB}, \gamma_{AB}, \alpha, \lambda)$ of reduced Cauchy data we can reconstruct the complete data $(P^{kl}, g_{kl}, \alpha, \lambda)$ by solving the four constraint equations. It can be easily verified, that due to our 2+1 –conformally–harmonic decomposition of $\Sigma$ the vector constraint (3) reduces to a single second order linear elliptic equation

$$q,_{33} + \tfrac{1}{2} \overset{2}{\Delta} q = \text{function of } p^{AB} \text{ and } \gamma_{AB} \tag{25}$$

for the unknown function

$$q(x^1, x^2, x^3) = \frac{1}{\overset{\circ}{\lambda}} \int_{-\infty}^{x^3} P^3{}_3(x^1, x^2, \rho) \, d\rho \tag{26}$$

together with the condition:

$$P^3{}_A = \overset{\circ}{\lambda} \, q,_A + \text{function of } p^{AB} \text{ and } \gamma_{AB} \tag{27}$$

( symbol $\overset{2}{\Delta}$ denotes the 2-dimensional laplacian defined by $\gamma$ ). To solve <u>uniquely</u> the linear equation (25) we <u>do not need</u> any boundary condition for q except the continuity condition for $P^{kl}$ at the center point $\underset{\circ}{x}$. This condition implies that $q \to 0$ exponentially for $\rho \to -\infty$ and it is sufficient to find <u>uniquely</u> the solution. The only lacking component of the complete Cauchy data is now the conformal factor f which can be found by solving the scalar constraint (4). As it is known (see e.g. [3]) it is a second order, elliptic equation for the function f. To find uniquely the solution we use the value $\lambda = \overset{\circ}{\lambda} f$ on $\Sigma$ which gives us the Dirichlet data for the equation.

## 4. Time evolution

In the paper [6] the fundamental formula has been proved:

$$- dH = \frac{1}{2\kappa} \int_{\Sigma} (\dot{P}^{kl} \, dg_{kl} - \dot{g}_{kl} \, dP^{kl}) + \frac{1}{\kappa} \int_{\partial\Sigma} (\dot{\lambda} \, d\alpha - \dot{\alpha} \, d\lambda) +$$

$$+ \frac{1}{2\kappa} \int_{\partial\Sigma} (\gamma \, dQ^{00} + \lambda \, ds - 2Q^{0}{}_{A} \, dn^{A} - \sigma^{AB} \, d\gamma_{AB}) \tag{28}$$

where the hamiltonian  H  is given by the formula

$$H = \frac{1}{2\kappa} \int_{\partial\Sigma} (Q^{00}\gamma + 2Q^{0}{}_{A} \, n^{A} - s \, \lambda) \tag{29}$$

The following notation has been used:  dots denote  $\partial_{0}$  (time derivative); $Q^{ij}$  is the A.D.M. "momentum" for the hypersurface  $\Lambda$  ( Q is defined in terms of extrinsic curvature  of $\Lambda$  in the same way as  P  was defined in terms of extrinsic curvature of  $\Sigma$ );

$$n^{A} = \tilde{\tilde{g}}^{AB} g_{0B} \tag{30}$$

where  $\tilde{\tilde{g}}$   is the inverse of the 2-metric  $g_{AB}$

$$\gamma = g_{00} - n^{A} n^{B} g_{AB} \tag{31}$$

$$s = \frac{1}{\lambda} g_{AB} S^{AB} \tag{32}$$

where   S   is the orthogonal projection of   Q   onto  $\partial\Sigma$  i.e.:

$$S^{AB} = Q^{AB} + Q^{0A} n^{B} + Q^{0B} n^{A} + Q^{00} n^{A} n^{B} \tag{33}$$

and finally  $\sigma$  is given by the traceless part of  S:

$$\sigma^{AB} = \frac{\lambda}{\overset{0}{\lambda}} (S^{AB} - \frac{1}{2} \tilde{\tilde{g}}^{AB} s \, \lambda) \tag{34}$$

The formula (28) is an analog of a definition of a hamiltonian vector field in classical mechanics, written in terms of symplectic relations:

$$- d H (p,q) = \dot{p} \, dq - \dot{q} \, dp \tag{35}$$

which is equivalent to

$$\dot{p} = - \frac{\partial H}{\partial q} \qquad ; \qquad \dot{q} = \frac{\partial H}{\partial p} \tag{36}$$

In field theory an additional non-evolutional boundary term is always present. This corresponds to the possibility of controlling not only Cauchy data but also boundary value  of the field. For example, the corresponding formula for the scalar field theory would be:

$$- dH (\pi, \phi) = \int_{\Sigma} (\dot{\pi}\, d\phi - \dot{\phi}\, d\pi) + \int_{\partial\Sigma} p^3\, d\phi \qquad (37)$$

where $\pi = p^0$ and $p^3$ are components of the momentum $p^\mu$ canonically conjugate to the scalar field $\phi$ (see [6], [7]). The formulae like (28), (35) or (37) have a precise mathematical meaning and can be written without coordinates (see e.g. [7]).

Intuitively, the formula (37) gives rise to the hamiltonian system $(\pi, \phi)$ provided the last term is "killed". This can be done by imposing the boundary value of $\phi$ on $\Lambda$, i.e. by restricting the class of the unknown fields to the subclass of those which fulfill the boundary conditions. Within this subclass we have $d\phi = 0$. This enables us (due to integration by parts) to write the formulae

$$\dot{\pi} = - \frac{\delta H}{\delta \phi} \qquad ; \qquad \dot{\phi} = \frac{\delta H}{\delta \pi} \qquad (38)$$

analogous to (36), i.e. to translate the field-evolution problem into the language of hamiltonian systems.

Physically, the choice of the boundary conditions means that we close the system composed of the field within $\Sigma$ by imposing at the boundary $\partial\Sigma$ an insulation from the outside world. Mathematically, this means that we choose an appropriate functional space (generally a Sobolev space) of functions satisfying boundary conditions as the configuration space of our system. The phase space will be defined as its cotangent bundle. This way we obtain a rigorously defined hamiltonian system.

Let us come back to the problem of gravitational field. The formula (28) implies the following procedure of solving the Einstein equations:

i) "Kill" the last term in (28) by imposing 6 boundary conditions on $\Lambda$. The particular choice of the first 4 of them, i.e. $Q^{00}$, s, $n^A$, is the choice of the tube $\Lambda$ in space-time (the choice of the family of boundary observers). The evolution will now be referred to this family. For any such a boundary reference we fix also the remaining two boundary conditions i.e. $\gamma_{AB}|\partial\Sigma$. The regularity conditions for the metric $g$ at the center $x_o$ give us also boundary conditions for $\gamma$ at the opposite end of the $x^3$-axis: $\partial_3\gamma_{AB}$ has to vanish exponentially for $x^3 \to -\infty$. These boundary conditions have to be included into the definition of the functional space which will be our configuration space (the phase space $\tilde{P}$ will be its cotangent bundle).

ii) Choose an element $(p^{AB}, \gamma_{AB}, \lambda, \alpha)$ in $\tilde{P}$ as an initial value of the field.

iii) The hamiltonian (29) has to be expressed in terms of legal variables. We use for this goal the identities

$$Q^0_{\ A} = P^3_{\ A} + \lambda\, \alpha,_A \tag{39}$$

and, for $q = sh\alpha$,

$$\partial_3\lambda = \frac{1}{\sqrt{1+q^2}} \left( q\, P^3_{\ 3} + \sqrt{g_{33}}\, \sqrt{|\gamma|}\, Q^{00} \right) \tag{40}$$

which the reader can easily verify. Finally, the first (most difficult one) term of the hamiltonian (29) becomes:

$$Q^{00}\gamma = -\frac{1}{f\,Q^{00}}\, [\, \overset{o}{\lambda}\, \sqrt{1+sh^2\alpha}\; \partial_3 f - P^3_{\ 3}\, sh\alpha\, ]^2 \tag{41}$$

The value of the hamiltonian depends on Cauchy data $(p^{AB}, \gamma_{AB}, \lambda, \alpha)$ and on the control parameters $(Q^{00}, s, n^A)$ at the boundary. If the latter are not constant in time the hamiltonian is explicitly time–dependent. The non–conservation of the energy is due to the interaction of our system with the outside world.

iv) Solve the initial value problem for the unconstrained hamiltonian system $(\tilde{P}, \omega)$ with the hamiltonian found above. This way we find the values of $p^{AB}$, $\gamma_{AB}$, $\lambda$ and $\alpha$ on different surfaces $\Sigma_t$ (possibly within a "thin sandwich $|t| < \epsilon$ only).

v) Solve constraint equations (25) and (4) on each $\Sigma_t$ separately. Use the value of $\lambda$ as the boundary condition for f. This way you know the complete Cauchy data on each $\Sigma_t$.

vi) To find lacking four components of the 4-metric $g_{\mu\nu}$ use gauge-conservation conditions. The reader can easily check that the condition $\partial_0(P^{kl}\, g_{kl}) = 0$ is equivalent to the elliptic equation for the lapse function $N = |g^{00}|^{-1/2}$ :

$$\Delta N = NR \tag{42}$$

where R is the scalar curvature of the 3-metric $g$. The boundary value of $N$ on $\partial\Sigma_t$ (necessary to find the solution) can be easily calculated from the equation (41).

Due to our 2+1 conformally harmonic decomposition of $\Sigma$ the conservation conditions for the gauge (10)–(11) reduce to the single elliptic linear equation for the radial component $N^3$ of the shift vector $N^k = \tilde{g}^{kl}\, g_{01}$ :

$$(f N^3,_3),_3 + \frac{1}{2}\, \overset{2}{\Delta}\, N^3 = \text{function of } P^{kl} \text{ and } g_{kl}. \tag{43}$$

plus the condition

$$N^A,_3 = -\gamma^{AB}\, N^3,_B + \text{function of } P^{kl} \text{ and } g_{kl}. \tag{44}$$

To solve uniquely (43) we use the boundary value for $N^3$ on $\partial\Sigma_t$ which is given by the already known function $\alpha$. The value of $N^3$ for $x^3 \to -\infty$ is also necessary. The choice of the latter means that we prescribe the space–time direction of the trajectory of the center $\underset{o}{x}$. The simplest way to do it is to decide that $\underset{o}{x}$ moves orthogonally with respect to $\Sigma_t$. This corresponds to homogeneous (vanishing) boundary condition at $-\infty$. However, all other choices are equally possible.

To integrate the condition (44) from the boundary towards the center we use the boundary value of the 2-dimensional vector field $N^A = n^A$ which we fixed on the whole $\Lambda$ at the beginning. This completes the information about $g_{\mu\nu}$.

## 5. Energy positivity

Due to our 2+1 conformally harmonic decomposition of $\Sigma$ the scalar constraint equation (4) can be rewritten in the following way:

$$-\frac{1}{2\kappa}\,\partial_r\,(\,r^2\overset{o}{\lambda}\phi,_r\,)\;-\;\frac{1}{2\kappa}\,\partial_A\,(\,\overset{o}{\lambda}\gamma^{AB}\phi,_B\,)\;+\;\frac{\overset{o}{\lambda}}{4\kappa}\,\overset{2}{\check{R}}\,(\gamma)\;-\;\frac{\overset{o}{\lambda}}{2\kappa}\;=$$

$$=\;\frac{1}{4\kappa f^2}\,P^k{}_l\,P^l{}_k\;+\;\frac{\overset{o}{\lambda}}{8\kappa}\,\gamma^{kl}\,\phi,_k\,\phi,_l\;+\;\frac{\overset{o}{\lambda}}{16\kappa}\,\gamma_{AC,3}\,\gamma_{BD,3}\,\gamma^{AB}\,\gamma^{CD} \tag{45}$$

where $\phi = \log \dfrac{f}{r^2}$ ; $r = \exp x^3$ and $\overset{2}{R}\,(\gamma)$ is the scalar curvature of the 2-metric $\gamma$. It turns out, that in the asymptotically flat case the tottal energy (i.e. the limit of the expression (29) for $R \to \infty$ ) equals:

$$M = \lim_{R \to \infty}\,(\,-\,\frac{1}{2\kappa}\,\int_{\partial K(0,R)}\,\overset{o}{\lambda}\,r^2\,\phi,_r\,) \tag{46}$$

The above quantity is, by the way, equal to the so called A.D.M. mass of the system. Integrating the left hand side of the equation (45) over asymptotically flat, non-compact Cauchy surface we obtain exactly the value of $M$ since the second term gives no contribution whereas the third and the fourth ones cancel each other due to the Gauss-Bonnet theorem. Therefore, M is equal to the integral of the manifestly positive right-hand side of (45 ).

In the case of non-empty space the right-hand side of (45) is modified by the matter energy density which is again positive. Therefore, our simple argument for the positivity of the energy remains valid.

Similar argument can also be used in topologically non-trivial case (the radial variable will run outside the horizon only!) but we do not discuss it in the present paper.

## 6. Final remarks

There are also different ways of controlling the boundary parameters. For example, one can perform a partial legendre transformation in (28) exchanging $\gamma$ with $Q^{00}$, $\lambda$ with $s$ but also $\dot{\lambda}$ with $\alpha$. This way we obtain the hamiltonian system which is different from the one discussed here. The 6 boundary control parameters of this system will be the 3-metric $g_{ij}$ on $\Lambda$. The energy of this system defined again as the hamiltonian differs from the energy (29) by the term coming from the Legendre transformation. The properties of such a system will be discussed in the forthcoming paper. Also the limit $R \to \infty$ will be discussed for both systems.

## REFERENCES


[1] Arnowitt R., Deser S., Misner.: The dynamics of general relativity. In: Gravitation – an introduction to current research (Witten L. ed.) N.Y. 1962, Wiley

[2] Ashtekar A.: Asymptotic properties of isolated systems: Recent developments. To appear in the Proceedings of GR 10, Padova, July 83

[3] Choquet-Bruhat Y., York J.W.: TheCauchy problem. In: General Relativity and Gravitation (Held A. ed.), N.Y. 1980, Plenum Press

[4] Ferraris M., Kijowski J.: Gen. Rel. Grav. 14 (1982) p.165

[5] Kijowski J.: Gen. Rel. Grav. 9 (1978) p.857

[6] Kijowski J.: Asymptotic Degrees of Freedom and Gravitational Energy. To appear in the Proceedings of Journees Relativistes, Torino, May 83

[7] Kijowski J., Tulczyjew W.M.: A Symplectic Framework for Field Theories. Springer Lecture Notes in Physics, vol 107 (1979)

[8] Regge T., Teitelboim C.: Ann. Phys. 88 (1974) p.286

[9] Tulczyjew W.M.: Symposia Mathematica 14 (1974) p. 247


Received January 18, 1985.

GEOMETRODYNAMICS PROCEEDINGS (1985), pp. 173-190
edited by A. Pràstaro
© 1985 by World Scientific Publishing Co.

# POLYNOMIAL IDENTITIES SATISFIED BY REALIZATIONS OF LIE ALGEBRAS

Mircea Iosifescu and Horia Scutaru

*Department of Theoretical Physics, Central Institute of Physics
Bucharest, P.O.B. MG-6, Romania.*

## 1. Introduction

Lie algebras can be realized as algebras of operators
- the Lie bracket of two operators is their commutator - and
as algebras of differentiable functions defined on a symplectic
manifold, the Lie bracket being, in this case, the Poisson
bracket associated to the symplectic structure on the manifold.
Realizations of this second type will be called in the following
Poisson ( or classical ) realizations ; realizations as operator
algebras will be refered to as quantum realizations.

Realizations of Lie algebras are encountered in Physics
as symmetry ( invariance ) algebras or as dynamical ( non -
-invariance ) algebras of Hamiltonian dynamical systems.

The considerations which follow have their starting point
in two "experimental" facts concerning realizations:

i) The first is a property which has been repeatedly
pointed out ( e.g. /1/-/7/) : the generators of a realization
of a Lie algebra satisfy algebraic identities.

ii) Let us fix a realization of a Lie algebra L and consider the set of linearly independent polynomials $P_i$, i=1,...,m
the vanishing of which gives rise to the identities $P_i = 0$
which characterize a Poisson realization of L. The other

174

"experimental" remark is that the set $P_i$ is closed with
respect to the extension of the adjoint action of L , i.e.
denoting by $x_i$ , $i=1,\ldots,n$ , the generators of the realization
of L , we have /8/

$$\{x_i,P_j(x_1,\ldots,x_n)\} = \sum_{k=1}^{m} \lambda_{ij}^k P_k(x_1,\ldots,x_n) \qquad (1.1)$$

where $\{\ ,\ \}$ denotes Poisson bracket.

Starting from these "experimental" facts, we shall point
out (section 2) a procedure /9/ of construction of polynomial
identities for realizations of Lie algebras. This procedure
associates polynomial identities of a realization of a Lie
algebra L to symmetric subrepresentations of the direct powers
of the adjoint representation of L. The procedure is valid
both for Poisson realizations and for quantum realizations of L.

The polynomial identities of a realization of a Lie algebra
L represent, for a dynamical system admitting L as its symmetry
or as its dynamical algebra, constraints imposed to the system
by its symmetry properties. Let indeed the Hamiltonian H of a
dynamical system depend on the generators $x_i$ of a realization
of the Lie algebra L, $H = H(x_1,\ldots,x_n)$ and let $P_k(x_1,\ldots,x_n)$
$= 0$ $(k=1,\ldots,m$ ) be the linearly independent identities satisfied by the generators $x_i$. We have, from (1.1),

$$\frac{dP_k(x_1,\ldots,x_n)}{dt} = \{P_k,H\} = -\sum_{j=1}^{n} \sum_{l=1}^{m} \frac{\partial H}{\partial x_j} \lambda_{jk}^l P_l(x_1,\ldots,x_n) = 0$$
$$(1.2)$$

i.e. the polynomials $P_k$ are constants of the motion ; the
identities $P_k = 0$ are constraints imposed to the motion by
its symmetry properties.

It is of course desirable to obtain all possible constraints which can be satisfied by a dynamical system with

symmetry. A procedure to solve this problem /10/ will be indicated in section 3 and exemplified on second-degree polynomial
constraints for orthogonal Lie algebras of arbitrary rank.

For reasons of simplicity, calculations will refer to
polynomials defined on the dual space $L^*$ of the Lie algebra L;
simplicity is however not the only reason : the polynomial
identities written in terms of coordinates in $L^*$ are archetypes
of the constraints imposed by the symmetry to the dynamical
system.

The presence of constraints in a dynamical system rises
the problem of their elimination. In the classical case, this
amounts to obtain explicit expressions of independent pairs
of canonically conjugated coordinates on the archetypal manifold defined in $L^*$ by the polynomial identities. The equivalent quantum problem is the determination of the minimum
number of pairs of bosonic operators in terms of which a given
realization of a Lie algebra can be expressed. The expression
of the generators of the realization in terms of these bosonic operators is usually called Holstein-Primakoff realization
and it has come to the attention of physicists in the last
years especially in connection with collective motion in nuclei
(e.g./11/) or with the study of 1/N expansions in quantum
mechanics or in field theory /12/,/13/.

We shall point out in section 5 (cf./14/) that Holstein-
Primakoff realizations of the Lie algebras sp(2d;R) and
so(2d;R) can be obtained from the polynomial identities of the
realizations.

Another result to which we want to call the attention

- and which is closely related to the polynomial identities -
is the possibility to construct an equivariant mapping $\mathcal{K}$ from
a G-invariant manifold $\mathcal{M}$ on which a Poisson realization is
defined, into a set of linear operators on a G-module V ( section 4 ) . This mapping, which     is     an analogue of the
moment map, allows to obtain the polynomial identities of the
realization and to associate them to G-orbits in EndV /15/.

The previous results allow the extension to the quantum
case of theorems obtained for the classical Gross-Neveu model
by Berezin /16/ and Papanicolaou /13/ (section 6).

## 2. Polynomial identities associated to subrepresentations of $(ad^{\otimes k})_s$

Let G be a Lie group, L be its Lie algebra and $L^*$ be
the dual space of L.

In what follows, we indicate how to obtain polynomials
defined on $L^*$ and satisfying (1.1) and show that a system of
polynomials with this property determines a symplectic submanifold of $L^*$ and thus a Poisson realization of L.

Let us introduce in L a basis $x_1, \ldots, x_n$ for the elements
of which the Lie product is

$$[x_i, x_j] = \sum_{k=1}^{n} c_{ij}^{k} x_k \qquad (2.1)$$

Let $\{u_1, \ldots, u_n\}$ be the basis in $L^*$ , dual with respect to this
basis of L, i.e. defined by $\langle u_i | x_j \rangle = \delta_{ij}$ (i,j=1,\ldots,n).
Elements $u \in L^*$ are then written as

$$u = \sum \xi_i(u) u_i \quad \text{with} \quad \xi_i(u) = \langle u | x_i \rangle \qquad (2.2)$$

The generator of the co-adjoint action $Ad^*(\exp t x_i)_u : L^* \rightarrow L^*$ , expressed in terms of the coordinate functions $\xi_i$, is

$$X_{x_i}(u) = \sum c_{ij}^k \xi_k(u) \left(\frac{\partial}{\partial \xi_j}\right)_u \tag{2.3}$$

The Kirillov-Kostant-Souriau (KKS) symplectic form $\omega$, defined by /17/

$$\omega(X_{x_i}, X_{x_j}) = -\langle u | [x_i, x_j] \rangle \tag{2.4}$$

determines, on any manifold $\mathcal{O}$ on which it is non-degenerate, a Poisson bracket

$$\{f, g\} = \omega(X_f, X_g) \tag{2.5}$$

where $f$, $g$ are differentiable functions on this manifold and $X_f$ is the Hamiltonian field defined by $df = X_f \lrcorner \omega$. In terms of the coordinate functions $\xi_i(u)$ of a point $u \in L^*$, this Poisson bracket has the expression

$$\{f, g\} = \sum_{i,j,k=1}^{n} c_{ij}^k \xi_k \frac{\partial f}{\partial \xi_i} \frac{\partial g}{\partial \xi_j} \tag{2.6}$$

In particular, if $f$ and $g$ are coordinate functions of a point $u \in \mathcal{O}$, we have

$$\{\xi_i, \xi_j\} = \sum_{k=1}^{n} c_{ij}^k \xi_k \tag{2.7}$$

i.e. the coordinate functions of a point $u \in \mathcal{O}$ generate a Poisson realization of the Lie algebra L.

We are now able to define the extension of the adjoint representation ad of L to a representation $\widetilde{ad}$ of L on the linear space $P(L^*)$ of the polynomials defined on $L^*$:

$$\widetilde{ad}(x_i)p = X_{x_i}p = \sum_{j,k=1}^{n} c_{ij}^k \xi_k \frac{\partial p}{\partial \xi_j} = \{\xi_i, p\} \qquad (p \in P(L^*)) \tag{2.8}$$

This representation is equivalent to the extension of the adjoint representation to the symmetric algebra S(L). There exists thus a natural number k for each irreducible sub-representation of $\widetilde{ad}$ such that it is equivalent with a subrepresentation of $(ad^{\otimes k})_s$ - the symmetric part of $ad^{\otimes k}$.

Let $\pi$ be a subrepresentation of $\widetilde{ad}$ . Let $p_1,\ldots,p_m$ be a basis of the subspace of $P(L^*)$ afforded by $\pi$ , i.e. such that

$$(X_{x_i} p_j)(u) = (\pi(x_i)p_j)(u) = \sum_{k=1}^{m} \pi_{jk}(x_i)p_k(u) \qquad (2.9)$$

Let us define for a subrepresentation $\pi$ an algebraic submanifold of $L^*$ by

$$\Theta_\pi = \left\{ u \in L^* \mid p_1(u) = \ldots = p_m(u) = 0 \right\} \qquad (2.10)$$

From Whitney's theorem /18/, it follows that the regular part of $\Theta_\pi$ , denoted $\Theta_\pi^o$, is a regular analytic submanifold of $L^*$ , the dimension of which is equal to $\dim L^* - s$ , where $s$ is the maximum value attained on $\Theta_\pi$ by the dimension of the linear space generated by $\left\{ (dp_1)_u,\ldots,(dp_m)_u \right\}$ , $u \in \Theta_\pi$ . From the definition of $\Theta_\pi$ and from the property (2.9) it follows that the vectors $X_{x_1}(u),\ldots,X_{x_n}(u)$ generate a linear subspace of the tangent space $T_u\Theta_\pi^o$ for any $u \in \Theta_\pi^o$ . The dimension of this subspace is given by $r(u) = \text{rank} \parallel \sum c_{ij}^k \xi_k(u) \parallel$. It is obvious that $\dim T_u\Theta_\pi^o = \dim L^* - s \geqslant r(u)$. Thus, in order to have $r(u) = \dim L^* - s$ , it is sufficient to prove that $r(u) \geqslant \dim L^* - s$ , for any $u \in \Theta_\pi^o$ .

Hence, if it is possible to prove that on $\Theta_\pi^o$ we have

$$\text{rank} \parallel \sum_{k=1}^{n} c_{ij}^k \xi_k \parallel \geqslant \dim \Theta_\pi^o \qquad (2.11)$$

then it results that $T_u\Theta_\pi^o$ is generated by the vectors $X_{x_1}(u),\ldots,X_{x_n}(u)$ and that it is possible to define on $\Theta_\pi^o$ a KKS symplectic form $\omega$ in a unique way by (2.4) and hence to define a Poisson bracket (2.6) associated to $\omega$ for any $f,g \in C^\infty(\Theta_\pi^o)$ . A preliminary analysis of the non-degeneracy of the manifold $\Theta_\pi$ defined by (2.10) i.e. a proof that $\dim \Theta_\pi > 0$ is necessary. A Poisson realization of the Lie algebra L can be thus associated to any subrepresentation $\pi$ of $(ad^{\otimes k})_s$ for which $\Theta_\pi$ is non-degenerate and satisfies the rank condition (2.11). ( For a one-dimensional subrepresentation,the polynomial relation is $I(u) = k$ , where $I(u)$ is an an invariant of the algebra and $k \in \mathbb{R}$ is arbitrary , fixed.)

### 3. Polynomial identities for semisimple Lie algebras

The relations $p_1(u) = \ldots = p_m(u) = 0$ in (2.10) are polynomial identities satisfied by the Poisson realization (2.7). The procedure outlined in the previous section has been applied in /9/ to the determination of polynomial identities for semisimple Lie algebras of ranks two and three. We shall point out,in the following, a method for the construction of polynomials belonging to subrepresentations of $(ad^{\otimes k})_s$ for an arbitrary semisimple Lie algebra and exemplify it for orthogonal algebras and for $k=2$ .

Let us remark that the adjoint representation of a Lie algebra L admits, on each of the following isomorphic linear spaces :

1. The space $P(L^*)$ of the polynomials on $L^*$ ,
2. The symmetric algebra $S(L)$,
3. The universal enveloping algebra $U(L)$,

unique isomorphic extensions. It is therefore sufficient to perform the reduction of the extension of the adjoint representation only in one of these three spaces. We shall perform the reduction in $P(L^*)$, where the extension $\widetilde{ad}$ of $ad$ is given by (2.9). The intertwining operator between the extensions 2 and 3 is given by the symmetrizer /19/. From the equivalence between extensions 1 and 2, it results that the restriction of $\widetilde{ad}$ to the space of polynomials of second degree is equivalent to the symmetrized direct square $(ad\otimes ad)_s$. To project out vectors which belong to the irreducible subrepresentation $\varsigma_j$ in $(ad\otimes ad)_s = \sum_{j=1}^{\nu} \oplus \varsigma_j$ is thus equivalent to project out vectors belonging to $\varsigma_j$ from the representation defined by the action of the linear operator $\widetilde{ad}$ (2.8) on second-degree polynomials in the indeterminates $\xi_i$ $(i=1,\ldots,\dim L)$.

The irreducible components $\varsigma_j$ can be labeled by the eigenvalues of the Casimir operator

$$C = \sum_{i=1}^{\dim L} \varsigma(x_i)\,\varsigma(x^i) \tag{3.1}$$

where $\varsigma = (ad\otimes ad)_s$ and $x^i \in L$ – the element dual to $x_i$ – is defined by

$$(x^i, x_j) = \delta_{ij} \tag{3.2}$$

the bracket $(\ ,\ )$ denoting a non-degenerate bilinear form on L ( if such a form exists ).

For the restriction of $\widetilde{ad}$ to second-degree homogeneous polynomials, the Casimir operator becomes

$$C = \sum_{i=1}^{\dim L} \left\{ \xi_i, \left\{ \xi^i, \cdot \right\} \right\} \tag{3.3}$$

where

$$\xi^i(u) = \langle u | x^i \rangle \tag{3.4}$$

It turns out, as a consequence of a result of Okubo /20/ that in the Clebsch-Gordan series of $(ad\otimes ad)_s$ the number of terms is $\nu = 4$, for all the following semisimple Lie algebras : $A_n\ (n \geqslant 3)$, $B_n\ (n \geqslant 2)$, $C_n\ (n \geqslant 2)$, $D_n\ (n \geqslant 5)$, the decomposition being multiplicity-free. The spectral decomposition of the

Casimir operator $C$ contains thus, for any of the semisimple Lie algebras quoted above, precisely four terms

$$C = \sum_{i=1}^{4} \lambda_i P_i \qquad (3.5)$$

where $P_i$ is the projection operator on the irreducible module $\mathcal{S}_i$ . The eigenvalues $\lambda_i$ (i=1,2,3,4) of the Casimir operator result, for any of these Lie algebras, as solutions of algebraic equations of fourth degree ; the expression of the equation for a given Lie algebra L are determined by finding the linear combinations of polynomials $C^q \xi_s \xi_t$ (q=0,1,2,3,4) which vanishes for any s,t=1,...,dimL , i.e. by finding the coefficients $\alpha_i$ such that $(\sum_{q=0}^{4} \alpha_q C^q) \xi_s \xi_t = 0$ for any s,t = 1,...,dimL. The projection operator associated to the eigenvalue $\lambda_\alpha$ , solution of the equation $\sum \alpha_q C^q = 0$, is

$$P_\alpha = \frac{\prod_{\beta \neq \alpha} (C - \lambda_\beta I)}{\prod_{\beta \neq \alpha} (\lambda_\beta - \lambda_\alpha)} \qquad (3.6)$$

and the vectors belonging to the irreducible representation $\rho_\alpha$ (characterized by $\lambda_\alpha$ ) of $C$ are obtained by applying $P_\alpha$ to the generic monomial $\xi_s \xi_t$ .

Let us exemplify on semisimple Lie algebras of types $B_n$ and $D_n$.

The generators $M_{ij}$ of the algebras so(2n+1,C) and so(2n,C) satisfy the structure equations

$$[M_{ij}, M_{kl}] = \delta_{il} M_{jk} + \delta_{jk} M_{il} - \delta_{ik} M_{jl} - \delta_{jl} M_{ik} \qquad (3.7)$$

where $i \neq j = 1,...,N$, $k \neq l = 1,...,N$, $M_{ji} = -M_{ij}$ with N=2n+1 and N=2n for so(2n+1) and so(2n) respectively.

The elements $M_{ij}$ and $M_{ji}$ are dual; the Casimir operator is

$$C = \sum_{i<j=1}^{N} \{M_{ij}, \{M_{ji}, \cdot\}\} \qquad (3.8)$$

182

Let $\Lambda_1,\ldots,\Lambda_n$ be the weights of the fundamental representations. Denoting by $\Lambda_0$ the weight of the adjoint representation, we have, for $B_n$ $(n \geqslant 3)$ and for $D_n$ $(n \geqslant 4)$ /21/ $\Lambda_0 = \Lambda_2$. The Clebsch-Gordan series for the reduction of the product $(\mathrm{ad} \otimes \mathrm{ad})_s$, for algebras of types $B_n$ and $D_n$ with $n \geqslant 4$, is /21/

$$( \Lambda_0 \otimes \Lambda_0 )_s = (0) \oplus (\Lambda_4) \oplus (2\Lambda_1) \oplus (2\Lambda_2) \qquad (3.9)$$

The equality (3.9) is valid for $B_2$ if $(\Lambda_4)$ is replaced with $(\Lambda_1)$ and for $B_3$ if $(\Lambda_4)$ is replaced by $(2\Lambda_3)$. The eigenvalues of the Casimir operator (3.8) for the representations $(\Lambda_4),(2\Lambda_1)$ and $(2\Lambda_2)$ in (3.9) are $8(N-4)$, $N$ and $8(N-1)$ respectively. This sequence of eigenvalues keeps valid for $B_2$ and $B_3$. (cf. also /22/).

The functions belonging to the irreducible subrepresentations of $(\mathrm{ad} \otimes \mathrm{ad})_s$ are, for the algebras $B_n$ and $D_n$ :

$$P_{(0)}\, M_{pq}\, M_{rs} = \frac{1}{N(N-1)} ( \delta_{qr}\, \delta_{ps} - \delta_{pr}\, \delta_{qs} ) \sum_{i,j=1}^{N} M_{ij}\, M_{ji} \qquad (3.10)$$

$$P_{(\Lambda_4)}\, M_{pq}\, M_{rs} = \tfrac{1}{3} ( M_{pq}\, M_{rs} + M_{ps}\, M_{qr} + M_{pr}\, M_{sq} ) \qquad (3.11)$$

$$P_{(2\Lambda_1)}\, M_{pq}\, M_{rs} = \frac{1}{N-2} \Big[ \, \delta_{qr} \Big( \sum_{i=1}^{N} M_{pi}\, M_{is} - \tfrac{1}{N} \delta_{ps} \sum_{i,j=1}^{N} M_{ij}\, M_{ji} \Big)$$
$$+ \, \delta_{ps} \Big( \sum_{i=1}^{N} M_{qi}\, M_{ir} - \tfrac{1}{N} \delta_{qr} \sum_{i,j=1}^{N} M_{ij}\, M_{ji} \Big)$$
$$- \, \delta_{pr} \Big( \sum_{i=1}^{N} M_{qi}\, M_{is} - \tfrac{1}{N} \delta_{qs} \sum_{i,j=1}^{N} M_{ij}\, M_{ji} \Big) \qquad (3.12)$$
$$- \, \delta_{qs} \Big( \sum_{i=1}^{N} M_{pi}\, M_{ir} - \tfrac{1}{N} \delta_{pr} \sum_{i,j=1}^{N} M_{ij}\, M_{ji} \Big) \Big]$$

$$P_{(2\Lambda_2)}\, M_{pq}\, M_{rs} = \tfrac{1}{3} ( 2 M_{pq}\, M_{rs} - M_{ps}\, M_{qr} - M_{pr}\, M_{sq} )$$
$$+ \frac{1}{N-2} \Big( - \delta_{qr} \sum_i M_{pi}\, M_{is} + \delta_{qs} \sum_i M_{pi}\, M_{ir} + \delta_{pr} \sum_i M_{qi}\, M_{is} - \delta_{ps} \sum_i M_{qi}\, M_{ir} \Big) \qquad (3.13)$$
$$+ \frac{1}{(N-1)(N-2)} ( \delta_{qr}\, \delta_{ps} - \delta_{qs}\, \delta_{pr} ) \sum_{i,j=1}^{N} M_{ij}\, M_{ji}$$

where $N=2n+1$ ( $N=2n$ ) for algebras of type $B_n$ ( $D_n$ ). The tensor (3.12) has been calculated also by Jarvis /21/.

## 4. <u>A moment map - like mapping</u>

Let, as before, L be the Lie algebra of the Lie group G
and $x_i$ be the generators of L. Let $x^i$ be the element of L
dual to $x_i$ with respect to the Cartan-Killing bilinear form.
Let $\mathcal{M}$ be a manifold on which an action of G and a Poisson
algebra $\mathcal{A}$ are defined. Let

$$R : \quad x_i \in L \longrightarrow f_{x_i}(m) \in C^\infty(\mathcal{M})$$
$$R : \quad [x_i, x_j] \longrightarrow \{f_{x_i}, f_{x_j}\}_m \qquad\qquad (4.1)$$

be a Poisson realization of L, defined on $\mathcal{M}$ . Let the action
of G on $\mathcal{A}$ be Hamiltonian /23/.

The moment map /24/ is a mapping $\phi: \mathcal{M} \rightarrow L^*$ from a point
$m \in \mathcal{M}$ to a linear form on L, defined by

$$\langle \phi(m) | x \rangle = f_x(m) \qquad\qquad (4.2)$$

for $m \in \mathcal{M}$ and $x \in L$. The moment map applies G-orbits on
into co-adjoint orbits ; it is equivariant, i.e.

$$\phi(g.m) = Ad_g^* \phi(m) \qquad\qquad (4.3)$$

where g.m denotes the action of $g \in G$ on $m \in \mathcal{M}$ .

Let V be a G-module and $\pi$ be a linear representation
of L afforded by V. We define a mapping $\mathcal{K} : \mathcal{M} \longrightarrow EndV$ by

$$\mathcal{K}(m) = \sum_{i=1}^{n} f_{x_i}(m) \pi(x^i) \qquad (m \in \mathcal{M}) \qquad (4.4)$$

The mapping $\mathcal{K}$ is equivariant /15/ i.e.

$$\mathcal{K}(g.m) = U(g^{-1}) \mathcal{K}(m) U(g) \qquad\qquad (4.5)$$

where U is a representation of G such that

$$\frac{d}{dt} U(\exp tx_i) = \pi(x_i) \qquad\qquad (4.6)$$

The mapping $\mathcal{K}$ is a generalization of a construction of

Carey and Hannabuss, related to orbits in twistor spaces /25/ and of a construction considered by Kramer in connection with collective motion in nuclei /26/,/27/.

Matrices of the type (4.4) are closely related to the matrix $A_L = \sum_{i=1}^{n} (\tau(x^i) \otimes x_i + \pi(x_i) \otimes x^i)$ considered in /6/, /7/ in order to explain the characteristic identities for Lie algebras pointed out in /4/,/5/. They are also related to the matrix $Q = \sum \pi(x_i) \otimes \rho(x^i)$ introduced in /28/ with the same purpose. ( $\pi$ and $\rho$ are linear representations of L ).

The matrix $\mathcal{K}$ introduced by (4.4) is useful in the determination of polynomial identities for Poisson realizations. indeed, the equivariance property of $\mathcal{K}$ is valid for polynomials in the indeterminate $\mathcal{K}$ as well. Thus, if a polynomial relation $P(\mathcal{K}(m)) = 0$ is valid for a given point m , it will be valid for any point g.m of the G-orbit through m. The matrix elements of this polynomial relation provide explicit polynomial identities for the generators of the Poisson realization (4.1), proving thus that any Poisson realization satisfies algebraic identities.

In /15/ we have used the properties of the mapping $\mathcal{K}$ to deduce some of the polynomial identities which have been obtained, for semisimple Lie algebras, by the method pointed out previously. Let us exemplify these results on the orthogonal algebras. For compact algebras of types $B_n$ and $D_n$ the matrix $\mathcal{K}$ has the expression

$$\mathcal{K} = \sum_{i,j} M_{ij} \pi(M_{ji}) = \sum_{i,j} M_{ij}(e_{ji} - e_{ij}) \qquad (4.7)$$

where $e_{ij}$ are elements of the Weyl basis. We have

$$\mathcal{K}^2 = \sum_{q,r} (4 \sum_{i} M_{ri} M_{iq}) e_{qr} \qquad (4.8)$$

The matrix elements of $\mathcal{K}^2$ are polynomials belonging to the subrepresentation $(2\Lambda_1) \oplus (0)$ of $(\mathrm{ad} \otimes \mathrm{ad})_s$, which have been determined previously. The expressions of the matrices $\mathcal{K}$ for non-compact algebras of types $B_n$ and $D_n$ are similar. In the particular case of the algebra so(4,2), the explicit expressions for the polynomial relations satisfied by the matrix $\mathcal{K}$ on four and six-dimensional orbits of SO(4,2) have been determined in /15/, using the classification of orbits obtained in /25/ and /29/.

### 5. An application : the derivation of the Holstein-Primakoff realization from polynomial identities

As stated in the introduction, the knowledge of the polynomial identities - i.e. of the constraints imposed to the dynamical system by its symmetry - can be used to eliminate these constraints. We shall indicate in the following how this elimination can be performed for realizations of the algebras so(2d;R) and sp(2d;R) , for any d. To do that, we first identify the second-degree polynomial identities satisfied by the generators of "collective" representations of so(2d;R) obtained in /13/ and of sp(2d;R) obtained in /11/. Taking then these identities as our starting point, we deduce the Holstein-Primakoff realizations obtained in /13/ and /11/ in a purely algebraic way. This approach can be applied to all representations (finite or infinite-dimensional, with maximal weight vector) the generators of which satisfy second-degree polynomial identities.

Let $F_\varepsilon$ be the Fock space of states for a Bose ($\varepsilon = -1$) or for a Fermi ($\varepsilon = +1$) system with Nd degrees of freedom /31/ on which the following canonical commutation ($\varepsilon = -1$) and anti-commutation ($\varepsilon = +1$) relations are defined :

$$[b_{is}, b_{jt}]_\varepsilon = [b_{is}^*, b_{jt}^*]_\varepsilon = 0 \quad (i,j=1,\ldots,d \ ; \ s,t=1,\ldots,N)$$
$$[b_{is}, b_{jt}^*]_\varepsilon = \delta_{ij} \cdot \delta_{st} \, I \tag{5.1}$$

In (5.1) I is the unity operator on $F_\varepsilon$ .

The generators of the skew-adjoint reducible representation of $sp(2d;R)$ ($\varepsilon = -1$) and of $so(2d;R)$ ($\varepsilon = +1$), defined on $F_\varepsilon$ by

$$A_{ij} = \sum_{s=1}^{N} b_{is}^* b_{js} - \varepsilon(N/2) \, \delta_{ij} I$$

$$B_{ij} = \sum_{s=1}^{N} b_{is}^* b_{js}^* \quad ; \quad O_{ij} = \sum_{s=1}^{N} b_{is} b_{js} \tag{5.2}$$

have the following nonvanishing Lie brackets

$$[A_{ij}, A_{kl}]_{-1} = \delta_{jk} A_{il} - \delta_{il} A_{kj}$$

$$[A_{ij}, B_{kl}]_{-1} = \delta_{jk} B_{il} - \varepsilon \delta_{jl} B_{ik}$$

$$[A_{ij}, C_{kl}]_{-1} = \varepsilon \delta_{il} C_{jk} - \delta_{ik} C_{jl} \tag{5.3}$$

$$[B_{ij}, C_{kl}]_{-1} = -\delta_{jk} A_{il} - \delta_{il} A_{jk} + \varepsilon \delta_{ik} A_{jl} + \varepsilon \delta_{jl} A_{ik}$$

To simplify notations we wrote $I$, $b_{ij}$, $b_{ij}$, $A_{ij}$, $B_{ij}$, $C_{ij}$ instead of $I^\varepsilon$, $b_{ij}^\varepsilon$, $b_{ij}^\varepsilon$, $A_{ij}^\varepsilon$, $B_{ij}^\varepsilon$, $C_{ij}^\varepsilon$ respectively.

Using the matrix notation $X = (X_{ij})$ ($i,j=1,\ldots,d$) and denoting by $X^t$ the transposed of X and by $X^*$ the matrix the elements of which are the adjoints of the corresponding elements of X, we have $B^t = -\varepsilon B$, $C^t = -\varepsilon C$, $A^t = A^*$, $B^* = C$, $C^* = B$ .

The "collective" space $F_\varepsilon^c$ is, by definition /11/,/13/, the subspace of $F_\varepsilon$ spanned by vectors of the form

$$B_{ij} B_{kl} \cdots B_{rs} v_\varepsilon \qquad (5.4)$$

where $v_\varepsilon$ is the vacuum state in $F_\varepsilon$ , defined by $b_{ir}v_\varepsilon = 0$
$i=1,\ldots,d$ , $r=1,\ldots,N$.

It can be proved that, in $F_\varepsilon^c$ , the generators (5.2) satisfy the second-degree polynomial identities

$$AB - BA^t + \varepsilon(AB - BA^t)^t = 0 \qquad (5.5)$$

$$CA - A^tC + \varepsilon(CA - A^tC)^t = 0 \qquad (5.6)$$

$$A^2 + ((A^t)^2)^t - BC - (CB)^t - (1/d)\mathrm{Tr}((A^2 + (A^t)^2)^t - BC - (CB)^t)I = 0$$
$$(5.7)$$

where, abusively, we denoted $I = ( \delta_{ij}I)$.

( The matrix elements of the identities (5.5-7) are equivalent to the polynomial identities associated to the subrepresentation $\Lambda_2$ of $(ad \otimes ad)_s$ of the algebra $sp(2d;R)$/15/ and to the subrepresentation $(2\Lambda_1)$ of $(ad \otimes ad)_s$ of $so(2d;R)$.)

The Holstein-Primakoff realizations for $sp(2d;R)$ /11/ and $so(2d;R)$ /13/ are of the form

$$A = -\varepsilon( a^*a + \tfrac{N}{2} I) , \quad B = Xa^* = a^*Y , \quad C = aX = Ya \qquad (5.8)$$

where $a^* = ( a_{ij}^*)$ ; $a = (a_{ij})$ $(i,j=1,\ldots,d)$ are $d \times d$ matrices the elements of which are bosonic creation and annihilation operators acting on a boson Fock space :

$$[a_{ij},a_{kl}]_{-1} = [a_{ij}^*,a_{kl}^*]_{-1} = 0 , \quad [a_{ij},a_{kl}^*]_{-1} = ( \delta_{ik}\delta_{jl} - \varepsilon\delta_{il}\delta_{jk})I$$
$$(5.9)$$

and $(a^*)^t = -\varepsilon a$ , $(a)^t = -\varepsilon a$; $X = (X_{ij})$, $Y = (Y_{ij})$ $(i,j=1,\ldots,d)$ are respectively vector and covariant vector operators for the $u(d)$ subalgebra.

The following statement holds /14/ :

The bosonic realizations (5.8) satisfy the identities (5.5-6) automatically and the identities (5.7) if and only if

$$X^2 = a^*a + (N + \varepsilon d - 1)I , \quad Y^2 = (a^*a)^t + NI \qquad (5.10)$$

### 6. Other possible applications

The covariant symbols of the generators of the algebras so(2d;R) and sp(2d;R) (/16/;(2.9)) generate Poisson realizations of these algebras /30/. The algebraic equations satisfied by these symbols are precisely those obtained by means of the moment map-like mapping $\mathcal{K}$.

The second-degree polynomial identities pointed out by Berezin and Papanicolaou in the classical case have a quantum analogue. The quantum identities (5.5-7 ; $\varepsilon$ =+1) can be obtained in the following way : the second-degree polynomials are obtained by symmetrization of the second-degree classical polynomials and it can be proved that their values on the sector considered in /13/ and /16/ are determined by their values on the vacuum state. This remarkable property is a consequence of the fact that these polynomials generate an irreducible module for the extension of the adjoint representation to the universal enveloping algebra. This last property implies also that the constraints remain invariant under the evolution generated by the Gross-Neveu Hamiltonian.

## References

1. N. Mukunda, J.Math.Phys. **8**, 1069 (1967)

2. G. Györgyi, Nuovo Cim. **53A**, 717 (1968)

3. A. O. Barut, A. Böhm, J.Math.Phys. **11**,2938 (1970)

4. A. J. Bracken, H. S. Green, J.Math.Phys. **12**,2106 (1971)

5. H. S. Green, J.Math.Phys. **12**,2106 (1971)

6. K. C. Hannabuss, Characterizing Equations for Semisimple Lie Groups, Preprint, Mathematical Institute, Oxford (1972)

7. D. M. O'Brien, A. Cant, A. L. Carey, Ann.Inst.H.Poincaré **26A**, 405 (1977)

8. M. Iosifescu, H. Scutaru, J.Math.Phys. **21**, 2033 (1980)

9. M. Iosifescu, H. Scutaru, J.Math.Phys. **25**, 2856 (1984)

10. M. Iosifescu, H. Scutaru, Polynomial Constraints Associated to Dynamical Symmetries, Preprint, Bucharest, FT-253 (1984)

11. J. Deenen, C. Quesne, J.Math.Phys. **23**, 2004 (1982)

12. L. D. Mlodinow, N. Papanicolaou, Ann.Phys.(N.Y.) **131**,1 (1984)

13. N. Papanicolaou, Ann.Phys. (N.Y.) **136**, 210 (1981)

14. M. Iosifescu, H. Scutaru, Boson Realizations and Quantum Constraints, Preprint, Bucharest, FT-255 (1984)

15. M. Iosifescu, H. Scutaru, A Moment Map-Like Mapping, Preprint, Bucharest, FT-254 (1984)

16. F. A. Berezin, Commun.Math.Phys. **63**, 131 (1978)

17. A. A. Kirillov, Elementy Teorii Predstavleniia, "Nauka" Publ. House, Moscow (1972)

18. V. S. Varadarajan, Lie Groups, Lie Algebras and their Representations, Prentice-Hall, Englewood Cliffs, N.J.(1974)

19. J. Dixmier, Algèbres Enveloppantes, Gauthier-Villars, Paris-Bruxelles-Montréal, (1974)

20. S. Okubo, J.Math.Phys. $\underline{20}$, 586 (1979)

21. P. D. Jarvis, J.Phys.A Math.Gen. $\underline{12}$, 1 (1979)

22. S. Okubo, J.Math.Phys. $\underline{23}$, 8 (1982)

23. V. Guillemin, S. Sternberg, Ann.Phys.(N.Y.), $\underline{127}$, 220 (1980)

24. J. M. Souriau, Structure des Systèmes Dynamiques, Dunod, Paris, (1970)

25. A. L. Carey, K. C. Hannabuss, Rep.Math.Phys. $\underline{13}$, 199 (1978)

26. P. Kramer, Ann.Phys. (N.Y.) $\underline{141}$, 254 (1982)

27. P. Kramer, Advances in the Theory of Collective Motion in Nuclei, in "Group Theoretical Methods in Physics", Lecture Notes in Physics vol. $\underline{201}$, 243 Springer (1984)

28. S. Okubo, J.Math.Phys. $\underline{18}$, 2382 (1977)

29. H. Zassenhaus, Can.Math.Bull. $\underline{1}$, 31,101,183 (1958)

30. H. Scutaru, Coherent States and Classical Limit of Algebraic Quantum Models, in "Nuclear Collective Dynamics, Lectures of the 1982 International Summer School of Nuclear Physics, Poiana Braşov, Romania" World Scientific, Singapore (1983)

31. In sections 4 and 5 the symbol N has different meanings.

Received January 18, 1985.

GEOMETRODYNAMICS PROCEEDINGS (1985), pp. 191-200
edited by A. Pràstaro
© 1985 by World Scientific Publishing Co.

# FREE MOTIONS IN MULTIDIMENSIONAL UNIVERSES

G. Marmo

Istituto di Fisica Nucleare,
Gruppo teorico, Sezione di Napoli
Dipartimento di Fisica,
Mostra d'Oltremare Pad. 19 - I 80125 NA

## Introduction

The quest for unification of the known "fundamental" forces has regained a good deal of interest for the early attempts by Kaluza and Klein[1] to unify electromagnetism and gravity.

Recent proposals are made to incorporate non Abelian "gauge fields" in the same way as Kaluza-Klein theory incorporates electromagnetism.

The main idea is that the universe, to begin with, is a huge manifold  E  and the only relevant quantities are geometrical.  As temperature goes down some kind of phase transition takes place and a fibered structure appears on  E.  Fibers are so "compact" that we cannot observe them directly. The only remnant being the gauge fields we see.  The gauge symmetry is a reflection of the symmetry of translation along the fiber.

Here, only gravity in higher dimensions is relevant, the gauge fields arise as part of the gravitational fields.

At the level of test particles we expect that geodetical motions in higher dimensions are able to represent Yang-Mills particles in four dimensions.  The classical motion of particles with internal structure was derived by Wong starting with Dirac-type equations[2].  These are first order differential equations in the "internal" variables and second order differential equations in the position.  Our main effort is to show how to go from second order geodetical motion in higher dimensions to the "mixed" system in four dimensions.

We consider two different "ansatz" on the metric tensor and discuss in which conditions we do get the "required" equations of motion.

### The "ansatz" on the metric tensor

On the manifold $E$ we consider coordinates $(x^M)$ and a metric tensor

$$\bar{g} = \bar{g}_{MN} \, dx^M \otimes dx^N \quad . \tag{1}$$

Geodetical motions are given by

$$\ddot{x}^M + \Gamma^M_{NP} \dot{x}^N \dot{x}^P = 0 \quad . \tag{2}$$

By using the fibered structure on $E$, and an appropriate "ansatz" for the metric tensor we would like to reduce (2) to the form

$$\begin{cases} \ddot{x}^\mu + \Gamma^\mu_{\nu\lambda}\dot{x}^\lambda \dot{x}^\nu = I_a F^a_{\mu\nu}\dot{x}^\nu g^{\lambda\mu} \\[2mm] \dot{I}_a + C^c_{ab}A^b_\mu I_c \dot{x}^\mu = 0 \end{cases}$$

where $(x^\mu)$ are space time coordinates, i.e. for the base manifold of the fibration

$$F \to E \to M$$

and $I_a$ are the "internal variables to be identified with isospin variables or non-abelian charges for the fiber $F$. By using a splitting of coordinates for $E$, say $(x^\mu, y^m)$, the metric tensor acquires the form

<table>
<tr><td></td><td align="center">4</td><td align="center">N</td></tr>
<tr><td align="center">4</td><td align="center">$\bar{g}_{\mu\nu}$</td><td align="center">$\bar{g}_{\mu n}$</td></tr>
<tr><td align="center">N</td><td align="center">$\bar{g}_{m\nu}$</td><td align="center">$\bar{g}_{mn}$</td></tr>
<tr><td></td><td align="center">4</td><td align="center">N</td></tr>
</table>

### A more appropriate parametrization is obtained by setting

$$\bar{g}_{\mu m} = B^n_\mu \, \bar{g}_{nm} \quad .$$

Because $\| \bar{g}_{mn} \|$ is invertible, this is no restriction.

We also define

$$\bar{g}_{\mu\nu} = g_{\mu\nu} - \bar{g}_{mn} B^m_\mu B^n_\nu \quad .$$

A computation of the Riemannian curvature furnishes

$$F^m_{\mu\nu} = B^m_{\mu;\nu} - B^m_{\nu;\mu}$$

$$B^m_{\mu;\nu} = B^m_{\mu;\nu} - B^m_\nu B^m_{\mu,b} \quad .$$

To turn $F^m_{\mu\nu}$ into a Yang-Mills field we need some additional input:

<u>We need a "gauge group" G, this is provided by vector fields along the internal manifold which close on the Lie algebra of the required Lie group.</u>

We set

$$Y_\alpha = Y^m_\alpha \frac{\partial}{\partial y^m}$$

with

$$|Y_\alpha, Y_\beta| = - C^\gamma_{\alpha\beta} Y_\gamma$$

and require

$$B^m_\mu = - Y^m_\alpha A^\alpha_\mu$$

where

$$\frac{\partial A_\mu}{\partial y^m} = 0 \quad ; \quad \frac{\partial Y^m_\alpha}{\partial x^\mu} = 0 \quad .$$

Now we get

$$F^m_{\mu\nu} = - Y^m_\alpha (A^\alpha_{\mu,\nu} - A^\alpha_{\nu,\mu} - C^\alpha_{\beta\gamma} A^\beta_\mu A^\gamma_\nu) \quad .$$

We require in addition that $g_{\mu\nu}$ is function only of $(x^\mu)$ to get, eventually

194

$$\left\| \begin{array}{c|c} g_{\mu\nu}(x) + A_\mu^\alpha(x)A_\nu^\beta(x)Y_\alpha^m(y)Y_\beta^n(y)\phi_{mn}(x,\,y) & \\ \hline & \\ -\phi_{mn}(x,\,y)Y_\alpha^m(y)A_\mu^\alpha(x) & \phi_{mn}(x,\,Y) \end{array} \right\|$$

where $\phi_{mn} = \bar{g}_{mn}$. At this point we can make two different requirements,
Vector fields $Y_\alpha$ are:

1. Killing vectors for the total metric;

2. Killing vectors for the metric restricted to the internal space.

Up to now the internal space is only a homogeneous space, i.e. the Lie group $G$ acts transitively. In principle it is diffeomorphic to $G/H$, where $H$ is some closed subgroup of $G$.

Early analysis of this theory by De Witt, R. Kerner, A. Trautmann, Cho-Freund, Orzalesi and many others[3] assumed $H = 1$, and $Y_\alpha$ to be killing vectors for the total metric. Recently E. Witten argued that a $G/H$ space (with $H$ a maximal proper subgroup) is a more realistic requirement[4].

Rougly speaking when we have a trivial bundle

$$E = M \times G$$

the group acting on $E$ is actually $G \times G$ (left action and right action), when we consider $G/H$, with $H$ acting on the left, the right action of $G$ passes to the quotient while only the subgroup (Normalizer of $H$ in $G$)/$H$ acts on the left in the quotient $G/H$.
These aspects will have important consequences when we go to analyze the geodetical motions on $E$.
Thus our starting point will be:
A fiber bundle $E \rightarrow M$ with structure group $G$, and a metric tensor

$$\bar{g} = g_{\mu\nu}(x)dx^\mu \otimes dx^\nu + \phi_{mn}(x,\,y)\Sigma^m \otimes \Sigma^n$$

where

$$\Sigma^m = dy^m - Y_\alpha^m(y)A_\mu^\alpha(x)dx^\mu$$

$$g_{\mu\nu}dx^{\mu} \otimes dx^{\nu}$$

defines a metric on  M

$$\tilde{g} = \phi_{mn}(x, y)dy^{m} \otimes dy^{n}$$

defines a metric along the fibers.

<u>Assumption:</u>

$$L_{Y_{\alpha}} \tilde{g} = 0$$

i.e.  $Y_{\alpha}$  are killing vectors for the metric when restricted to the fibers.

<u>Geodetical motions and Lagrangian function</u>

The motion takes place on  TE,  the tangent bundle of the total space  E;  $TE \xrightarrow{\pi_E} E$.

The Lagrangian function is taken to be  $\mathcal{L}\bar{g}$:

$$2 \mathcal{L}\bar{g} = g_{\mu\nu}\dot{x}^{\mu}\dot{x}^{\nu} + \phi_{mn}\dot{\Sigma}^{m}\dot{\Sigma}^{n}$$

where

$$\dot{\Sigma}^{m} = \dot{Y}^{m} - Y_{\alpha}^{m} A_{\mu}^{\alpha} \dot{x}^{\mu} \quad .$$

For computational purposes it is better to use some coordinate free notation.

A second order vector field on  $TE$[5]  will be denoted by  $\Delta$:

$$\Delta = \dot{x}^{A} \frac{\partial}{\partial x^{A}} + \Delta^{A} \frac{\partial}{\partial \dot{x}^{A}}$$

in canonical coordinates for  TE.

The Lagrangian can be written in the form

$$2 \mathcal{L}\bar{g} = g(\Delta, \Delta) + \tilde{g}(\Delta, \Delta)$$

where some abuse of notation is being used, we should consider  $\pi_E^{*}g$  rather than  g.

Euler-Lagrange equations can be written in the intrinsic form

$$\left(\frac{d}{dt} \frac{\partial \mathcal{L}}{\partial \dot{x}} - \frac{\partial \mathcal{L}}{\partial x}\right)dx^{A} + \frac{d}{dt} \left(\frac{\partial \mathcal{L}}{\partial \dot{x}} dx^{A}\right) - d\mathcal{L} \equiv L_{\Delta}\theta_{\mathcal{L}} - d\mathcal{L}$$

196

where

$$\theta_{\mathcal{L}} = \frac{\partial \mathcal{L}}{\partial \dot{x}} \, dx^A$$

and $\frac{d}{dt}$ is being replaced by the Lie derivative $L_\Delta$.

It is more <u>convenient to "project"</u> $L_\Delta \theta_{\mathcal{L}} - d\mathcal{L} = 0$ <u>along non holonomic directions</u> if these are globally defined.

If $Y_\alpha = Y_\alpha \frac{\partial}{\partial x}$ on $E$ we define

$$\dot{Y}_\alpha = Y_\alpha^A \frac{\partial}{\partial x^A} + \left(\frac{d}{dt} Y_\alpha^A\right) \frac{\partial}{\partial \dot{x}^A} \qquad \text{on} \quad TE \quad .$$

Simple results are the following:

1.  If the set of vector fields $\{Y_\alpha\}$ close on some Lie algebra $g_G$, the set of vector fields $\{Y_\alpha\}$ close on the same Lie algebras;

2.  $|\dot{Y}_\alpha, \Delta|$ contains only vertical terms, i.e. $< [\Delta, \dot{Y}_\alpha] | \pi^* df > = 0$.

From

$$i_{\dot{Y}_\alpha} L_\Delta \theta_{\mathcal{L}} = i_{\dot{Y}_\alpha} d\mathcal{L}$$

we get

$$L_\Delta i_{\dot{Y}_\alpha} \theta_{\mathcal{L}} = L_{\dot{Y}_\alpha} \mathcal{L} \quad .$$

Setting

$$P_\alpha = Y_\alpha^A \frac{\partial \mathcal{L}}{\partial \dot{x}^A}$$

we can replace $\mathcal{E} \cdot \mathcal{L} \cdot$ equations with

$$\frac{d}{dt} P_\alpha = L_{\dot{Y}_\alpha} \mathcal{L}$$

as long as vector field $Y_\alpha$ are enough to generate the tangent space at each point.

The main advantage is that we get equations expressed in terms of global quantities if directions are globally defined.

When the Lagrangian is associated with a metric tensor,

$$2 \mathcal{L}_{\bar{g}} = \bar{g}(\Delta, \Delta)$$

we get

$$\frac{d}{dt} P_\alpha = \frac{1}{2} \left[ (L_{\dot{Y}_\alpha} \bar{g})(\Delta, \Delta) + \bar{g}([\dot{Y}_\alpha, \Delta], \Delta) + \bar{g}(\Delta, [\dot{Y}_\alpha, \Delta]) \right]$$

but $|\dot{Y}_\alpha, \Delta|$ is vertical, so

$$\frac{d}{dt} P_\alpha = \frac{1}{2} (L_{\dot{Y}_\alpha} \bar{g})(\Delta, \Delta) \ .$$

Going back to our situation we get:

1. If $Y_\alpha$ are killing vectors for the full metric, associated "momenta" $P_\alpha$ are constants of the motion.

2. If $Y_\alpha$ are killing vectors for the metric "along" the internal space the situation is different.

We consider

$$\mathcal{L}\bar{g} = \frac{1}{2} (g(\Delta, \Delta) + \phi_{mn}\dot{\Sigma}^m\dot{\Sigma}^n)$$

defining

$$\Delta'' \equiv \Delta - \dot{Y}_\alpha A^\alpha(\Delta)$$

we get

$$\mathcal{L}\bar{g} = \frac{1}{2} (g(\Delta, \Delta) + \tilde{g}(\Delta', \Delta')) \quad .$$

Then

$$L_{\dot{Y}_\alpha} \mathcal{L}\bar{g} = \frac{1}{2} (\tilde{g}([\dot{Y}_\alpha, \Delta'], \Delta') + \tilde{g}(\Delta', [\dot{Y}_\alpha, \Delta']))$$

$$= \tilde{g}(-[Y_\alpha, Y_\beta]A^\beta(\Delta), \Delta') \quad .$$

From

$$P_\alpha = \phi_{mn}Y_\alpha^m\dot{\Sigma}^n = \tilde{g}(Y_\alpha, \Delta')$$

we get

$$\dot{P}_\alpha = - C^\gamma_{\alpha\beta}A^\beta(\Delta)P_\gamma = - C^\gamma_{\alpha\beta}A^\beta_\mu P_\gamma\dot{x}^\mu \quad .$$

This equation suggests the identification of $P_\alpha$ with the isospin variable $I_\alpha$ up to a dimensional factor.

<u>Remark:</u> In dimension $4+1$ the momentum along the fifth dimension is identified with the electric charge.

We have to check now for the remaining equations, these require a little bit more algebra.

We obtain[6]

$$\ddot{x}^\mu + \Gamma^\mu_{\nu\lambda}\dot{x}^\mu\dot{x}^\lambda = P_\alpha F^\alpha_{\nu\lambda}\dot{x}^\nu g^{\lambda\mu} - \frac{1}{2}\partial^\mu\tilde{g}^{mn}(x)\hat{P}_m\hat{P}_n \quad .$$

Here we have been obliged to introduce the momenta associated with right invariant vector fields $\hat{Y}_\alpha$, this to get the scalar field $\tilde{g}^{mn}$ independent of y. Obviously on $G/H$ these "momenta" are not globally defined. To get the "usual" Wong's equations we are forced to set

$$\partial^\mu\tilde{g}^{mn}(x) = 0 \quad .$$

Anyhow we can investigate in which cases we get a "projectable" system of differential equations. We consider some simple situations.

1. $\partial^\mu\tilde{g}^{mn}(x) = 0$ .

   Equations of motion can be projected onto the bundle $TM \otimes g^*_G$.

2. $\tilde{g}^{mn} = \phi(x)K^{mn}$ , $\qquad K^{mn} \in \mathbf{R}$

   $$\partial^\mu\tilde{g}^{mn} = (\phi^{-1}\partial^\mu\phi)\tilde{g}^{mn}$$

   and we get

   $$\ddot{x}^\mu + \Gamma^\mu_{\lambda\nu}\dot{x}^\nu\dot{x}^\lambda = P_a F^a_{\mu\lambda}\dot{x}^\nu g^{\lambda\mu} - \frac{1}{2}(\phi^{-1}\partial^\mu\phi)P^2$$

   $$P^2 = \hat{g}^{ab}P_a P_b = \tilde{g}^{mn}\hat{P}_m\hat{P}_n \quad .$$

   Equations can be projected onto $TM \otimes g^*_G$.

## Conclusions

The recent discovery of abelian Kaluza-Klein monopoles requires $\partial^\mu\phi_{mn}(x) \neq 0$[7].

If the pure Einstein theory in higher dimensions contains non-abelian monopoles, it seems reasonable that $\partial^\mu\phi_{mn}(x)$ must be different from zero. If this is the case, we have shown that in the geometrical setting of $G/H$ fibers, we do not get Wong's equations from geodetical motions. Then non-abelian monopole solutions seem to create problems already at the classical level. Recently[8] it has been argued that interpretation problems arise already at the level of abelian Kaluza-Klein monopoles.

It seems that in higher-dimensions one has to abandon the idea of using a pure Einstein theory to unify fundamental forces.


## Acknowledgement

Results presented here have been obtained in collaboration with C.C. Chiang and S.C. Lee.


## References


[1]  Th. Kaluza, Sitzber. Preuss. Akda. Wiss. (1921), 966;
     O. Klein, Z. Phys. 37 (1926) 895;
     P.G. Bergman, "An Introduction to the Theory of Relativity", Prentice
     Hall, New York (1942).

[2]  S.K. Wong, Nuovo Cimento 65A, (1970), 1635;
     S. Sternberg, Proc. Nat. Acad. Sci. 74 (1977) 5253;
     R. Giachetti, R. Ricci and E. Sorace, J. Math. Phys. 22 (1981) 1703;
     Ch. Duval and P. Horvathy, Ann. Phys. (N.Y.) 142 (1981) 10;
     A.P. Balachandran, G. Marmo, B.S. Skagerstam and A. Stern, "Gauge
     Symmetries and Fibre Bundles" Lecture Notes in Physics 188, Springer
     (1983);
     C.A. Orzalesi and M. Pauri, Nuovo Cimento B67 (1982) 193.

[3]  B. De Witt, "Dynamic Theory of Groups and Fields" Gordon & Breach,
     New York (1965);
     A. Trautman, Rep. Math. Phys. 1 (1970) 29;
     Y.M. Cho, J. Math. Phys. 16 (1976) 235;
     Y.M. Cho and P.G.O. Freund, Phys. Rev. D12 (1975) 1711;
     R. Kerner, Ann. Ist. H. Poincaré 9A (1968) 143;
     C. Destri, C.A. Orzalesi and P. Rossi, Ann. Phys. (N.Y.) 147 (1983) 321.

[4]  E. Witten, Nucl. Phys. B186 (1981) 412;
     A. Salam and J. Strathdee, Ann. Phys. (N.Y.) 141 (1982) 316;
     S. Weinberg, Phys. Lett. B125 (1983) 265;
     R. Percacci and S. Randjbar-Daemi, J. Math. Phys. 24 (1983) 807.

[5]  R. Abraham and J. Marsden, "Foundations of Mechanics", Benjamin,
     Reading, Mass (1978);
     A. Trautman, "Differential Geometry for Physicists" Bibliopolis,
     Naples, (1985).

[6]  C.C. Chiang, S.C. Lee and G. Marmo, "Lagrangian Dynamics on Higher
     Dimensional Spaces with applications to Kaluza-Klein Theories",
     I.P. AS TP-31-84 (to appear in J. Math. Phys. 1985).

[7]  R. Sorkin, Phys. Rev. Lett. 51 (1983) 87;
     D.J. Gross and M.J. Perry, Nucl. Phys. B226 (1983) 29;
     S.M. Barr, Phys. Lett. 129B (1983) 303;

H.M. Lee and S.C. Lee, Phys. Lett. $\underline{149B}$ (1984) 95;
S.C. Lee, Phys. Lett. $\underline{149B}$ (1984) 98, 100.

[8]  J. Gegenberg and G. Kunstatter, "The Motion of Charged Particles in
Kaluza-Klein Space-Time", Preprint University of Toronto - April 1984.

Received March 16, 1985.

GEOMETRODYNAMICS PROCEEDINGS (1985), pp. 201-213
edited by A. Pràstaro
© 1985 by World Scientific Publishing Co.

# HARMONICALLY IMMERSED LORENTZ SURFACES

Tilla Milnor

*Mathematics Department, Rutgers University*
*New Brunswick, NJ 08903, USA*

I'll be talking today about the hyperbolic analog of an elliptic problem which has received a great deal of attention from mathematicians. (See [2].) At the heart of the discussion is an example well known to physicists (see [3], [13] or [14]) which I first heard about at a lecture given by the Chinese mathematician C.H. Gu. (See [4].) Involved is a harmonic map

$$M^2 \to \mathcal{S}^2$$

from the Minkowski 2-plane $M^2$ to the standard, round unit 2-sphere $\mathcal{S}^2$. By choosing coordinates $u,v$ appropriately on $M^2$, the induced metric I obtained by using the map to pull back to $M^2$ the metric of the sphere takes on the form

$$I = du^2 + 2 \cos\omega du dv + dv^2,$$

while the Minkowski metric $dx^2 - dy^2$ on $M^2$ takes the form

$$dx^2 - dt^2 = \sigma \, du dv$$

for some function $\sigma = f(u)g(v)$. Here $\omega$ is the angle between the $u,v$ coordinate curves as measured by I, with $0 < \omega < \pi$. Moreover, since the unit sphere has constant curvature 1, the intrinsic curvature $K(I)$ of the induced metric I is 1. The exact expression of this fact is the sine-Gordon equation

$$\omega_{uv} = -\sin\omega.$$

Anyone acquainted with classical surface theory is bound to be intrigued by the situation just described. For, at the turn of the century, Hilbert [5] studied surfaces of constant negative Gauss curvature sitting in ordinary Euclidean 3-space $E^3$. He showed that there are local coordinates $u,v$ anywhere on such a surface $S$ in terms of which

$$I = du^2 + 2\cos\omega\, du\, dv + dv^2$$
$$II = 2\sin\omega\, du\, dv$$

where $I$ is the metric induced on $S$ by the surrounding space, and $II$ is the second fundamental form (which gives the deviation of $S$ from its own tangent plane at any point). Once again, $\omega$ measures the angle between coordinate curves, with $0 < \omega < \pi$. Moreover, the fact that Gauss and intrinsic curvature $K(I)$ both equal $-1$ is expressed by the sine-Gordon equation

$$\omega_{uv} = \sin\omega.$$

Hilbert used these observations to show (modulo a subtle error later corrected by Holmgren [6]) that <u>there is no complete surface in $E^3$ with constant negative Gauss curvature</u>. Put another way, Hilbert's theorem states that the hyperbolic 2-plane cannot be isometrically immersed in $E^3$. (See Appendix 1 of [7] for an updated exposition of the proof.)

The striking resemblance of the example cited by Gu to Hilbert's local description of surfaces in $E^3$ with $K(I) \equiv -1$ raises obvious questions. How does one account for the similarity? Do harmonic maps from general Lorentz surfaces into arbitrary pseudo-Riemannian manifolds have the same local behavior?  If so, is there an analog of Hilbert's theorem for Lorentz surfaces harmonically immersed in arbitrary pseudo-Riemannian manifolds?

In the course of this talk I will answer these questions. Our results indicate that for all their dissimilarities, harmonically immersed Lorentz surfaces have much in common with harmonically immersed surfaces with a Riemannian prescribed metric.  In particular, we'll see that except in degenerate situations, harmonically immersed Lorentz surfaces possess a natural Riemann surface structure, and on this Riemann surface a certain quadratic differential defined by the harmonic map is forced to be holomorphic.  Thus the equation characterizing the harmonic immersion of a Lorentz surface (which is basically the wave equation) generally implies the Cauchy Riemann equations.

In order to proceed, we need some definitions.  Let  S denote an oriented 2-dimensional manifold, that is, an oriented surface.  On  S,  prescribe a metric

$$h = h_{ij} dx^i dx^j \qquad i,j = 1,2$$

which is pseudo-Riemannian, so that  $\det h \neq 0$.  At many points below we take  $\det h < 0$,  making  $(S,h)$  into a Lorentz surface. The symbol $\mathfrak{M}$ denotes an n-dimensional manifold on which we prescribe a pseudo-Riemannian metric $\mathfrak{g}$.  Since our concern is with mappings

$$f: S \to \mathfrak{M}$$

which are immersions, there is no loss in taking $n \geq 2$.  We assume  $C^\infty$  smoothness throughout.  Any map  $f: S \to (\mathfrak{M}, \mathfrak{g})$

induces on  S  the metric

$$I = f^* \mathcal{G} = g_{ij} dx^i dx^j$$

which is pseudo-Riemannian only if  det I ≠ 0.  Even if  f  is an immersion, there is no guarantee that  det I ≠ 0.  However, if  det I ≠ 0,  then  f  must be an immersion, which means that the Jacobian map  $f_*$  of  f  has rank 2  at every point, making  f  locally an imbedding of  S  in $\mathcal{M}$.

The map f:(S,h) → $(\mathcal{M}, \mathcal{G})$  is <u>harmonic</u> if and only if it is extremal for the energy (or action) integral

$$E = \int_S e \, dA_h$$

where the function

$$e = \frac{1}{2} tr_h I = \frac{h_{11}g_{22} + h_{22}g_{11} - 2h_{12}g_{12}}{2 \det h}$$

is the energy function of  f,  and where  $dA_h$  is the element of area for the prescribed metric h  on  S.  Note that the value of  E  depends only upon the conformal class of  h,  which means that  E  is the same for prescribed metric  λh  as it is for prescribed metric h,  given any function  λ ≠ 0  on  S.  It follows that  f:(S,λh) → $(\mathcal{M}, \mathcal{G})$  is harmonic if and only if f:(S,h) → $(\mathcal{M}, \mathcal{G})$  is harmonic.  (This statement is false for maps from a higher dimensional manifold.)

In practical terms, one says that a map  f:(S,h) → $(\mathcal{M}, \mathcal{G})$  is harmonic if and only if  f  satisfies the Euler-Lagrange equation for  E,  which states that the tension vector field

$$\tau = tr_h(\nabla f_*)$$

must vanish identically. Here $\nabla f_*$ is the vector valued 2-form
on S which harmonic map people call the second fundamental
form of f. One defines $\nabla f_*$ by giving its value

$$\nabla f_*(X,Y) = {}_{g}\nabla_{f_*X} f_*Y - f_*({}_{h}\nabla_X Y)$$

for any pair X,Y of $C^\infty$ vector fields on S, with ${}_{g}\nabla$ and
${}_{h}\nabla$ denoting covariant differentiation associated with $g$ and
h respectively, and with $f_*$ denoting the Jacobian of f
which carries vectors tangent to S to vectors tangent to the
image of S in $\mathcal{M}$.

We note in passing that most differential geometers refer
to the form B defined by

$$B(X,Y) = {}_{g}\nabla_{f_*X} f_*y - f_*({}_{I}\nabla_X Y)$$

as the second fundamental form of f. Of course, B is only
defined when $\det I \neq 0$ so that covariant differentiation ${}_{I}\nabla$
makes sense. Then

$$\nabla f_*(X,Y) = f_*({}_{I}\nabla_X Y - {}_{h}\nabla_X Y) + B(X,Y)$$

with B(X,Y) the normal component of $\nabla f_*(X,Y)$. In terms of
local coordinates in S and $\mathcal{M}$,

$$\tau = h^{ij}({}_{I}\Gamma^k_{ij} - {}_{h}\Gamma^k_{ij})\frac{\partial f}{\partial x^k} + tr_h B \qquad i,j,k = 1,2$$

where ${}_{I}\Gamma^k_{ij}$ and ${}_{h}\Gamma^k_{ij}$ are the Christoffel symbols for I and
h respectively. The normal component of $\tau$ is $tr_h B$. It is
convenient to think of $tr_h B$ as the h-mean curvature vector

field of  f,  since  $\mathrm{tr}_I B$  is the ordinary mean curvature vector field of  f.

In terms of local coordinates indexed by  i,j,k  on  S  and by  $\alpha,\beta,\gamma$  on  $\mathcal{M}$,  the condition  $\tau \equiv 0$  becomes

$$_h\Delta f^\alpha + g\Gamma^\alpha_{\beta\gamma} \frac{\partial f^\beta}{\partial x^i} \frac{\partial f^\gamma}{\partial x^j} h^{ij} = \tau^\alpha = 0 \qquad \alpha = 1,2,\cdots,n$$

where

$$_h\Delta = h^{ij}\{\frac{\partial^2}{\partial x^i \partial x^k} - {}_h\Gamma^k_{ij} \frac{\partial}{\partial x^k}\}$$

is the Laplace Beltrami operator for  h.  If the symbols  $g\Gamma^\alpha_{\beta\gamma}$  all vanish (as they do when  $\mathcal{M}$  is Euclidean or Minkowski's n-space), then by using local coordinates  x,y  on  S  in terms of which  $h = \lambda(dx^2 \pm dy^2)$  for some function  $\lambda$,  the condition  $\tau \equiv 0$  reduces to Laplace's equation for each  $f^\alpha$  when  det h>0 and to the wave equation for each  $f^\alpha$  when  det h<0.  If the  $\Gamma^\alpha_{\beta\gamma}$  do not all vanish, one can achieve the same effect <u>at any one point  p</u>  by using geodesic normal coordinates in  $\mathcal{M}$  centered at  f(p).  It is in this sense that one thinks of the equation  $\tau \equiv 0$  characterizing the harmonic immersion of a Lorentz surface as being basically the wave equation.

One more bit of notation is needed before we can begin to answer the questions raised at the beginning of this talk. Given the map  $f:(S,h) \to (\mathcal{M},g)$,  let

$$K = \det I/\det h \qquad H = \tfrac{1}{2}\mathrm{tr}_h I$$

$$H' = \sqrt{H^2-K}$$

where we agree to take  H'>0  when  $H^2>K$  and  iH'<0  when  $H^2<K$. Note that  H  is just another name for the energy function  e

of f. Our first theorem shows that the behavior of the particular harmonic map cited by Gu is quite typical for harmonically immersed Lorentz surfaces. Proofs of the various results stated below can be found in [11] or [12].

<u>Theorem 1</u>. If f is harmonic with det h<0, det I $\neq$ 0 and H$^{\cdot}$ $\neq$ 0 then local coordinates u,v are available anywhere on S in terms of which h = $\sigma$dudv for some function $\sigma$ while

$$I = \pm du^2 + 2 \left\{ \begin{array}{c} \cos\omega \\ \cosh\omega \\ \sinh\omega \end{array} \right\} dudv \pm dv^2,$$

corresponding to the cases

$$\left\{ \begin{array}{l} \det\ I>0 \\ \det\ I<0,\ H^2>K \\ \det\ I<0,\ H^2<K \end{array} \right\}.$$

In all three cases, $\omega$ can be interpreted as the angle between the u,v coordinate curves as measured by I, with $0<\omega<\pi$ in the first case, and $0<\omega$ in the second case. If $K(I) \equiv$ constant $\neq$ 0, then in the cases indicated, $\omega$ satisfies

$$\left\{ \begin{array}{l} \omega_{uv} = c\ \sin\omega \\ \omega_{uv} = c\ \sinh\omega \\ \omega_{uv} = c\ \cosh\omega \end{array} \right\}$$

for some constant $c\neq 0$.

In the first case distinguished above, I has no null directions. In the second case, the two null directions of I are not separated by the null directions of h. In the third case, the two null directions of I are separated by the null directions of h. The degenerate case H$'$ = 0 occurs when at least one null direction of  I  coincides with at least one null direction of  h  (as when  I∝h,  so that  f  is conformal).

The appearance of the cosh-Gordon equation in Theorem 1 makes me wonder whether the equation is of importance elsewhere.  Nobody I have spoken with to date knows of any physical applications of the cosh-Gordon equation.  Moreover, every geometric occurrence of the equation which I know of can be explained (ultimately) by Theorem 1.

The following analog of Hilbert's theorem is obtained by combining Theorem 1 with a result due to Wissler.  (See [15].)

<u>Theorem 2</u>.  If  f  is harmonic with  det h<0  and  I  a complete Riemannian metric, then  K(I)  cannot be bounded away from zero.

Theorem 2 applies whenever a Lorentz surface is harmonically immersed as a complete, spacelike surface in some pseudo-Riemannian manifold.  By Fact 1 below, Theorem 2 yields Hilbert's theorem as a special case.  Note that in Facts 1 and 2,  I  and  II  are the classically defined first and second fundamental forms of an immersed surface.  Because of Fact 1 and Theorem 1, solutions of the sine-Gordon, sinh-Gordon and cosh-Gordon equations can be used to generate (locally) surfaces in Euclidean or Minkowski 3-space with  det II<0  and  K(I) constant.  (See [1].)

<u>Fact 1</u>.  A surface  S  immersed in Euclidean or Minkowski 3-space with  det II ≠ 0  has constant Gauss curvature if and only if the identity map  (S,II) → (S,I)  is harmonic.

Fact 1 is best appreciated by comparison with the following statement.

<u>Fact 2.</u>  A surface  S  immersed in Euclidean or Minkowski 3-space with  det I $\neq$ 0  has constant mean curvature if and only if the identity map  (S,I) $\rightarrow$ (S,II)  is harmonic.

When  H'det I $\neq$ 0  on a harmonically immersed Lorentz surface  (S,h)  there is a Riemann surface  R  defined on  S  by the immersion.  The easy way to obtain  R  is by using the local coordinates  u,v  provided by Theorem 1 to generate complex parameters  w = u+iv  which (one easily checks) determine a unique complex analytic structure  R  on  S.  Slightly more complicated definitions of this same structure  R  give more insight into properties of the immersion, as we shall see below.

In order to proceed, consider the metrics  h' I'  and  W defined by  f  on  S  as follows.  Wherever  H' $\neq$ 0,  set

$$H'h' = I-Hh, \quad H'I' = HI-Kh$$

and wherever  det I $\neq$ 0  let

$$\sqrt{|det I|} \; W = \begin{vmatrix} dy^2 & -dxdy & dx^2 \\ h_{12} & h_{12} & h_{22} \\ g_{11} & g_{12} & g_{22} \end{vmatrix}$$

for arbitrary local coordinates  x,y  on  S.  If  $H^2 > K$,  W  is indefinite, while  h'  (respectively  I')  is indefinite when  h (respectively  I)  is definite, and definite when  h  (respectively  I)  is indefinite.  If  $H^2 < K$,  all the forms  h, h', I and  I'  are indefinite, but  W  (known to classical geometers as the lines-of-curvature form) is definite.

Of course, any definite metric $g$ on $S$ determines the structure of a Riemann surface $R_g$ on $S$. When the prescribed metric $h$ on $S$ is definite, the natural Riemann surface to consider on $S$ is $R_h$. If the prescribed metric $h$ on $S$ is indefinite and $H^2 > K$ for $f$, the natural Riemann surface to consider on $S$ is $R_h'$. If the prescribed metric $h$ on $S$ is indefinite and $H^2 < K$ for $f$, the natural Riemann surface to consider on $S$ is $R_W$. The Riemann surface $R$ defined previously on any harmonically immersed Lorentz surface for which $H'$ det $I \neq 0$ coincides with $R_h'$ when $H^2 > K$ and with $R_W$ when $H^2 < K$.

In order to extract the Cauchy Riemann equations from the condition $\tau \equiv 0$, we need one more bit of notation. Any real quadratic form $\Lambda$ on $S$ can be expressed in terms of coordinates $x,y$ which are isothermal for a definite metric $g$, yielding functions $A$, $B$, $C$, and $\lambda$ such that

$$g = \lambda(dx^2 + dy^2)$$
$$\Lambda = A dx^2 + 2B dx dy + C dy^2.$$

The function $\phi = A - C - 2iB$ changes like the coefficient of a quadratic differential $\phi dz^2$ on $R_g$ for different choices of the complex parameter $z = x+iy$ on $R_g$. We name this quadratic differential

$$\Omega = \Omega(\Lambda, R_g) = \phi dz^2.$$

One calls $\Omega$ holomorphic provided that $\phi$ is an analytic function of $z$ for each choice of a complex parameter $z$ on $R_g$.

Breaking the tension field $\tau$ into its components $\tan \tau$

and norm $\tau$, it is the condition $\tan \tau \equiv 0$ which is equivalent to the Cauchy Riemann equations in the next three results.

<u>Theorem 3</u>. $f:(S,h) \to (\mathcal{M},\mathcal{g})$ with $\det h > 0$ and $\det I \neq 0$ is harmonic if and only if $\text{norm } \tau \equiv 0$ and $\Omega(I,R_h)$ is holomorphic.

<u>Theorem 4</u>. $f:(S,h) \to (\mathcal{M},\mathcal{g})$ with $\det h > 0$, $\det I \neq 0$ and $H^2 > K$ is harmonic if and only if $\text{norm } \tau \equiv 0$ and $\Omega(I',R_h')$ is holomorphic.

<u>Theorem 5</u>. $f:(S,h) \to (\mathcal{M},\mathcal{g})$ with $\det h < 0$, $\det I \neq 0$ and $H^2 < H$ is harmonic if and only if $\text{norm } \tau \equiv 0$ and $\Omega(H'h',R_W)$ is holomorphic.

These results have the following consequence.

<u>Corollary</u>. If $f:(S,h) \to (\mathcal{M},\mathcal{g})$ is harmonic with $H' \det I \neq 0$ then the metrics $H'h$, $H'h'$ and $W$ are all flat on $S$. (Moreover, one of these flat metrics is definite.)

Having noted numerous echoes of classical surface theory in the study of harmonically immersed surfaces, we include a final observation which underlies this phenomenon. For surfaces immersed in Euclidean or Minkowski 3-space, the second fundamental form $II$ must satisfy the Codazzi-Mainardi equations with respect to the first fundamental form $I$, a property we denote by $Cod(I,II)$. Using this notation, a new characterization of the condition $\tan \tau \equiv 0$ is found in the next two results.

<u>Theorem A</u>. $f:(S,h) \to (\mathcal{M},\mathcal{g})$ with $H \cdot K \neq 0$ is harmonic if and only if $\text{norm } \tau \equiv 0$ and $Cod(\Gamma,I)$ where $\Gamma = Hh$.

<u>Theorem B</u>.  $f:(S,h) \to (\mathcal{M}, \mathcal{G})$  with  $HK \neq 0$  is harmonic if and only if  norm $\tau \equiv 0$  and  $Cod(I,II)$  where  $II = |K|^{1/2} h$.

For uses of the metric  $\Gamma$  defined in Theorem A, see [8], [9] and [10].  For uses of the metric $II$  defined in Theorem B, see [11].

References


[1]  S.S. Chern, "Geometrical interpretation of the sinh-Gordon equation," <u>Ann Polon Math</u> 39 (1980), 74-80.

[2]  J. Eells and L. Lemaire, "A report on harmonic maps," <u>Bull. London Math. Soc.</u> 10 (1978), 1-68.

[3]  M. Forger, "Instantons in nonlinear $\sigma$-models, gauge theories and general relativity," <u>Lecture Notes in Physics</u> 139 (1981).

[4]  C.H. Gu, "On the Cauchy problem for harmonic maps defined on two-dimensional Minkowski space," <u>Comm. Pure Appl. Math</u> 33 (1980), 727-738.

[5]  D. Hilbert, "Ueber Flächen von constanter Gausschér Krümmung," <u>Trans. AMS</u> 2 (1901), 87-99.

[6]  E. Holmgren, "Sur les surfaces a courbure constante négative," <u>C.R. Acad. Sci. Paris</u> 134 (1902), 740-743.

[7]  T.K. Milnor, "Efimov's theorem about complete immersed surfaces of negative curvature," <u>Advances In Math.</u> 8 (1972), 474-543.

[8]  _____________, "The energy-1 metric on harmonically immersed surfaces," <u>Michigan Math. J.</u> 28 (1971), 341-346.

[9]  _____________, "Classifying harmonic maps of a surface in $E^n$," Supplementary volume <u>Amer. J. Math.</u> 103 (1981) honoring P. Hartman, 211-218.

[10]  _____________, "Are harmonically immersed surfaces at all like minimally immersed surfaces?" <u>Seminar on Minimal Submanifolds</u>, E. Bombieri editor, Princeton U. Press (1983), 99-110.

[11]  _____________, "Harmonic maps and classical surface theory in Minkowski 3-space," <u>Trans. AMS</u> 280 (1983), 161-185.

[12]  _____________, "Intrinsic curvature of the induced metric on harmonically immersed surfaces," <u>Proc. AMS</u>, to appear.

[13] C.W. Misner, "Harmonic maps as models for physical
     theories," Phys. Rev. D 18 (1978), 4510-4524.

[14] K. Pohlmeyer, "Integrable Hamiltonian systems and inter-
     actions through quadratic constraints," Comm. Math. Phys.
     46 (1976), 207-221.

[15] C. Wissler, "Globale Tschebyscheff-Netze auf Riemannschen
     Mannigfaltkeiten und Fortsetzung von Flächen konstanter
     negativer Krümmung," Comm. Math. Helv. 47 (1972), 348-372.


This work was partially supported by NSF grant R11-8310315.

Received April 22, 1985.

GEOMETRODYNAMICS PROCEEDINGS (1985), pp. 215-228
edited by A. Pràstaro


# ON THE SYMMETRY PROPERTIES OF CONSTRAINED HAMILTONIAN SYSTEMS

Mihail Mintchev

*Istituto Nazionale di Fisica Nucleare, Sezione di Pisa, Pisa, Italy.*

## 1. Formulation of the Problem

In this note we are concerned with one aspect of the problem of describing the symmetry properties of constrained Hamiltonian systems (see e.g. [1] ). We start by considering a Hamiltonian system $h = \{ \Gamma , \omega , H \}$ with finite degrees of freedom. $\Gamma$, $\omega$, and $H$ are the phase space, the symplectic form and the Hamiltonian respectively. We denote by $C^{\infty}(\Gamma)$ the space of smooth real valued functions on $\Gamma$ and assume that the Hamiltonian vector field $X_H$ is complete. Following [2], we call a submanifold $\mathcal{M} \subset \Gamma$ compatible with $f \in C^{\infty}(\Gamma)$ if $X_f$ is tangent to $\mathcal{M}$.

Suppose now that $\Gamma$ admits a nontrivial submanifold $\mathcal{M}$ which is compatible with H. Clearly, any point on $\mathcal{M}$ is not allowed by time evolution to leave $\mathcal{M}$. Therefore $\mathcal{M}$ defines a new dynamical system $h_c(\mathcal{M})$ – the constrained system generated by $\mathcal{M}$. As well known, $h_c(\mathcal{M})$ is a Hamiltonian system, whose basic structures $\Gamma_c$, $\omega_c$ and $H_c$ are determined (see e.g. [2] ) in terms of $\Gamma$, $\omega$, H and $\mathcal{M}$.

Consider furthermore a constant of the motion I of h and denote by $\mathcal{M}(I)$ the maximal submanifold of $\mathcal{M}$ which is compatible both with H and I. The following possibilities occur:

$$\mathcal{M}(I) = \mathcal{M} \quad , \qquad (1.1)$$

$$\mathcal{M}(I) \subsetneq \mathcal{M} , \quad \begin{cases} \mathcal{M}(I) \neq \phi , & (1.2a) \\[2em] \mathcal{M}(I) = \phi . & (1.2b) \end{cases}$$

In the case (1.1), which has been already investigated in the literature, I gives rise to a constant of the motion $I_c$ for all motions of $h_c(\mathcal{M})$. The study of the cases (1.2a,b), which appear quite frequently in the applications, is the main topic of this note. In particular, it is shown below that if (1.2a) takes place, then I gives rise to a constant of the motion not for all but only for certain motions of $h_c(\mathcal{M})$. These motions can be completely characterized in terms of $\mathcal{M}(I)$, whose construction is given in Sect. 2. From (1.2b) it follows that in general $h_c(\mathcal{M})$ keeps no trace of all symmetries of h. This fact is not surprising since, as well known, the system h in which $h_c(\mathcal{M})$ is "imbedded", is to a certain extent arbitrary.

Finally, it is worth mentioning that quantization in general may affect $\mathcal{M}(I)$. Indeed, as shown in Sect. 3.b, if a classical conservation law I of the type (1.1) is subject to anomalies on quantum level, for the quantum law $I_q$ it may happen that $\mathcal{M}(I_q)$ is of the type (1.2).

## 2. Construction of the Submanifold $\mathcal{M}(I) \subset \mathcal{M}$

The submanifold $\mathcal{M}$ , being compatible with H, is defined locally by fixing the values of a number of constants of the motion of h. For simplicity we will limit ourselves to considering the case when there exists one constant of the motion $I_1$ of h, which defines $\mathcal{M}$ globally, i.e.

$$\mathcal{M} = \left\{ (p_1,\ldots,p_n,q^1,\ldots,q^n) \equiv (p,q) \in \Gamma \; : \; I_1(p,q) = \nu \right\} \quad (2.1)$$

for fixed $\nu \in \mathbb{R}$. Let I be a constant of the motion of h and let

$$I_2 \equiv \left\{ I , I_1 \right\} , \qquad (2.2)$$

where $\{\cdot,\cdot\}$ is the Poisson Bracket (PB) defined by $\omega$ . Due to the Jacobi identity, $I_2$ is also a constant of the motion of h. Consider the restriction $I_{2|\mathcal{M}}$ of $I_2$ to $\mathcal{M}$. If $I_{2|\mathcal{M}} = 0$, one has that $\mathcal{M}$ is compatible with I . Then from the definition of $\mathcal{M}(I)$ it follows that $\mathcal{M}(I) = \mathcal{M}$ , i.e. one gets (1.1).

Assume now that $\mathcal{M}$ is not compatible with I, i.e.

$$I_{2|\mathcal{M}} \neq 0 \quad . \tag{2.3}$$

Consider

$$\mathcal{M}_1 \equiv \mathcal{M} \cap \{(p,q) \in \Gamma : \ I_2(p,q) = 0\} \tag{2.4}$$

and suppose that $\mathcal{M}_1$ is a submanifold of $\mathcal{M}$. If $\mathcal{M}_1 = \phi$ , then there exist no nontrivial submanifold of $\mathcal{M}$ , which is compatible both with H and I. Therefore $\mathcal{M}(I) = \mathcal{M}_1 = \phi$ and the case (1.2b) takes place. Let $\mathcal{M}_1 \neq \phi$ and let us consider the constant of the motion $I_3$ of h defined by

$$I_3 \equiv \{I , I_2\} \quad . \tag{2.5}$$

There are two possibilities:

$$I_3|_{\mathcal{M}_1} = 0 \tag{2.6}$$

and

$$I_3|_{\mathcal{M}_1} \neq 0 \quad . \tag{2.7}$$

In the case (2.6), $\mathcal{M}_1$ is compatible with I. The compatibility of $\mathcal{M}_1$ with H follows simply from the fact that $I_1$ and $I_2$ are constants of the motion of h. Thus $\mathcal{M}(I) = \mathcal{M}_1$. Comparing (2.7) with (2.3), one deduces that in the case (2.7) one has to repeat the above operation with the substitution $I_2 \longmapsto I_3$ and $\mathcal{M} \longmapsto \mathcal{M}_1$. This leads to the definition of a submanifold $\mathcal{M}_2 \subsetneq \mathcal{M}_1$ and so on. In such a way, applying the above algorithm, one obtains the chain

$$\mathcal{M} \supset \mathcal{M}_1 \supset \dots \supset \mathcal{M}(I) \quad , \tag{2.8}$$

which ends up with the submanifold $\mathcal{M}(I)$ we are looking for.

Now let us comment on the case when $h_c(\mathcal{M})$ (and a fortiori h) has infinite degrees of freedom. The underlying structures in constructing (2.8) are the same, but for the fact that now they are based on an infinite dimensional manifold $\Gamma$ (see e.g. [3]), modelled on a locally convex topological space.

Finally, we note that in general h admits constants of the motion leading to infinite chains (2.8). For such constants of the motion, the above algorithm is not powerful from the constructive point of view. Fortunately however, there exist physically interesting cases for which (2.8) is finite. The rest of this note is devoted to the analysis of some of these cases, occuring in gauge theories.

## 3. Applications
### 3a. Classical Gauge Theories

In this subsection we are dealing with a classical gauge theory with a compact gauge group G. Without loss of generality one may assume that G has no proper connected invariant subgroups. We adopt the family of local covariant gauges, which provide a clear and sipmle framework for implementing the constraints. In these gauges the dynamics is described by the nondegenerate Lagrangian density[1]

$$\mathcal{L} = \mathcal{L}_{YM} + \mathcal{L}_{GF} + \mathcal{L}_{FP} + \mathcal{L}_{M} \, . \tag{3.1}$$

Here

$$\mathcal{L}_{YM} = (-1/4)\underline{F}_{\mu\nu}\cdot\underline{F}^{\mu\nu} \equiv (-1/4)F^{a}_{\mu\nu}F^{a\,\mu\nu} \, , \tag{3.2a}$$

$$\underline{F}_{\mu\nu} = \partial_\mu \underline{A}_\nu - \partial_\nu \underline{A}_\mu + g\underline{A}_\mu \wedge \underline{A}_\nu \quad , \quad g \in \mathbb{R}$$

---

[1] Color vectors $\{V^a : a = 1,\ldots,\dim G\}$ are denoted compactly by $\underline{V}$.

$$(\underline{A}_\mu \wedge \underline{A}_\nu)^a = f^{abc} A^b_\mu A^c_\nu$$

$$\mathcal{L}_{GF} = -\underline{A}_\mu \cdot (\partial^\mu \underline{B}) + (\xi/2)\underline{B} \cdot \underline{B} \quad , \quad \xi \in \mathbb{R} \tag{3.2b}$$

$$\mathcal{L}_{FP} = -i(\partial^\mu \underline{\bar{c}}) \cdot (D_\mu \underline{c}) \quad , \quad D_\mu \underline{c} = \partial_\mu \underline{c} + g\underline{A}_\mu \wedge \underline{c} \quad , \tag{3.2c}$$

where $f^{abc}$ are the structure constants of G and $\mathcal{L}_M$, whose explicit form is not needed in what follows, involves in general several kinds $\alpha = 1,\ldots,N$ of matter fields $\varphi^{(\alpha)}_k$, $k = 1,\ldots,n_\alpha$. The fields $\underline{A}_\mu$, $\underline{B}$, $\underline{c}$ and $\underline{\bar{c}}$ transform under the adjoint representation of G, while the matter fields $\varphi^{(\alpha)}_k$ transform under an, in general reducible, $n_\alpha$ dimensional representation of G, whose Hermitian generators will be denoted by $T^{(\alpha)a}_{jk}$. The hermiticity assignment for the fields is that of Ref.4.

Being nondegenerate, $\mathcal{L}$ defines unambiguously a Hamiltonian system h. For canonical coordinates on the phase space $\Gamma$ of h one may take $\underline{A}_i$, $\underline{B}$, $\underline{c}$, $\underline{\bar{c}}$ and $\varphi^{(\alpha)}_k$. Then the canonically conjugate momenta, denoted respectively by $\underline{\pi}^i_A$, $\underline{\pi}_B$, $\underline{\pi}_c$, $\underline{\pi}_{\bar{c}}$ and $\pi^{(\alpha)}_k$ are given by:

$$\underline{\pi}^i_A = \underline{F}^{io} \quad , \quad \underline{\pi}_B = -\underline{A}^o \quad , \quad \underline{\pi}_c = i\partial_o \underline{c} \quad , \tag{3.3a}$$

$$\underline{\pi}_{\bar{c}} = -i(D_o \underline{c}) \quad , \quad \pi^{(\alpha)}_k = \frac{\partial \mathcal{L}_M}{\partial(\partial_o \varphi^{(\alpha)}_k)} \tag{3.3b}$$

Canonical equal time PB's are postulated and the Faddeev-Popov fields $\underline{c}$ and $\underline{\bar{c}}$, as well as the Fermi matter fields, are considered as generators of an infinite dimensional Grassmann algebra. The Hamiltonian is defined by

$$H = \int d^3x\, \Theta^{oo}(x^o,\underline{x}) \quad , \tag{3.4}$$

where $\Theta^{\mu\nu}(x)$ is the Hilbert energy-momentum tensor [5]

$$\Theta^{\mu\nu}(x) = -2\,\Delta(x)\,\frac{\delta S}{\delta g_{\mu\nu}(x)}\Big|_{g_{\mu\nu}(x)\,=\,g_{\mu\nu}} \quad . \tag{3.5}$$

Here $\Delta(x) = \{-\det[g_{\mu\nu}(x)]\}^{1/2}$ and $S = \int d^4x\, \Delta(x)\, \mathcal{L}(x)$. Explicitely one has

$$\Theta^{00} = \frac{1}{2}(\, \underline{\pi}_A^i \cdot \underline{\pi}_A^i + \,^*\underline{F}^{i0} \cdot \,^*\underline{F}^{i0}\,) + \underline{A}_i \cdot (\,\partial^i \underline{B}\,) +$$

$$+ \underline{\pi}_B \cdot \left[\, \partial_i \underline{\pi}_A^i + g\underline{A}_i \wedge \underline{\pi}_A^i + g\,\underline{\pi}_C \wedge \underline{c} - ig(\sum_{\alpha=1}^{N} \underline{T}_{jk}^{(\alpha)} \varphi_k^{(\alpha)}\,)\, \pi_j^{(\alpha)} \right] +$$

$$+ i(\,\partial_i \underline{\bar{c}}\,) \cdot (D^i \underline{c}) - i\,\underline{\pi}_c \cdot \underline{\pi}_{\bar{c}} - \frac{\xi}{2}\,\underline{B} \cdot \underline{B} + \Theta_M^{00} \quad, \tag{3.6}$$

where $\,^*\underline{F}_{\mu\nu} = \frac{1}{2}\varepsilon_{\mu\nu\varrho\sigma}\,\underline{F}^{\varrho\sigma}$ and $\Theta_M^{00}$ is given by (3.5) with the replacement $S \longmapsto S_M = \int d^4x\, \Delta(x)\, \mathcal{L}_M(x)$.

Now let us summarise some facts concerning the symmetry properties of the system $\mathcal{L}$. There are two types of conserved currents - dynamically and topologically conserved currents. The conservation of the former is a consequence of the equations of motion, while the latter are conserved independently of dynamics. We firstly list some dynamical currents needed for our discussion below. They are

$$\underline{J}_\mu^N = g\left[\underline{A}^\nu \wedge \underline{F}_{\nu\mu} + \underline{A}_\mu \wedge \underline{B} + \underline{j}_\mu - i\underline{\bar{c}} \wedge D_\mu \underline{c} + i(\partial_\mu \underline{\bar{c}}) \wedge \underline{c}\right] , \tag{3.7a}$$

$$J_\mu^B = \underline{B} \cdot (D_\mu \underline{c}) - (\partial_\mu \underline{B}) \cdot \underline{c} + \frac{i}{2} g(\partial_\mu \underline{\bar{c}}) \cdot (\underline{c} \wedge \underline{c}) - \partial^\nu (\underline{F}_{\nu\mu} \cdot \underline{c}) . \tag{3.7b}$$

$$J_\mu^5 = \sum_{k=1}^{n} \bar{\Psi}_k \gamma_\mu \gamma^5 \Psi_k \quad , \tag{3.7c}$$

where

$$\underline{j}_\mu = -i \sum_{\alpha=1}^{N} \underline{T}_{kj}^{(\alpha)} \varphi_j^{(\alpha)} \frac{\partial \mathcal{L}_M}{\partial(\partial^\mu \varphi_k^{(\alpha)})}$$

and we have assumed that the family of matter fields $\varphi_k^{(\alpha)}$ involves the massless Dirac fields $\Psi_k$, minimally coupled to $\underline{A}_\mu$ and transforming under a n-dimensional representation of G. The color Noether current $\underline{J}_\mu^N$ and the Becchi-Rouet-Stora (BRS) current $J_\mu^B$ are conserved as a consequence of the invariance of $\mathcal{L}$ under global (x-independent) gauge transformations and BRS transformations [6] respectively. The conservation of the axial vector current (3.7c) reflects the invariance of $\mathcal{L}$ under $\gamma^5$ transformations.

Let us comment briefly on the BRS invariance of $\mathcal{L}$, which plays the fundamental role in defining the constrained system we are interested in. Firstly we recall that the BRS transformations are in some sense the substitute of the local (x-dependent) gauge transformations.[‡2] However, differently from the latter, which form an infinite dimensional group, the BRS transformations are a one parameter family defined entirely in terms of the dynamical variables of the theory. The parameter in question is a Grassmann variable, anticommuting with $\underline{c}$, $\underline{\bar{c}}$ and the Fermi matter fields. This condition implies that the BRS transformations are nilpotent - another substantial difference in comparison with the local gauge transformations. Moreover it is the source of the graded character of the charge algebra (3.11) below.

We need in what follows also the topological currents

$$\underline{J}_\mu^F = g\partial^\nu(\underline{F}_{\nu\mu} \wedge \underline{c}) \quad . \tag{3.8a}$$

$$J_\mu^T = (g/4\pi)^2 \varepsilon_{\mu\nu\rho\sigma} \partial^\nu \left[ \underline{c} \cdot (\partial^\rho \underline{A}^\sigma) \right] \tag{3.8b}$$

which being divergences of antisymmetric tensor fields are identically conserved.

In order to introduce charges corresponding to the dynamical and topological currents considered above, we will assume that the behaviour at spatial infinity of the basic fields $\underline{A}_\mu$, $\underline{B}$, $\underline{c}$, $\underline{\bar{c}}$ and $\varphi_k^{(\alpha)}$ is such that the integrals

$$Q_Y = \int d^3x J_0^Y(x^\circ,\underline{x}) \quad , \quad \underline{Q}_Z = \int d^3x \underline{J}_0^Z(x^\circ,\underline{x}) \quad , \tag{3.9}$$

where $Y = B$, $T$, $5$ and $Z' = F$, $N$, exist. We stress that these conditions do not imply that the basic fields necessarily vanish at spatial

---

infinity. Since in nonabelian gauge theories the basic fields are not observable, any assumption of trivial boundary conditions at spatial infinity for them looks artificial and in some cases excludes physically interesting configurations. For example the Yang-Mills-Higgs system in the Prasad-Sommerfield limit admits (see e.g. [7]) stationary solutions of finite energy, whose Higgs components approach nonzero constants as $|\underline{x}| \to \infty$ .

The main dynamical input for our considerations below are the Lagrange-Euler equations obtained by varying $\mathcal{L}$ with respect to $\underline{A}_\mu$ . These equations can be given the form [4]

$$\partial^\nu \underline{F}_{\nu\mu}(x) + \underline{J}^N_\mu(x) = \left\{ Q_B , iD_\mu \underline{\bar{c}}(x) \right\}_+ . \qquad (3.10)$$

Here and in what follows, the subscript $(-)+$ means that the corresponding PB is (anti)symmetric. Eqs. (3.10) combined with the canonical PB's and the nilpotency of $Q_B$ lead to the graded algebra [8]:

$$\left\{ Q^a_N , Q^b_Z \right\}_- = gf^{abc} Q^c_Z , \qquad Z = N, F \qquad (3.11a)$$

$$\left\{ Q_B , Q^a_N \right\}_- = Q^a_F , \qquad (3.11b)$$

$$\left\{ Q_B , Q_5 \right\}_- = 0 , \qquad (3.11c)$$

$$\left\{ Q_B , Q_B \right\}_+ = \left\{ Q_B , Q^a_F \right\}_+ = \left\{ Q_B , Q_T \right\}_+ = 0 , \qquad (3.11d)$$

The constrained system $h_c(\mathcal{M})$ which has physical relevance is generated by the submanifold $\mathcal{M} = \left\{ \Gamma : Q_B = 0 \right\} \subset \Gamma$ consisting of all points of $\Gamma$ for which $Q_B = 0$. $\mathcal{M}$ is compatible with H since $Q_B$ is a constant of the motion of h. Therefore $h_c(\mathcal{M})$ is well defined and its physical meaning is transparent. Roughly speaking, the observables of $h_c(\mathcal{M})$ are given by the BRS invariant observables of h.

Now we are in the position to discuss some symmetry properties of

$h_c(\mathcal{M})$ stemming from h. Applying the algorithm developed in the previous section, one may easily construct the manifolds $\mathcal{M}(\underline{Q}_F)$, $\mathcal{M}(Q_5)$ and $\mathcal{M}(\underline{Q}_N)$. Using (3.11c,d) one gets

$$\mathcal{M}(\underline{Q}_F) = \mathcal{M}(Q_5) = \mathcal{M} \qquad (3.12a)$$

while (3.11b,d) imply

$$\mathcal{M}(\underline{Q}_N) = \mathcal{M} \cap \{\Gamma : \underline{Q}_F = 0\} \subset \mathcal{M} \ . \qquad (3.12b)$$

Note that, because of (3.11a), the chain (2.8) corresponding to $\underline{Q}_N$ interrupts after the first step. Moreover, by choosing appropriate initial data (for the details see [8]), one may show that $\mathcal{M}(\underline{Q}_N) \neq \phi$ and $\mathcal{M}(\underline{Q}_N) \neq \mathcal{M}$ .

According to the general discussion in Sect.1, $\underline{Q}_F$ and $Q_5$ give rise to constants of the motion for all physical motions, i.e. for all motions of $h_c(\mathcal{M})$. On the contrary, $\underline{Q}_N$ are constants for a physical motion if and only if $\underline{Q}_F = 0$ for this motion. This fact confirms (from the point of view of constrained Hamiltonian dynamics) the observation, made by several authors [9], that the color Noether charges cannot be defined unambiguously for any physical motion.

Finally we draw the reader's attention to the interplay between dynamical and topological conservation laws which takes place in determining the symmetry properties of $h_c(\mathcal{M})$. Indeed, from the analisys above it follows that for physical motions the topological charges $\underline{Q}_F$ control the observability of the color charges $\underline{Q}_N$ which, in turn, are dynamical.


### 3b. Quantum Gauge Theories


After the above background on classical gauge theories, we turn to the quantum case. The discussion of the quantum analogue of the system h is relatively simple because of a remarkable correspondence [10] between the basic structures on classical and quantum levels.

The quantum counterpart of $\Gamma$ is a complex linear space D, equipped with a nondegenerate sesquilinear form $\langle \cdot , \cdot \rangle$. The energy-momentum tensor $\Theta^{\mu\nu}(f)$ and the currents $J_\mu^Y(f)$ and $\underline{J}_\mu^Z(f)$ (Y = B, T, 5; Z = F, N), where f is a suitable test function, become $\langle \cdot , \cdot \rangle$-symmetric operators acting on D. All these currents beside $J_\mu^5(f)$ are assumed to be conserved. Concerning $J_\mu^5(f)$, due to the Adler-Bell-Jackiw anomaly [11], one has

$$\partial^\mu J_\mu^5(f) = (g^2/32\pi^2):\underline{F}_{\mu\nu}{}^*\underline{F}^{\mu\nu}:(f) \quad , \tag{3.13}$$

where :...: is a suitably defined normal operator product [12]. In absence of monopoles one gets

$$\partial^\mu J_\mu^5(f) = \partial^\mu C_\mu(f) \quad ,$$

$C_\mu$ being the so called Chern-Simons characteristic class

$$C_\mu = (g/4\pi)^2 \varepsilon_{\mu\nu\varsigma\sigma}:\left[\underline{A}^\nu \cdot (\partial^\varsigma \underline{A}^\sigma) + (g/3)\underline{A}^\nu \cdot (\underline{A}^\varsigma \wedge \underline{A}^\sigma)\right]:(f) \quad .$$

Following [11] one defines the current

$$\tilde{J}_\mu^5(f) = J_\mu^5(f) - C_\mu(f) \quad , \tag{3.14}$$

which according to (3.13) is conserved. It is the zero-component of $\tilde{J}_\mu^5(f)$ which on quantum level generates the $\gamma^5$ transformations. Let H, $Q_Y$, $\tilde{Q}_5$ and $\underline{Q}_Z$ be the Hamiltonian and the charges defined by $\Theta^{oo}(f)$, $J_o^Y(f)$, $\tilde{J}_o^5(f)$ and $\underline{J}_o^Z(f)$ with the usual space-time smearing [13] adopted for defining charge operators in quantum field theory. All the just mentioned objects are related in the following way. Let $D' = \{\Phi \in D : Q_B \Phi = 0\}$. Then one postulates:

(1) $\langle \cdot , \cdot \rangle$ is nonnegative on D';

(2) The nonabelian Maxwell equations hold in the mean on D' [14], i.e.

$$\langle \Phi , [\partial^\nu \underline{F}_{\nu\mu}(f) + \underline{J}^N_\mu(f)]\Psi \rangle = 0 \quad , \qquad (3.15)$$

for any $\Phi$ , $\Psi \in D'$. In this sense D' is the quantum analogue of $\mathcal{M}$.

(3) The Hamiltonian H commutes with $Q_Y$ (Y = B, T), $\widetilde{Q}_5$ and $\underline{Q}_Z$ which, in turn, satisfy (3.11a,b,d) with the replacement

$$\{ \cdot \ , \ \cdot \}_{\underline{+}} \longmapsto -i [ \cdot \ , \ \cdot ]_{\underline{+}} \ ,$$

$[ \cdot \ , \ \cdot ]_{\underline{+}}$ being the (anti)commutator. Due to the Adler-Bell-Jackiw anomaly, on quantum level (3.11c) is replaced by

$$[Q_B \ , \ \widetilde{Q}_5]_- = Q_T \qquad (3.16)$$

Now some general remarks concerning the above assumptions are in order. First of all, the quantum counterpart of (3.11d) implies that $Q_B$ is nilpotent. Secondly, $\langle \cdot , \cdot \rangle$ cannot be nonnegative on the whole D, since $\langle \cdot , \cdot \rangle$ is nondegenerate and there are $\langle \cdot , \cdot \rangle$-symmetric nontrivial nilpotent operators (e.g. $Q_B$) acting on D. Moreover although nondegenerate on D, $\langle \cdot , \cdot \rangle$ becomes degenerate on D'. This meets analogy with the classical case, where the restriction of the PB $\{ \cdot , \cdot \}$ to $\mathcal{M}$, which is well defined since $\mathcal{M}$ is compatible with H, is also degenerate [2].

Let $D'' = \{ \Phi \in D' : \langle \Phi , \Phi \rangle = 0 \}$. Then the Hilbert state space $\mathcal{H}$ of the constrained system $h_c(\mathcal{M})$ (called also the physical space of h) is given by $\overline{D'/D''}$, where the completion of the factor space D'/D'' is taken with respect to the scalar product $(\cdot , \cdot)$ induced on D'/D'' by $\langle \cdot , \cdot \rangle$ (for details see e.g. [15] ). Clearly $\mathcal{H}$ is the quantum counterpart of $\Gamma_c$.

Now we are going to construct the quantum analogues $D'(\underline{Q}_F)$, $D'(\widetilde{Q}_5)$ and $D'(\underline{Q}_N)$ of $\mathcal{M}(\underline{Q}_F)$, $\mathcal{M}(\underline{Q}_5)$ and $\mathcal{M}(\underline{Q}_N)$. Because of the commutation relations of these charges with $Q_B$ (assumption (3) above) one gets

$$D'(\underline{Q}_F) = D' \quad , \tag{3.17a}$$

$$D'(\tilde{Q}_5) = \{ \Phi \in D' : Q_T \Phi = 0 \} \subsetneq D' \quad , \tag{3.17b}$$

$$D'(\underline{Q}_N) = \{ \Phi \in D' : \underline{Q}_F \Phi = 0 \} \subsetneq D' \quad . \tag{3.17c}$$

which have to be compared with (3.12). Let us consider $\mathcal{H}'(\tilde{Q}_5) = \overline{D'(\tilde{Q}_5)/D''(\tilde{Q}_5)}$, where $D''(\tilde{Q}_5) = \{ \Phi \in D'(\tilde{Q}_5) : \langle \Phi , \Phi \rangle = 0 \}$. At this point the question arises how $\mathcal{H}'(\tilde{Q}_5)$ and $\mathcal{H}$ are related. The answer is given by the first part of Lemma 1 in Ref.15, where it is proven that there exists a subspace $\mathcal{H}(\tilde{Q}_5) \subset \mathcal{H}$ , which is isomorphic to $\mathcal{H}'(\tilde{Q}_5)$. From the cited Lemma one has that $\mathcal{H}(\tilde{Q}_5)$ is the image of the mapping $\mathcal{E} : \mathcal{H}'(\tilde{Q}_5) \longrightarrow \mathcal{H}$ , constructed as follows. Consider the mapping $E : D'(\tilde{Q}_5)/D''(\tilde{Q}_5) \longrightarrow D'/D''$ defined by $E : \underset{\sim}{\Phi} \longmapsto \Phi \longmapsto \tilde{\Phi}$ where $\underset{\sim}{\Phi}$ and $\tilde{\Phi}$ are the equivalence classes of $\Phi$ in $D'(\tilde{Q}_5)/D''(\tilde{Q}_5)$ and $D'/D''$ respectively. As easily seen $E$ is independent of the choice of $\Phi \in \underset{\sim}{\Phi}$ . Therefore $E$ is well defined. Moreover one can show that $E$ is linear and continuous. Finally, $\mathcal{E}$ is obtained by extending $E$ by continuity to the whole $\mathcal{H}'(\tilde{Q}_5)$. In the same way one may construct $\mathcal{H}(\underline{Q}_N) \subset \mathcal{H}$ .

In conclusion, it follows from the above discussion that $\tilde{Q}_5$ and $\underline{Q}_N$ give rise to well defined conserved quantum numbers not for all states in $\mathcal{H}$ , but only on certain subspaces $\mathcal{H}(\tilde{Q}_5)$ and $\mathcal{H}(\underline{Q}_N)$ of $\mathcal{H}$. Concerning $\underline{Q}_N$ this result has to be expected already on classical grounds. The new phenomenon above is the inclusion $\mathcal{H}(\tilde{Q}_5) \subsetneq \mathcal{H}$, which illustrates how anomalies may affect the symmetry properties of constrained quantum Hamiltonian systems.

Acknowledgements: The author is grateful to Prof. A. Di Giacomo and to all the members of the theoretical group of the University of Pisa for the warm hospitality. Discussions with Prof. R. Ferrari, Dr. E. d'Emilio and Dr. G. Paffuti are also kindly acknowledged.

REFERENCES

[1]  P. A. M. Dirac, Canad. J. Math. 2 (1950) 125.

P. A. M. Dirac, Proc. R. Soc. Lond. A 246 (1958) 326.

L. D. Faddeev, Theor. Math. Phys. 1 (1969) 1.

A. J. Hanson, T. Regge, C. Teitelboim, Constrained Hamiltonian

Systems (Accademia Nazionale dei Lincei, Roma, 1976).

Concerning the symmetry properties of constrained Hamiltonian

systems we refer to G. Marmo, N. Mukunda, J. Samuel, La

Rivista del Nuovo Cimento 6 (1983) 2 and references therein.

[2]  N. Woodhouse, Geometric Quantization (Clarendon Press, Oxford,

1980), Chapter 2.

[3]  Y. Choquet-Bruhat, C. DeWitt-Morette, M. Dillard-Bleik, Analysis,

Manifolds and Physics (North-Holland, Amsterdam, 1982).

[4]  T. Kugo, I. Ojima, Suppl. Prog. Theor. Phys. 66 (1979) 1.

[5]  D. Hilbert, Königl. Gesel. d. Wiss. Göttingen, Nachr. Math.

Phys. Kl. (1915) 395.

[6]  C. Becchi, A. Rouet, R.Stora, Ann. Phys. 98 (1976) 287.

[7]  C. H. Taubes, Commun. Math. Phys. 86 (1982) 257, 299.

[8]  M. Mintchev, Phys. Lett. 146B (1984) 207.

[9]  B. D. Bramson, Proc. R. Soc. Lond. A 341 (1975) 463.

S. Schlieder, Nuovo Cimento A63 (1981) 137.

J. Tafel, A. Trautman, J. Math. Phys. 24 (1983) 1087.

L. F. Abbott, S. Deser, Phys. Lett. 116B (1982) 259.

K. P. Tod, Proc. R. Soc. Lond. A 389 (1983) 369.

For the case when monopoles are present see:

P. Nelson, A. Monohar, Phys. Rev. Lett. 50 (1983) 943.

A. Balachandran et al, Phys. Rev. Lett. 50 (1983) 1553.

P. Nelson, S. Coleman, Nucl. Phys. B 237 (1984) 1.

[10] E. d'Emilio, M. Mintchev, Phys. Lett. 123B (1983) 119.

[11] S. Adler, Phys. Rev. 177 (1969) 2426;

J. Bell, R. Jackiw, Nuovo Cimento 60A (1969) 43.

[12] W. Zimmerman, Brandeis Lectures 1970, (M.I.T. Press, Cambridge 1970).

[13] M. Requardt, Commun. Math. Phys. 50 (1976) 259.

[14] The possible realizations of Maxwell-type equations and the related problem of locality are discussed by E. d'Emilio, M. Mintchev, Fortschr. Phys. 32 (1984) 503, Section 1.

[15] M. Mintchev, E. d'Emilio, J. Math. Phys. 22 (1981) 1267.

Received January 18, 1985.

GEOMETRODYNAMICS PROCEEDINGS (1985), pp. 229-236
edited by A. Pràstaro


# ON A PROPERTY OF HIGHER ORDER POINCARÉ-CARTAN FORMS IN THE CONSTRUCTIVE APPROACH

Jaime Munoz Masqué
*Universidad de Salamanca*

In [2] we described a method which permits a global and
effective construction of Poincaré-Cartan forms for all variational problems of arbitrary order. Briefly, this method may be
summarized as follows: Let $p: E \longrightarrow X$ be a submersion on a manifold
$X$ oriented by a volume element $\eta$. Let us fix a linear connection
$\nabla_0$ of the base manifold $X$ and a derivation law $\nabla$ in the vertical
bundle $V(E)$ of the fibered manifold $E$. By means of the pair $\nabla, \nabla_0$
it is possible to associate to each $r$-order Lagrangian density
$L\eta$, $L \epsilon C^\infty(J^r E)$, an ordinary $n$-form $\Theta = \Theta(\nabla, \nabla_0)$ on $J^{2r-1}(E)$, $n = \dim X$,
which can be globally written as

$$\Theta = \theta^{(1)} \wedge \bar{\Omega}_1 + \ldots + \theta^{(r)} \wedge \bar{\Omega}_r - L\eta \, ,$$

for certain valued forms $\bar{\Omega}_1, \ldots, \bar{\Omega}_r$ and where $\theta^{(k)}$ is the $k$-th
structure 1-form associated to the pair of derivation laws $(\nabla, \nabla_0)$.
Furthermore, the form $\Theta$ satisfies the following condition:

(1) $\qquad d\Theta \equiv \theta^{(1)} \wedge \mathcal{E} \quad$ modulo 2-contact forms,

where $\mathcal{E}$ is a $V(E)^*$-valued $n$-form on $J^{2r-1}(E)$ explicitly given in
[2] and which will be called the Euler-Lagrange form associated
to the variational problem defined by $L\eta$ with respect to the pair
$\nabla, \nabla_0$. The above congruence may be expressed in a much more precise
form (see [2], fundamental theorem on page 143). It is known that
the Poincaré-Cartan form $\Theta$ thus constructed does not depend on the
derivation laws chosen $\nabla$ and $\nabla_0$ when $r \leqslant 2$ and arbitrary $n$ or $n=1$
and arbitrary $r$, and it coincides in these cases with the classi-
cal expressions of this form given by several authors using different methodologies (see [1] or [3] for $r=1$, [4] for $r=2$ and [5]
for $n=1$). However, this is no longer true in the general case; i.e.,
when $r>2$ and $n>1$.

The aim of this paper is to prove the following general result:

**Theorem.** *The Poincaré-Cartan form constructed by the above mentioned method does not depend on the vertical derivation law chosen. That is,*

$$\Theta(\nabla, \nabla_o) = \Theta(\nabla', \nabla_o) \ .$$

Thus, the derivation law $\nabla$ defined on the vertical bundle is merely technical. It is only necessary in order to be able to calculate the exterior differentials of the valued forms which appear in the theory. On the other hand, the use of a linear connection on the base manifold X is quite natural and, in fact, in classical Field Theory the formulation of a variational problem in most cases determines a linear connection associated to the problem, so that by choosing this connection it is possible to obtain a Poincaré-Cartan form canonically associated to the variational problem under consideration.

Before passing on to the proof of the theorem, we shall give some preliminary results (in particular the local expressions of the forms $\theta^{(k)}$ and $\bar{\Omega}_k$) which will be used in the proof of the theorem.

**a)** By using the fundamental recurrence relationship, $\theta^{(k)} = L\,\theta^{(k-1)}$, explained in [2], and the formal properties of the operator L, it follows that there exist unique functions

$$\overset{o}{a}_{\beta\alpha} \in C^{\infty}(X) \ (|\beta| \leqslant |\alpha|), \ a^{hi}_{\alpha} \in C^{\infty}(J^{|\alpha|}) \ ,$$

defined on an open coordinate domain, such that:

$$(2) \qquad L^u \theta^h_o = \frac{1}{u!} \sum_{|\beta| \leqslant u} \sum_{|\alpha| = u} \overset{o}{a}_{\beta\alpha} \theta^h_\beta \otimes (dx)^\alpha$$

$$(3) \qquad L^v(\partial/\partial y_h) = \frac{1}{v!} \sum_{i} \sum_{|\alpha| = v} a^{hi}_{\alpha} (dx)^\alpha \otimes (\partial/\partial y_i)$$

$$(4) \qquad \theta^{(k)} = \frac{1}{(k-1)!} \sum_{h,i} \sum_{|\beta| \leqslant k-1} \sum_{|\sigma| = k-1} A^{hi}_{\beta\sigma} \theta^h_\beta \otimes (dx)^\sigma \otimes (\partial/\partial y_i)$$

where

$$(5) \qquad A^{hi}_{\beta\sigma} = \sum_{\alpha \leqslant \sigma, \ |\beta| \leqslant |\alpha|} \binom{|\sigma|}{|\alpha|} \overset{o}{a}_{\beta\alpha} a^{hi}_{\sigma-\alpha} \ .$$

Here, $(dx)^{\alpha}$ stands for the symmetric product $(dx_1)^{\alpha_1}\ldots(dx_n)^{\alpha_n}$, $\alpha$ being the multi-index $\alpha=(\alpha_1,\ldots,\alpha_n)$, and $\theta^i_\alpha$ is the standard 1-contact form: $\theta^i_\alpha = dy^i_\alpha - \sum_j y^i_{\alpha+(j)} dx_j$. Note also that functions $\overset{o}{a}_{\beta\alpha}$ do not depend on the index h.

Furthermore, these functions are completely determined by the following recurrence relations:

$$(6i) \qquad \overset{o}{a}_{oo} = 1$$

$$(6ii) \qquad \overset{o}{a}_{\beta\alpha} = \sum_j \left[\, \partial \overset{o}{a}_{\beta,\alpha-(j)}/\partial x_j \;+\; \overset{o}{a}_{\beta-(j),\alpha-(j)} \right]$$
$$- \sum_{j,k,\ell} (1+\alpha_\ell-\delta_{j\ell}-\delta_{k\ell})\,\overset{o}{\Gamma}{}^\ell_{jk}\,\overset{o}{a}_{\beta,\alpha+(\ell)-(j)-(k)}$$

$$(7i) \qquad a^{hi}_o = \delta_{hi} \quad \text{and} \quad a^{hi}_{(j)} = \Gamma^i_{jh} + \sum_\nu y^\nu_{(j)} \Gamma^i_{\nu h}$$

$$(7ii) \qquad a^{hi}_\alpha = \sum_j \left[ D_j a^{hi}_{\alpha-(j)} + \sum_g a^{hg}_{\alpha-(j)} a^{gi}_{(j)} \right]$$
$$- \sum_{j,k,\ell} (1+\alpha_\ell-\delta_{j\ell}-\delta_{k\ell})\,\overset{o}{\Gamma}{}^\ell_{jk}\,a^{hi}_{\alpha+(\ell)-(j)-(k)}$$

where $\overset{o}{\Gamma}{}^\ell_{jk}$ and $\Gamma^h_{ji}$, $\overline{\Gamma}^a_{hi}$ are the local coefficients of $\overset{o}{\nabla}$ and $\nabla$, respectively, and $D_j$ stands for "the vector field" in $J^\infty$ given by

$$D_j = \partial/\partial x_j + \sum_i \sum_{|\alpha|=o}^\infty y^i_{\alpha+(j)} (\partial/\partial y^i_\alpha) \,.$$

Moreover, from the above formulas we obtain:

$$(8i) \qquad \overset{o}{a}_{\beta\alpha} = \delta_{\beta\alpha}|\alpha|!/\alpha! \quad \text{for} \quad |\beta|=|\alpha|$$

$$(8ii) \qquad A^{hi}_{\beta\sigma} = \delta_{hi}\delta_{\beta\sigma}|\sigma|!/\sigma! \quad \text{for} \quad |\beta|=|\sigma|$$

Note also that $A^{hi}_{\beta\sigma}$ is a differentiable function on $J^{|\sigma|-|\beta|}$.

b) According to $[2]$, forms $\overline{\Omega}_k$ are defined by the following formulas:

$$\bar{\Omega}_k = \Omega_k + \sum_{i=1}^{r-k} (-1)^i c_{1,1}^k \, c \prod_{j=1}^{i-1} (c_{1,1}^{k+j} L) \, d\Omega_{k+i} \quad (k=1,\ldots,r) \ ,$$

where $\Omega_k$ is the contraction of the first covariant index with the first contravariant one in $\eta \otimes F_k$ , and sections $F_o, F_1, \ldots, F_r$ are obtained from the vertical covariant development of the 1-form $-d\mathcal{L}$ (see [2] for notations and further information). With the same notations as in the preceding formulas, locally we set:

$$F_k = \sum_i \sum_{|\alpha|=k} F_\alpha^i (\partial/\partial x)^\alpha \otimes dy_i \quad (k=o,1,\ldots,r) \ ,$$

where the functions $F_\alpha^i$ satisfy the following equations:

(9i) $\qquad |\beta|! F_\beta^h = - \partial\mathcal{L}/\partial y_\beta^h \qquad \text{for } |\beta|=r$

(9ii) $\qquad |\beta|! F_\beta^h + \sum_i \sum_{|\alpha|=|\beta|+1} \alpha! A_{\beta\alpha}^{hi} F_\alpha^i = - \partial\mathcal{L}/\partial y_\beta^h \quad (|\beta|=o,\ldots,r-1)$

Choosing coordinates $(x_j)$ in $X$ so that $\eta = dx_1 \wedge \ldots \wedge dx_n$ and writing

$$\eta_j = dx_1 \wedge \ldots \wedge \widehat{dx_j} \wedge \ldots \wedge dx_n \ ,$$

we obtain:

$$\Omega_k = \sum_{i,j} \sum_{|\alpha|=k-1} (-1)^{j-1} (1+\alpha_j) F_{\alpha+(j)}^i \eta_j \otimes (\partial/\partial x)^\alpha \otimes dy_i$$

Then, from the first formula of this section we also have that

$$\bar{\Omega}_k = \sum_{i,j} \sum_{|\alpha|=k-1} (-1)^{j-1} (1+\alpha_j) \bar{F}_{\alpha+(j)}^i \eta_j \otimes (\partial/\partial x)^\alpha \otimes dy_i$$

where functions $\bar{F}^i_\alpha$ fulfil the following conditions:

(10) $\bar{F}^i_\alpha = F^i_\alpha$ for $|\alpha| = r$, and for $|\alpha| = 1, \ldots, r-1$, we have,

$$\bar{F}^i_\alpha = F^i_\alpha - \sum_j (1+\alpha_j) \left[ D_j \bar{F}^i_{\alpha+(j)} - \sum_h a^{ih}_{(j)} \bar{F}^h_{\alpha+(j)} \right]$$

$$- \sum_{j,k,\ell} (1+\alpha_k - \delta_{k\ell})(1+\alpha_j + \delta_{jk} - \delta_{j\ell}) \overset{o}{\Gamma}{}^{\ell}_{jk} \bar{F}^i_{\alpha+(j)+(k)-(\ell)}$$

Equations (9) and (10) determine the functions $F^h_\beta$ and $\bar{F}^i_\alpha$ by descending recurrence on $|\beta|$ and $|\alpha|$, respectively. Finally, we also obtain the local expression of Poincaré-Cartan form:

$$\Theta = \sum_{k=1} \theta^{(k)} \wedge \bar{\Omega}_k - \mathcal{L}\eta = \sum_{h,j} \sum_{|\beta|=o}^{r-1} (-1)^{j-1} f^h_{\beta j} \theta^h_\beta \wedge \eta_j - \mathcal{L}\eta ,$$

where

(11) $\qquad f^h_{\beta j} = \sum_i \sum_{|\alpha|=|\beta|}^{r-1} (\alpha+(j))! A^{hi}_{\beta\alpha} \bar{F}^i_{\alpha+(j)} \qquad (|\beta|=o, \ldots, r-1)$ .

**Proof of the theorem.** Let $\Theta$ be the Poincaré-Cartan form constructed with the pair $(\nabla, \nabla_o)$ and let $\Theta'$ be the form constructed with the same linear connection $\nabla_o$ of the manifold X and another vertical derivation law $\nabla'$ in $V(E)$. First we shall prove some properties of functions $f^h_{\beta j}$ which are of course also true for the corresponding coefficients $f'^h_{\beta j}$ of the form $\Theta'$. We set:

$$G^h_{\alpha j} = \sum_i \sum_{\alpha \leqslant \sigma, \, |\sigma| \leqslant r-1} (\alpha+(j))! \binom{|\sigma|}{|\alpha|} a^{hi}_{\sigma-\alpha} \bar{F}^i_{\sigma+(j)} \qquad (|\alpha|=o, \ldots, r-1) .$$

Hence,

$$f^h_{\beta j} = \sum_{|\alpha|=|\beta|}^{r-1} \overset{o}{a}_{\beta\alpha} G^h_{\alpha j} \qquad (|\beta|=o, \ldots, r-1) ,$$

as follows from the definition of the functions $A^{hi}_{\beta\sigma}$ (formula (5)).

Functions $G^h_{\alpha j}$ satisfy the following property: If $\alpha+(j) = \alpha'+(j')$, then $G^h_{\alpha j} = G^h_{\alpha' j'}$. We can therefore define functions $\bar{G}^h_\alpha$ for $|\alpha| = 1, \ldots, r$ , such that,

$$G^h_{\alpha j} = \bar{G}^h_{\alpha+(j)} \qquad (|\alpha|=o, \ldots, r-1) .$$

In fact, if $\alpha + (j) = \alpha' + (j')$, we have $\alpha = \tau + (j')$, $\alpha' = \tau + (j)$, for a certain multi-index $\tau$, and then from the above formulas we obtain:

$$G^h_{\alpha j} = G^h_{\tau + (j'),j} = \sum_i \sum_{\substack{|\sigma| \leqslant r-1 \\ \tau \leqslant \sigma - (j')}} (\sigma + (j))! \binom{|\sigma|}{|\tau|+1} a^{hi}_{\sigma - (j') - \tau} \bar{F}^i_{\sigma + (j)}$$

$$(\text{by setting } \sigma' = \sigma - (j')) = \sum_i \sum_{\substack{|\sigma'| \leqslant r-2 \\ \tau \leqslant \sigma'}} (\sigma' + (j) + (j'))! \binom{|\sigma'|+1}{|\tau|+1} a^{hi}_{\sigma' - \tau} \bar{F}^i_{\sigma' + (j) + (j')}$$

$$(\text{by setting } \sigma'' = \sigma' + (j)) = \sum_i \sum_{\substack{|\sigma''| \leqslant r-1 \\ \tau \leqslant \sigma'' - (j)}} (\sigma'' + (j'))! \binom{|\sigma''|}{|\tau|+1} a^{hi}_{\sigma'' - (j) - \tau} \bar{F}^i_{\sigma'' + (j')}$$

$$= G^h_{\tau + (j),j'} = G^h_{\alpha' j'} \ .$$

On the other hand, since $d\Theta \equiv \theta^{(1)} \wedge \mathcal{E}$ modulo 2-contact forms (formula (1)), and there exists a valued $(n-1)$-form $\Phi$ such that $\mathcal{E}' = \mathcal{E} + \Phi \wedge \theta^{2r-1}$ (see [2] on page 144), we obtain

$$i_{D_1} \cdots i_{D_n} d(\Theta' - \Theta) = 0 \ ,$$

where $D_j$ is the vector field in $J^\infty$ previously defined. By writing the above equation in local coordinates, we have:

$$(12) \qquad \sum_j D_j (f'^h_{oj} - f^h_{oj}) = 0$$

$$(13) \qquad \sum_j D_j (f'^h_{\beta j} - f^h_{\beta j}) + \sum_j (f'^h_{\beta - (j),j} - f^h_{\beta - (j),j}) = 0 \qquad (|\beta| = 1, \ldots, r-1)$$

$$(14) \qquad \sum_j (f'^h_{\beta - (j),j} - f^h_{\beta - (j),j}) = 0 \qquad (|\beta| = r) \ .$$

We shall now prove by descending recurrence on $k = o, \ldots, r-1$ that,

$$(*) \qquad f^h_{\beta j} = f'^h_{\beta j} \quad \text{and} \quad G^h_{\alpha j} = G'^h_{\alpha j} \quad \text{for } |\alpha| = |\beta| = k \ .$$

For $k = r-1$ from (11), bearing in mind (8ii), (10) and (9i), it follows that

$$f^h_{\beta j} = f'^h_{\beta j} = - \frac{1 + \beta_j}{r} (\partial L / \partial y^h_{\beta + (j)}) \qquad (|\beta| = r-1) \ ,$$

and for $|\alpha|=r-1$ we have $G^h_{\alpha j} = (\alpha!/|\alpha|!)f^h_{\alpha j}$ . Hence, in this case $G^h_{\alpha j} = G'^h_{\alpha j}$ . Let us assume that conditions (*) are fulfilled for $r-1,\ r-2,\ldots,\ k>o$ . Then, equation (13) for $|\beta|=k$ reduces to the following:

$$0 = \sum_j (f'^h_{\beta-(j),j} - f^h_{\beta-(j),j}) = \sum_j \sum_{|\alpha|=k-1} \overset{o}{a}_{\beta-(j),\alpha}(\bar{G}'^h_{\alpha+(j)} - \bar{G}^h_{\alpha+(j)})$$

$$= \left(\sum_j \frac{(k-1)!}{(\beta-(j))!}\right)(\bar{G}'^h_\beta - \bar{G}^h_\beta) = (k!/\beta!)(\bar{G}'^h_\beta - \bar{G}^h_\beta) .$$

Hence, $\bar{G}'^h_\beta = \bar{G}^h_\beta$ , or in other words, $G^h_{\alpha j} = G'^h_{\alpha j}$ for $|\alpha|=k-1$ . Furthermore, for $|\beta|=k-1$ we have:

$$f'^h_{\beta j} = \sum_{|\alpha|=|\beta|}^{r-1} \overset{o}{a}_{\beta\alpha}G'^h_{\beta j} = \frac{(k-1)!}{\beta!}\ G'^h_{\beta j} + \sum_{|\alpha|=k}^{r-1} \overset{o}{a}_{\beta\alpha}G^h_{\alpha j}$$

$$= \frac{(k-1)!}{\beta!}\ G^h_{\beta j} + \sum_{|\alpha|=k}^{r-1} \overset{o}{a}_{\beta\alpha}G^h_{\alpha j} = \sum_{|\alpha|=|\beta|}^{r-1} \overset{o}{a}_{\beta\alpha}G^h_{\alpha j} = f^h_{\beta j} .$$

Thus (*) is proved for $|\alpha|=|\beta|= k-1$, and the proof is complete.


# References


[1] P.GARCIA. The Poincaré-Cartan Invariant in the Calculus of Variations, Symposia Math.,14, Academic Press (London, 1974), 219-246.

[2] P.GARCIA,J.MUÑOZ. On the Geometrical Structure of Higher Order Variational Calculus, Proc.IUTAM-ISIMM Symp. on Modern Developments in Analyt.Mechanics, Turin, June 7-11,1982, Vol.I-Geometrical Dynamics,Tecnoprint(Bologna,1983) 127-147.

[3] H.GOLDSCHMIDT,S.STERNBERG, The Hamilton-Cartan Formalismo in the Calculus of Variations, Ann.Inst.Fourier 23, 203-267 (1973).

[4] D.KRUPKA, Lepagean Forms in Higher Order Variational Theory,
     Proc.IUTAM-ISIMM Symp. on Modern Developments in Analyt.
     Mechanics, Turin, June 7-11, 1982, Vol.I-Geometrical
     Dynamics, Tecnoprint(Bologna,1983) 197-238 .

[5] S.STERNBERG, Some Preliminary Remarks on the Formal
     Variational Calculus of Gel'fand and Dikii. Lec.Not.
     in Math. 676, Springer-Verlag (Berlin,1978), 399-407 .

Received February 20, 1985.

GEOMETRODYNAMICS PROCEEDINGS (1985), pp. 237-247
edited by A. Pràstaro


# COVARIANT CANONICAL FORMALISM FOR GRAVITY THEORIES

J.E.Nelson and T.Regge

*Istituto Nazionale di Fisica Nucleare, Sezione di Torino,
Istituto di Fisica Teoriaca, Università di Torino,
10125 Torino, Italy.*

## 1. INTRODUCTION

In this talk I will present a piece of work that has been developed in the last year in Turin in collaboration with Alessandro D'Adda and Tullio Regge [1] that concerns our mutual interests, namely those of constructing a consistent quantum theory of gravity and gravitational theories i.e.theories of the coupled gravitational and matter fields. We are concerned with the canonical formalism for constrained systems, of which gravity is a superb example, as developed by Dirac [2] and others [3], because it forms a basis for quantisation and has, so far, given us greater insight into the structure of the gravitational field. However, the most common criticism of this method is that it is not covariant, that it necessarily separates space and time and destroys the manifest covariance of Einstein's theory, although it allows us to understand which are the true physical, dynamical degrees of freedom and which are only the (apparent) gauge degrees of freedom.

Therefore we have developed a canonical formalism for the exterior calculus of differential forms which is naturally covariant, and can be shown, when we choose a gauge, to correspond to the standard theory [4,5] in canonical form, where by 'standard' we mean written in terms of all field components.

In order to develop this formalism it was necessary to extend the concepts of velocity, momentum, Poisson Bracket, etc., familiar from classical mechanics, to differential forms, and we found some surprising results. The final form for the Hamiltonian is remarkably simple, and is first class. I must point out that some of the calculations are non - trivial and require heavy algebra, although it was in the details of the theory that many problems were solved. We have developed the formalism for first - order tetrad ( or vierbein ) gravity because the method of Dirac [2] works particularly well for systems which possess extra gauge invariance [5] and also because we wish to extend the theory to include supergravity theories. Here I will discuss pure gravity only in 4 space - time dimensions where the group is the Poincare' group $P$ , but we expect unified theories to be represented by extensions of the Poincare' group in higher dimensions.

I will denote the vierbein ( 1-form ) by $V^a$ , with components $L^a{}_\mu$ i.e.

$$V^a = L^a{}_\mu \, dx^\mu$$

and the spin connection 1-form by $\omega^{ab} = - \omega^{ba}$ i.e.

$$\omega^{ab} = \omega^{ab}{}_\mu \, dx^\mu$$

Lower case latin indices a,b,c,.... are used in tangent space, whereas greek indices denote space - time tensors. Both kinds of indices take values 0..3. I also use a tangent space compound index A,B,C... in the range 1....10 to label elements in the Lie algebra of $P$ , which have a vector range $A = a$ and a

where

$$V_b = \eta_{ab} V^a \qquad\qquad \omega_t{}^b = \eta_{at} \omega^{ab}$$

$$\eta_{ab} = \eta^{ab} = (-1, +1, +1, +1)$$

Note that we may add on to the 4-form Lagrangian density $L$ any exterior derivative of a 3-form functional of $V^a, \omega^{ab}$ only, without changing the variational equations. The effect for the canonical formalism would be to change the definitions of momenta and therefore the form of the Hamiltonian by a canonical transformation, and is what we choose to do here. It is analogous to the removal of second time derivatives in second-order formalism and will make the correspondence between component and covariant formalism so much more transparent. We rewrite

$$L = \tfrac{1}{2} R^{ab} \wedge V^c \wedge V^d \, \epsilon_{abcd}$$

$$= \omega^{ab} \wedge dV^c \wedge V^d \, \epsilon_{abcd}$$

$$- \tfrac{1}{2} \omega^{at} \wedge \omega_t{}^b \wedge V^c \wedge V^d \, \epsilon_{abcd}$$

$$+ d\left( \tfrac{1}{2} \omega^{ab} \wedge V^c \wedge V^d \, \epsilon_{abcd} \right)$$

and ignore the total derivative. Thus we take as our action

$$S = \int \left( \omega^{ab} \wedge dV^c - \tfrac{1}{2} \omega^{at} \wedge \omega_t{}^b \wedge V^c \right) \wedge V^d \, \epsilon_{abcd}$$

We need to specify what we mean by 'velocity' corresponding to our field variables. In component formalism [4,5] the velocity $\dot{q} = dq/dt =$ time derivative of $q$, where time $t$ is already specified (commonly $t = x^o$ ). Here, instead we define 2-form velocities as

$$dq^A = \partial q^A{}_\alpha / \partial x^\beta \; dx^B \wedge dx^\alpha$$

240

tensor range $A = ab$ i.e. the compound 1-form is denoted $q^A$ where

$$q^A = (V^a, \omega^{ab})$$

Repeated indices are to be understood as summed over their range with the convention

$$M^A L_A = M^a L_a + \tfrac{1}{2} M^{ab} L_{ab}$$

## 2. ACTION PRINCIPLE AND CONSTRAINTS

We consider the well-known Einstein-Hilbert action for pure gravity in four dimensions, written in terms of differential forms [6]

$$S = \int L = \int \tfrac{1}{2} R^{ab} \wedge V^c \wedge V^d \, \epsilon_{abcd}$$

with the curvature 2-forms

$$R^{ab} = d\omega^{ab} - \omega^{at} \wedge \omega_t{}^b$$

Variation of this action a' la Palatini (i.e. with $V^a, \omega^{ab}$ as independent variables ) yields

$$R^{ab} \wedge V^c \, \epsilon_{abcd} = 0$$

$$R^a \wedge V^b \, \epsilon_{abcd} = 0$$

$$( \Rightarrow R^a = 0 \quad \text{for non-singular } V^b )$$

with the torsion

$$R^a = d V^a - \omega^{ab} \wedge V_b$$

(with of course $d^2 x^\alpha = 0$)

and, correspondingly, define conjugate momenta as the variations of the action with respect to these velocities. We denote by $\Pi_a$, $\Pi_{ab}$ the 2-form momenta conjugate to $V^a$, $\omega^{ab}$ respectively (written compoundly as $\Pi_A$, $q^A$) and find the set of primary constraints, denoted $\phi_a$, $\phi_{ab}$

$$\phi_{ab} = \Pi_{ab} \approx 0$$

$$\phi_a = \Pi_a - \omega^{cd} \wedge V^b \, \epsilon_{abcd} \approx 0$$

(note there is no factor of $\tfrac{1}{2}$ since $d\omega^{ab} = -d\omega^{ba}$ appears twice in the action)

or, written compoundly

$$\phi_A = \Pi_A - \tfrac{1}{2} q^B \wedge q^C \, M_{CBA} \approx 0$$

and
$$M_{ABC} = 0 \quad \text{if} \quad \text{(i) } C \text{ is tensor, and/or}$$
$$\text{(ii) } B, A \text{ both tensor or both vector}$$

ie $M_{a\;bc\;d} = -2\,\epsilon_{abcd}$ , all others zero.

We write the constraints $\phi_A \approx 0$ (weakly zero) meaning that they define a submanifold of space – time, and any linear functional of the $\phi_A$ will similarly be written as $\approx 0$ .


## 3. THE HAMILTONIAN, FORM BRACKETS

The canonical Hamiltonian (4-form) is easily calculated as

$$H = dq^A \wedge \Pi_A - L + \Lambda^A \wedge \phi_A$$
$$= \tfrac{1}{2} \omega^{at} \wedge \omega_t{}^b \wedge V^c \wedge V^d \, \epsilon_{abcd} + \Lambda^a \wedge \phi_a + \tfrac{1}{2} \Lambda^{ab} \wedge \phi_{ab}$$

242

where we have added on the primary constraints $\phi_A$ with arbitrary 2-form multipliers $\Lambda^A$.

In order to look for secondary constraints we need now to define a bracket operation for differential forms, in analogy with the Poisson bracket of classical mechanics. We define

$$\left(q^A, \Pi_B\right) = \delta^A{}_B$$

or

$$\left(V^a, \Pi_b\right) = \delta^a{}_b$$

$$\left(\omega^{ab}, \Pi_{cd}\right) = \delta^{ab}{}_{cd}$$

$$= \delta^a{}_c \, \delta^b{}_d - \delta^a{}_d \, \delta^b{}_c$$

with all others zero. This 'form bracket' defines the map

$$F^a \times F^b \longrightarrow F^{a+b-3}$$

where $F^a$ is the set of a-forms. It can be extended to a generic polynomial in the $q^A, \Pi_B$ by repeated use of the combination rules:

(1) $\quad (A,B) = (-1)^{ab} \, (B,A)$

(2) $\quad (A, BC) = (A,B)C + (-1)^{b(a+1)} B(A,C)$

(3) $\quad (AB, C) = (-1)^{bc} (A,C) B + (-1)^a A(B,C)$

(4) $\quad (-1)^{c(b+1)} (C, (A,B)) + (-1)^{a(c+1)} (A, (B,C))$

$$+ (-1)^{b(a+1)} (B, (C,A)) = 0$$
$$\text{(Jacobi identity)}$$

In (1) – (4) a,b,c, denote the degrees of the forms A,B,C respectively. This form bracket turns out to be fundamental for the canonical formalism of the exterior calculus of forms. It combines forms defined at the same point of space – time, but can be related to the Poisson bracket. The first use we make of it is in the identification of exterior derivative.  We define

$$dA = (A, H) + \partial A$$

for any form A, where the operator $\partial$ acts only on external fields, such as the components $\Lambda_{BC}{}^{A}$ of $\Lambda^{A}$, which are c-numbers.  i.e. $\partial q^{A} = \partial \pi_{A} = 0$ .  Using this definition of $d$ we immediately find

$$dV^{a} = \Lambda^{a}$$
$$d\omega^{ab} = \Lambda^{ab}$$

i.e $dq^{A} = \Lambda^{A}$

and for the constraints

$$d\phi_{a} \approx \left( \omega^{ct} \wedge \omega_{t}{}^{d} - \Lambda^{cd} \right) \wedge V^{b} \, \epsilon_{abcd}$$

$$d\phi_{ab} \approx \omega_{b}{}^{t} \wedge V^{c} \wedge V^{d} \, \epsilon_{atcd} - \omega_{a}{}^{t} \wedge V^{c} \wedge V^{d} \, \epsilon_{btcd}$$
$$- 2 \Lambda^{c} \wedge V^{d} \, \epsilon_{abcd}$$

There are now two possibilities.  Either the $d\phi_{A}$ are weakly zero or give us new (secondary) constraints.  In fact we can solve directly for the $\Lambda^{A}$ as

$$\Lambda_{BC}{}^{A} = R_{BC}{}^{A} - C_{BC}{}^{A} \qquad\qquad \Lambda^{A} = -\tfrac{1}{2} q^{B} \wedge q^{C} \, \Lambda_{CB}{}^{A}$$

where $C_{BC}{}^{A}$ are the Poincaré' structure constants (see Appendix) and

$$R_{BC}{}^{A} = - R_{CB}{}^{A}$$

are the curvature components

$$R_{BC}{}^{a} = 0 \qquad\qquad R_{CD}{}^{ab} = 0 \ \text{if} \ C \text{ or } D \text{ is tensor.}$$

In that case we can write

$$d\phi_{a} \approx - R^{cd} \wedge V^{b} \, \epsilon_{abcd}$$

$$d\phi_{ab} \approx - 2 R^{c} \wedge V^{d} \, \epsilon_{abcd}$$

which are just the variational equations, i.e. there are no secondary constraints. We can now write the Hamiltonian as

$$H = \Lambda^A \wedge \phi_A - \frac{1}{4} q^D \wedge q^C \wedge q^B \wedge q^A \, C_{AB}{}^T Q_{CTD}$$

$$\text{with} \quad Q_{ABC} = M_{BAC} + M_{CAB} \qquad \Lambda^A = R^A - C^A$$

and it satisfies

$$dH = \partial H + \frac{1}{2}(H,H) = DR^A \wedge \phi_A \approx 0$$

$$\text{with} \quad DR^A = dR^A + R^B \wedge q^C C_{CB}{}^A = 0$$

i.e. the Bianchi identities for the curvatures $R^A$.


## 4. DISCUSSION

As a consequence of our definition of exterior derivative $d$ we have the following formula for $d^2$

$$d^2 A = (H, \partial A) + \partial(H, A) + (\partial H, A) - (K, A) + \partial^2 A$$

for any polynomial $A$ of $q^A$ and $\pi_B$, where

$$K = dH = 0 \qquad \partial H = -\frac{1}{2} q^D \wedge q^C \wedge q^B \Lambda_{BC,D}{}^A \wedge \phi_A$$

In particular we find that

$$d^2 q^A = DR^A = 0$$

$$d^2 \phi_A = -\frac{1}{2} q^B \wedge q^C \wedge \Lambda_{CB,A}{}^D \wedge \phi_D \approx 0$$

$$d^2 f = \frac{1}{2} q^C \wedge q^B \left( f_{,BC} - f_{,CB} - \Lambda_{BC}{}^A f_{,A} \right) = 0$$

where $f$ is an external field 0-form (e.g. $\Lambda_{BC}{}^A$). Note that the operator $d^2$ is only weakly zero when acting on the constraints $\phi_A$.

Any quantity is said to be first class if its form bracket with all the $\phi_A$ is weakly zero. Thus

$$(H, \phi_A) = d\,\phi_A$$
$$= -q^B \wedge \Lambda_{BA}{}^C \wedge \phi_C$$
$$\approx 0$$

i.e. our Hamiltonian is first class, although not all of the constraints $\phi_A$ are. This can be seen from the bracket

$$(\phi_A, \phi_B) = q^C Q_{CAB}$$

which is not weakly zero. However, the combinations $R^A \wedge \phi_A$ and $d^2 \phi_A$ are first class since

$$(R^A \wedge \phi_A, \phi_B) \approx R^A \wedge q^C Q_{CAB} = 0$$
$$(d^2 \phi_A, \phi_S) \approx -\tfrac{1}{2} q^B \wedge q^C \wedge \Lambda_{CB,A}{}^T \wedge q^D Q_{DTS}$$
$$= -\tfrac{1}{2} q^D \wedge q^B \wedge q^C \left( \Lambda_{CB}{}^T Q_{DTS} \right)_{,A} = 0$$

when we use the Einstein variational equations (see Appendix)

We have therefore a canonical framework for the exterior calculus of forms. The form bracket is fundamental; we expect it to be useful in the discussion of unified theories, like supergravity. The form bracket and the relationship between this formalism and the component formalism will be discussed elsewhere.

# APPENDIX

The Einstein variational equations can be written as

$$R^A \wedge q^C \varphi_{CAB} = 0$$

or in components

$$R_{AB}{}^T \varphi_{CTD} + R_{BC}{}^T \varphi_{ATD} + R_{CA}{}^T \varphi_{BTD} = 0$$

The alternating tensor $\epsilon_{abcd}$ satisfies

$$\epsilon_{0123} = -1 \qquad \epsilon^{0123} = +1$$

$$\epsilon_{abcd}\, \epsilon^{arst} = 6\, \delta^{[r}{}_b\, \delta^s{}_c\, \delta^{t]}{}_d$$

where $[ab]$ means $ab - ba$

The Poincaré structure constants $C_{BC}{}^A$ are

$$C_b{}_{cd}{}^a = \delta^a{}_c\, \eta_{bd} - \delta^a{}_d\, \eta_{bc}$$

$$C_{cd}{}_{ef}{}^{ab} = -\delta^a{}_c\, \delta^b{}_f\, \eta_{de} + \delta^a{}_d\, \delta^b{}_f\, \eta_{ce}$$

$$-\delta^a{}_d\, \delta^b{}_e\, \eta_{cf} + \delta^a{}_c\, \delta^b{}_e\, \eta_{df}$$

$$- \text{(exchange of } a \text{ with } b)$$

with all others zero. They satisfy the Jacobi identity

$$C_{AB}{}^F C_{FE}{}^T + C_{BE}{}^F C_{FA}{}^T + C_{EA}{}^F C_{FB}{}^T = 0$$

## REFERENCES


[1] A.D'Adda, J.E.Nelson and T.Regge,submitted to Ann.  Phys.

[2] P.A.M.Dirac, Canad.  J.  Math.  2 (1950) 129.

[3] See, for  example,  A.Hanson,  T.Regge  and  C.   Teitelboim 'Constrained  Hamiltonian  Systems',  Accademia  Nazionale  dei Lincei, Roma 1976.

[4]  R.Arnowitt,  S.Deser  and  C.Misner,  in"Gravitation:   An Introduction  to  Current  Research"  (L.Witten,ed.)  Wiley,  NY (1962).

[5] J.E.Nelson and C.Teitelboim, Phys.Lett.69B (1977),81; Ann.Phys.116 (1978),1.

[6] See, for example, T.Regge "The Group  Manifold  Approach  to Unified  Gravity"  -  Les Houches Summer School, Session XV 1983 (North Holland 1984).


Received January 18, 1985.

GEOMETRODYNAMICS PROCEEDINGS (1985), pp. 249-266
edited by A. Pràstaro


# INVARIANT DIFFERENTIATION TECHNIQUES

Albert Nijenhuis

*University of Pennsylvania*


1. Introduction  In recent years theoretical physicists have started
to use some differential operations on tensors that are independent
of any connection, or Riemann metric.
Although most of these applications lie in the areas of classical
mechanics and relativity theory, the operations should be seriously
considered in the study of phenomena that are not affected by gravitational fields.

This paper does not discuss specific applications of the connection
independent differential operations to physics.  Rather, we attempt to
give a reasonably up-to-date listing of those that are known, thus providing access for those who might be less familiar with the area.

There is no way in which we could provide an actually complete
listing: although Terng[1] has established algebraic conditions for differential invariants, at present it is not possible to use these conditions effectively.

In view of this situation it is not surprising that we can present
here another class of operations that is believed to be new.

Several of the operations fit very neatly into the context of graded
Lie algebras, thus providing some element of unity.  Before discussing
those, however, we first give a quick listing, and a discussion of
invariance properties.

2. A quick listing. The oldest connection independent differential
operators on tensors are no doubt the Lie bracket $[u,v]$ of two tangent (contravariant) vector fields u, v;  and the exterior derivative
(curl) dw of a cotangent (covariant) vector field w:

$$(2.1) \qquad [u,v]^{\lambda} = u^{\mu}\partial_{\mu}v^{\lambda} - v^{\mu}\partial_{\mu}u^{\lambda} \;\; ; \;\; (dw)_{\lambda\mu} = \partial_{\lambda}w_{\mu} - \partial_{\mu}w_{\lambda} \; .$$

It should be emphasized that, in the absence of a Riemannian metric, a
strict distinction between co- and contravariant vectors must be maintained. Contravariant vectors typically denote directions of flow, velocities, etc., while the covariant vectors include gradients of scalar
functions, such as forces. The two are "dual" to each other, in the
sense that a contraction, such as $w_{\lambda}v^{\lambda}$ is a scalar. Thus, a covariant
vector is a scalar valued linear function on the contravariant vectors,
and conversely. The notation  w(v)  is an index-free expression of this
fact. The directional derivative $u^{\lambda}\partial_{\lambda}s$ of a scalar is such a linear
function; it is denoted  ds(u);  thus  ds  is a covariant vector; also
us  is used, indicating that a contravariant vector may be seen as a
(directional) derivative operator on the scalars.

In a similar way, differential forms (alternating covariant tensors)
can be viewed as alternating multilinear functions on the tangent vectors.
Thus, dw is defined in index-free form by

$$(2.2) \qquad dw(u,v) = uw(v) - vw(u) - w([u,v]) \; .$$

It would be hard to place a date on the birth of the concept of vector:
it existed informally long before it was formally recognized. The
distinction between co- and contravariant vectors is of much more
recent date, of course.

The Lie derivative $[u,T]$ of a tensor field  T  with respect to a vector field  u  dates back to Ślebodziński  (1931) $[16]$ ,  though Lie was no doubt aware of it in a less formal version.  It generalizes the bracket of (2.1), produces a tensor of the same type as  T,  and satisfies the usual Leibnitz rule for derivatives of products.

Another invariant was obtained by Schouten $[14]$  as follows. Take any Lagrangian (density)  $L$  , depending on a tensor field  P  of any type, and its derivatives up to a certain order.  Now, the integral of  $L$  over a region does not change under an infinitesimal displacement of  $L$  (under suitable boundary conditions), and the Lie derivative of  $L$  can be expressed through its dependence on   P.  Performing the usual integrations by parts that go with variational problems, one can thus obtain an identity involving the Lagrangian derivative  $[L]$   of  $L$ and derivatives of   P, of the type of a covariant vector density. So far not much new.

Now replace $[L]$ by a tensor density  $T$  of the same type.  The identity will no longer hold, but the expression which was equal to zero is now a differential invariant of  $T$  and   P,  of covariant vector density type:

$$T^{\;\cdot\;\cdots\;\cdot\,\lambda_1\ldots\lambda_q}_{\;x_1\ldots x_p}\{\partial_\mu P^{x_1\ldots x_p}_{\cdot\;\;\;\cdot\,\lambda_1\ldots\lambda_q}+A^{x_1}_{\mu}\partial_\varrho P^{\varrho x_2\ldots x_p}_{\cdot\;\;\;\;\;\;\cdot\,\lambda_1\ldots\lambda_q}+\cdots -\partial_{\lambda_1} P^{x_1\ldots x_p}_{\cdot\;\;\;\cdot\,\mu\lambda_2\ldots\lambda_q}-\cdots\}$$

(2.3)

$$+P^{\varrho x_2\ldots x_p}_{\cdot\;\;\;\;\;\;\cdot\,\lambda_1\ldots\lambda_q}\partial_\varrho T^{\;\cdot\;\;\cdot\;\;\cdot\,\lambda_1\ldots\lambda_q}_{\;\mu x_2\ldots x_p}+\cdots -P^{x_1\ldots x_p}_{\cdot\;\;\;\cdot\,\mu\lambda_2\ldots\lambda_q}\partial_{\lambda_1} T^{\;\cdot\;\cdots\;\cdot\,\lambda_1\ldots\lambda_q}_{\;x_1\ldots x_p}-\cdots$$

To my knowledge this differential invariant is not commonly used.  Of course, it is not a strong one, with only  n  components.

I have no way to write (2.3) in an index-free form.  To my feeling this
means that I don't really understand it as well as operations that can
be reduced to the most elementary ones (2.1).

Better known are Schouten's differential invariants of purely contravariant tensors, either symmetric or alternating [12] .
These were the results of an intended systematic and complete search for
differential invariants of tensor  fields, around 1940.  Schouten never
studied them in depth.  The symmetric one,

$$(2.4) \qquad [P^{(p+1)}, Q^{(q+1)}]^{\varkappa_1 \ldots \varkappa_p \lambda \mu_1 \ldots \mu_q} = (p+1)\, P^{\nu(\varkappa_1 \ldots \varkappa_p} \partial_\nu Q^{\lambda \mu_1 \ldots \mu_q)} - {} \\ {} - (q+1)\, Q^{\nu(\mu_1 \ldots \mu_q} \partial_\nu P^{\lambda \varkappa_1 \ldots \varkappa_p)} ,$$

produces a symmetric tensor of degree p+q+1; it is directly related to
Poisson bracket of scalar functions on the cotangent bundle [8] : the
coefficients in the power series expansion of these functions in the
coordinates of the fiber are symmetric controvariant tensors in the
base space.

The alternating Schouten bracket is defined by a similar formula:

$$(2.5) \qquad (p+1)\, P^{\nu[\varkappa_1 \ldots \varkappa_p} \partial_\nu Q^{\lambda \mu_1 \ldots \mu_q]} - (q+1)\, Q^{\nu[\mu_1 \ldots \mu_q} \partial_\nu P^{\lambda \varkappa_1 \ldots \varkappa_p]} .$$

Deviating from [8]  we denote $(-1)^P$  times this expression by
$[P^{(p+1)}, Q^{(q+1)}]^{\varkappa_1 \ldots \varkappa_p \lambda \mu_1 \ldots \mu_q}$ ,  as a result of which we have

$$(2.6) \qquad [P^{(p+1)}, Q^{(q+1)}] = (-1)^{pq+1}\, [Q^{(q+1)}, P^{(p+1)}] .$$

This bracket satisfies an identity which generalizes (2.2), and
which could be used as a definition of  $[P,Q]$ .  Let  w  be a (p+q+1)-
form, and let  $(w_P)_{\lambda_1 \ldots \lambda_q} = w_{\lambda_1 \ldots \lambda_q \mu_0 \ldots \mu_p}^{\mu_0 \ldots \mu_p}$  be a "partial
substitution", then

$$(2.7) \qquad (dw)(P \wedge Q) = (dw_Q)(P) + (-1)^{(p+1)(q+1)}(dw_P)(Q) - (-1)^{p+1} w([P,Q]) .$$

In 1951, while assistant of J.A.Schouten, the author found another
invariant of tensor fields. It was motivated by an old problem: what are
the conditions that the eigenvectors of a symmetric tensor field $h_{\lambda\mu}$
in a Riemannian manifold be locally integrable; i.e., that the p-planes
spanned by any p of them be tangent to a family of p-dimensional
submanifolds? (The eigenvalues were assumed distinct.) An answer
was given in [11], p.196, but it contained the eigenvectors explicitly,
though in undifferentiated form. In 1949, A.Tonolo [18] put the integrability conditions in a form which no longer required the calulation
of any eigenvalues or eigenvectors - but only for 3-dimensional space.
In 1951 Schouten [13] generalized Tonolo's solution to n dimensions,
at which time I was shown the result. I responded that, then, the problem could be made independent of the metric by considering the eigen-
vectors of $h_\lambda{}^\nu = h_{\lambda\mu}g^{\mu\nu}$. Assuming his search for differential invariants
in 1940 had been exhaustive, Schouten did not believe this [15] , and
thus left the area for me to explore. The result was a new invariant [7]:

$$(2.8) \qquad H_{\mu\lambda}^{\bullet\bullet\kappa} = 2h_{[\mu}{}^{\bullet\rho}\partial_{\rho]}h_\lambda^{\bullet\kappa} - 2h_\rho^{\bullet\kappa}\partial_{[\mu}h_{\lambda]}^{\bullet\rho} \qquad ,$$

in terms of which Tonolo's generalized problem could be solved.
Almost simultaneously, Eckmann and Frölicker [2] found the torsion
of an almost-complex structure $(h_\lambda^{\bullet\kappa}h_\mu^{\bullet\lambda} = -\delta_\mu^\kappa)$, but since they wrote
the last term in (2.8) as $2h_{[\lambda}{}^{\bullet\rho}\partial_{\mu]}h_\rho^{\bullet\kappa}$, their expression was not in
general a tensor. In 1955 Haantjes [5] found what is probably the
ultimate form of the solution to Tonolo's problem; it also involved the
invariant (2.8).

The formula (2.8) can be made bilinear, and yields what is known as
the torsion tension [h,k]; it is defined for two vector 1-forms, and
yields a vector 2-form (tensor once contravariant, twice covariant and
alternating). It satisfies the formula (which can also be used as its
definition)

$$(2.9) \quad [h,k](u,v) = [hu,kv] + [ku,hv] - k[hu,v] - h[ku,v]$$
$$-k[u,hv] - h[u,kv] + kh[u,v] + hk[u,v],$$

which clearly displays the chracter of alternating function of two tangent vectors whose value is a tangent vector. Note that all brackets on the right side are Lie brackets of tangent vectors.

The bracket $[h,k]$ can be generalized to vector forms of any degree; it then includes Lie derivatives of vector forms as a special case) [8]:

$$(2.10) \quad [M,N]^{\kappa}_{.\mu_1\dots\mu_q\nu_1\dots\nu_r} = M^{\tau}_{.[\mu_1\dots\mu_q}\partial_{|\tau|}N^{\kappa}_{.\nu_1\dots\nu_r]} - N^{\tau}_{.[\nu_1\dots\nu_r}\partial_{|\tau|}M^{\kappa}_{.\mu_1\dots\mu_q]} -$$
$$- q\, M^{\kappa}_{.\tau[\mu_1\dots\mu_q}\partial_{\mu_1}N^{\tau}_{.\nu_1\dots\nu_r]} + r\, N^{\kappa}_{.\tau[\nu_1\dots\nu_r}\partial_{\nu_1}M^{\tau}_{.\mu_1\dots\mu_q]}$$

Another related expression [7] :

$$(2.11) \quad \Sigma^{\kappa}_{\mu\lambda} = h_{\mu}{}^{.\rho}\partial_{\rho}k_{\lambda}{}^{.\kappa} - k_{\lambda}{}^{.\rho}\partial_{\rho}h_{\mu}{}^{.\kappa} - h_{\rho}{}^{.\kappa}\partial_{\mu}k_{\lambda}{}^{.\rho} + k_{\rho}{}^{.\kappa}\partial_{\lambda}h_{\mu}{}^{.\rho}$$

consists of half of the terms in $[h,k]$ . It is a tensor if and only if h and k commute, and has been extensively used by Yano [21]. It leads to invariants

$$(h_{\rho}{}^{.\kappa}k_{\lambda}{}^{.\rho} - k_{\rho}{}^{.\kappa}h_{\lambda}{}^{.\rho})\,\partial_{\mu}v^{\lambda} + \Sigma^{\kappa}_{\mu\lambda}\,v^{\lambda},$$
$$(2.12)$$
$$\partial_{\mu}(h_{\rho}{}^{.\kappa}k_{\lambda}{}^{.\rho} - k_{\rho}{}^{.\kappa}h_{\lambda}{}^{.\rho})\,w_{\kappa} - \Sigma^{\kappa}_{\mu\lambda}\,w_{\kappa}$$

and can be generalized to a differential operator for tensors of any type, but becomes very complicated.

More promising is the following operator, because it can so simply be extended to all tensors:

$$\mathcal{D}_{\mu}v^{\kappa} = (k_{\lambda}^{\rho}h_{\mu}^{\lambda} - h_{\lambda}^{\rho}k_{\mu}^{\lambda})\partial_{\rho}v^{\kappa}$$
$$(2.13)$$
$$+ (2h_{\mu}^{\rho}\partial_{[\rho}k_{\lambda]}^{\kappa} + 2h_{\sigma}^{\kappa}\partial_{[\lambda}k_{\mu]}^{\sigma} + k_{\mu}^{\rho}\partial_{\lambda}h_{\rho}^{\kappa} - k_{\lambda}^{\rho}\partial_{\rho}h_{\mu}^{\kappa})v^{\lambda}\;.$$

It is believed to be new, and is further discussed in section 3.

Another, somewhat similar-looking expression by A.G.Walker [11] , defines a differential operator for any tensor, given an almost complex structure. His <u>torsional derivative</u> is

$$(2.14) \qquad T^{i}_{j\cdots\|rs} = H^{p}_{rs}\,\partial_{p}\,T^{i\cdots}_{j\cdots} - T^{p\cdots}_{j\cdots}\,h^{i}_{prs} - \ldots - T^{i\cdots}_{p\cdots}\,h^{p}_{jrs} - \ldots,$$

$$h^{i}_{jrs} = -\tfrac{1}{2}\,\partial_{j}\,H^{i}_{rs} - \tfrac{1}{2}\,h^{i}_{p}\,(h^{q}_{j}\partial_{q}H^{p}_{rs} - H^{q}_{rs}\partial_{q}h^{p}_{j} - H^{p}_{qs}\partial_{r}h^{q}_{j} - H^{p}_{qr}\partial_{s}h^{q}_{j}).$$

As Willmore [20] has shown, for contravariant  v  it can be written as

$$(2.15) \qquad v^{i}_{\|rs} = \tfrac{1}{2}\left(h\,[hv,\,H] - [v,\,H]\right)^{i}_{rs} ,$$

and can be generalized.

The torsion of an almost-tensor product structure on a manifold is given by Hangan [6] :

$$(2.16) \qquad M^{i}_{jk} = S^{uv}_{[jk]}\partial_{r}S^{ir}_{uv} - S^{uv}_{[j|r|}\partial_{k]}S^{ir}_{uv} - S^{uv}_{r[k}\partial_{|v|}S^{ir}_{j]u} + S^{uv}_{[j|r}\partial_{v}S^{ir}_{u|k]} .$$

M  is a tensor only if the structure tensor S  satisfies some appropriate algebraic conditions.

3. __Invariance properties.__  In order to show that an expression such
as (2.1) or (2.4) is a tensor, one must, in principle, determine its
behavior under a change of coordinates.  More coveniently, one
usually replaces all partial derivatives by covariant ones and determines if the $\Gamma$-terms thus added vanish (assuming a symmetric connection).
The latter method is rather indirect, in a way; why introduce a connection to prove independence of it.

In some cases, such as (2.3), one starts with a tensor, and the
nature of the problem guarantees that the tensor character is maintained.

In the index-free methods a different approach is often used.  The
differentiable structure on the manifold is not viewed as determined
by the coordinate systems, but by the (ring of) differentiable
functions, $\phi$ .  A tangent vector field is now defined as a linear
mapping  u:  $\phi \to \phi$ (linear over the constants) which satisfies
Leibnitz's rule:

$$(3.1) \qquad u(fg) = (uf)g + f(ug),$$

i.e. , it is a __derivation__.  A product fu  and the bracket $[u,v]$ ,
defined by (fu)g=f(ug)  and $[u,v]f = u(vf)-v(uf)$  are also tangent
vectors fields because, as a simple calculation shows, they are derivations.

A 1-form (covariant vector field)  is defined to be a map of the
tangent vector fields to $\phi$ , which is not just linear over the
constants, but linear over  $\phi$  : w(fu)=fw(u).  This linearity over
the functions guarantees that  w  is defined at each point.
Tensors are similarly defined as multilinear maps  over  $\phi$  .

To show that  dw ,  defined by (2.2) , is a 2-form, one must prove
linearity of dw(u,v)  in u,v  over $\phi$ :  dw(fu,v)=fdw(u,v).  (The
linearity over the constants is obvious, by inspection.)

Note that $[u,v]$ is not linear over $\Phi$ in u and v, so some calculation has to be done.

The tensor character of (2.8) can be shown using connections, but for (2.9) one would show the linearity in u and v over $\Phi$ . (Of course, $[h,k]$ is not linear in h over $\Phi$ !). Actually, for (2.9) there is a simpler way to prove linearity in v over $\Phi$ , using Lie derivatives:

$$(3.2) \qquad [h,k](u,v) = [hu,k]v + [ku,h]v - k[u,h]v - h[u,k]v .$$

Note that (3.2) contains v only in undifferentiated form.

A related line of argument $[15]$ led Schouten to his bracket operators (2.4) and (2.5): let T be any tensor field, and u,v, vector fields; write down the expression involving Lie derivatives

$$(3.3) \qquad u \otimes [v,T] + [u,T] \otimes v;$$

it resemles an operation, namely $[.,T]$, applied to the product $u \otimes v$ by a Leibnitz-type rule. Question: is this expression equal to some $[u \otimes v,T]$, which would then be some generalized Lie derivative? Answer: yes, if and only if there is a special kind of linearity over $\Phi$: the expression (3;3) applied to fu,v or applied to u,fv should give the same results. A calculation shows that this is not the case in general. However, by restricting T to be purely controvariant and by applying some (skew-) symmetrizations the offending terms can be removed. (Of course, Schouten used connections and lots of subscripts to do the above, but the idea is the same.)

Let us apply the same idea to the expression

$$(3.4) \qquad [hu,k] - h[u,k]$$

where h, k are vector 1-forms. We want to check if (3.4) is linear in u over the functions. Note that $[hu,.]$ and $[u,.]$ denote

258

Lie derivatives.  We use the rule

(3.5)
$$[fu,k] = f[u,k] + k(u\otimes df) - (u\otimes df)k;$$

the presence of  df  indicates the absence of linearity in u over $\Phi$.
Apply this to  (3.4):

(3.6)
$$[hfu,k] - h[fu,k] - f[hu,k] + fh[u,k]$$
$$= k(hu\otimes df) - (hu\otimes df)k - hk(u\otimes df) + h(u\otimes df)k$$
$$= (kh - hk)(u\otimes df)$$

This shows that (3.4)  is not linear in u  over $\Phi$ , but that symmetrization over h,k  leads to a  $\Phi$-linear operation; that is (2.9).  The unsym-
metrized form leads to (2.11).

There is a "product rule" involving terms of the form  $[h,k\ell]$  etc.  Due
to the remainder term in (3.6), which required a symmetrization, it
is a bit more complicated that one might wish:

(3.7)
$$[h,k\ell] + [h\ell,k] = h[k,\ell] + k[h,\ell] + [h,k]\tilde{\wedge}\,\ell;$$

the last term is defined by

(3.8)
$$([h,k]\tilde{\wedge}\,\ell)(u,v) = [h,k](\ell u,v) + [h,k](u,\ell v) .$$

(3.7) follows rather easily from another form of (2.9):

(3.9)
$$[h,k](u,v) = [hu,k]v - [hv,k]u - h[u,k]v + h[v,k]u$$

in which the brackets on the right are again Lie derivatives, so a Leibnitz
rule holds for k.

The formula (3.6) also gives rise to (2.11).  In view of  $[fu,v]=f[u,v] - (u\otimes df)v$  we have

(3.10)
$$[(kh - hk)fu,v] - f[(kh - hk)u,v] = -(kh - hk)(u\otimes df)v$$

so, by adding the corresponding sides of (3.10) and (3.6)  we see that

$$(3.11) \qquad \mathcal{D}_u v = \left[(kh - hk)u, v\right] + \left(\left[hu, k\right] - h\left[u, k\right]\right)v$$

is a differential invariant of h,k and v, but is $\Phi$-linear in u.
This also follows by re-arranging terms:

$$(3.12) \qquad \mathcal{D}_u v = \left[v, hk\right]u - \left[v, k\right]hu - \left[kv, h\right]u$$

The expressions (3.11,12) show $\mathcal{D}_u v$ as just a sum of Lie derivatives,
but it has a special property:  defining $\mathcal{D}_u f = df(kh - hk)u$,  we have
$\mathcal{D}_u(fv) = (\mathcal{D}_u f)v + f\mathcal{D}_u v$ .  That means that $\mathcal{D}_u$  can be extended to act
on all tensors, and obeys a Leibnitz rule for products.  In fact,

$$(3.13) \qquad \mathcal{D}_u T = \left[(kh - hk)u, T\right] + \left(\left[hu, k\right] - h\left[u, k\right]\right)*T$$

where   $h*T$, for a vector 1-form  h  and any tensor T  is defined by

$$(3.14) \qquad (h*T)_{\lambda_1 \cdots \lambda_p}^{\phantom{\lambda_1 \cdots \lambda_p} k_1 \cdots k_q} = \sum_{i=1}^{q} h_{\rho}^{k_i} T_{\lambda_1 \cdots \lambda_p}^{\phantom{\lambda_1 \cdots \lambda_p} k_1 \cdots \rho \cdots k_q} - \sum_{j=1}^{p} h_{\lambda_j}^{\rho} T_{\lambda_1 \cdots \rho \cdots \lambda_p}^{\phantom{\lambda_1 \cdots \rho \cdots \lambda_p} k_1 \cdots k_q};$$

this is the usual  "infinitesimal" action.

The form of (2.8) is somewhat similar to that for the curvature of a
connection.  It is therefore plausible to expect a kind of Bianchi
identity; i.e., a formula which is linear in h and H, containing first
order derivatives of each.  It turns out (with a good deal of work)
that such a formula exists:

$$(3.15) \qquad H^{\tau}_{.[\nu\mu}\partial_{|\tau|}h^{\kappa}_{.\lambda]} - h^{\tau}_{.[\nu}\partial_{|\tau|}H^{\kappa}_{.\mu\lambda]} + h^{\kappa}_{.\tau}\partial_{[\nu}H^{\tau}_{.\mu\lambda]} + 2H^{\kappa}_{.\tau[\nu}\partial_{\mu}h^{\tau}_{.\lambda]} = 0.$$

It is now tempting to replace  H  in this formula by an arbitrary
vector 2-form.  The result is no longer zero, of course, but is
it a tensor?  The answer turns out to be positive.  This observation
led to the discovery of the invariant (2.10).  Next came a escond
look at (3.15), which can be written as  $\left[h, \left[h, h\right]\right] = 0$.    It suggests
the existence of a trilinear identity for the bracket (2.10).
More about this in section 4.

**4. Graded Lie algebras.** In one of their numerous joint papers, K.Kodaira and D.C.Spencer felt the need to refer to a source for the concept of graded Lie algebra (GLA), and chose the obvious, encyclopedic one: "Homological Algebra" by Eilenberg and MacLane. Actually, no such thing is defined in that book, though probably it should be. The concept and the name are so obvious that you know one when you see one. Or, you should. In any case, the concept was there in [8], but the name had to wait till [3].

A graded algebra of any kind consists of a system of modules $\{A_n\}$, indexed by an abelian group, usually the integers $\mathbb{Z}$, or the integers modulo 2, $\mathbb{Z}_2$, in which bilinear products $A_n \times A_m \to A_{n+m}$ are defined. Graded commutativity means $ba=(-1)^{mn}ab$ if $a \in A_m$, $b \in A_n$; associativity is defined as usual. An example of a graded commutative, associative algebra is the differential forms, with $A_m$ consisting of those of degree m, and exterior multiplication. The alternating contravariant tensors are a similar example.

A graded Lie algebra is either of two things: 1. a graded algebra which is an ordinary Lie algebra, or 2. a graded algebra in which the product, usually denoted $[,]$, satisfies $[b,a]=(-1)^{mn+1}[a,b]$, while also a Jacobi identity holds $(c \in A_p)$:

$$(4.1) \quad (-1)^{mp}[[a,b],c] + (-1)^{nm}[[b,c],a] + (-1)^{pn}[[c,a],b] = 0.$$

A GLA of the first type is given by the Schouten bracket (2.4) with $A_m$ consisting of the contravariant symmetric tensors of degree m+1. Usually, GLA's are understood to be of the second type. For example, the alternating Schouten bracket (2.5), with signs as indicated, and with the same shift in degrees as above, is one. See [8] for details.

Applications of GLA's to Physics are discussed in [1]. These are of a very specific nature, and are not discussed here.  When, these algebras are graded by $\mathbb{Z}_2$, they are sometimes referred to as superalgebras.  The part of even degree is a common Lie algebra.

In a GLA, elements of odd degree commute:  $[a,b] = [b,a]$ , and $[a,a]$ ia usually not identically zero.  In fact, the property $[a,a] = 0$ for some $a \in A_1$ is frequently the key part of a theory, see $[9]$, which includes elaborate cohomology structures.

One way to generate  a GLA  is by considering the derivations of a graded algebra.  A linear map  D  of this algebra into itself is a derivation of degree  r   if $D(A_m) \subset A_{m+r}$  and

(4.2)  $$D(ab) = (Da)b + (-1)^{rm} a(Db).$$

Let  $B_r$  denote the module of these, then for  $D_1 \in B_r$,  $D_2 \in B_s$ it is easy to see that

(4.3)  $$[D_1, D_2] = D_1 D_2 - (-1)^{rs} D_2 D_1$$

is a derivation of degree  r+s;  equipped with this product,  $\{B_r\}$  is a GLA.

We give two similar examples.  V denotes a finite dimensional vector space.  In the first example, $A_m$  consists of all m-linear maps of V into the scalar field, with multiplication by tensor product.  In the second example, $A_m$  consists only of the alternating m-linear scalar-valued functions, with exterior product.  Both graded algebras are associative, the latter is graded commutative.  In both cases the algebra is generated by  $A_1 = V^*$ .  Let  $e_1, e_2, \ldots$  be a basis of V, and  $\varepsilon_1, \varepsilon_2, \ldots$  a dual basis of $A_1$.  Let  $D \in B_r$,  and set

262

$$(4.4) \qquad L = \sum e_i \otimes D\varepsilon_i$$

then we have $Da = a\,\bar{\circ}\,L$ in the first example, and $Da = a\,\bar{\wedge}\,L$ in the second, where [4,10]

$$(4.5) \qquad a\,\bar{\circ}\,L(v_1,\ldots,v_{m+r}) = \sum_{i=1}^{m} (-1)^{ir}\, a(u_1,\ldots,u_{i-1},L(u_i,\ldots,u_{i+r}),\ldots,v_{m+r})$$

$$a\,\bar{\wedge}\,L(v_1,\ldots,v_{m+r}) = \sum \operatorname{sgn}(\pi)\, a(L(v_{\pi_1},\ldots,v_{\pi_{r+1}}),v_{\pi_{r+2}},\ldots,v_{\pi_{r+m}})$$

The second summation extends over a set of $\binom{m+r}{m-1}$ permutations $\pi$ that cause the expressions to alternate in $v_1,\ldots,v_{m+r}$.

Let $C_{r+1}$ denote the module of all $(r+1)$-linear maps of V to V, in the first example, and the alternating ones in the second.

Then for $L \in C_{r+1}$ the maps $a \to a\,\bar{\circ}\,L$ or $a \to a\,\bar{\wedge}\,L$ are derivations, which we denote $i_L$. Define $M\,\bar{\circ}\,L$ and $M\,\bar{\wedge}\,L$ in analogy to (4.5), then the following "commutative-associative law" holds for $\bar{\wedge}$ , and an analogous one for $\bar{\circ}$ ($L \in C_{r+1}$, $M \in C_{s+1}$):

$$(4.6) \qquad (a\,\bar{\wedge}\,L)\,\bar{\wedge}\,M - a\,\bar{\wedge}\,(L\,\bar{\wedge}\,M) = (-1)^{rs}\left\{ (a\,\bar{\wedge}\,M)\,\bar{\wedge}\,L - a\,\bar{\wedge}\,(M\,\bar{\wedge}\,L)\right\}$$

The commutators, $[L,M]^{\bullet} = L\,\bar{\circ}\,M - (-1)^{rs}M\,\bar{\circ}\,L$ , $[L,M]^{\wedge} = L\,\bar{\wedge}\,M - (-1)^{rs}M\,\bar{\wedge}\,L$ , however, do satisfy the Jacobi identity for a reduced grading, and $[i_L, i_M] = i_{[L,M]^{\bullet}}$ or $i_{[L,M]^{\wedge}}$ .

These GLA's are important in the deformation theory of algebras. Let $\mu \in C_2$, then in the first example $\mu\,\bar{\circ}\,\mu = \frac{1}{2}[\mu,\mu]^{\bullet}$ vanishes if and only if $\mu : V \times V \to V$ is the product map of an associative algebra, [4] . In the second example, $\mu\,\bar{\wedge}\,\mu = \frac{1}{2}[\mu,\mu]^{\wedge}$ vanishes if and only if $\mu$ is the product map of a Lie algebra structure on V [10].

In 1956 Frölicher and Nijenhuis [3] identified the bracket (2.10) as being related to derivations. Let $\Phi_p$ denote the totality of differential forms of degree $p$ on a manifold; in particular $\Phi_o = \Phi$. Then, $\Phi_p$ is an infinite dimensional vector space over the constants, and a finitely generated module over $\Phi$. Also, $\Phi_o$ and $\Phi_1$ generate $\{\Phi_p\}$. This implies that the derivations of $\{\Phi_p\}$ are uniquely determined by their action on $\Phi_o$ and $\Phi_1$. Denote by $\Psi_p$ the vector p-forms over the manifold, then the second example discussed above identifies a whole class of derivations on $\{\Phi_p\}$ : let $L$ be a vector 1-form, then $i_L$ is a derivation: $i_L w = w \bar{\wedge} L$. This derivation is linear over $\Phi$ in $w$, and has therefore trivial (zero) action on $\Phi_o$. These derivations ("of type $i_*$") are the only ones that have this property.

The exterior derivative is another derivation, of degree 1, and $d^2 = 0$. Further derivations are the commutators of d and the derivations of type $i_*$ ; these are said to be of type $d_*$ ;

$$(4.7) \qquad d_L w = [i_L, d] w = dw \bar{\wedge} L + (-1)^r d(w \bar{\wedge} L), \qquad L \in \Psi_r.$$

These derivations are characterized by the property $[d_L, d] = 0$.

Every derivation $D$ of $\{\Phi_p\}$ of degree $r$ is of the form $D = i_L + d_M$, where $L \in \Psi_{r+1}$, $M \in \Psi_r$, and the decomposition is unique.

There are now three types of commutators of derivations: $[i_L, i_M]$ , $[d_L, d_M]$ and $[i_M, d_L]$. The first of these is, of course, $i_{[L,M]^{\bar{\wedge}}}$, the second is $d_{[L,M]}$, where $[L,M]$ is the differential invariant (2.10). The last one is

$$(4.8) \qquad [i_M, d_L] = d_{L \bar{\wedge} M} + (-1)^p i_{[M,L]}, \qquad L \in \Psi_p.$$

There are four kinds of Jacobi identities. If all terms are of type $i_*$, the Jacobi identity for $[\ ,\ ]^{\bar{\wedge}}$ results. If all are of type d$_*$ , we find the Jacobi identity for (2.10) , and if two of the derivations are of one type, and the third is of the other type, then another identity for the bracket (2.10) is obtained:

$$(4.9) \quad \begin{aligned} &[L \bar{\wedge} N, M] + (-1)^{p(n-1)} [L, M \bar{\wedge} N] - [L, M] \bar{\wedge} N = \\ &(-1)^{m(p-1)} L \bar{\wedge} [N, M] + (-1)^{p-1} M \bar{\wedge} [N, L] \end{aligned} \qquad \begin{aligned} &L \in \Psi_p \\ &M \in \Psi_m \\ &N \in \Psi_n \end{aligned}$$

This identity generalizes (3.7).

As stated before, the GLA of vector forms is applied to almost complex structures. If h is the structure tensor (hh=-I) then $[h,h]$ , the torsion, vanishes if and only if h is compatible with complex coordinate systems. In the study of deformations of complex structures extensive use is made of this GLA and its related cohomology.

Bibliography

[1] L.Corwin, Y.Ne'eman and S.Sternberg, *Graded Lie algebras in mathematics and physics*. Rev.Mod.Phys.,47(1975),575.

[2] B.Eckmann and A.Frölicher, *Sur l'intégralité des structures presque complexes*. C.R.Paris 232(1951),2284-2286.

[3] A.Frölicher and A.Nijenhuis, *Theory of vector-valued differential forms I*. Kon.Nederl.Akad.Wetensch.Proc. A59(1956)338-359.

[4] M.Gerstenhaber, *The cohomology structure of an associative ring*. Annals of Math. 78(1963)267-288.

[5] J.Haantjes, *On $X_m$-forming sets of eigenvectors*. Kon.Nederl.Akad.Wetensch. Proc. A 58 (1955)158-162.

[6] T.Hangan, *Sur l'intégralité de certaines G-structures tensorielles*. C.R.Paris A 288(1979)835-837.

[7] A.Nijenhuis, *$X_{n-1}$-forming sets of eigenvectors*. Kon.Nederl.Akad. Wetensch. Proc. A 54 (1951) 200-212.

[8] A.Nijenhuis, *Jacobi-type identities for bilinear differential concomitants of certain tensor fields*. Kon.Nederl.Akad.Wetensch.Proc. A 58(1955) 390-403.

[9] A.Nijenhuis and R.W.Richardson, *Cohomology and deformations in graded Lie algebras*. Bull.Amer.Math.Soc. 72 (1966)1-29.

[10] A.Nijenhuis and R.W.Richardson, *Deformations of Lie algebra structures*. J.Math.Mech. 17 (1967) 89-106.

[11] J.A.Schouten, *Der Ricci Kalkül*. Springer, Berlin, 1924.

[12] J.A.Schouten, *Ueber Differentialkomitanten zweier kontravarianter Gröszen*. Kon.Nederl.Akad.Wetensch.Proc. 43(1940) 449-452.

[13] J.A.Schouten, *Sur les tenseurs de $V_n$ aux directions principales $V_{n-1}$-normales*. Coll.de géom.diff. Louvain (1951) 67-70.

[14] J.A.Schouten, *Ricci Calculus*, Springer, Berlin, 1954.

[15] J.A.Schouten, *On the differential operators of first order in tensor calculus*. Conv.Geom.Diff. Ed.Cremonese, Roma (1954).

[16] W.Ślebodziński, *Sur les equations de Hamilton*. Bull.Acad.Roy. Belg. 17(1931) 864-870.

[17] C.L.Terng, *Natural vector bundles and natural differential opertaors*. Amer.J.Math. 100(1978) 775-828.

[18] A.Tonolo, *Sopra una classe di operazioni finite*. Ann.Mat.Pura Appl. 29 (1949) 29-53.

[19] A.G.Walker, *Dérivation torsionnelle et seconde torsion pour une structure presque complexe*. C.R.Paris 245 (1957) 1213-1215.

[20] T.J.Willmore, *Deriavations and vector 1-forms*. J.Lond.Math.Soc. 35 (1960) 425-432.

[21] K.Yano and M.Ako, *On certain operations associated with tensor fields*. Kodai Math.Sem.Rep. 26(1968) 414-436.

Received January 19, 1985.

GEOMETRODYNAMICS PROCEEDINGS (1985), pp. 267-281
edited by A. Pràstaro
© 1985 by World Scientific Publishing Co.

# A UTIYAMA TYPE THEOREM IN THE C-K-S GAUGE APPROACH TO GRAVITY I

A. Pérez-Rendòn

*Universidad de Salamanca (Spain)*

## 1. Introduction

As is well known, the Yang-Mills theory [1] allows us to obtain the interaction equations of a *matter field* (defined on a vector bundle associated with a principal bundle $p:P \longrightarrow X$) with a *gauge field* (defined on the bundle $K$ of connections on $P$) by postulating:

1) The Lagrangian $L$ of the matter field is invariant under a *group of internal symmetries* (= the structure group $G$ of the bundle $P$).

2) The Lagrangian $V$ of the gauge field is *gauge-invariant* (gauge group = {automorphisms of $P$ trivial on the base manifold $X$}).

Thus if we wish to know which are all the possible gauge fields interacting with a given matter field, *we must* previously *characterize the gauge-invariant functions defined on* $J^1K$ (= bundle of 1-jets of connections).

This problem was first resolved by Utiyama in his well-known paper [2], where he showed that:

**Utiyama's theorem.** A function on $J^1K$ is gauge-invariant when:

a) The gauge potentials $\Gamma_i^a$ (= components of connection 1-forms) and their derivatives $\dfrac{\partial \Gamma_i^a}{\partial x_j}$ appear in $V$ only through the particular combinations:

$$R_{ij}^a = \frac{\partial \Gamma_j^a}{\partial x_i} - \frac{\partial \Gamma_i^a}{\partial x_j} - \frac{1}{2} c_{be}^a (\Gamma_i^b \Gamma_j^e - \Gamma_j^b \Gamma_i^e)$$

(= components of the curvature 2-forms).

b) Furthermore, $V$ should fulfill the equations:

$$\frac{1}{2} c_{be}^a R_{ij}^e \frac{\partial V}{\partial R_{ij}^a} = 0 \quad \text{(for } b = 1,\ldots,\dim G)$$

A more geometric statement and proof of this theorem was formulated by Garcia in [3], and may be briefly summarized by:

Let $R$ be the epimorphism of bundles on $X$ :

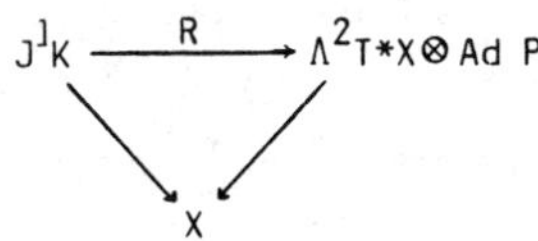

given by $R(j_x^1 \sigma) = (R_\sigma)_x$ , where $R_\sigma$ is the curvature 2-form of the connection $\sigma$ and Ad P is the *adjoint bundle* of $P$ (see below Proposition 2.2).

A function $V$ on $J^1K$ is gauge-invariant if and only if

$$J^1 K \xrightarrow{\;R\;} \Lambda^2 T^* X \otimes \mathrm{Ad}\, P \xrightarrow{\;\hat{V}\;} \mathbf{R} \;,$$
$$\underbrace{\phantom{J^1 K \xrightarrow{\;R\;} \Lambda^2 T^* X \otimes \mathrm{Ad}\, P \xrightarrow{\;\hat{V}\;} \mathbf{R}}}_{V}$$

$\hat{V}$ being a gauge invariant function on $\Lambda^2 T^* X \otimes \mathrm{Ad}\, P$ . (This result by García is included in the book by Bleecker [4] and in Eck's thesis [5])

The present paper is the first of a series dedicated to generalizing these results by giving a Utiyama type theorem in the Cartan-Kibble-Sciama (C-K-S) gauge approach to gravity; i,e., a theorem which establishes *which conditions should be fulfilled by the Lagrangians of the gravitational field*.

In order to do so, in this paper we shall concentrate on two problems:

**a)** We give a geometrically clear formulation of the C-K-S theory by means of the fibre bundle language.

**b)** We shall study the interaction equations of a matter field with a gravitational field, whose Lagrangian densities are $L\omega$ and $V\omega$ , in the following cases:

**1)** When both Lagrangians are *arbitrary*.

**2)** $V$ is *invariant under all the automorphisms of the principal bundle* $p : P \longrightarrow \mathbf{R}^4$ (where $P$ is the Poincaré group) with the Lorentz group as the structure group and $L$ is arbitrary.

**3)** When we also assume $L$ to be *Poincaré-invariant*.

The matter fields are defined over vector bundles associated with the bundle $P$ and the gravitational field over the fibred product $Y = F(\mathbf{R}^4) \times_{\mathbf{R}^4} K$ ; $\pi_2 : F(\mathbf{R}^4) \longrightarrow \mathbf{R}^4$ is the *bundle of orthonormal frames of the Minkowski space-time* $(\mathbf{R}^4, T_2)$ and $\pi_3 : K \longrightarrow \mathbf{R}^4$ is the *bundle of connections* on $P$ .

Following the results partly presented in a joint work with D.H. Ruipérez (cf.[8]), we devote Section 3 to formulating a *minimal coupling principle*, showing that the *interaction modified Lagrangian* (given by Kibble in [13]) may be obtained without assuming any invariance under some symmetry group of the Lagrangian of the initial field.

In Section 4, we establish which equations should be fulfilled by the Lagrangian of the gravitational field in order to be *aut P-invariant* (Aut $P$ = {group of auto-morphisms of the bundle $p : P \longrightarrow \mathbf{R}^4$}) . However, we leave for a forthcoming publication the study of the more general expression of the Lagrangians which are solutions of these equations.

Section 5, devoted to the study of the Lagrangians of a matter field, gives the definitions of *P-invariant* and *aut P-invariant* Lagrangians and shows that a Lagrangian is aut P-invariant if and only if it is P-invariant.

Finally in Section 6 we demostrate that the equations of a gravitational field coupled to a P-invariant matter field are very similar to those given by Hehl and Von der Heyde in the PG field theory (see [14]).

A future paper will also deal with the problem of whether it is possible to give a generalization for C-K-S-theory of the results of Mangiarotti and Modugno [7].

## 2. Preliminaries: algebra of infinitesimal automorphisms

Let $P$ be the Poincaré group and $p: P \longrightarrow \mathbf{R}^4$ be the trivial principal bundle with the Lorentz group $L$ as the structure group and base manifold $\mathbf{R}^4 = P/L$ .

With $g_P$ and $g_L$ we shall denote the Lie algebras of $P$ and $L$ , respectively, and with $A$ the ring of $C^\infty$-differentiable functions on $\mathbf{R}^4$ .

**Proposition 2.1.** The vector bundle $T_L P$ of *right $L$-invariant* vector fields on $P$ is isomorphic to the vector bundle associated with $p: P \longrightarrow \mathbf{R}^4$ whose typical fibre is $g_P$ . Accordingly, on the $A$-module of sections of $T_L P$ it is possible to define trivially a Lie algebra structure, fulfilling:

$$\text{aut } P: = \Gamma(T_L P) \simeq A \otimes_R g_P \quad \text{only as } A\text{-modules}$$

**Definition 2.1.** aut $P$ is the algebra of all *infinitesimal automorphisms* of $p: P \longrightarrow \mathbf{R}^4$ .

Let $\{\varepsilon_i, \tau_N\}$ (for $i = 1, \ldots, 4$ ; $N =$ pair of indices $(i,j)$ where $i < j \leq 4$) be a basis of $g_P$ such that

$$(2.1) \quad \begin{cases} \textbf{a) } \{\tau_N\} \text{ is a basis of } g_L . \\[2mm] \textbf{b) } [\varepsilon_i, \varepsilon_j] = 0 \; ; \; [\tau_N, \varepsilon_i] = c_{Ni}^k \varepsilon_k \\[2mm] \quad [\tau_N, \tau_M] = c_{N\,M}^{Q} \tau_Q \quad \text{and} \quad \tau_{ij} = -\tau_{ji} \end{cases}$$

The *right $P$-invariant* vector fields on $P$ , associated to $\{\varepsilon_i, \tau_N\}$ , form a basis of the $A$-module aut $P$ , which we shall call $(R\varepsilon_i, R\tau_N)$ .

If $(x_i, \varphi_{hj})$ is a coordinate system on $P$ , we can write:

$$(2.2) \quad \begin{cases} R\varepsilon_i = \dfrac{\partial}{\partial x_i} \\[4mm] R\tau_N = x_i \dfrac{\partial}{\partial x_j} - x_j \dfrac{\partial}{\partial x_i} + \varphi_{ih} \dfrac{\partial}{\partial \varphi_{jh}} - \varphi_{jh} \dfrac{\partial}{\partial \varphi_{ih}} \quad ; \quad \text{if } N = i,j . \end{cases}$$

**Proposition 2.2.** The subbundle $V_L P \subset T_L P$ of *vertical vector fields* is isomorphic to the *adjoint bundle* of $P$ , Ad $P$ ; that is, to the vector bundle associated with $P$ with typical fibre $g_L$ . We now have:

$$\text{gau } P: = \Gamma(V_L P) \simeq A \otimes_R g_L \quad , \text{ as } A\text{-algebras}$$

**Definition 2.2.** gau $P$ is the *gauge algebra* of the principal bundle $P$ .

If we take $\varphi_{ih} \dfrac{\partial}{\partial \varphi_{jh}} - \varphi_{jh} \dfrac{\partial \cdot}{\partial \varphi_{ih}} = \tilde{\tau}_N$ , we will obtain the expression in coordinates of the vertical, $L$-*invariant*, vector field on $P$ associated to $\tau_N$ . Thus, all elements $\gamma \varepsilon$ aut $P$ could be written as:

$$\gamma = \xi_i(x)\,\frac{\partial}{\partial x_i} + h_N(x)\,\tilde{\tau}_N \quad ; \quad \xi_i(x)\,,\ h_N(x) \in A \ .$$

**Proposition 2.3.** aut $P$ and gau $P$ are the "Lie algebras" of the groups:

Aut $P$ = {automorphisms of the principal bundle $P$} .

Gau $P$ = {automorphisms of $p:P \longrightarrow \mathbf{R}^4$ with a trivial action on $\mathbf{R}^4$} .

(see [9]).

### 3. The structure form $\bar{\vartheta}$

### 3.1. The canonical structure form on $J^1E$

Let $\pi_1:E \longrightarrow \mathbf{R}^4$ be a vector bundle associated with $P$, whose typical fibre is $F$ . We can define on $E$ the vector bundle

$$\bar{\pi}_1:\bar{E} = \mathrm{Hom}(\pi_1^*TR^4,VE) \longrightarrow E \ ,$$

$\pi_1^*TR^4$ being the pull-back of $TR^4$ on $E$ .

$\bar{E}$ is the vector bundle associated with the affine bundle of 1-jets $\tilde{\pi}_1:J^1E \longrightarrow E$.

*On $J^1E$ there is a canonical structure 1-form $\vartheta$ with values in the module of sections of $\tilde{\pi}_1^*VE$ (= pull-back of $VE$ on $J^1E$) :*

Let $x$ be an arbitrary point of $\mathbf{R}^4$ and $s:\mathbf{R}^4 \longrightarrow E$ a section of $E$ . If $j_x^1 s$ is the 1-jet of $s$ at $x$ and $\bar{D}$ is a tangent vector to $J^1E$ at $j_x^1 s$ , then

$$(3.1) \qquad\qquad \vartheta_{j_x^1 s}(\bar{D}) = d_x^V s(D)$$

where $D = \tilde{\pi}_{1*}(\bar{D})$ and $d_x^V s$ (= the *vertical differential* of $s$ at $x$) is the projector of $TE_{s(x)}$ on $VE_{s(x)}$ given by the exact sequence

$$0 \longrightarrow TR_x^4 \xrightarrow{\ ds_x\ } TE_{s(x)} \xrightarrow{\ d_x^V s\ } VE_{s(x)} \longrightarrow 0 \ .$$

*On $\bar{E}$ the structure form is not unique:*

Given a connection $\sigma$ on $E$ , we can define on $\bar{E}$ the structure form $\vartheta^\sigma$ :

$$\vartheta^\sigma_{\bar{y}}(\bar{D}) = D_y - \sigma_x(D') + \psi(D_y') \ ,$$

$\bar{D}$ is any tangent vector at $\bar{y} = (y,\psi)$ , $\psi \in \mathrm{Hom}(\pi_1^*TR^4,VE)_y$ and, if $x = \pi_1(y)$ , then, $D_x'$ , $D_y$ are the projections of $\bar{D}_{\bar{y}}$ on $\mathbf{R}^4$ and $E$ , respectively.

Therefore, if $\sigma$ , $\sigma'$ are two different connections, we have the following relation between the structure forms $\vartheta^\sigma$ , $\vartheta^{\sigma'}$ :

$$\vartheta^\sigma_{\bar{y}}(\bar{D}) = \vartheta^{\sigma'}_{\bar{y}}(\bar{D}) + (\sigma' - \sigma)_x D'$$

### 3.2. Coordinate expression of $\vartheta^\sigma$

Since $E$ is a vector bundle associated with $P$ , there exists a canonical morphism $\zeta_1$ of A-algebras (see [8] and [10]):

$$\zeta_1 : \text{aut } P \longrightarrow \Gamma(T_{\pi_1} E)$$

with values on $\pi_1$-*projectable* vector fields and consistent with the projections on $R^4$ :

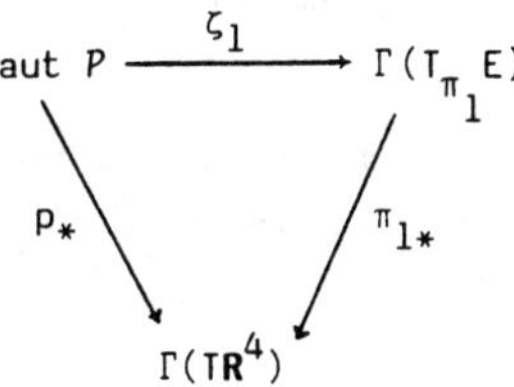

So, we also have:

$$\zeta_1 : \text{gau } P \longrightarrow \Gamma(VE) \quad .$$

And, by making use of the fact that $\text{gau } P \simeq A \otimes_R \mathfrak{g}_L$ , we finally obtain a representation of Lie algebras:

$$\zeta_1 : \mathfrak{g}_L \longrightarrow \Gamma(VE) \quad .$$

If we take a coordinate system on $E$ , $(x_i, z_j)$ — $i = 1, \ldots, 4$ ; $j = 1, \ldots, n = \dim F$ — where the $z_j$ are linear functions on $F$, and $(x_1, \ldots, x_4)$ is a coordinate system on $R^4$ such that $T_2 = -dx_1^2 - dx_2^2 - dx_3^2 + dx_4^2$ , then, we can write:

$$\zeta_1(\gamma) = \zeta_1(\xi_i \frac{\partial}{\partial x_i} + h_N \tilde{\tau}_N) = \xi_i \frac{\partial}{\partial x_i} + h_N a_{kj}^N z_k \frac{\partial}{\partial z_j} \quad ,$$

the constants $a_{kj}^N$ depending on the representation of $L$ into $F$ , given to construct the associated vector bundle $E$ .

Let $\sigma_o$ be the flat connection on $E$ and $\vartheta^{\sigma_o}$ be the structure form associated to $\sigma_o$ . We shall now define on $\bar{E}$ a coordinate system $(x_i, z_j, p_{ij})$ , where $p_{ij}(\bar{y}) = - [\vartheta_{\bar{y}}^{\sigma_o}(\frac{\partial}{\partial x_i})] z_j$ . From (3.1), we have:

$$(3.2) \qquad \vartheta^{\sigma_o} = (dz_j - p_{ij}\, dx_i) \circ \frac{\partial}{\partial z_j}$$

and any other structure form will be written:

$$(3.3) \qquad \vartheta^\sigma = (dz_j - (p_{ij} + \Gamma_i^N(x)\, a_{kj}^N\, dz_k) dx_i) \circ \frac{\partial}{\partial z_j} \quad ,$$

where $\Gamma_i^N(x)\, a_{kj}^N\, z_k\, dx_i \circ \frac{\partial}{\partial z_j} = \Gamma_i^N(x)\, dx_i \circ \zeta_1(\tau_N)$ is the coordinate expression of the $(VE)$-valued 1-form on $\mathbf{R}^4$ associated with the connections $(\sigma,\sigma_o)$ .

Now, if instead of using the vectors $(\frac{\partial}{\partial x_i})_x$ to determine the functions $p_{ij}$ , we take at each point $x_o \in \mathbf{R}^4$ an orthonormal frame $(D_i)_{x_o}$ (= a *tetrad*) and we introduce

$$\bar{p}_{ij}(\bar{y}) = -[\vartheta_{\bar{y}}^{\sigma_o}(D_i)]z_j \ ,$$

we will have that $\bar{p}_{ij}(\bar{y}) = e_i^k(x)p_{kj}(\bar{y})$ , when $(D_i)_{x_o} = e_i^k(x_o)(\frac{\partial}{\partial x_k})_{x_o}$ .
That is:

*Each section* $u$ *of the bundle* $\pi_2 : F(\mathbf{R}^4) \longrightarrow \mathbf{R}^4$ *of orthonormal frames of* $(\mathbf{R}^4, T_2)$ *allows us to define a coordinate system* $(x_i, z_j, \bar{p}_{ij})$ *on* $\bar{E}$ .

In this new coordinate system we have:

$$(3.4) \qquad \vartheta^\sigma = (dz_j - (e_i^\ell(x)\bar{p}_{\ell j} + \Gamma_i^N(x)a_{kj}^N\, z_k)dx_i) \circ \frac{\partial}{\partial z_j} \ ;$$

$e_i^\ell(x)\, , \Gamma_i^N(x) \in A$ and $(e_i^\ell(x_o))$ denote the inverse matrix of $(e_\ell^i(x_o))$ .

### 3.3. The form $\bar{\vartheta}$

Let $\pi_3 : K \longrightarrow \mathbf{R}^4$ be the *vector bundle of connections* on $P$ , i.e.,

$$K = \mathrm{Hom}(T\mathbf{R}^4, \mathrm{Ad}\ P)$$

and let $Y$ denote the fibred product $Y = F(\mathbf{R}^4) \times_{\mathbf{R}^4} K$ .

In [8] we showed that on $\bar{E} \times_{\mathbf{R}^4} J^1 Y$ it is possible to define a 1-form of structure $\bar{\vartheta}$ with the following property:

*Every section* $(u,\sigma)$ *of* $Y$ *induces a section* $i_{u,\sigma}$ *of the bundle* $\bar{E} \times_{\mathbf{R}^4} J^1 Y \longrightarrow \bar{E}$ *such that*

$$\bar{\vartheta}\big|_{\mathrm{Im}(i_{u,\sigma})} = \vartheta^\sigma \text{ ,"written in the coordinate system defined by } u \text{ on } \bar{E}\text{"}$$

**Remark.** The bundle of connections on an arbitrary principal bundle $p : P \longrightarrow X$ is (see Garcia [11]) *the space of orbits* $J^1P/G$ *of the structure group* $G$ *of* $P$ *operating on the 1-jet space* $J^1P$ and is thus an *affine bundle.*

The bundle $K$ is the *vector bundle associated with* $J^1P/L$ , taking as section zero the flat connection on $P$ . Every section of $K$ is given by a pair of connections $(\sigma,\sigma_o)$ but, since $\sigma_o$ is always fixed, we may identify the section $(\sigma,\sigma_o)$ with the connection $\sigma$ .

In the above-mentioned paper [ 8 ] , we also showed that on the fibred product $\bar{E} \times_{\mathbf{R}^4} J^1 Y$ it is possible to define a coordinate system $(x_i, z_j, e_\ell^h, \Gamma_k^N, p_{ij}, H_{i\ell}^h, B_{ik}^N)$ such that, with reference to it, the form $\bar{\vartheta}$ may be written:

$$(3.5) \quad \bar{\mathfrak{I}} = [dz_j - (e_i{}^\ell p_{\ell j} + \Gamma_i^N a_{kj}^N z_k)dx_i] \circ \frac{\partial}{\partial z_j} + (de_\ell{}^h - H_{i\ell}^h \, dx_i) \circ \frac{\partial}{\partial e_h{}^\ell} +$$

$$+ (d\Gamma_k^N - B_{ik}^N \, dx_i) \circ \frac{\partial}{\partial \Gamma_k^N} \quad .$$

If this expression is compared with equation (3.4), we see that the $e_i{}^\ell$ and the $\Gamma_k^N$ are no longer functions of the ring $A$ , but rather independent variables defined, respectively, on $F(\mathbf{R}^4)$ and $K$ , in the following way:

If $u_x$ is an orthonormal frame at $x$ , $e_\ell{}^h(x,u_x) = e_h^*(u_x^{-1}(\frac{\partial}{\partial x_\ell}))$ , where $(e_h)$ is the usual basis of $\mathbf{R}^4$ and $(e_h^*)$ is its dual basis.

$$\Gamma_k^N(x,(\sigma-\sigma_0)_x) = \tau_N^*((\sigma-\sigma_0)_x \frac{\partial}{\partial x_k}) \quad .$$

Finally, for every $\bar{z} = (\bar{y}, j_x^1 u, j_x^1 (\sigma-\sigma_0)) \in \bar{E} \times_{\mathbf{R}^4} J^1 Y$ , we have

$$H_i{}^h{}_\ell(\bar{z}) = -[\vartheta'_{j_x^1 u} (\frac{\partial}{\partial x_i})]e_\ell{}^h \quad ; \quad B_{ih}^N(\bar{z}) = -[\vartheta''_{j_x^1(\sigma-\sigma_0)} (\frac{\partial}{\partial x_i})]\Gamma_k^N$$

where $\vartheta'$ , $\vartheta''$ are the structure forms canonically defined on $J^1 F(\mathbf{R}^4)$ and $J^1 K$ .

### 3.4. A minimal coupling principle

Let $L$ and $V$ be two functions on the manifolds $\bar{E}$ and $J^1 Y$ , respectively.

**Definition.** A *minimal coupling between two classical fields, whose Lagrangian densities are* $L\omega$ *and* $V\omega$ , ($\omega = e\,dx_1 \wedge \ldots \wedge dx_4$ *and* $e = \mathrm{Det}(e_i{}^j)$ ) *is given by the variational problem* $(\bar{\mathfrak{I}},(L+V)\omega)$ , *defined on* $\bar{E} \times_{\mathbf{R}^4} J^1 Y$ . Its Euler-Lagrange equations are the *interaction equations* of both fields.

**Note.** For a complete exposition of the $1^{st}$ order variational calculus (= variational problems whose Lagrangians involve at most first order derivatives) on 1-jet bundles, *using the structure form* , see references [12] and [10].

**Remark 1.** In the coordinate system $(x_i, z_j, e_\ell{}^h, \ldots, B_{ik}^N)$ defined above, the form $\bar{\mathfrak{I}}$ and the Lagrangians $L$ and $V$ will be written:

$$\bar{\mathfrak{I}} \text{ in agreement with equation (3.5)}$$

$$L = L(x_i, z_j, p_{ij}) \quad \text{and} \quad V = V(x_i, e_\ell{}^h, \Gamma_k^N, H_{i\ell}^h, B_{ik}^N) \quad ,$$

both *arbitrary* functions.

However, if we take on $\bar{E} \times_{\mathbf{R}^4} J^1 Y$ a new coordinate system $(x_i, \ldots, p'_{ij}, \ldots, B_{ik}^N)$ where:

$$p'_{ij} = e_i{}^\ell p_{\ell j} + \Gamma_i^N a_{kj}^N z_k \quad ,$$

274

we then have that $\bar{\vartheta}$ , $L$ and $V$ will be given by:

$$\bar{\vartheta} = (dz_j - p'_{ij}\, dx_i) \circ \frac{\partial}{\partial z_j} + (de_h{}^{\ell} - H^{\ell}_{ih}\, dx_i) \frac{\partial}{\partial e_h{}^{\ell}} +$$

$$+ (d\Gamma^N_k - B^N_{ik}\, dx_i) \circ \frac{\partial}{\partial \Gamma^N_k}$$

$$L = L(x_i, z_j, e^i{}_{\ell}(p'_{ij} - \Gamma^N_i\, a^N_{kj}\, z_k)) \; ; \; V = V(x_i, e_{\ell}{}^h, \Gamma^N_k, H^h_{i\ell}, B^N_{ik}) \quad .$$

That is, the expression of $L$ in this second coordinate system coincides (except for the sigh) with the *interaction modified Lagrangian* given by Kibble in [13].

**Remark 2.** In the next sections 4 and 5, we shall set which conditions $V$ and $L$ should fulfill in order to be, respectively, the Lagrangian of a *gravitational field defined on the fibre bundle* $Y$ and the Lagrangian of a *matter field over the vector bundle* $E$ .

Finally, we shall finish this section giving the expression in coordinates of the *Euler-Lagrange equations* of the principle of minimal interaction defined on $E \times_{R^4} J^1 Y$ by the forms $(\bar{\vartheta}, (L+V)\omega)$ :

$$\textbf{A.-} \quad \frac{\partial}{\partial x_i}\left(\frac{\partial L}{\partial\left(\frac{\partial z_j}{\partial x_i}\right)}\right) + \Gamma^N_i\, a^N_{jk}\, \frac{\partial L}{\partial\left(\frac{\partial z_k}{\partial x_i}\right)} = \frac{\partial L}{\partial z_j}$$

$$\textbf{B.-} \quad \frac{\partial}{\partial x_i}\left(\frac{\partial V}{\partial\left(\frac{\partial \Gamma^A_h}{\partial x_i}\right)}\right) - \frac{\partial V}{\partial \Gamma^A_h} = -a^A_{kj}\, z_k\, \frac{\partial L}{\partial\left(\frac{\partial z_j}{\partial x_h}\right)}$$

$$\textbf{C.-} \quad \frac{\partial}{\partial x_i}\left(\frac{\partial Ve_{\ell}{}^h}{\partial\left(\frac{\partial e_h{}^h}{\partial x_i}\right)}\right) - \frac{\partial Ve_{\ell}{}^h}{\partial e_h} = e^h{}_{\ell}L - e^i{}_{\ell}\left(\frac{\partial z_j}{\partial x_i} - \Gamma^A_i\, a^A_{kj}\, z_k\right) \frac{\partial L}{\partial\left(\frac{\partial z_j}{\partial x_h}\right)}$$

## 4. Utiyama's Theorem

In Yang-Mills theory, all possible Lagrangians of a gauge field are functions on $J^1 K$ ($K$ is now the bundle of connections on an arbitrary principal bundle) which are *invariant under the gauge group*.

It is natural to generalize this definition by saying:

*The Lagrangians of the gravitational field* $\pi : Y \longrightarrow R^4$ *are all functions* $V$ *on* $J^1 Y$ *which are invariant under the group* Aut $P$ *of automorphisms of* $p : P \longrightarrow R^4$ .

Consequently, the first problem to solve in this section is: *How does the group* Aut $P$ *act on* $Y$?; or more specifically, which is the infinitesimal version of the action of Aut $P$ on $Y$ ?.

### 4.1. Transformation properties of the gravitational potentials

Let $\zeta_2$ and $\zeta_3$ be the morphisms of $\text{aut}\,P$ into $\Gamma(T_{\pi_2}F(\mathbf{R}^4))$ and $\Gamma(T_{\pi_3}K)$ defined as follows:

**a)** If $\gamma = \xi_i(x)\dfrac{\partial}{\partial x_i} + h_N(x)\tilde{\tau}_N$ is an element of $\text{aut}\,P$, then $\zeta_2(\gamma)$ is given by $D_1^\gamma + D_2^\gamma$, where:

**1.-** $D_1^\gamma = \xi_i\dfrac{\partial}{\partial x_i} - \dfrac{\partial \xi_\ell}{\partial x_j}\,e_\ell{}^m\,\dfrac{\partial}{\partial e_j{}^m}$ is the only vector field on $F(\mathbf{R}^4)$ verifying,

$$(\pi_2)_* D_1^\gamma = \xi_i\frac{\partial}{\partial x_i} = P_*(\gamma) \quad \text{and} \quad L_{D_1^\gamma}(\eta) = 0$$

$\eta$ being the *soldering form* on $F(\mathbf{R}^4)$.

**2.-** $D_2^\gamma = -c^m_{Nk}\,h_N\,e_j{}^k\,\dfrac{\partial}{\partial e_j{}^m} = h_N\left(e_j{}^i\,\dfrac{\partial}{\partial e_j{}^\ell} - e_j{}^\ell\,\dfrac{\partial}{\partial e_j{}^i}\right)$ is the image of $h_N\tilde{\tau}_N$ by the natural isomorphism between the bundles $P$ and $F(\mathbf{R}^4)$ ; $N = i,\ell$.

Therefore:

$$(4.1) \qquad \zeta_2(\gamma) = \xi_i\frac{\partial}{\partial x_i} - \left(\frac{\partial \xi_\ell}{\partial x_j}\,e_\ell{}^m + c^m_{Nk}\,h_N\,e_j{}^k\right)\frac{\partial}{\partial e_j{}^m}$$

**b)** The coordinate expression of $\zeta_3(\gamma)$ is:

$$(4.2) \qquad \zeta_3(\gamma) = \xi_i\frac{\partial}{\partial x_i} + \left(\frac{\partial h_N}{\partial x_i} - \Gamma^N_k\frac{\partial \xi_k}{\partial x_i} - c^N_{MA}\,h_M\,\Gamma^A_i\right)\frac{\partial}{\partial \Gamma^N_i}\ .$$

(see [8], page 64 for an intrinsic definition of $\zeta_3$).

From formulae (4.1) and (4.2) we infer that *the transformation properties of the gravitational potentials* $e_j{}^m$ and $\Gamma^N_i$ are given by the following equations:

$$\delta e_j{}^m = -\left(\frac{\partial \xi_\ell}{\partial x_j}\,e_\ell{}^m + c^m_{Nk}\,h_N\,e_j{}^k\right)$$

$$\delta \Gamma^N_i = \frac{\partial h_N}{\partial x_i} - \Gamma^N_k\frac{\partial \xi_k}{\partial x_i} - c^N_{MA}\,h_M\,\Gamma^A_i$$

which agree with the transformation properties of the "vierbein components" $(h^M_k)$ and of the "local affine connections" $(A^{ij}_\mu)$ introduced by Kibble in [13] (see page 216, formulae (4.11) and (4.9)).

Because $Y = F(\mathbf{R}^4)\times_{\mathbf{R}^4} K$, the morphisms $\zeta_2$ and $\zeta_3$ allow us to define a new morphism $\zeta'$ of $\text{aut}\,P$ into $\Gamma(T_\pi Y)$ in the following way:

$$(4.3) \qquad \zeta'(\gamma) = \xi_i\frac{\partial}{\partial x_i} - \left(\frac{\partial \xi_\ell}{\partial x_j}\,e_\ell{}^m + c^m_{Nk}\,h_N\,e_j{}^k\right)\frac{\partial}{\partial e_j{}^m} +$$

$$+ \left(\frac{\partial h_N}{\partial x_i} - \Gamma^N_k\frac{\partial \xi_k}{\partial x_i} - c^N_{MA}\,h_M\,\Gamma^A_i\right)\frac{\partial}{\partial \Gamma^N_i}$$

### 4.2. Lagrangians of the gravitational field

**Definition.** The Lagrangians of the gravitational field $\pi : Y \longrightarrow \mathbb{R}^4$ are the functions $V$ on $J^1 Y$ such that

$$L_{\overline{\zeta'(\gamma)}} (V\omega) = 0 \quad ,$$

for any $\gamma \in \operatorname{aut} P$, $\overline{\zeta'(\gamma)}$ being the 1-jet prolongation to $J^1 Y$ of $\zeta'(\gamma)$, $\omega = e\,dx_1 \wedge \ldots \wedge dx_4$ and $e = \operatorname{Det}(e_i{}^j)$ .

**Theorem.** *All Lagrangians $V$ are characterized by a finite number of conditions:*

$$\frac{\partial V}{\partial B^A_{ik}} + \frac{\partial V}{\partial B^A_{ki}} = 0 \quad ; \quad \frac{\partial V}{\partial H^m_{ij}} + \frac{\partial V}{\partial H^m_{ji}} = 0 \quad ; \quad \frac{\partial V}{\partial x_k} = 0 \quad ;$$

$$(4.4) \qquad \frac{\partial V}{\partial \Gamma^N_i} = c^A_{NM}\, \Gamma^M_j\, \frac{\partial V}{\partial B^A_{ij}} + c^m_{Nk}\, e_j{}^k\, \frac{\partial V}{\partial H^m_{ij}} \quad ;$$

$$(4.5) \qquad \frac{\partial V}{\partial e_i{}^n} = R^A_{kn}\, \frac{\partial V}{\partial B^A_{ik}} + T^m_{jn}\, \frac{\partial V}{\partial H^m_{ij}} \quad ;$$

$$c^M_{NA}\, \Gamma^A_j\, \frac{\partial V}{\partial \Gamma^M_j} + c^m_{Nk}\, e_j{}^k\, \frac{\partial V}{\partial e_j{}^m} + c^A_{NM}\, B^M_{ij}\, \frac{\partial V}{\partial B^A_{ij}} + c^m_{N\ell}\, H^\ell_{ij}\, \frac{\partial V}{\partial H^m_{ij}} = 0 \quad .$$

where

$$R^A_{kn} := e^\ell{}_n (B^A_{k\ell} - B^A_{\ell k} - c^A_{NM}\, \Gamma^N_\ell\, \Gamma^M_k)$$

$$T^m_{jn} := e^\ell{}_n (H^m_{j\ell} - H^m_{\ell j} - c^m_{Nk}\, \Gamma^N_\ell\, e_j{}^k)$$

**Proof.** As the 1-jet prolongation of $\zeta'(\gamma)$ to $J^1 Y$ is

$$\overline{\zeta'(\gamma)} = \zeta'(\gamma) + \left( \frac{\partial^2 h_N}{\partial x_i \partial x_k} - \frac{\partial^2 \xi_\ell}{\partial x_i \partial x_k}\, \Gamma^N_\ell - c^N_{MA}\, \Gamma^A_i\, \frac{\partial h_M}{\partial x_i} - B^N_{\ell k}\, \frac{\partial \xi_\ell}{\partial x_i} - \right.$$

$$\left. - B^N_{ih}\, \frac{\partial \xi_h}{\partial x_k} - c^N_{MA}\, h_M\, B^A_{ik} \right) \frac{\partial}{\partial B^N_{ik}} - \left( \frac{\partial^2 \xi_\ell}{\partial x_i \partial x_k}\, e_\ell{}^m + c^m_{Nj}\, \frac{\partial h_N}{\partial x_i}\, e_k{}^j + \right.$$

$$\left. + H^m_{nk}\, \frac{\partial \xi_n}{\partial x_i} + H^m_{i\ell}\, \frac{\partial \xi_\ell}{\partial x_k} + c^m_{N\ell}\, h_N\, H^\ell_{ik} \right) \frac{\partial}{\partial H^m_{ik}} \quad ,$$

$L_{\overline{\zeta'(\gamma)}} (V\omega) = 0$ , for an arbitrary $\gamma$ , whenever the coefficients of $\dfrac{\partial^2 h_N}{\partial x_i \partial x_k}$ , $\dfrac{\partial^2 \xi_\ell}{\partial x_i \partial x_k}$ , $\dfrac{\partial h_N}{\partial x_i}$ , $\dfrac{\partial \xi_\ell}{\partial x_i}$ , $h_N$ , $\xi_\ell$ vanish, i.e. whenever $V$ fulfills all the conditions given in the statement.

As was already mentioned in the Introduction, the study of the functions $V$ which will fulfill all these conditions together with the generalization, for gauge

theories with a "non-Internal" group, of the geometric results given by Garcia in [3],
will be dealt in a forthcoming publication.

We shall now finish the section by seeing how equations B and C, given in
3.4, can be written when $V$ is the Lagrangian of a gravitational field.

If we insert (4.4) and (4.5), we find:

$$\mathbf{B'.-}\quad \frac{\partial}{\partial x_i}\left(\frac{\partial V}{\partial\left(\frac{\partial\Gamma^A_h}{\partial x_i}\right)}\right) - c^N_{AM}\,\Gamma^M_j\,\frac{\partial V}{\partial B^N_{hi}} - c^m_{Ak}\,e^k_j\,\frac{\partial V}{\partial H^m_{hj}} = -a^A_{kj}\,z_k\,\frac{\partial L}{\partial\left(\frac{\partial z_j}{\partial x_h}\right)}$$

$$\mathbf{C'.-}\quad \frac{\partial}{\partial x_i}\left(\frac{\partial V}{\partial\left(\frac{\partial e^\ell_h}{\partial x_i}\right)}\right) - e^h_\ell V - R^A_{k\ell}\,\frac{\partial V}{\partial B^A_{hk}} - T^m_{j\ell}\,\frac{\partial V}{\partial H^m_{hj}} =$$

$$= e^h_\ell L - e^i_\ell\,\left(\frac{\partial z_j}{\partial x_i} - \Gamma^A_i\,a^A_{mj}\,z_m\right)\frac{\partial L}{\partial\left(\frac{\partial z_j}{\partial x_h}\right)}$$

which are the equations of the gravitational field minimally coupled to a matter field
with an arbitrary Lagrangian $L$.


## 5. Lagrangians of the matter field

Because $\text{aut}\,P \simeq A \otimes_R g_P$, the morphism $\zeta_1$ of $\text{aut}\,P$ in $\Gamma(T_{\pi_1}E)$, given in
section 3.2, allows us furthermore to define a representation $\zeta_1$ of the Lie algebra
$g_P$ into $\Gamma(T_{\pi_1}E)$.


**Definition 5.1.** We shall say that a matter field of Lagrangian $L$, defined on
$\bar{E}$, is *P-invariant* when

$$(5.1)\qquad\qquad \bar{X}_i(L) = \bar{Z}_N(L) = 0 \ ,$$

where $\bar{X}_i$ and $\bar{Z}_N$ are the $\overset{\sigma}{\vartheta}{}^o$-prolongations to $\bar{E}$ of the vector fields
$X_i := \zeta_1(\epsilon_i)$ ; $Z_N := \zeta_1(\tau_N)$ .

[By the $\overset{\sigma}{\vartheta}{}^o$-*prolongation to $\bar{E}$ of a vector field* $D$ given on $E$ we shall
refer to the unique vector field $\bar{D}$ on $\bar{E}$, projectable on $D$ and such that
$L_{\bar{D}}\overset{\sigma}{\vartheta}{}^o = B \cdot \overset{\sigma}{\vartheta}{}^o$ , cf. [12], [10]].

In the coordinate system $(x_i, z_j, p_{ij})$ we shall have that

$$\bar{X}_i = \frac{\partial}{\partial x_i}$$

$$\bar{Z}_N = x_i\,\frac{\partial}{\partial x_j} - x_j\,\frac{\partial}{\partial x_i} + a^N_{hk}\left(z_h\,\frac{\partial}{\partial z_k} + p_{\ell h}\,\frac{\partial}{\partial p_{\ell k}}\right) + c^h_{N\ell}\,p_{hm}\,\frac{\partial}{\partial p_{\ell m}} \ ; \ (N = i,j) \ .$$

278

Accordingly, conditions (5.1) will be expressed as:

$$(5.2) \quad \begin{cases} \dfrac{\partial L}{\partial x_i} = 0 \\[2ex] a^N_{hk}\left(z_h \dfrac{\partial L}{\partial z_k} + p_{\ell h}\dfrac{\partial L}{\partial p_{\ell k}}\right) + c^k_{N\ell}\, p_{km}\dfrac{\partial L}{\partial p_{\ell m}} = 0 \quad . \end{cases}$$

By using the morphisms $\zeta_1$ , $\zeta_2$ and $\zeta_3$ it is also possible to construct a morphism $\zeta$ of aut $P$ into vector fields on $E \times_{R^4} Y$ , projectable on $R^4$ , in the following way:

$$\text{If} \quad \gamma = \xi_i \frac{\partial}{\partial x_i} + h_N\, \tilde{\tau}_N \in \text{aut } P, \text{ then, } \zeta(\gamma) = h_N\, a^N_{kj}\, z_k \frac{\partial}{\partial z_j} + \zeta'(\gamma) \quad .$$

**Definición 5.2.** We shall say that a matter field with Lagrangian density $L\omega$ , defined on $E \times_{R^4} J^1 Y$ , is aut $P$-*invariant* when for every $\gamma \in \text{aut } P$ ,

$$L_{\overline{\zeta(\gamma)}}(L\omega) = 0$$

where $\overline{\zeta(\gamma)}$ is the $\bar{\mathfrak{F}}$-prolongation to $\bar{E} \times_{R^4} J^1 Y$ of the vector field $\zeta(\gamma)$ .

**Theorem.** A *matter field is* aut P-*invariant if and only if it is* P-*invariant.*

**Proof.** Simple calculus shows that

$$\overline{\zeta(\gamma)}\,(p_{\ell m}) = h_N\,(a^N_{km}\, p_{\ell k} + c^k_{N\ell}\, p_{km}) \quad .$$

Consequently

$$L_{\overline{\zeta(\gamma)}}(L\omega) = \overline{\zeta(\gamma)}\,(L)\omega + L(\overline{\zeta(\gamma)}e)dx_1 \wedge \ldots \wedge dx_4 + Le\, L_{\overline{\zeta(\gamma)}}(dx_1 \wedge \ldots \wedge dx_4) =$$

$$= (\zeta(\gamma)L + h_N(a^N_{km}\, p_{\ell k} + c^k_{N\ell}\, p_{km})\frac{\partial L}{\partial p_{\ell m}})\omega - L\frac{\partial \xi_\ell}{\partial x_\ell}\omega + L\frac{\partial \xi_\ell}{\partial x_\ell}\omega = 0 \quad .$$

Since $L = L(z_j, p_{ij})$ and also fulfills conditions (5.2).

## 6. Gravitational fields coupled to P-invariant matter fields

Let now suppose that we have a minimal coupling principle defined on $\bar{E} \times_{R^4} J^1 Y$ by the forms $(\bar{\mathfrak{F}}, (L+V)\omega)$ , where $L$ *is the Lagrangian of a* P-*invariant matter field and* $V$ *is one of the possible Lagrangians of the gravitational field*, i.e., $L$ fulfills conditions (5.2) and $V$ satisfies all the equations given in the Theorem of section 4.2.

Let we furthermore suppose that $s' = (e_\ell{}^h(x), \Gamma^N_\ell(x))$ is a section of $\pi : Y \longrightarrow R^4$ and a solution of the equations $B'$ and $C'$ ($\S4.2$), and let $\gamma^{s'}_n = e^i_n(x)(\frac{\partial}{\partial x_i} + \Gamma^N_i(x)\tilde{\tau}_N) =$

$$= e^i_n(x)\frac{\partial}{\partial x_i} + (\sigma - \sigma_0)(e^i_n(x)\frac{\partial}{\partial x_i}) \text{ , where } \sigma - \sigma_0 = \Gamma^N_\ell(x)dx_\ell\tilde{\tau}_N \text{ . Obviously, } \gamma^{s'}_n \in \text{aut } P$$

and so, using the morphism $\zeta$ , we obtain

$$D_n^{s'} := \zeta(\gamma_n^{s'}) = e^i_{\ n}(x)(\frac{\partial}{\partial x_i} + \Gamma^A_i(x)a^A_{kj}\, z_k\, \frac{\partial}{\partial z_j}) +$$

$$+ (e^i_{\ n}(x)\,\frac{\Gamma^N_i(x)}{\partial x_j} + \Gamma^N_i(x)\,\frac{\partial e^i_{\ n}(x)}{\partial x_j} - \Gamma^N_i\,\frac{\partial e^i_{\ n}(x)}{\partial x_j} - c^N_{MA}\,\Gamma^A_j\, e^i_{\ n}(x)\Gamma^M_i(x))\,\frac{\partial}{\partial\Gamma^N_j} -$$

$$- (\frac{\partial e^i_{\ n}(x)}{\partial x_j}\, e_i^{\ m} + c^m_{Nk}\, e_j^{\ k}\, e^\ell_{\ n}(x)\,\Gamma^N_\ell(x))\,\frac{\partial}{\partial e_j^{\ m}}$$

Moreover, if $\bar{D}_n^{s'}$ is the $\bar{\mathfrak{I}}$-prolongation of $D_n^{s'}$ to $\bar{E}\times_{R^4} J^1 Y$ , we have $L_{\bar{D}_n^{s'}}((L+V)\omega) = 0$ since both $L$ and $V$ are aut $P$-invariant.

This means that $\bar{D}_n^{s'}$ is an *infinitesimal automorphism* of the variational problem $(\bar{\mathfrak{I}},(L+V)\omega)$ and its corresponding *Noether invariant* $y_n^{s'}$ can be calculated, using the, *Poincaré-Cartan form* $\bar{\theta}$ :

$$y_n^{s'} = i\bar{D}_n^{s'}\cdot\bar{\theta}$$

where $\bar{\theta} = \mathfrak{I}\wedge\Omega_{L+V} - (L+V)\omega$ and $\Omega_{L+V}$ is the *Legendre-transformation form* (cf. [12], [6], [10]).

Let $\bar{s}$ now be the $\bar{\mathfrak{I}}$-prolongation to $\bar{E}\times_{R^4} J^1 Y$ of a stationary section $s = (z_j(x), e_\ell^{\ h}(x), \Gamma^N_\ell(x))$ of the variational problem we are dealing with, such that $(e_\ell^{\ h}(x), \Gamma^N_\ell(x)) = s'$ .

[The $\bar{\mathfrak{I}}$-prolongation to $\bar{E}\times_{R^4} J^1 Y$ of a section $s:R^4\longrightarrow E\times_{R^4} Y$ is the only section $\bar{s}:R^4\longrightarrow \bar{E}\times_{R^4} J^1 Y$ satisfying the conditions:

$$\mathfrak{I}|_{\bar{s}} \equiv 0 \quad ; \quad \bar{\pi}\circ\bar{s} = s \ ] \quad .$$

Let us calculate $y_n^{s'}|_s$ :

$$i\bar{D}_n^{s'}\,\bar{\theta}|_{\bar{s}} = [dz_j - (e_i^{\ \ell}p_{\ell j} + \Gamma^A_i\, a^A_{kj}\, z_k)dx_i](D_n^{s'})(-1)^h\, e^h_{\ m}\,\frac{\partial L}{\partial p_{mj}}\, e\omega_h|_{\bar{s}} +$$

$$+ (d\Gamma^A_h - B^A_{ih}\, dx_i)(D_n^{s'})(-1)^\ell\,\frac{\partial V}{\partial B^A_{\ell h}}\, e\omega_\ell|_{\bar{s}} +$$

$$+ (de_h^{\ \ell} - H^\ell_{ih}\, dx_i)(D_n^{s'})(-1)^k\,\frac{\partial V}{\partial H^\ell_{kh}}\, e\omega_k|_{\bar{s}} - i\bar{D}_n^{s'}\,(L+V)\omega|_{\bar{s}} =$$

$$= -(-1)^h\, e^i_{\ n}(x)(\frac{\partial z_j}{\partial x_i} - \Gamma^A_i(x)\, a^A_{kj}\, z_k(x))\,\frac{\partial L}{\partial(\frac{\partial z_j}{\partial x_h})}\, e\omega_h +$$

$$+ (-1)^h\, [e^i_{\ n}(x)(\frac{\partial\Gamma^N_i}{\partial x_j} - c^N_{MA}\,\Gamma^M_i(x)\,\Gamma^A_j(x)) - e^i_{\ n}(x)\,\frac{\partial\Gamma^N_j}{\partial x_i}]\,\frac{\partial V}{\partial(\frac{\partial\Gamma^N_j}{\partial x_h})}\, e\omega_h +$$

$$+ -(-1)^h \left[ \frac{\partial e^i_n}{\partial x_j} \, e_i{}^m(x) + c^m_{Nk} \, e_j{}^k(x) \, e^\ell{}_n(x) \, \Gamma^N_\ell(x) + e^i{}_n(x) \frac{\partial e_j{}^m}{\partial x_i} \right] \cdot$$

$$\cdot \frac{\partial V}{\partial \left( \frac{\partial e_j{}^m}{\partial x_h} \right)} e\omega_h - (-1)^{h+1} \, e^h{}_n(x)(L+V)\omega_h e \, , \quad \text{where} \quad \omega_h = dx_1 \wedge \cdots \wedge \widehat{dx_h} \wedge \cdots \wedge dx_4 \, .$$

Using the fact that $\dfrac{\partial e^i_n}{\partial x_j} \, e_i{}^m = -e^i{}_n \dfrac{\partial e_i{}^m}{\partial x_j}$ , we have:

$$y^{s'}_{n|\bar{s}} = (-1)^{h+1} \left[ e^i{}_n \left( \frac{\partial z_j}{\partial x_i} - \Gamma^A_i \, a^A_{kj} \, z_k \right) \frac{\partial L}{\partial \left( \frac{\partial z_j}{\partial x_h} \right)} - e^i{}_n \left( \frac{\partial \Gamma^N_i}{\partial x_j} - \frac{\partial \Gamma^N_j}{\partial x_i} - c^N_{MA} \, \Gamma^M_i \, \Gamma^A_j \right) \cdot \frac{\partial V}{\partial \left( \frac{\partial \Gamma^N_j}{\partial x_h} \right)} - \right.$$

$$\left. - e^i{}_n \left( \frac{\partial e_i{}^m}{\partial x_j} - \frac{\partial e_j{}^m}{\partial x_i} - c^m_{Nk} \, \Gamma^N_i \, e_j{}^k \right) \frac{\partial V}{\partial \left( \frac{\partial e_j{}^m}{\partial x_h} \right)} - e^h{}_n(L+V) \right] \omega_h e =$$

$$= (-1)^{h+1} \left[ e^i{}_n \left( \frac{\partial z_j}{\partial x_i} - \Gamma^A_i \, a^A_{kj} \, z_k \right) \frac{\partial L}{\partial \left( \frac{\partial z_j}{\partial x_h} \right)} - R^N_{jn} \frac{\partial V}{\partial \left( \frac{\partial \Gamma^N_j}{\partial x_h} \right)} \right.$$

$$\left. - \Gamma^m_{jn} \frac{\partial V}{\partial \left( \frac{\partial e_j{}^m}{\partial x_h} \right)} - e^h{}_n \, (L+V) \right] \omega_h e = -y^{s'}_{nh} \, \omega_h e \, |_{\bar{s}} \, .$$

And, so the set $C'$ of equations of the gravitational field can be written as:

$$\frac{\partial}{\partial x_i} \left( \frac{\partial V}{\partial \left( \frac{\partial e_h{}^n}{\partial x_i} \right)} \right) = y^{s'}_{nh}$$

Let us now consider the vector field $\tilde{\tau}_N \in \text{aut } P$ and write $D_N = \zeta(\tilde{\tau}_N)$ . In coordinates we have:

$$D_N = a^N_{jk} \, z_j \frac{\partial}{\partial z_k} - c^M_{NA} \, \Gamma^A_i \frac{\partial}{\partial \Gamma^M_i} - c^m_{Nk} \, e_j{}^k \frac{\partial}{\partial e_j{}^m}$$

The corresponding Noether invariant is

$$y_{N|\bar{s}} = iD_N \cdot \bar{\Theta}|_{\bar{s}} = y_{Nh} \, e\omega_h|_{\bar{s}} = (-1)^h \left[ a^N_{kj} \, z_k \frac{\partial L}{\partial \left( \frac{\partial z_j}{\partial x_h} \right)} - \right.$$

$$\left. - c^M_{NA} \, \Gamma^A_i \frac{\partial V}{\partial \left( \frac{\partial \Gamma^M_i}{\partial x_h} \right)} - c^m_{Nk} \, e_j{}^k \frac{\partial V}{\partial \left( \frac{\partial e_j{}^m}{\partial x_h} \right)} \right] \omega_h e \, .$$

Hence, the set  B'  of equations of the gravitational field becomes:

$$\frac{\partial}{\partial x_i}\left(\frac{\partial V}{\partial(\frac{\partial \Gamma^N_h}{\partial x_i})}\right) = - \,{}^y N_h$$

## References


[1]  C.N. Yang and R.L. Mills - Conservation of isotopic spin and isotopic gauge invariance, Phys. Rev. 96 (1954), 191-195.

[2]  R. Utiyama - Invariant theoretical interpretation of interaction, Phys. Rev. 110 (1956), 1597-1607.

[3]  P.L. García - Gauge algebras, curvature and symplectic structure, J. Differential Geom. 12 (1977), 209-227.

[4]  D. Bleecker - Gauge theory and variational principles, Addison Wesley (1981), Reading, Mass.

[5]  D.J. Eck - Gauge-natural bundles and generalized gauge theories - Mem. Amer. Math. Soc. Vol. 33, nº 247 (1981).

[6]  P.L. García and A. Pérez-Rendón - Reducibility of the symplectic structure of minimal interaction, In "Differential Geometrical Methods in Mathematical Physics", Bonn 1977, Lecture Notes in Math. nº 676, 409-433.

[7]  L. Mangiarotti and M. Modugno - On the geometric structure of gauge theories. Preprint. U. Firenze.

[8]  A. Pérez-Rendón and D.H. Ruipérez - Gauge theories on homogeneous manifolds, In "Differential Geometric Methods in Mathematical Physics", Math. Phys. Stud. nº 6, D. Reidel, Dordrecht (1984), 55-74.

[9]  S. Sternberg - On the role of field theories in our physical conception of geometry, In "Differential Geometrical Methods in Mathematical Physics", Bonn 1977, Lecture Notes in Math. nº 676, 1-80.

[10]  A. Pérez-Rendón - Principles of minimal interaction, In "Proc. of the Intern. Meeting on Geometry and Physics", Florence 1982, Pitagora Editrice Bologna (1983), 185-216.

[11]  P.L. García - Connections and 1-jet fibre bundles, Rend. Sem. Mat. Univ. Padova 47 (1972), 227-242.

[12]  P.L. García and A. Pérez-Rendón - "Symplectic approach to the theory of quantized fields I and II - Comm. Math. Phys. 13 (1969), 24-44; and Arch. Rational Mech. Anal. 43 (1971), 101-124.

[13]  T.W.B. Kibble - Lorentz invariance and gravitational field, J. Mathematical Phys. 2 (1961), 212-221.

[14]  F.W. Hehl - Four Lectures on Poincaré gauge field theory.- In "Cohomology and Gravitation. Spín, Torsion, Rotation and supergravity", Proc. of the Erice-School 1979, Plenum, New York (1980).


Received April 13, 1985.

GEOMETRODYNAMICS PROCEEDINGS (1985), pp. 283-420
edited by A. Pràstaro
© 1985 by World Scientific Publishing Co.

# DYNAMIC CONSERVATION LAWS

Agostino Pràstaro

Dipartimento di Matematica, Università della Calabria
87036 Arcavacata di Rende (CS), Italy.

# 1.- INTRODUCTION

The historian merit of the Noether's theorem is to give a general framework
to find dynamic conservation laws for Lagrangian dynamic systems, (see refs.
[1] ). But, when the system is not Lagrangian the Noether's framework fails
and, however, even if for Lagrangian systems one has cases where conservation
laws can be checked which are not of Noetherian type.

In this paper we have further developed a geometric framework, first introduced in ref. [12] , to discover conservation laws of a dynamic system
which is more general of the Noetherian one.  This is obtained by considering
the differential invariants associated to any finite sub-group H of the symmetry group of a continuum system, that are numerical functions $f : J D^k(C) \to \mathbb{R}$
on the space of the states $J D^k(C)$ such that $X.f = 0$ for any infinitesimal generator X of the natural action of H on $J D^k(C)$.  In fact  f  can be considered
as a k-order Lagrangian conserved under the group H and so by using the
Noether's theorem for such Lagrangians we can obtain closed differential
forms  $\gamma \equiv (j^\infty c)^* \beta$  for any dynamic configuration c of the continuum
system (CS) which is also an extremal for f.

On the other hand, it is well known that some entities that are conserved on
the classical level cannot be conserved on the microscopic scale, where
quantum effects are no negligible.  So, a correct unitary description of
dynamic conservation laws in a field theory must be placed in a framework
large enough to take into account also the quantum phenomena.

Thus, we are conduced to ask us which formulation of quantum mechanics is
today suitable to interpretate the microscopic physical phenomena.  We think
that despite of the indesputable success of the classical quantum theory
(see e.g. refs.   [3] )   there are serious obstructions in order to
adopt that point of view in a larger framework such as the unified field
theory.  We belive that these difficulties cannot be removed also taking
into account some more recent geometric reformulations, (see e.g. refs. [4,
5] ).  In fact, the most serious obstruction in adopting the usual quantization
point of view (=Dirac's approach) in field theory is to ascribe to the
combination of non-linearity with the effects of fully covariance.

However, the essential meaning of quantum effects in field theory is on one side the discretization of energies and charges (quanta) and on the other the possibility for a system to change from a classical configuration to another. Now, the discretization of charges can be seen as a phenomen connected with the topology of the basis space, while the quantum jumps are to ascribe to a "microdynamics", non-relevable on macroscopic level. This microdynamics is essentially responsable of the microchanges of configuration of a conti- nuum system. So, in order to take into account the quantum effects we shall introduce another geometric structure to add to that characterizing a classical continuum system. This is the quantum situs. Roughly speaking a quantum situs is a collection of equivalence classes of configurations, two configurations being equivalent if together coincide on two 3-dimensional submanifolds with two dynamic configurations. Then, by using a path-integration point of view we can calculate the transition amplitude between two dynamic configurations even if this transition is strictly forbidden classically, as the configurations that interpolate the initial and final dynamic configurations, are not neces- sarily solutions of the field equation.

The classic Schroedinger equation (that has not a covariant meaning),is now substituted by another equation (S) , called with the same name, that instead plaies the role to identify an internal constraint into the quantum situs. In practice, this should correspond to the artful, (and not well justified, from the mathematical point of view) proceeding of"renormalization"used in the classic quantum field theory).

Finally, we reformulate the concept of Dirac's quantization by considering the representation of physical fields by means of linear operators on suita- ble Hilbert spaces such that the corresponding restriction on the set of solutions of tne dynamic equation gives a (anti)-hermitian representation. Then, a direct relation by the quantum-cobordism is obtained by using the concept of spectral measure on a suitable subset ot the quantum situs. In this way the concept of quantization is not more negatively influenced by the non-linearity of the system and it appears as a direct expression of the formal properties of the dynamic equation.

## 2.-NOTATIONS

We shall always consider differentiable manifolds of finite dimension and class $C^\infty$ and maps of class $C^\infty$.

. $\mathcal{D}^k$: <u>derivative functor</u> of order k [10] .

. $D^k f:V \to \mathcal{D}^k(V,W)$: <u>k-derivative of a map</u> f:V$\to$W between differentiable manifolds. $D^k s:M \to \mathcal{D}^k(W) \subset \mathcal{D}^k(V,W)$: k-derivative of a section s of $\pi$:W$\to$M.

. $J\mathcal{D}^k(W)$: <u>jet-derivative space of order k of the fiber bundle</u> $\pi$:W$\to$M. One has $D^k s:M \to J\mathcal{D}^k(W)$. $\pi_k:J\mathcal{D}^k(W) \to M$ and $\pi_{k,k'}:J\mathcal{D}^k(W) \to J\mathcal{D}^{k'}(W)$, $0 \leqslant k' \leqslant k$, fibered structures of $J\dot{\mathcal{D}}^k(W)$.

. $\Pi\mathcal{D}^k(W)$: open subbundle of $J\mathcal{D}^k(W)$ of k-derivative of sections with maximal rank.

. $\widehat{J\mathcal{D}}^k(W,W)$: subbundle of $J\mathcal{D}^k(W,W)$ of k-derivative of fiber bundle morphisms f:W$\to$W over differentiable maps $\underline{f}$:M$\to$M of $\pi$:W$\to$M.

. $\overset{\smile}{J\mathcal{D}}{}^k(W)$: <u>k-order sesquiholonomic prolongation</u> of W: defined by the kernel of the following double flèche: $J\mathcal{D}J\mathcal{D}^{k-1}(W) \longrightarrow J\mathcal{D}J^k\bar{\mathcal{D}}^2(W)$
$$\downarrow$$
$$J\mathcal{D}^{k-1}(W).$$

. $\overline{J\mathcal{D}}^k(W)$: <u>k-order semiholonomic prolongation</u> of W: defined by the kernel of the following double flèche: $J\mathcal{D}\overline{J\mathcal{D}}^{k-1}(W) \longrightarrow J\mathcal{D}\overline{J\mathcal{D}}^{k-2}(W)$ ; $\overline{J\mathcal{D}}(W)=J\mathcal{D}(W)$.
$$\downarrow$$
$$J\mathcal{D}^{k-1}(W)$$

One has : (a) $J\mathcal{D}^k(W) \subset \overset{\smile}{J\mathcal{D}}{}^k(W) \subset \overline{J\mathcal{D}}^k(W)$; (a) $\overline{J\mathcal{D}}(W)=J\mathcal{D}(W)=\overset{\smile}{J\dot{\mathcal{D}}}(W)$; (c) $\overline{J\mathcal{D}}^2(W)=\overset{\smile}{J\mathcal{D}}^2(W)$.

. $\Pi^k(M,N)$ ,(dimM=dimN): open subbundle of $J\mathcal{D}^k(M,N)$ of invertible transformations
$$M \to N.$$

. $\Pi^k(M)$: open subbundle of $J\mathcal{D}^k(M,M)$ with $\Pi^k(M)=\Pi^k(M,M)$. $p_k:\Pi^k(M) \to M$ and $p_{k,k'}:\Pi^k(M) \to \Pi^{k'}(M)$, $0 \leqslant k' \leqslant k$, $\bar{p}_k:\Pi^k(M) \to M$ natural projections: $p_k$=source projection; $\bar{p}_k$=target projection.

. $\widehat{\Pi}^k(W)$: open subbundle of $\Pi^k(W)$ of invertible fibered transformations of $\pi$:W$\to$M. $\hat{p}_k:\widehat{\Pi}^k(W) \to W$ , $\hat{p}_{k,k'}:\widehat{\Pi}^k(W) \to \widehat{\Pi}^{k'}(W)$, $0 \leqslant k' \leqslant k$ , $\hat{\bar{p}}_k:\widehat{\Pi}^k(W) \to W$ natural projections: $\hat{p}_k$=source projection; $\hat{\bar{p}}_k$=target projection.

. TX: <u>tangent bundle</u> of a manifold X; $T^*X$: <u>cotangent bundle</u> of X. T: tangent functor ; $T^*$: cotangent functor.

. $C^\infty(W)$: space of $C^\infty$ sections of a fiber bundle $\pi$:W$\to$M.

. $C^\infty(X)$: space of $C^\infty$ sections of a fiber bundle $\pi$:W$\equiv$X$\times \mathbb{R} \to$X.

. $\hat{C}^\infty(TW)$: space of $C^\infty$-vector fields W$\to$TW on $\pi$:W$\to$M , $\pi$-related with vector fields on M.

. $J\mathcal{D}^k(TW)$: subbundle of $J\mathcal{D}^k(TW)$ of k-order jet-derivatives of vector fields on W $\pi$-related with vector fields on W.

. $\overset{(k)}{X}$ : <u>k-prolongation of a vector field</u> X:W→TW  $\pi$-related with a vector field X:M→TM  $[10]$  .

. $\Lambda^o_p X$: <u>bundle of differentiable p-forms</u> on X.

. $S^r_s X$ : <u>bundle of symmetric tensors of type (r,s)</u> on X.

. d$\alpha$: <u>differential of a differential form</u> $\alpha$.

. $\delta\alpha$: <u>codifferential</u> of $\alpha$ .

. $\partial f$ : <u>infinitesimal variation</u> of a map f:ℝ x X → Y   $[10]$.

. $f^* s$ : <u>pull-back of a superfield of geometric objects</u> s by means of a morphism f.

. $f^* W$: <u>pull-back of a fiber bundle</u>, $\pi$:W→M  by means of a map f:X→M.

. $T^r_s M \otimes T^p_q M \cong d^*(T^r_s M \otimes T^p_q M)$,  d:M↪MxM is the diagonal map.

. $T^p_q M \otimes W \cong d^*(T^p_q M \otimes W)$,  $\pi$:W→M  is a fiber bundle.

. vTW: <u>vertical tangent bundle</u> of  $\pi$:W→M; T=vertical tangent functor. One has the short exact sequence of vector bundles over W:  $0 \to vTW \to TW \overset{\beta}{\longrightarrow} \pi^* TM$,  where $\beta =(p,T(\pi))$ being p the canonical projection TW→W.  One has the canonical isomorphism:  $vTJ\mathcal{D}^k(W) \cong J\mathcal{D}^k(vTW)$.

. $\overset{v}{d}f$: <u>vertical differential</u> of a real-valued function defined on W .

. A sequence($\blacklozenge$)  $...\to E_{r-1} \overset{\psi_r}{\longrightarrow} E_r \overset{\psi_{r+1}}{\longrightarrow} E_{r+1} \to$ ... of vector bundles $E_r$ over M and vector fiber bundles morphisms  $\psi_r$ is formally exact at $E_r$ if we have the exact sequences, s $\geqslant$ 0,  $...\to J\mathcal{D}^s(E_{r-1}) \overset{J\mathcal{D}^s(\psi_r)}{\longrightarrow} J\mathcal{D}^s(E_r) \overset{J\mathcal{D}^s(\psi_{r+1})}{\longrightarrow}_{r+1} J\mathcal{D}^s(E_{r+1})$ at $J\mathcal{D}^s(E_r)$. ($\blacklozenge$) is <u>locally exact</u> at $E_r$ if for any section h of $E_r$ over U⊆M such that $\psi_{r+1}$(h)=0, there exists a section s of $E_{k+1}$ over U ⊆ M such that $\psi_r$(s)=h on U. ($\blacklozenge$) is <u>exact</u> (at $E_r$) if it is both formally and locally exact at any $E_r$ (at $E_r$).

. E ---→W $\overset{\pi}{\longrightarrow}$ M  : <u>affine bundle</u>  $\pi$:W→ M  modelled on the vector bundle $\bar\pi$:E→ M. One has the sequence:  $\pi^*_{k-1,o}(S^o_k M \otimes vTW) \dashrightarrow J\mathcal{D}^k(W) \overset{\pi_{k,k-1}}{\longrightarrow} J\mathcal{D}^{k-1}(W)$.

. $W_p$ : fiber over  p ∈ M of the fiber bundle  $\pi$:W→M.

. N($\phi$) :<u>normal bundle</u> of a morphism  $\phi$:W →W' of fibered manifolds  $\pi$:W→M , $\pi'$:W'→ M', defined by one of the following short exact sequences of vector bundles over W: 1)  $0 \to TW \to \phi^* TW' \to N(\phi) \to 0$ ; 2) $0 \to vTW \to \phi^* vTW' \to N(\phi) \to 0$.

. An <u>exact sequence</u> of fibered manifolds over M and morphisms : $W \overset{\phi}{\to} W' \overset{\psi}{\to} W''$ is such that there exists a section f" of W' such that :(a) im($\phi$)=ker$_{f"}$($\psi$) as sets; (b) The vertical sequence of vector bundles pulled back over W by reciprocal images is exact: $vTW \to \phi vTW' \to (\psi \circ \phi)^* vTW''$.  In particular, if O denotes the fiber bundle id$_M$ :M→ M we get the following exact sequence:

$0 \longrightarrow \ker_{f'}(\phi) \longrightarrow W \underset{f'}{\overset{\phi}{\rightrightarrows}} \overset{\pi}{\longrightarrow} W'$ , whenever $f'(M) \subset \text{im}(\phi)$. A sequence of affine bundles over M $W \overset{\phi}{\longrightarrow} W' \overset{\psi}{\longrightarrow} W''$ is exact iff the associated sequence $vTW \overset{\phi^*}{\longrightarrow} vTW' \overset{(\psi \cdot \phi)^*}{\longrightarrow} vTW''$ is exact.

. $\text{coker}(\phi)$ : If $\phi:W \longrightarrow W'$ is a homomorphism of affine bundles of locally constant rank, then there exist an affine bundle $\text{coker}(\phi)$ and an epimorphism of affine bundles $\psi:W' \longrightarrow \text{coker}(\phi)$ such that the sequence $W \overset{\phi}{\longrightarrow} W' \overset{\psi}{\longrightarrow} \text{coker}(\phi) \longrightarrow 0$ is exact. We can identify $\text{coker}(\phi)$ with $\text{coker}(vT(\phi))$.

. $\underline{F}$ : <u>sheaf of sections</u> of a fiber bundle $p:F \longrightarrow X$.

. $\mathbf{D}$: <u>Non-Linear Spencer operator</u> (over $J\mathcal{D}^{k-1}(W)$): $\mathbf{D}:J\mathcal{D}(J\mathcal{D}^k(W)) \longrightarrow T^*M \otimes J\mathcal{D}^{k-1}(vTW)$, $\mathbf{D}.f \equiv \mathbf{D} \cdot Df = D(\pi_{k,k-1} \cdot f) - \beta_{1,k-1} \cdot f$ , $\forall f \in C^\infty(J\mathcal{D}^k(W))$, where $\beta_{1,k-1}$ is the canonical embedding $J\mathcal{D}^k(W) \longrightarrow J\mathcal{D}(J\mathcal{D}^{k-1}(W))$. Then, the following sequence $0 \longrightarrow \underline{W} \overset{D^k}{\longrightarrow} J\mathcal{D}^k(W) \underset{0}{\overset{D}{\rightrightarrows}} (T^*M \otimes J\mathcal{D}^{k-1}(vTW))$ is locally exact. $0$ is the zero section of $T^*M \otimes J\mathcal{D}^{k-1}(vTW)$ over $J\mathcal{D}^k(W)$ by reciprocal image.

If $W \equiv E$ is a vector bundle over M, we have that $\mathbf{D}$ identifies a linear operator (<u>linear Spencer operator</u>) $\mathbf{D}:J\mathcal{D}(J\mathcal{D}^k(E)) \longrightarrow T^*M \otimes J\mathcal{D}^{k-1}(E)$. So, we obtain the following exact complex (<u>linear Spencer complex</u>):

$$0 \longrightarrow \underline{E} \longrightarrow J\mathcal{D}^k(E) \overset{\mathbf{D}}{\longrightarrow} T^*M \otimes J\mathcal{D}^{k-1}(E) \overset{\mathbf{D}}{\longrightarrow} \overset{\circ}{\Lambda}_2 M \otimes J\mathcal{D}^{k-2}(E) \overset{\mathbf{D}}{\longrightarrow} \ldots \overset{\mathbf{D}}{\longrightarrow} \overset{\circ}{\Lambda}_n M \otimes J\mathcal{D}^{k-n}(E) \longrightarrow 0$$

where $J\bar{\mathcal{D}}^m(E)=0$ for $m < 0$, and $\mathbf{D}.(\omega \wedge u) = d\omega \wedge (\pi_{k,k-1} \circ u) + (-1)^j \omega \wedge \underline{D}.u$, for $\omega \in \overset{\circ}{\Lambda}_j M$, $u \in \Lambda M \otimes J\mathcal{D}^k(E)$.

The restriction of $-\mathbf{D}$ to $\overset{\circ}{\Lambda}_j M \otimes \varepsilon_k (\overset{\circ}{S}_k M \otimes E)$ , ($\varepsilon_k$ =canonocal monomorphism $\overset{\circ}{S}_k M \otimes E$ $J\mathcal{D}^k(E)$, cames from a morphism $\delta:\overset{\circ}{\Lambda}_j M \otimes \overset{\circ}{S}_k M \otimes E \longrightarrow \overset{\circ}{\Lambda}_{j+1} M \otimes \overset{\circ}{S}_{k-1} M \otimes E$ of vector bundles (<u>Spencer morphism</u>, see sect.3).

. $L_X s$: <u>Lie derivative of a super-field of geometric objects</u> s with respect to a vector field X on the basis bundle, projectable on a vector field on the basis manifold $\left[12\right]$. (Sometimes the symbol $\mathscr{L}$ is used also).

. $C(W)$: <u>category whose objects are subbundles</u> of $\pi:W \longrightarrow M$ and whose morphisms are the local fiber bundle automorphisms between those objects.

. <u>k-referment</u> at the point $p \in M$ : linear referment of the vector space $J\mathcal{D}^{k-1}(TM)_p$.

. <u>k-coreferment</u> at the pint $p \in M$: linear referment of the vector space $J\mathcal{D}^{k-1}(T^*M)_p$.

. $\mathsf{T}^k_m(M) \equiv J\mathcal{D}^k(\mathbb{R}^m,M)_{(o,-)} \equiv \pi_k^{-1}(o)$; $\mathsf{T}^{*k}_m(M) \equiv J\mathcal{D}^k(M,\mathbb{R}^m)_{(-,o)} \equiv \bar{\pi}_k^{-1}(o)$, where $\pi_k:J\mathcal{D}^k(\mathbb{R}^m,M) \longrightarrow \mathbb{R}^m$ and $\bar{\pi}_k:J\mathcal{D}^k(M,\mathbb{R}^m) \longrightarrow \mathbb{R}^m$ are the canonical projections. Example: $\mathsf{T}^1_1(M)=TM$; $\mathsf{T}^{*1}_1(M)=T^*M$.

. <u>Bundle of k-referments</u> on M: $H^k(M) \equiv \bigcup_{p \in M} H^k(M)_p \equiv \bigcup_{p \in M} \Pi^k(\mathbb{R}^n, M)_{(o,p)}$ , (n=dimM),

. <u>Bundle of k-coreferments</u> on M: $H_k(M) \equiv \bigcup_{p \in M} H_k(M)_p \equiv \bigcup_{p \in M} \Pi^k(M, \mathbb{R}^n)_{(p,o)}$.

. $H^k(M)$ (resp. $H_k(M)$) has the structure of principal fiber bundle over M with structure group $L_n^k = \Pi^k(\mathbb{R}^n)_{(o,o)}$. The right action of $L_n^k$ on $H^k(M)$ is defined by $D^k f(o).D^k \alpha(o) = D^k(f \cdot \alpha^{-1})(\alpha(o))$. Note that $L_n^k$ has a natural

structure of Lie group with the following product $D^k f_1(o).D^k f_2(o) = D^k(f_2 \bullet f_1)(o)$. In the canonical basis of $\mathbb{R}^n$, $D^k f(o)$ is represented by a k-ple of matrices: $D^k f(o) = (A_j^i, A_{js}^i, \ldots, A_{j_1 \ldots j_k}^i)$ with $A_{j_1 \ldots j_\alpha}^i \equiv (\partial x_{j_\alpha} \ldots \partial x_{j_1}.f^i)(o)$. Then, the product law is obtained by derivation of composition law.

<u>Example</u>. (a) (k=1): $L_n^1 = GL(n,R)$, $(A_j^i)(B_s^r) = (C_s^i \equiv A_j^i B_s^j)$;

(b) (k=2): $L_n^2 = \{(A_j^i, A_{ij}^k)\}$ : $(A_j^i, A_{ij}^k)(B_j^i, B_{ij}^k) = (C_i^i, C_{ij}^k) = (A_k^i B_i^k, A_{ij}^k B_k^s + A_i^k B_{kj}^r)$.

$H^k(M)$ (resp. $H_k(M)$) identifies a covariant (resp. controv.) functor from the category of diffrentiable manifolds of dimension n to the category of the $L_n^k$-principal fiber bundles.

$G_n^k$ : <u>k-(jet derivative) prolongation of a Lie group G</u> : is the following extension of Lie groups: $1 \longrightarrow T_n^k(G) \longrightarrow G_n^k \longrightarrow L_n^k \longrightarrow 1$ , where: (a) $T_n^k(G)$ is endowed with a natural structure of Lie group with moltiplication $D^k f(o).D^k h(o) = D^k(f \bullet h)(o)$. One has a natural representation of $L_n^k$ in $T_n^k(G)$: $r:L_n^k \longrightarrow Aut(T_n^k(G))$, $r(D^k f(o))(D^k h(o)) = D^k(h \bullet f^{-1})(o)$; (b) $G_n^k \equiv T_n^k(G) \times_r L_n^k$ = semi-direct product with respect to r. One has the canonical maps:

(a') $i:T_n^k(G) \longrightarrow G_n^k$ , $i(a) = (a,e)$; (b') $s:L_n^k \longrightarrow G_n^k$, $s(b) = (e,b)$; (c) $p:G_n^k \longrightarrow L_n^k$, $p(a,b) = b$. s is a section of this extension.

$T_n^k P$ : <u>k-(jet derivative)-prolongation of the principal fiber bundle $(P,M,\pi,G)$</u> $= T_n^k(\pi)^* H^k(M) \cong JD^k(P) \times_M H^k(M)$. $T_n^k P$ has a structure of principal fiber bundle over M with structure group $G_n^k$. In particular $T_n^k M = H^k(M)$ with respect to the principal fibration $(M,M,id_M;G=\{e\})$. In general $JD^k(P)$ is not a principal fiber bundle! If F is a G-manifold, $T_n^k(F)$ is a $G_n^k$-manifold with action map $\phi:G_n^k \times T_n^k(P) \longrightarrow T_n^k(P)$ given by: $((D^k g(o), D^k \alpha(o)); D^k \mu(o)) \mapsto D^k(g.\mu \alpha^{-1})(o)$ Then, if $(P,M,\pi;G)$ is a principal fiber bundle and $\pi:W \to M$ is a fiber bundle with fiber F associated to P, then $JD^k(W)$ is a fiber bundle with fiber-type $T_n^k(F)$ associated with $T_n^k P$.

Set $E(P) \equiv TP/G$ = space of tangent vector of P mod the right translation of G. Then $E(T_n^k P) \cong JD^k(E(P))$. In particular $E(H^k(M)) \cong JD^k(TM)$.

$P \subset H^k(M)$: <u>G-structure of order k on M</u>: sub-principal fiber bundle $P$ of $H^k(M)$, with structure group $G \subset L_n^k$ [1] . (In this case $JD^q(P) \longrightarrow P^k$ (q $\leq$ k) is a principal fiber bundle with structure group $T_n^k(G)$ $(n=\dim M)$.

$P^k$ is integrable at the order k+q in a point $p \in M$ if $\forall u \in P_p^k$, $JD^q(P)_u \cap (H_{P^k}^{k+q}(M))_u \neq \emptyset$ [+]

If $P^k \subset H^k(M)$ is integrable at the order k+q, in any point $p \in M$ $P^k$ is said to be integrable at the order k+q . In this case $P^{k+q} \equiv JD^q(P^k) \cap H_{P^k}^{k+q}(M) \longrightarrow M$ is a sub-principal fiber bundle of $H^{k+q}(M)$ and one has $(P^{k+q})^{+q'} = P^{k+q+q'}$. $P^{k+q}$ is called the <u>q-prolongation</u> of $P^k$.

- $\overset{\circ}{\Lambda}{}_p^o(\pi_E)$ : space of horizontal p-forms on $\pi : W \to M$.

- $S(E_k)$ : set of solutions of a differential equation $E_k \subset J\mathfrak{D}^k(W)$.

---

(+) $H_{P^k}^{k+q}(M)$ is the restriction to $P^k$ of the $L_{n,k}^{k+q}$-principal fiber bundle $H^{k+q}(M) \to H^k(M)$ , where $L_{n,k}^{k+q} \equiv \ker(L_n^{k+q} \to L_n^k)$.

# 3. - BASIC RESULTS ON THE FORMAL THEORY OF PDE

A <u>non-linear differential equation of order k</u> on a fiber bundle $\pi : W \to M$ is a fibered submanifold $E_k \subset J\mathcal{D}^k(W)$ of the k-th order jet-derivative bundle $J\mathcal{D}^k(W)$ over $\pi$. If $\bar{\pi} : E \to M$ is a vector bundle, a <u>linear differential equation</u> of order k on $\bar{\pi}$ is a vector subbundle $E_k \subset J\mathcal{D}^k(E)$. The <u>r-prolongation</u> of $E_k$ is the subset $E_{k+r} \equiv J\mathcal{D}^r(E_k) \cap J\mathcal{D}^{k+r}(W)$ of $J\mathcal{D}^{k+r}(W)$. Note, that the non-linear Spencer operator restricted to the first prolongation $E_{k+1}$ of $E_k$ gives a first order operator $\mathbb{D} : E_{k+1} \to T^*M \otimes vTE_k$ over $E_k$. So, a k+1-differential equation $E'_{k+1} \subset J\mathcal{D}^{k+1}(W)$ is contained in $E_{k+1}$ iff: (1) $\pi_{k+1,k}(E'_{k+1}) \; E_k$; (2) $\mathbb{D}(E'_{k+1}) \; T^*M \otimes vTE_k$, (where $\mathbb{D}$ is just the non-linear Spencer operator).

A differential equation $E_k \subset J\mathcal{D}^k(W)$ is <u>formally compatible</u> if $\pi_{k+r} : E_{k+r} \to M$ is surjective $\forall r \geqslant 0$. A linear PDE is always compatible. A differential equation $E_k \subset J\mathcal{D}^k(W)$ is <u>formally transitive</u> if $\pi_{k+r,o} : E_{k+r} \to W$ is surjective $\forall r \geqslant 0$.

The <u>symbol</u> of $E_k \subset J\mathcal{D}^k(W)$ is the family of vector spaces over $E_k$: $g_k \equiv vTE_k \cap \pi^*_{k,o}(S^o M \otimes vTW)$. Set $F_o \equiv N(i_k)$, where $i_k$ is the embedding $E_k \subset J\mathcal{D}^k(W)$. ($\dim F_o = \dim J\mathcal{D}^k(W) - \dim E_k$). We have the following commutative and exact diagram of families of vector spaces over $E_k$:

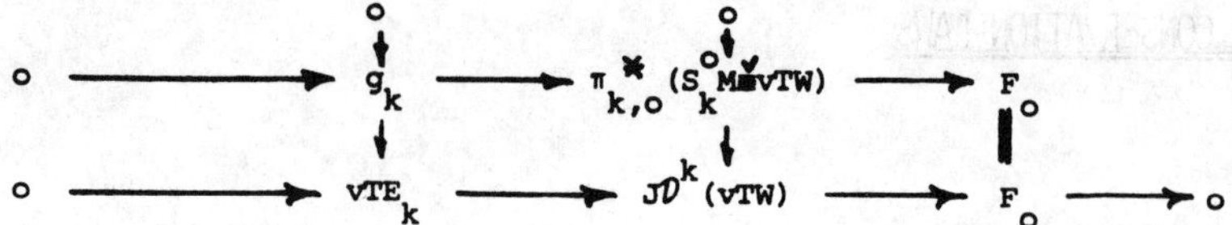

The <u>r-prolongation</u> of the symbol $g_k$ of $E_k$ is the family of vector spaces $g_{k+r}$ over $E_k$ given as the kernel of the composite morphism: $\pi^*_{k,o}(S^o_{k+r} M \otimes vTW) \rightarrow \pi^*_{k,o}(S^o_r M \otimes S^o_k M \otimes vTW) \rightarrow S^o_r M \otimes F_o$. So, we have the following exact sequences of family of vector spaces over $E_k$: $o \rightarrow g_{k+r} \rightarrow \pi^*_{k,o}(S^o_{k+r} M \otimes vTW) \rightarrow S^o_r M \otimes F_o$. From which we try the following other exact sequence: $o \rightarrow g_{k+1} \rightarrow \pi^*_{k,o}(S^o_{k+1} M \otimes vTW) \overset{\psi}{\rightarrow} T^* M \otimes F_o \rightarrow F_1 \rightarrow o$, where $F_1 \equiv \operatorname{coker}(\psi)$ is a family of vector spaces over $E_k$. If $g_{k+1}$ is a vector bundle over $E_k$, then $F_1$ is a vector bundle over $E_k$ and $\dim F_1 = n.\dim F_o - m \dfrac{(n+k)!}{(k+1)!(n-1)!} + \dim g_{k+1}$. Of course we have the following exact sequence of vector spaces over $E_{k+r}$: (▶) $o \rightarrow \pi^*_{k+r,k} g_{k+r} \rightarrow vTE_{k+r} \overset{\pi^*_{k+r,k+r-1}}{\rightarrow} vTE_{k+r-1}$

<u>Definition  .1</u> A differential equation $E_k \subset JD^k(W)$ is <u>formally integrable</u> if for $r \geqslant o$, $g_{k+r+1}$ is a vector bundle over $E_k$ and the map $\pi_{k+r+1,k+r} : E_{k+r+1} \rightarrow E_{k+r}$ is surjective.

<u>Proposition  .1</u> The following propositions are equivalent:

(1) Any prolongation $E_{k+r}$ ,$o \leqslant r \leqslant h$, is an affine subbundle of $JD^{k+r}(W)|_{E_{k+r-1}}$ modeled on the vector bundle $\pi^*_{k+r,k+r-1} g_{k+r}$ over $E_{k+r-1}$;

(2) $g_{k+r+1}$ is a vector bundle over $E_k$ and $E_{k+r+1} \rightarrow E_{k+r}$ is surjective for $o \leqslant r \leqslant h$;

Further, if any one of these assertions is satisfied, then $(E_{k+1})_{+r} = (E_k)_{+r+1}$.

Then, we try that the formal integrability of a PDE $E_k \subset JD^k(W)$ is equivalent to the following commutative diagram, $\forall r \geqslant 1$,

$$
\begin{array}{ccc}
o \longrightarrow E_{k+r-1} & \longrightarrow & JD^{k+r-1}(W) \\
\Vert \qquad\quad \uparrow & & \uparrow \\
o \longrightarrow E_{k+r} & \longrightarrow & JD^{k+r}(W) \\
\Vert \qquad\quad \uparrow & & \uparrow \\
o \longrightarrow \pi^*_{k+r,k+r-1} g_{k+r} & \longrightarrow & \pi^*_{k+r,k+r-1}(S^o_{k+r} M \otimes vTW)
\end{array}
$$

where the horizontal lines are exact, $\forall r \geqslant 1$.

Associated to any differential equation $E_k \subset JD^k(W)$ we recognize a family of vector spaces $\{H^{m-j,j}_q\}_{q \in E_k}$ ,$m \geqslant k$, $j,k \in \mathbb{N}$, where $H^{m-j,j}_q$ is the cohomology at $\Lambda^o_j M \otimes g_{m-j}$ (<u>Spencer cohomology</u>) of the following complex (<u>first Spencer comlex</u>): (▽) $g_m \overset{\delta}{\rightarrow} T^* M \otimes g_{m-1} \overset{\delta}{\rightarrow} \Lambda^o_2 M \otimes g_{m-2} \overset{\delta}{\rightarrow} \cdots \rightarrow \Lambda^o_r M \otimes g_{m-r} \rightarrow \cdots \rightarrow \Lambda^o_{m-k} M \otimes g_k \overset{\delta^{m-k}}{\rightarrow}$

$$\rightarrow \Lambda^o_{m-k+1} M \otimes S^o_{k+1} M \otimes vTW$$

with the coboundaries $\delta^m$ the vector morphisms canonically induced by the coboundaries $d^o: S_m^o V \to V^* \otimes S_{m-1}^o V$; $d^j: \Lambda_j^o V \otimes S_{k-j}^o V \to \Lambda_{j+1}^o V \otimes S_{k-j-1}^o V$, $d^j(\omega \otimes u) = (-1)^j \omega \wedge d^o(u)$, being V a vector space. Note that we put $g_{k+r-p} = S_{k+r-p}^o M \otimes vTW$ if $r < p$, and $g_{k+r-p} = 0$ if $k+r-p < 0$. If $H^{m,j} = 0$ for $m > k$ and $j \geq o$, $E_k$ is called <u>involutive</u>. If $H^{m,j} = 0$ for $m \geq k$ and $0 \leq j \leq r$, $E_k$ is called r-acyclic. If $g_{k+r} = 0$ for some $r \geq 0$, $E_k$ is said to be <u>finite</u>. (Of course if $g_k = o$ $E_k$ is involutive). Note, that $g_k$ is always 1-acyclic, that is the sequences

$$0 \to g_{k+r} \longrightarrow T^* M \otimes g_{k+r-1} \longrightarrow \Lambda_2^o M \otimes g_{k+r-2}$$

are exact for $r \geq 1$. Further, if $g_k$ is involutive and of finite type then must be $g_k = o$.

One can see that there exists an integer $k_o > k$ depending on the dimension of M, the dimension of W and k such that $E_{k_o}$ is involutive.

<u>Proposition</u> .2 (<u>Criterion of involutivness</u>) Let $E_k \subset JD^k(W)$ be a fifferential equation. $(g_k)_q \subset ( \pi_{k,o}^* (S_k^o M \otimes vTW))_q$ $\forall q \in E_k$, is involutive iff there exists a local coordinate system in which we have $\dim(g_{k+1})_q = \sum_{i=o}^{n-1} \dim(g_k^{(i)})$, where $g_k^{(i)} = \{ X \in (g_k)_q \mid X(v_1) = \ldots = X(v_i) = 0 \}$, where $\{v_1, \ldots, v_n\}$ is the natural basis of $T_{\pi_k(q)} M$ induced by the system of coordinates (n=dimM).

Let $E_k \subset JD^k(W)$ be a differential equation. We can associate to $E_k$ the following structures:

(A)    Set $C^j = \{ \Lambda_j^o M \otimes_{E_k} vTE_k \} / \delta(\Lambda_{j-1}^o M \otimes_{E_k} G_{k+1})$ , where $G_{k+r} = g_{k+r}$ if $E_k \subset JD^k(W)$ and $G_{k+r} = S_{k+r}^o M \otimes_{E_k} vTW$ if $E_k = JD^k(W)$, $1 \leq j \leq n = \dim M$ . $C^j$ are families of vector spaces over $E_k$. One has $C^1 = JD(E_k)/E_{k+1}$.

(B)   We have the exact sequence of vector spaces over $E_k \subset JD^k(W)$:

$$0 \longrightarrow G_{k+2} \longrightarrow S_2^o M \otimes_{E_k} vTE_k \xrightarrow{\sigma_1} T^* M \otimes_{E_k} C^1 \xrightarrow{\tau} C^2 \longrightarrow 0,$$

where $\tau$ is the epimorphism of vector spaces over $E_k$ induced by the multiplication map $T^* M \otimes T^* M \to \Lambda_2^o M$ and $\sigma_1$ is given by composition $S_2^o M \otimes_{E_k} vTE_k \to T^* M \otimes T^* M \otimes_{E_k} vTE_k \xrightarrow{\sigma} \to T^* M \otimes C^1$ being $\sigma$ the natural projection.

Then, one can prove the following

<u>Theorem</u> .1 If $E_k \subset JD^k(W)$ is a differential equation such that the map $E_{k+1} \to E_k$ is surjective and $g_{k+1}$ is a vector bundle over $E_k$ then there is a morphism over $E_k$: $\kappa(E_k): E_{k+1} \to C^2 = \{ \Lambda_2^o M \otimes_{E_k} vTE_k \} / \delta(T^* M \otimes g_{k+1})$ such that one has the following exact sequence: ($\heartsuit$) $E_{k+2} \to E_{k+1} \underset{o \cdot \pi_{k+1,k}}{\overset{\kappa(E_k)}{\rightrightarrows}} C^2$ .

$\kappa(E_k)$ is called the <u>curvature map</u> of $E_k$. More precisely $\kappa(E_k)$ has values into Spencer cohomology spaces $H^{k,2} = \ker\{ \delta: \Lambda_2^o M \otimes g_k \to \Lambda_3^o M \otimes S_{k-1}^o M \otimes vTW \} / \delta(T^* M \otimes g_{k+1}) \subset C^2$ .

Further, iff $g_{k+2}$ is a vector bundle over $E_k$ one has the following exact sequence of vector bundles over $E_k$: $0 \to g_{k+2} \to T^*M \underset{E_k}{\otimes} g_{k+1} \to \Lambda_2^o M \otimes vTE_k \to C^2$.
We have the following important:

<u>Theorem</u> .2  If $E_k \subset JD^k(W)$ is a differential equation such that the map $E_{k+1} \to E_k$ is surjective, $g_{k+1}$ is a vector bundle over $E_k$ and $g_k$ is 2-acyclic, then $E_k$ is formally integrable.

<u>Proof</u>. Since $g_k$ is 2-acyclic we have that $H^{k,2}=0$, so the exactness of sequence ($\heartsuit$) proves that $E_{k+1} \to E_k$ is surjective.  Further, one can prove that if $g_k$ is 2-acyclic and $g_{k+1}$ is a vector bundle then $g_{k+r}$ $(r \geq 1)$ is a vector bundle over $E_k$.  Then, proceeding by induction and by using ($\nabla$) we conclude that $E_k$ is formally integrable.

<u>Theorem</u> .3 (<u>Criterion of formal integrability</u>) Let $E_k \subset JD^k(W)$ be a differential equation.  Then, there is an integer $k_o = k+h \geq k$ , depending only on k the dimension of M and dimension of W, such that $E_{k_o}$ is involutive and such that if $g_{k+r+1}$ is a vector bundle over $E_k$ and $\pi_{k+r+1,k+r}: E_{k+r+1} \to E_{k+r}$ is surjective for $o \leq r \leq h$, then $E_k$ is formally integrable.

It is also useful, in order to decide the surjectivity of the map $E_{k+1} \to E_k$ , to consider some other geometric structures associated to $E_k$.

<u>Definition</u> .2   Let $E_k \subset JD^k(W)$ be a differential equation on $\pi: W \to M$. Define the families $F_o,\ldots,F_n$ of vector spaces over $E_k$ by the formula $F_r = (\Lambda_r^o M \underset{E_k}{\otimes} F_o)/\delta(\Lambda_{r-1}^o M \underset{E_k}{\otimes} G'_1)$, where $G'_r = im(S_{k+r}^o M \underset{E_k}{\otimes} vTW \to S_r^o M \underset{E_k}{\otimes} F_o)$, n=dimM.
We have the following

<u>Proposition</u> .3  1.  One has the following short exact sequences:
$$0 \to g_{k+r} \to S_{k+r}^o M \underset{E_k}{\otimes} vTW \to G'_r \to 0 .$$
2.  $H^{k+r}(g_k)=H^{r+1,s-1}(G'_1)$, $\forall r \geq o$ , $\forall s \geq 2$.

3.  $G'_{k+1}$ is the r-prolongation of $G'_1$ iff $g_k$ is 2-acyclic.  In that case $G'_1$ is (s-1)-acyclic iff $G_k$ is s-acyclic. In particular $G'_1$ is involutive iff $G_k$ is involutive.

4.  If $g_k$ is involutive and if $g_{k+1}$ is a vector fiber bundle over $E_k$ then $F_r$ is a vector bundle over $E_k$ for any r=1,...,n.

__Theorem__ .4  If $E_k \subset JD^k(W)$ is a differential equation on $\pi:W \to M$ and if $g_{k+1}$ is a vector bundle over $E_k$ then $F_1$ is a vector bundle over $E_k$ and a section $\kappa$ of this vector bundle exists such that we have the exact sequence: $E_{k+1} \xrightarrow{\pi_{k+1,k}} E_k \overset{\kappa}{\underset{o}{\rightrightarrows}} F_1$ .

__Example.__  The Spencer equation $(S_p)=\ker \mathbf{D} \subset JD(JD^k(W))$ is an involutive equation but it is not formally integrable.

__Note.__  In Physics differential equations are generally obtained as kernels of differential operators.  So, if f is a fixed section of a fiber bundle $\pi_K: \mathbb{K} \to M$, and $E_k = \ker_f K \subset JD^k(W)$, where $K: JD^k(W) \to \mathbb{K}$ is a fiber bundle morphism characterizing a differential operator $K.:C^\infty(W) \to C^\infty(K)$.  Then, $E_{k+q} = \ker K^{(q)} \subset JD^{k+q}(W)$, where $K^{(q)}$ is the q-prolongation of K : $K^{(q)} \equiv \beta_{k,q} \overset{D^q f}{\bullet} JD^q(K)$, being $\beta_{k,q}$ the embedding $JD^{k+q}(W) \to JD^q(J^k(W))$. The symbol $g_k$ of $E_k$ is related to K by the formula $g_k = \ker \sigma_k(K)$, where $\sigma_k(K) \equiv vT(K) \bullet \mu_{(k)} : S_k^o M \bar{\otimes} vTW \to vT\mathbb{K}$ is the symbol of K, being $\mu_{(k)}$ the vector bundle morphism on $JD^k(W)$ given by the following exact sequence of vector bundles over $JD^k(W)$:

$$(\mathbf{\mathsf{\Sigma}}) \quad 0 \to \pi^\ast_{k,o}(S_k^o M \bar{\otimes} vTW) \xrightarrow{\mu_{(k)}} vTJD^k(W) \longrightarrow \pi^\ast_{k,k-1} vTJD^{k-1}(W) \longrightarrow 0$$

Moreover, $g_{k+r} = \mu_{(k+r)} \bullet \pi^\ast_{k+r,o}(S_{k+r}^o M \bar{\otimes} vTW) \cap vTE_{k+r}$.  If W=E is a vector bundle over M and $E_k \subset JD^k(E)$ is a linear equation, we can always write $E_k = \ker(\phi)$, where $\phi: JD^k(E) \to F_o = JD^k(E)/E_k$ is a morphism of vector bundles over M.  We can always consider that $\phi$ is an epimorphism. We get the identification $g_{k+q} \subset S_{k+q}^o M \bar{\otimes} E$ for the symbol of $E_{k+q}$. If $\phi^{(q)}: JD^{k+q}(E) \to JD^q(F_o)$, $\forall q \geq 0$, is a morphism of costant rank, then $E_{k+q}$ is a vector bundle ($\forall q \geq 0$). In this case $E_k$ is said to be regular. (But could be non-formally integrable). It is useful the following

__Theorem__ .5  Let $E_k \subset JD^k(E)$ be a linear PDE on E.  If we restrict $E_k$ to a sufficiently small open set $U \subset M$, or if $E_k$ is sufficiently regular, then there exist integers $s \geq 0$, $r \geq 0$ such that $E_{k+r}^{(s)} = \pi_{k+r+s,k+r}(E_{k+r+s}) \subset E_{k+r}$ is an involutive formally integrable differential equation with the same solutions as $E_k$ and such that $(E_{k+r+s}^{(s)})_{+q} = E_{k+r+s+q}^{(s)}$ , $\forall q \geq 0$.

__Proof.__  See ref. $\left[1/n\right]$ .  $\square$

So, we can assume that $E_k$ is an involutive formally integrable differential equation in the following.

__Theorem .6__ Let $E_k = \ker(\phi) \subset J\mathcal{D}^k(E)$. Then, to $E_k$ we can canonically associate a formally exact finite length sequence (__Janet sequence__):

$$(\spadesuit) \quad 0 \longrightarrow \Theta \longrightarrow \underline{E} \xrightarrow{\;\mathcal{D}\;} \underline{F}_o \xrightarrow{\;\mathcal{D}_1\;} \underline{F}_1 \xrightarrow{\;\mathcal{D}_2\;} \underline{F}_2 \longrightarrow \cdots \xrightarrow{\;\mathcal{D}_n\;} \underline{F}_n \longrightarrow 0$$

where $\Theta$ is the sheaf of solutions of $E_k$, $\mathcal{D}_r \equiv \psi^r \bullet b$, $\psi^r$ is the canonical epimorphism $\psi^r : J\mathcal{D}(F_{r-1}) \longrightarrow F_r = \operatorname{coker}(\psi^{r-1})^{(1)}$ and $\psi^o = \phi$. Moreover, we may have $F_r \equiv \Lambda^o_r M\boxtimes F_o / \delta(\Lambda^o_{r-1} M\boxtimes G'_1)$.

__Remark__. The linear Spencer complex gives by restriction to $R_k$ the linear Spencer complex for $R_k$:

$$(\bullet) \quad 0 \longrightarrow R_m \xrightarrow{\;D\;} T^* M\boxtimes R_{m-1} \xrightarrow{\;D\;} \Lambda^o_2 M\boxtimes R_{m-2} \xrightarrow{\;D\;} \cdots \xrightarrow{\;D\;} \Lambda^o_n M\boxtimes R_{m-n} \longrightarrow 0 \quad, \text{ where } R_m = J\mathcal{D}^m(TM),$$

for $m \leqslant k$. The cohomology of $\underline{E} \xrightarrow{\;\mathcal{D}\;} \underline{F}_o \xrightarrow{\;\mathcal{D}_1\;} \underline{F}_1$ depends only on $R_k$ and is isomorphic to the cohomolohy $H^{m-1,1}$ of $(\bullet)$ at $\underline{T^* M\boxtimes R}_{m-1}$, for $m \geqslant k+1$. The operator $\mathcal{D}_1$ is the compatibility conditions for $\mathcal{D}$ and express all the formal obstructions to solving the inhomogeneous equation $\mathcal{D}.s=f$, $f \in \underline{F}_o$. Therefore, the cohomology $H^{m-1,1}$ is independent of $m$ for $m \geqslant k+1$ and is the obstruction to the local solvability of the differential operator $\mathcal{D}$. $H^{m-1,1}$ is called the __first(linear) Spencer cohomology group of $R_k$__.

__Proposition .4__ Set $C_r(E) \equiv (\Lambda^o_2 M\boxtimes J\mathcal{D}^k(E))/\delta(\Lambda^o_{r-1} M\boxtimes S^o_{k+1} M\boxtimes E)$, $C_o(E) \equiv J\mathcal{D}^k(E)$. Then, there is a formally exact sequence (__Second linear Spencer sequence__):

$$(\Downarrow) \quad 0 \longrightarrow \Theta \longrightarrow \underline{C}_o \xrightarrow{\;D_1\;} \underline{C}_1 \xrightarrow{\;D_2\;} \cdots \longrightarrow \underline{C}_n \longrightarrow 0 \quad.$$

__Proof.__ It is an application of Theorem 3.6 with $\phi \equiv D^k : \underline{E} \longrightarrow \underline{J\mathcal{D}^k}(E)$, as $D^k$ is an involutive operator with zero solution sheaf and zero symbol. $\square$

__Note.__ The relation between $(\spadesuit)$ and $(\Downarrow)$ is given by the following commutative diagram:

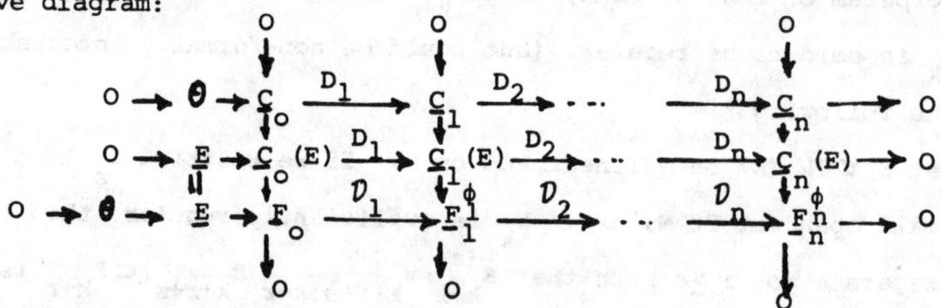

The epimorphisms $\phi_r$ are induced inductively by $\phi \equiv \phi_o$.

A remarkable application of Theorem 3.3 is an alternative form of the Frobenius theorem for r-distributions $\mathbb{E} \subset TM$ over a manifold M.

__Theorem .7__ Let $\mathbb{E} \subset TM$ be a r-distribution on M. Let $\{\xi_\alpha\}_{\alpha=1,\ldots,r}$ be r-unconnected vector fields defined on $U \subset M$ which span the distribution at each point of $U \subset M$. Then, $\mathbb{E}$ is completely integrable ($\equiv$involutive) iff the following first order linear differential system $\xi_\rho.\phi \equiv \xi^i_\rho(\partial x_i.\phi)=0$ is involutive formally integrable, ($\phi \in C^{\infty}(M)$).

<u>Definition</u> .3  A differential equation $E_k \subset JD^k(W)$ is <u>completely integrable</u> if for any $q \in E_k$ there exists a local solution $s:U \to W$, such that $D^k s(x)=q$, where $\pi_k(q)=x$, and $D^k s(U) \subset E_k$.

This implies that the map $E_{k+l} \longrightarrow E_k$, $\forall l \geq 0$, is surjective.

The formal integrability assures the completely integrability in the analytic case.

Of particular interest are differential equations $E_k \subset JD^k(W)$ diffeomorphic to $\pi_{k,k-1}(E_k) \cong E_{k-1}$ via the projection $\pi_{k,k-1}$. These equations are related to the so-called k-connections on a fiber bundle $\pi:W \to M$.

<u>Theorem</u> .8  Let $E_k \subset JD^k(W)$ be a differential equation diffeomorphic to its projection $\pi_{k,k-1}(E_k) \cong E_{k-1} \subset JD^{k-1}(W)$. Then, $E_k$ is completely integrable iff $E_{k+1} = \overset{v}{E}_{k+1} \equiv JD(E_k) \cap \overset{v}{JD}{}^{k+1}(W)$.

<u>Proof.</u>  Note, that as $E_k \cong E_{k-1}$ one has also that $\overset{v}{E}_{k+1} \cong E_k$. Then, since $E_{k+1} \subset \overset{v}{E}_{k+1}$, $E_{k+1} \to E_k$ is surjective iff $\overset{v}{E}_{k+1} = E_{k+1}$. On the other hand, the diffeomorphism $E_{k-1} \cong E_k$ implies that $E_k$ can be written in local coordinates as $y^j_{i_1 \ldots i_k} = F^j_{i_1 \ldots i_k}(x^\alpha, y^j, \ldots, y^j_{i_1 \ldots i_{k-1}})$. So, $E_k$ identifies a distribution $\overset{}{E}$ which since $E_{k+1} \to E_k$ is surjective and $g_k = 0$, Theorem .3 assures to be completely integrable. $\square$

<u>Note.</u>  A <u>k-connection</u> on W is a k-order differential equation $C_k \subset JD^k(W)$ diffeomerphic to $JD^{k-1}(W)$ by means of the projection $\pi_{k,k-1}:JD^k(W) \to JD^{k-1}(W)$. $C_k$ is called <u>flat</u> if $C_k$ is completely integrable. In this case we say that the fiber bundle W is <u>flat</u>. A k-connection $C_k \subset JD^k(W)$ can be identified by means of a section $\daleth:JD^{k-1}(W) \longrightarrow JD^k(W)$ of the affine fiber bundle $\pi_{k,k-1}:JD^k(W) \to JD^{k-1}(W)$ modelled on the vector bundle $\pi^*_{k-1,o}(S^o_k M \overset{v}{\otimes} vTW) \to JD^{k-1}(W)$. So, to $C_k$ we can associate the curvature map $_\daleth R \equiv pr \bullet \daleth^{(1)}|_{C_k}:C_k \to \overset{v}{JD}{}^{k+1}(W)/J\overset{v}{D}{}^{k+1}(W)$, where $pr:\overset{v}{JD}{}^{k+1}(W) \to \overset{v}{JD}{}^{k+1}(W)/JD^{k+1}(W)$ is the canonical projection on the quotient manifold and $\daleth^{(1)}$ is the first prolongation of the section $\daleth$ defining $C_k$. Thus, by means of Theorem 3.8 $C_k$ is flat iff $C_k \subset \ker(_\daleth R)$.

## 4. - PSEUDOGROUPS AND PDE

In this section we shall study the symmetry properties of differential equations. This is obtained by means of the so-called pseudogroups. So, first we shall consider this important geometric structure and the corresponding problem of integrability. Further, we will analize the relation between pseudogroups and super-bundles of geometric objects. Finally, we shall obtain methods of formal integration by means of study of pseudogroups on the manifolds characterizing the differential equations.

<u>Definition</u> .1 Let X be a topological space. Then, a <u>pseudogroup</u> $G$ of <u>homeomorphisms</u> of X is a set $\{f_\mu\}_{\mu \in I}$ of homeomorphisms with domain $U_\mu$ and range $V_\mu$ open subsets of X, such that: (a) $id_X \in G$ ; (b) $U_\upsilon \subset U_\mu \Rightarrow f_\mu | U_\upsilon \in G$; $f_\mu \in G \Rightarrow f_\mu^{-1} \in G$ ; (d) $f_\mu, f_\upsilon \in G$, $V_\mu \cap U_\upsilon \neq \phi \Rightarrow f_\upsilon \bullet f_\mu \in G$ ; (e) If $f_\mu \in G$ for $\mu \in I_1 \subset I$ , $f_\mu : U \to V_\mu$ and $f_\mu | U_\mu \cap U_\upsilon = f_\upsilon | U_\mu \cap U_\upsilon$ , for $(\mu,\upsilon)$ $I_1 x I_1$, and if the mapping $f: \bigcup_{\mu \in I_1} U_\mu \to \bigcup_{\mu \in I_1} V_\mu$ defined by $f | U_\mu = f_\mu$ is a homeomorphism then $f \in G$.

$G$ is called <u>transitive</u> if in addition it satisfies: (f) For any two points $p, p' \in X$ there is an element $f \in G$ such that $p' = f(p)$. Thus, $G$ is transitive iff the orbit of any element $p \in X$, that is $X_p \equiv \{ f_\mu(p) \}_{\mu \in I} \subset X$, is X. If X is a differentiable manifold of class $C^k$ we can consider pseudogroups of local diffeomorphisms of X of class $C^k$.

<u>Definition</u> .2 A <u>groupoid</u> is a set $J$ together with a law of composition defined for some pairs of elements such that: (a) If eitheir $(\alpha\beta)\gamma$ or $\alpha(\beta\gamma)$ is defined, then both are, and they are equal; (b) If $\alpha\beta$ and $\beta\gamma$ are defined and $\beta$ is an <u>identity</u> , (that is $\alpha\beta = \alpha$, $\beta\gamma = \gamma$ ), then $\alpha\gamma$ is defined; (c) for each $\alpha$ there are identities $\varepsilon_L$ and $\varepsilon_R$ such that $\varepsilon_L \alpha$ and $\alpha \varepsilon_R$ are defined; (d) For each $\alpha$ there is an $\alpha^{-1}$ such that $\alpha^{-1}\alpha$ is a right identity for $\alpha$ and $\alpha \alpha^{-1}$ is a left identity for $\alpha$ .

$J$ is called <u>connected</u> (or <u>transitive</u>) if it also satisfies: (e) Given any identities $\varepsilon$ and $\varepsilon'$, there is an element $\alpha$ such that $(\varepsilon\alpha)\varepsilon'$ is defined. The set M of identities of $J$ is called <u>basis</u> of $J$ . One has the following applications: $\pi_R: J \to M$ (<u>right identity</u>); $\pi_L: J \to M$ (<u>left identity</u>). $J$ is connected iff $(\pi_R, \pi_L): J \to M \times M$ is surjective. If $M = \{e\}$ then the composition law in $J$ is always defined and one has a group.

A <u>topological groupoid</u> (resp. <u>differential groupoid</u>) is a groupoid with a topology (resp. <u>differential structure</u>) such that composition and taking of inverse are continuous (resp. differentiable). Furthermore, the basis M is a subspace (resp. sub-manifold) of $J$.

A <u>Lie groupoid</u> is a transitive differentiable groupoid. Then one can prove that a Lie groupoid is locally trivial, that is for any couple $(x,y) \in M \times M$ there exists an open neighbourhood U of y in M and a differentiable map $s: U \to J$ such that $\pi_R(s(y'))=x$, $\pi_L(s(y'))=y$, $\forall\, y' \in U$.

The relation between Definition 4.1 and 4.2 is given by the following

<u>Proposition</u> .1 To any pseudogroup $G$ of homeomorphisms of X it is associated a groupoid $J(G)$ that characterizes G.

<u>Proof.</u> $J(G)$ is the groupoid of the germs of elements of $G$ with the composition law: $g_x \cdot f_y = (g \circ f)_y$, which makes sense iff $f_y(y)=x$. $J(G)$ has the topology generated by these set, of the form $\{f_x\}$, where f is a fixed element of $G$ and x varies over some open set contained in the domain of f. $J(G)$ characterizes G in the sense that a homeomorphism between open subsets of X belongs to $G$ if its germ at each point belongs to $J(G)$. $G$ is transitive iff $J(G)$ is transitive. $\square$

<u>Remark.</u> A pseudogroup $G$ over a space X can be used to construct manifolds M locally modelled on X by requering that the transiction functions should be elements of $G$. Such manifolds are called <u>manifolds with a $G$-structure.</u> Furthermore, if $G$ is a transitive pseudogroup on $\mathbb{R}^n = X$ and $G_o \equiv \{\, a \in GL(n, \mathbb{R}) \mid a = (\partial x_i \cdot f^j)(o), f \in G, f(o)=o \,\}$, the closure $G(G)$ of $G_o$ is called Lie group associated to $G$. Then, if a manifold modelled on $\mathbb{R}^n$ has such a $G$-structure, then the group of its tangent bundle is reducible to $G(G)$. Note, that the inverse is false. (This problem is related to the integrability of a Lie equation).

Let us give some other useful definitions on a groupoid.

**Definition .3** Let $J$ be a differentiable groupoid. An <u>invertible section</u> of $J$ is a local section $s:M \quad J$ for the map $\pi_R:J \to M$ such that $\pi_L \bullet s:M \to M$ is a local diffeomorphism of the basis M.

**Proposition .2** The set $\Theta$ of invertible sections of the differentiable groupoid $J$ is a pseudogroup for the following composition law: $(s,s')=s"$, where $s"$ is the section $x \mapsto s"(x)=s'(\pi_R(x)).s(x)$.

**Definition .4** 1. Let $J$ be a differentiable groupoid. The <u>k-prolongation</u> of $J$ is the space $J^k \subset J\!D^k(J)$ of k-derivatives of invertible sections of $J$

2. Similarly we have the definitions of <u>k-semiholonom$^{ic}$ prolongation</u> $\bar{J}^k \subset \bar{J\!D}^k(J)$ and the definition of <u>k-sesquiholonom$^k$ prolongation</u> $\check{J}^k \subset J\!D^k(J)$.

**Proposition .3** 1. If $J$ is a Lie groupoid $J^k, \bar{J}^k, \check{J}^k$ are so.

2. For any $k'$, $0 \leqslant k' \leqslant k$, one has the projections: $J^k \to J^{k'}$, $\bar{J}^k \to \bar{J}^{k'}$, $\check{J}^k \to \check{J}^{k'}$.

**Theorem .1** (Relation between Lie groupoids and principal fiber bundles).

1. Let $J$ be a Lie groupoid (+) with basis M; the set $J_x \equiv \pi_R^{-1}(x)$ is a principal fiber bundle $(J_x, M, \pi_L; G_x)$ where the structure group $G_x = (\pi_R, \pi_L)^{-1}(x,x)$.

2. Let $(P,M,\pi;G)$ be a principal fiber bundle with structure group G. Then, the quotient space $J \equiv P \times P/\!\sim$, where $\sim$ is the equivalence relation given by $(za,z'a) \sim (z,z')$, $\forall$ a $\in$ G, has a natural structure of Lie groupoid with basis M. $J \cong J((Aut(P)))$. Furthermore $J_x \cong P$. Then, one has the correspondence: $\Theta \Longleftrightarrow Aut(P)$.

3. If $J$ is the groupoid associated to $(P,M,\pi;G)$, $J^k$ is the groupoid associated to $T_n^k P$.

**Definition .5** A <u>k-order differential system</u> for a Lie groupoid $J$ is a (Lie) sub-groupoid $\Psi_k$ of $J^k$: $\Psi_k \subset J^k$. So, $\Psi_k$ is a differential equation on $J$ of order k. $\Psi_k$ is <u>completely integrable</u> if the projection $\Psi_{k+q} \equiv J\!D^q(\Psi_k) \cap J^{k+q} \to \Psi_k$ is surjective for any $q \geqslant 0$. The solutions of $\Psi_k$ are the invertible sections $s$ of $J$ such that $D^k s:M \to \Psi_k$.

**Definition .6** Let $J$ be a differentiable groupoid. An <u>infinitesimal displacment</u> of $J$ is a $\pi_R$-vertical tangent vector of $J$ applied in a point of $M \subset J$.

---

(+) If $J$ is not transitive the choice of x is not indifferent. So, we shall assume that $J$ is a Lie groupoid.

Let $\mathbb{D}(J)$ be the set of infinitesimal displacments of $J$ . One has the following diffeomorphism: $\mathbb{D}(J)\cong M\underset{i\times\bar{\pi}}{x} v\overset{R}{T}J$ , where $i:M \to J^R$ is the canonical injection, $v\overset{R}{T}J$ is the $\pi_R$-vertical tangent space of $J$ and $\bar{\pi}:v\overset{R}{T}J \to J$ is the canonical projection.

We list some usegul propositions regarding $\mathbb{D}(J)$.

<u>Proposition</u> .4 1. If $J$ is the groupoid associated to a G-principal bundle $P \to M$, there exists a canonical diffeomorphism $\mathbb{D}(J)\cong E(P)\cong TP/G \to TP\cong\mathbb{D}(J)\underset{M}{x} P$.

2. If $P$ and $P'$ are two principal fiber bundles with a same associated groupoid then $E(P)=E(P')$.

3. $\mathbb{D}(J)$ is a Lie algebroid, namely the sheaf of sections of $\mathbb{D}(J)$ is a sheaf of Lie algebras and the surjective morphism of vector bundles $r:\mathbb{D}(J) \to TM$ induces a morphism of Lie algebras. Furthermore, the kernel of $r$ is a fiber bundle of Lie algebras.

4. If $J$ is a Lie group, $\mathbb{D}(J)$ is the usual Lie algebra of $J$.

5. Any Lie algebroid is isomorphic to the algebroid of a $\pi_R$-simply connected groupoid (that is $\pi_R^{-1}(x)$ is simply connected for any $x \in M$).

6. $\mathbb{D}(J)$ identifies a functor between the category of principal fiber bundles and the category of Lie algebroids.

7. If $(P,M,\pi;G)$ is a principal fiber bundle with connected basis, then for any Lie sub-algebroid $E'$ of $\mathbb{D}(J)\cong E(P)$, there exists a connected sub-principal fiber bundle $P'$ (connected in the sense of sub-manifold of $P$), such that $E(P')$ $=E'$. Thus, any other connected principal sub-fiber bundle $P''$ such that $E(P'')=E'$ is deduced from $P'$ by action of some element $s$ of G: $z \mapsto z.s$.

8. Let $J$ be a Lie groupoid with connected basis M. If $E'$ is a Lie sub-algebroid of $\mathbb{D}(J)$, then there exists a unique sub-groupoid $\pi_R$-connected $J'$ such that $\mathbb{D}(J')=E'$.


Let us now specialize these general considerations to the pseudogroup Aut(X) of local diffeomorphisms of the differentiable manifold X.

The groupoid $J(\text{Aut}(X))$ aysociated to Aut(X) coincides with $\Pi^o(X)\cong X\times X$ and we get: 1) $\pi_R$=source projection$\cong p_o$; 2) $\pi_L$=target projection$\cong\bar{p}_o$; 3) M=X.

The k-prolongation $(\Pi^o(X))^k$ of $\Pi^o(X)$ is just $\Pi^k(X)$. As $\Pi^o(X)$ is a Lie groupoid so is $\Pi^k(X)$ with the following canonical projections; (a) target projection: $\bar{p}_k : \Pi^k(X) \to X$; (b) source projection: $p_k : \Pi^k(X) \to X$. The composition law is the following: $\gamma_k : \Pi^k(X) \times_X \Pi^k(X) \to \Pi^k(X)$, $(D^k f(p), D^k g(p)) \mapsto D^k(f \bullet g)(p)$. The identities are $D^k id_X(p)$, $\forall p \in X$ and the inverse of $D^k f(p)$ is $D^k f^{-1}(p)$. Furthermore, $p_{k,o} \equiv (p_k, \bar{p}_k) : \Pi^k(X) \to X \times X$ is an epimorphism. $\Pi^k(X)$ is the groupoid associated to the $L_n^k$-principal bundle $H^k(X)$, (n=dimX).

If $G \subset Aut(X)$ is a differential subpseudogroup of $Aut(X)$, we have a canonical injective application $j_k : J(G)^k \to \Pi^k(X)$ sanding $D^k(f_x) \mapsto (D^k f)(x)$. Set $I^k(G) \equiv im(j_k) \subset \Pi^k(G)$. In general $j_k$ is not an embedding, so $I^k(G)$ is not in general a sub-manifold of $\Pi^k(X)$. Furthermore, if $G$ is transitive so is $I^k(G)$.

A Lie equation of order k on X is a k-order differential system $R_k$ for the Lie groupoid $\Pi^o(X)$: $R_k \subset \Pi^k(X)$. So, the composition law $\gamma_k$ of $\Pi^k(X)$ restricts to $R_K$. $R_k$ is a Lie subgroupoid of $\Pi^k(X)$ if $R_k$ is a transitive differential equation. The set $G$ of solutions of $R_k$ is a subpseudogroup of $Aut(X)$ such that $I^k(G) \subseteq R_k$. The equality holds iff $R_k$ is completely integrable. In this case $G$ is said to be a Lie pseudogroup of order k on X. $G$ is a transitive Lie pseudogroup of order k iff the corresponding Lie equations is formally transitive (as well completely integrable).

Any G-structure of order k on a manifold X (n=dimM), $G \subset L_n^k$, can be identified with a Lie equation of order k on X. In fact, as $\Pi^k(X)$ is associated to $H^k(X)$, we can associate to $P \subset H^k(X)$ a subgroupoid $R_k$ of $\Pi^k(X)$ which is a Lie equation. The local authomorphisms of a G-structure $P \subset H^k(X)$, that are local diffeomorphisms $f \in Aut(X)$ such that $JD^k(f) \in Aut(P)$, are solutions of the groupoid $R_k \subset \Pi^k(X)$ associated to P. The G-structure is called transitive and locally homogeneous if $R_k$ is completely integrable. In this case for any $(z,z') \in P \times P$ there exists a local diffeomorphism transforming z into z'. Then, the pseudogroup $Aut(P)$ is a Lie pseudogroup. (Note that if P is integrable $R_k$ is so, but the inverse is false).

The set $\mathbb{D}(\Pi^k(X))$ of infinitesimal displacments of the groupoid $\Pi^k(X)$ can be identified wit $J\mathcal{D}^k(TX)$: $\mathbb{D}(\Pi^k(X)) \cong J\mathcal{D}^k(TX) \equiv D^k(id_X) \vee TJ\mathcal{D}^k(XxX)$. In particular, for any Lie equation $R_k \subset \Pi^k(X)$ we have $\mathbb{O}(R_k) \equiv D^k(id_X) \vee TR_k \equiv R_k \subset J\mathcal{D}^k(TX)$. We call $R_k$ the <u>linearized</u> of $R_k$ . $R_k$ is a Lie algebroid with respect to the Lie bracket $[,]$ for vector fields on $R_k$: $[R_k,R_k] \subset R_k$. (Any vector subbundle $R_k$ of $J\mathcal{D}^k(TX)$ such that is closed for the bracket $[,]$ is called an <u>infinitesimal Lie equation of order k</u> on X).

$R_k$ is transitive iff $R_k$ is a transitive groupoid. In this case we have the short exact sequence: $0 \to R_k^o \to R_k \to TX \to 0$ $(\blacksquare)$ , where $R_k^o \equiv R_k \cap J\mathcal{D}_o^k(TX)$, and $J\mathcal{D}_o^k(TX) = \ker(\pi_{k,o}) \subset J\mathcal{D}^k(TX)$. Furthermore, $G$ is a transitive Lie pseudogroup of order k, with finite equation $R_k$ and linearized equation $R_k$ iff $R_k$ is a formally transitive differential equation completely integrable. Note that on $J\mathcal{D}^k(TX)_x$ is defined also a canonical algebraic bracket $\{,\}$ :

$(\heartsuit)$ $\{,\}_x : J\mathcal{D}^k(TX)_x \boxempty J\mathcal{D}^k(TX)_x \to J\mathcal{D}^{k-1}(TX)_x$ , $\{D^k\xi(x), D^k\eta(x)\} = D^{k-1}[\xi,\eta](x)$,

where $\xi, \eta$ are vector fields defined on a neighbourhood of $x \in X$. This bracket (as well $[,]$) satisfies the Jacobi identity and gives to $J\mathcal{D}^\infty(TX)_x \equiv$ proj lim $J\mathcal{D}^k(TX)_x$ the structure of a Lie algebra. One has the formula:
$[D^k\xi, D^k\eta] = \{D^{k+1}\xi, D^{k+1}\eta\} + \xi \lrcorner (\mathbb{D}.(D^{k+1}\eta)) - \eta \lrcorner (\mathbb{D}.(D^{k+1}\xi))$.

From Theorem 4.1/1 we get that to a Lie groupoid $R_k \subset \Pi^k(X)$ it is associated the following principal fiber bundle over X: $(P \equiv p_o^{-1}(x), X, \bar{p}_o; G \equiv p_{k,o}^{-1}(x,x))$, for any $x \in X$. Further, from Proposition 4.4/1 we get $R_k \cong TP/G$. So, the sequence $(\blacksquare)$ can be written as follows: $0 \to vTP/G \to TP/G \to TX \to 0$ , where $P \equiv p_o^{-1}(x)$.

<u>Definition .7</u> Let $R_k \subset \Pi^k(X)$ be a Lie equation of order k on X. A numerical function $\phi: J\mathcal{D}^k(XxX) \to \mathbb{R}$ is called a <u>differential invariant of</u> $R_k$ if $\phi(D^k g(x).D^k f(x)) = \phi(D^k f(x))$, $\forall D^k g(x) \in R_k$.

<u>Definition .8</u> Let $X' \equiv X$ . We define the map $\blacksquare: J\mathcal{D}^k(TX') \to vTJ\mathcal{D}^k(XxX')$ by $\blacksquare: D^k\eta \to \blacksquare(D^k\eta) \equiv \bar{\eta}$.

We have $[\blacksquare(D^k\xi), \blacksquare(D^k\eta)] = \blacksquare([D^k\xi, D^k\eta])$, $\forall \xi, \eta \in C^\infty(TX')$.

<u>Proposition .5</u> 1. If $\phi$ is a differential invariant of $R_k$ it follows that $\phi$ is an invariant of the distribution $\blacksquare R_k$ on $J\mathcal{D}^k(XxX')$, that is

$\blacksquare(D^k\eta).\phi=0, \forall\, D^k\eta \in C^\infty(R_k)$.

2.  This distribution is involutive, so, at least locally, on an open neighbourhood of $D^k(id_X)(X) \subset \Pi^k(X)$, we can obtain a maximum and functionally independent set $\{\phi^\tau\}$ of differential invariant of order k.  Such a set is called  <u>fundamental set of order k.</u>  If $\dim\mathcal{JD}^k(X\times X')=s$ and the fiber dimension of $R_k$ is q, one has at most (s-q-n) non-trivial differential invariants in the fundamental set of order k (n=dimX).  (Note that as $R_k=(id_k)^*vTR_k \subset vT\mathcal{JD}^k(X\times X'),(id_k\equiv D^k(id_X))$, it follows that between the differential invariants there are always also the coordinate functions $x^i$ on X, but these are trivial differential invariants of $R_k$).

3.  $R_k$ can be locally defined in a neighbourhood of $id_k(X)$ by the following equations $\phi^\tau(y)=s^\tau(x)$, where $s^\tau(x)\equiv\phi^\tau(D^k id_X(x))$.

4.  If $\phi$ is a differential invariant of $R_k$ then $(\square_{i_1} \ldots \square_{i_q}.\phi)$ is a differential invariant of $R_{k+q}$, $\forall\, q\geq 0$. (See Appendix for the definition of $\square_i$).

Let us now investigate on the relation between Lie pseudogroups and fiber bundles.  We shall prove that Lie pseudogroups are strictly related with a particular category of super-bundles of geometric objects $[12,13]$.  Let us, first introduce the following fundamental

<u>Definition  .9</u>  A <u>reduced Lie super-bundle of geometric objects of order k</u> over a manifold X is a super-bundle of geometric objects $S\equiv(W,B;\,\mathbb{B})$, being $W\equiv\{\,\pi:W\rightarrow X\}$ the total bundle, $B\equiv\{\,\pi_B:B\rightarrow X\}$ the basis bundle and $\mathbb{B}$ is a  covariant functor  $C(B)\longrightarrow C(W)$ such that : (1) $\mathbb{B}$ depends only on the sub-category $C_B(X)$ of $C(X)$, image of the natural projection-functor R: $C(B)\rightarrow C(X)$; (2) the s-isotropy pseudogroup $G_s$ associateg to any section s of $W$ : $G_s\equiv\{\,f\equiv(f_W,f_M)\in G_W|f_W\cdot s\cdot f_M^{-1}\equiv f^*s =s\}\subset G_W$ , is a Lie pseudogroup of order k on X, where $G_W\equiv\mathbb{B}(\mathrm{Hom}(C(B)))\subset\mathrm{Hom}(C(W))$.

Then, we have the following important

<u>Theorem  .2</u>  A reduced Lie super-bundle of geometric objects of order k on X identifies : 1) a natural  action of $R_k$ (=Lie equation defining $G_s$) on W: $\gamma_k:R_k\underset{X}{\times} W \longrightarrow W$ with the inverse $\gamma_k^{-1}:W\underset{X}{\times} R_k\rightarrow W$ such that the following diagrams commute:

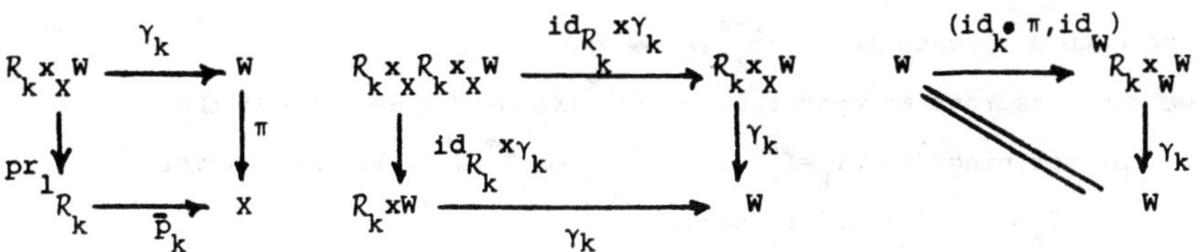

2) A natural action of $C^\infty(R_k)$ on $C^\infty(W)$ defined by the following commutative diagram:

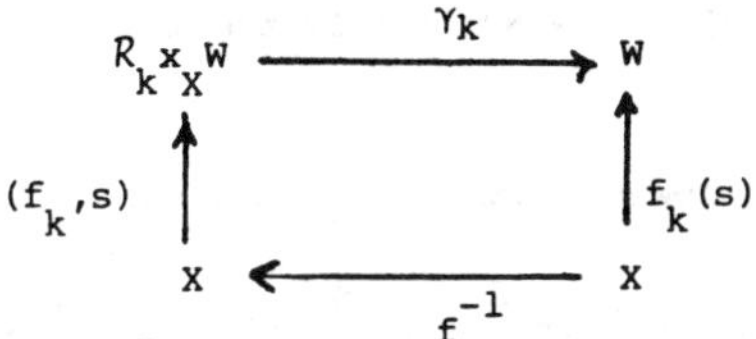

which is associative.

Further, any solution $\rho$ of $R_k$ can be lifted on W in a fiber bundle diffeomorphism $(\overset{\vee}{\rho},\rho)$ , such that the following diagram

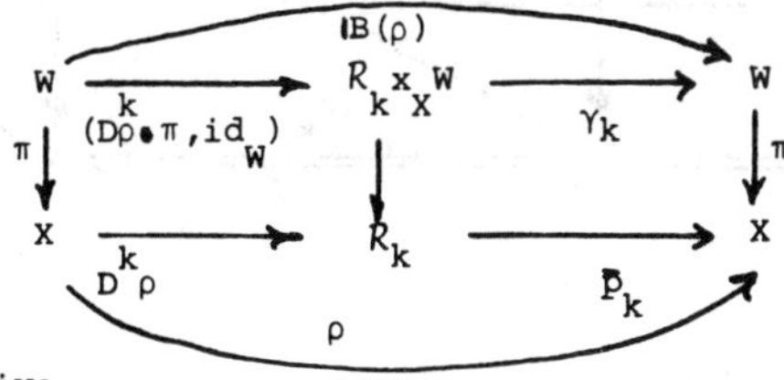

is commutative.

Viceversa, a k-order Lie equation $R_k \subset \Pi^k(X)$ on X identifies a reduced Lie super-bundle of geometric objects of order k on X with isotropy equation $K_k$.

Furthermore, a natural epimorphism $\Phi:\Pi^k(X) \to W$ and section s of W exist such that $R_k = \ker_s \Phi$ ; namely the following sequence : $0 \to R_k \longrightarrow \Pi^k(X) \overset{\Phi}{\underset{s\bullet p_k}{\rightrightarrows}} W$ is exact.

<u>Proof.</u>    In fact $\gamma_k$ is given by $\gamma_k(D^k f(x),s(x)) = ( \mathbb{B}(f)\bullet s \bullet f^{-1})(f(x))$.

Viceversa, given $R_k \subset \Pi^k(X)$ we canonically obtain a reduced Lie super-bundle of geometric objects of order k given by :

(a) basic bundle: $\pi_B:B \equiv R_k \cap \bar{p}_k^{-1}(x) \to X$; $\pi_B$ is the target projection of $\Pi^k(X)$. B is a principal fiber bundle with structure group $G \equiv R_k \cap p_{k,o}^{-1}(x,x)$;

(b) total bundle: $\pi:W \equiv p_k^{-1}(x)/G$; $\pi$ is the source projection of $\Pi^k(X)$. W is a bundle of homogeneous spaces with tipical fiber $F \equiv p_{k,o}^{-1}(x,x)/G$ ;

(c) covariant functor: $\mathbb{B}:C(\mathcal{B}) \to C(\mathcal{W})$    given by: (i) $\mathbb{B}(B|U) = W|U$;

(ii) (note that $\text{Hom}(C(\mathcal{B})) \cong \Theta =$ set of invertible sections of $R_k \equiv \{f_k\}$),

$\mathbb{B}(f_k)( [D^k g(p)] ) = [D^k(f\bullet g\bullet f^{-1})(f(p))]$ .

Furthermore, one can see that $C^\infty(R_k) = \{ f \in \text{Hom}(C(\mathcal{B}))|\ f_k(s) = s \}$    , where s is a section of W identified by means of the image in W of $R_k \cap \bar{p}_k^{-1}(x)$ by means of the natural projection $\Phi: \bar{p}_k^{-1}(x) \to W$.

Finally, we can construct an epimorphism $\Phi: \Pi^k(X) \to \cdot W$ that extends the previous $\Phi$ by defining $\Phi \circ f_k = f_k^{-1}(s)$, $\forall f_k \in C^\infty(\Pi^k(X))$. Furthermore, one can see that $f_k \in C^\infty(R_k)$ iff $f_k(s)=s$. $\square$

<u>Proposition</u> .6 For the linearized $R_k$ of a Lie equation $R_k \subset \Pi^k(X)$ we can write $R_k = \ker(Ls)$, where $Ls: J\mathcal{D}^k(TX) \to s^* vTW$ is the Lie derivative associated to s $[10,12]$.

<u>Proof.</u> In fact we have the following commutative diagram:

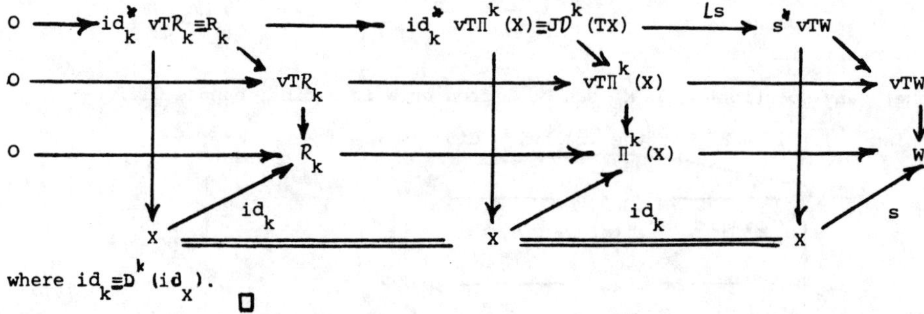

where $id_k \equiv D^k(id_X)$. $\square$

Let us consider, now, the integrability of a Lie equation.

<u>Proposition</u> .7 1.If $R_k \subset \Pi^k(X)$ is a transitive Lie equation, the symbol $M_{k+r}$ of the r-prolongation of the linearied equation $R_k$ , $r \geqslant 0$, is a vector bundle and we have explicit isomorphisms: $R^o_{k,x_o} \cong R^o_{k,x}$ ; $M_{k+r,x_o} \cong M_{k+r,x}$; $(H^{k+r,s})_{x_o} \cong (H^{k+r,s})_x$, $\forall$ $x_o, x \in X$, where $H^{\bullet,\bullet}$ is the Spencer cohomology of $R_k$. So, if $R_k$ is a transitive Lie equation, $R_k$ is formally integrable iff $R_{k+\ell+1} \to R_{k+\ell}$ is surjective with $0 \leqslant \ell \leqslant h$, being $R_{k+h}$ involutive.

2. Let $R_k \subset \Pi^k(X)$ be a transitive Lie equation. If $R_{k+r}$ is a vector bundle we have $R_{k+r} \cong (D^{k+r} id_X)^* vTR_{k+r}$.

3. If $R_k \subset \Pi^k(X)$ is a Lie equation such that $R_{k+r}$ is a fibered submanifold of $J\mathcal{D}^{k+r}(X \times X)$, then $R_{k+r}$ is a Lie equation $\forall$ $r \geqslant 0$.

4. If $R_k \subset J\mathcal{D}^k(TX)$ is an infinitesimal Lie equation such that $R_{k+r}$ is a vector subbundle of $J\mathcal{D}^{k+r}(TX)$, then $R_{k+r}$ is a system of infinitesimal Lie equation , $\forall$ $r \geqslant 0$.

5. Let $R_k \subset \Pi^k(X)$ be a Lie equation. The Spencer cohomolgy $H^{k+r,s}$ of $R_k$ does not depend on the section s of W such that $R_k = \ker \Phi_s$ .

__Theorem .3__ Let $R_k \subset \Pi^k(X)$ be a Lie equation over a manifold X. Let $M_k$ the symbol of $R_k$. Let us define the following natural vector bundles of order k over the natural bundle W associated to $R_k \equiv \ker_s \Phi$, $\Phi : \Pi^k(X) \to W$:

$W_r \equiv \Lambda^o_r X \otimes vTW / \delta(\Lambda^o_{r-1} X \otimes N_1)$, $W^o_o \equiv vTW$, $N_r = \mathrm{im}(\sigma_r(vT(\Phi^{(r)}))$ in the following exact sequence of vector bundles over X associated to $M_k$:

$$0 \longrightarrow M_{k+r} \longrightarrow S^o_{k+r} X \otimes TX \xrightarrow[\sigma_r(vT(\Phi^{(r)}))]{} S^o_r X \otimes vTW \ .$$

1. Then, $R_k$ is formally integrable iff its defining section s verifies $\dim(s^* W_1)$ analytic integrability conditions that constitute a natural system $B_1(s|c) \equiv \ker_c I \subset J\mathcal{D}(W)$, being c a suitable section $W \to W_1$ and I is an affine natural epimorphism $I : J\mathcal{D}(W) \to W_1$ over W given by: $I(q) = \mathrm{pr}(q-a)$, $\forall q \in J\mathcal{D}(W)$, where pr is the natural projection $T^* X \otimes vRW \to W_1$ and $a = \Phi^{(1)}(f_{k+1}(x))$, being $f_k \in C^\infty(\Pi^k(X)$, $\Phi(f_k(x)) = \pi_{1,o}(q) \in W$. The natural section locally depends on a finite number of constants called __structure constants__.

2. Furthermore, $B_1(s|c) \subset J\mathcal{D}(W)$ is analytic and formally integrable iff the section c is a solution of a differential equation $J(c) \subset J\mathcal{D}(W_1)$ (=__Jacobi conditions__).

__Proof.__ 1. Let $R_k$ be formally integrable with a 2-acyclic symbol. Let $B_1$ be the image of $\Phi^{(r)}$. $B_1$ is a fibered submanifold of $J\mathcal{D}(W)$ and coincides with the kernel of I. Then, we have the following commutative and exact diagram of natural affine bundles over W:

$$
\begin{array}{ccccccccc}
0 & \longrightarrow & N_1 & \longrightarrow & T^* X \otimes W_o & \longrightarrow & W_1 & \longrightarrow & 0 \\
& & \downarrow & & \downarrow & & \downarrow & & \\
0 & \longrightarrow & B_1 & \longrightarrow & J\mathcal{D}(W) & \xrightarrow{I} & W_1 & \longrightarrow & 0 \\
& & \downarrow & & \downarrow & & \downarrow o & & \\
& & W & = & W & = & W & &
\end{array}
$$

where $o : W \to W_1$ is the zero section of the vector bundle $W_1 \to W$.

We have also the following commutative and exact diagram of affine bundles:

$$
\begin{array}{ccccccccc}
0 & \longrightarrow & M_{k+1} & \longrightarrow & S^o_{k+1} X \otimes TX & \xrightarrow{\sigma_1(vT(\Phi^{(1)}))} & TX \otimes W_o & \xrightarrow{\tau} & W_1 & \longrightarrow & 0 \\
& & \downarrow & & \downarrow & \Phi^{(1)} & \downarrow & I & \downarrow & & \\
0 & \longrightarrow & R_{k+1} & \longrightarrow & \Pi^{k+1}(X) & \longrightarrow & J\mathcal{D}(W) & \xrightarrow{I} & W_1 & & \\
& & \downarrow & & \downarrow & \Phi & \downarrow & & & & \\
0 & \longrightarrow & R_k & \longrightarrow & \Pi^k(X) & \longrightarrow & W & \longrightarrow & 0 &
\end{array}
$$

According to results of section 2, we have an exact sequence:

$R_{k+1} \longrightarrow R_k \xrightarrow[\text{o}]{\kappa} p_k^*(s^*W_1)$. Now, we are able to compute the map $\kappa$ by using I. In fact, for any $f_{k+1} \in C^\infty(\Pi^{k+1}(X))$ over $f_k \in C^\infty(R_k)$ we have $\kappa \circ f_k = \tau(f_{k+1}(Ds)-Ds)=f_k(I\bullet Ds)-I(Ds)$. Therfore, $\kappa \circ f_k = 0$ iff $f_k(I\bullet Ds)=I\bullet Ds$, whenever $f_k(s)=s$. Now, the above condition is satisfied iff a natural section $c:W \longrightarrow W_1$ exists such that $I\bullet Ds = c\bullet s$.

2. As the symbol $M_k$ of $R_k$ is 2-acyclic, we have the following commutative diagram in which the upper row is an exact sequence of natural vector bundles over W:

$$
\begin{array}{ccccccccc}
\text{O} & \to & N_2 & \longrightarrow & S_2^o X \boxtimes W_o & \longrightarrow & T\, X \boxtimes W_1 & \longrightarrow & W_2 \to \text{O} \\
 & & \downarrow & & \downarrow & & \downarrow & & \downarrow \\
\text{O} & \to & B_2 & \longrightarrow & J\!D^2(W) & \longrightarrow & J\!D(W_1) & & \\
 & & \downarrow & & \downarrow & & \downarrow & & \\
\text{O} & \to & B_1 & \longrightarrow & J\!D(W) & \longrightarrow & W_1 & & \\
 & & \downarrow & & \downarrow & & \downarrow & & \\
 & & W & = & W & = & W & &
\end{array}
$$

We shall proceed by steps: (a) We are able to construct a natural morphism $J: J\!D(W_1) \longrightarrow W_2$; (b) The restriction of $J\bullet J\!D(c):J\!D(W) \longrightarrow W_2$ to $B_1(s|c)$ is the curvature map $\kappa$ of $B_1(s|c)$; (c) the condition in order to have the surjectivity of $B_1(s|c)_{+1} \longrightarrow B_1(s|c)$ by means of the exact sequence $B_1(s|c)_{+1} \longrightarrow B_1(s|c) \xrightarrow{\kappa} W_2$ is given by $J(c)$: $\kappa \bullet Ds = J\bullet J\!D(c)\bullet Ds = 0$ that , for fixed s, gives a condition on c. $\square$

<u>Remark.</u> We can alternatively examine the integrability conditions on the linearized system $R_k$ (see e.g. ref. $[1/n]$).

<u>Example.</u> If $G$ is a Lie group G acting simply and transitively on a manifold M (dimG=dimM=n), the general equations $R_1$ and $R_1$ are expressed by means of the canonical Maurer-Cartan invariant A(G)-valued 1-form: $s \equiv \omega = \omega_i^h(x)dx^i \boxtimes Z_h$, where $\{Z_h\}$ is a basis of the Lie algebra A(G) of G:

$$R_1 \qquad \omega_k^h(y)y_i^k - \omega_i^k(x) = \text{o}$$

$$R_1 \qquad \omega_k^h(x)\,\xi_i^k + \xi^r(\partial x_r.\omega_i^h(x))=\text{o}.$$

The equation $B_1(\omega|c)$ is given by $\alpha_k^i(x)\alpha_e^j(x)\left[(\partial x_i.\omega_j^h)(x)-(\partial x_j.\omega_i^h)(x)\right]-C_{ke}^h=\text{o}$, where $\alpha_h^i(x)\omega_k^h(x)=\delta_k^i$. The symbol of this non-linear system is involutive and we can see that this system is completely integrable iff the structure constants $C_{ke}^h$ satisfy the Jacobi conditions. These coincide with the equation $J(c)$ that justifies the name given to $J(c)$.

Let us study the problem of integrability of a Lie equation by using the Spencer operator.

<u>Theorem</u> .4  1.  Let $R_k \subset \Pi^k(X)$ be a Lie equation of order k on X. Then, one has the following commutative diagrams:

① (First non-linear Spencer sequences):

$$
\begin{array}{ccccccccc}
0 & \longrightarrow & \mathrm{Aut}(X) & \xrightarrow{\;D^{k+1}\;} & \Pi^{k+1}(X) & \xrightarrow{\;\bar{D}\;} & T^*X\otimes JD^k(TX) & \xrightarrow{\;\bar{D}'\;} & \Lambda^o_2 X\otimes JD^{k-1}(TX) \\
 & & \uparrow & & \uparrow & & \uparrow & & \uparrow \\
0 & \longrightarrow & G & \longrightarrow & R_{k+1} & \xrightarrow{\;\bar{D}\;} & T^*X\otimes R_k & \xrightarrow{\;\bar{D}'\;} & \Lambda^o_{-2} X\otimes JD^{k-1}(TX) \\
 & & \uparrow & & \uparrow & & \uparrow & & \uparrow \\
 & & 0 & & 0 & & 0 & & 0
\end{array}
$$

where the rows are locally exact.  $\bar{D}$ is obtained by the non-linear Spencer operator tensoring by $id_{T^*X}$ and $\bar{D}'$ is defined by the following formula:
$$\bar{D}'.\sigma\,(\xi,\eta)=\bar{D}\,\sigma(\xi,\eta)-\{\,\sigma(\xi)\,\sigma\,\sigma(\eta)\}\quad ,\forall\,\xi,\eta\in \underline{TX}. \quad (\bar{D}' \text{ is not involutive}).$$

② (Second non-linear Spencer sequences):

$$
\begin{array}{ccccccccc}
0 & \longrightarrow & \mathrm{Aut}(X) & \xrightarrow{\;D^k\;} & \Pi^k(X) & \xrightarrow{\;\bar{D}_1\;} & C_1(TX) & \xrightarrow{\;\bar{D}_2\;} & C_2(TX) \\
 & & \uparrow & & \uparrow & & \uparrow & & \uparrow \\
0 & \longrightarrow & G & \xrightarrow{\;D^k\;} & R_k & \xrightarrow{\;\bar{D}_1\;} & C_1 & \xrightarrow{\;\bar{D}_2\;} & C_2 \\
 & & \uparrow & & \uparrow & & \uparrow & & \uparrow \\
 & & 0 & & 0 & & 0 & & 0
\end{array}
$$

where the rows are locally exact.  $C_r(TX)$ and $C_r$ are the following vector bundles: $C_r(TX)\equiv(\Lambda^o_r X\otimes JD^k(TX))/\delta\,(\Lambda^o_{r-1} X\otimes S^o_{k+1} X\otimes TX)$; $C_r\equiv\Lambda^o_r X\otimes R_k / \delta(\Lambda^o_{r-1} X\otimes g_{k+1})$.
$\bar{D}_1$ is defined by the following formula: $\bar{D}_1.f_k=(\bar{D}.f_k)\bullet A^{-1}$, being $A\equiv id_{T^*X}+\bar{D}f_1$.
Finally $\bar{D}_2$ is defined by the following commutative diagram:

$$
\begin{array}{ccccc}
\Pi^{k+1}(X) & \xrightarrow{\;\bar{D}\;} & T^*X\otimes JD^k(TX) & \xrightarrow{\;\bar{D}'\;} & \Lambda^o_2 X\otimes JD^{k-1}(TX) \\
\Big\| & & \Big\downarrow{\scriptstyle(\Lambda^o_2(B)\otimes id_{JD^k(TX)})_\bullet} & & \Big\downarrow{\scriptstyle(\Lambda^o_2(B)\otimes id_{JD^{k-1}(TX)})_\bullet} \\
\Pi^{k+1}(X) & \xrightarrow{\;\bar{D}_1\;} & T^*X\otimes JD^k(TX) & \xrightarrow{\;\bar{D}_2\;} & \Lambda^o_2 X\otimes JD^{k-1}(TX)
\end{array}
$$

where $B=A^{-1}$.  ($\bar{D}_1$ is involutive).

2.  Whenever $R_k$ is formally integrable with a 2-acyclic (involutive) symbol, the restricted non-linear operator $\bar{D}_1:R_k \to C_1$ is formally integrable with a 2-acyclic (involutive) symbol.

3.  Whenever $R_k$ is formally integrable with a 2-acyclic (involutive) symbol, $\bar{D}_2:C_1 \to C_2$ is a formally integrable (involutive) first order non-linear operator.

4. From ① and ② we can try , by the proceeding of linearization, the corresponding linear Spencer sequences.

5. The cohomology of $T^*X \boxtimes R_k$ of the first non-linear Spencer sequences and the cohomology at $C_1$ of the second non-linear Spencer sequences are isomorphic to the cohomology at W of the non-linear Janet sequence:

$$(\text{Janet sequence}) \qquad O \longrightarrow G \longrightarrow \text{Aut}(X) \xrightarrow{\Phi \cdot D^k} \underline{W} \underset{O}{\overset{I \cdot D}{\rightrightarrows}} \underline{W}_1 \quad , \ R_k \equiv \ker_s \Phi \ .$$

<u>Remark.</u> Above costructions of sequences 1 , 2 can be related to a $R_k$-valued form on $R_k$. Let $\omega'_{k+1} : J\mathcal{D}^{k+1}(W) \to T^*J\mathcal{D}^{k+1}(W) \boxtimes J\mathcal{D}^k(vTW)$ be the canonical $vTJ\mathcal{D}^k(W)$-valued 1-form on $J\mathcal{D}^{k+1}(W)$ defined by $\omega'_{k+1} \circ T(f_{k+1}) = \mathbf{D} \cdot f_{k+1}$. The fundamental <u>Maurer-Cartan form</u> for the groupoid $R_k \subset \Pi^k(X)$ is the $R_k$-valued differential 1-form $\bar{\omega}$ on $R_{k+1}$, $\bar{\omega} : R_{k+1} \to T^*R_{k+1} \boxtimes R_k$ defined by restriction of $\omega'_{k+1}$ to $R_{k+1}$ on $J\mathcal{D}^{k+1}(W)$ with $W \rightleftharpoons X \times X \to X$ and taking the pull-back by $D^{k+1}(\text{id}_X)$. Then, $\bar{\omega}$ satisfies the following eaquation:
$$\mathbf{D} \cdot \bar{\omega} - \frac{1}{2} \ \{\bar{\omega}, \bar{\omega}\} \quad = O.$$
The non-linear first order differential operator $\bar{\mathbf{D}} : \underline{\Pi^{k+1}(X)} \to \underline{T^*X \boxtimes J\mathcal{D}^k(TX)}$ of Theorem 4.4 can be also defined by means of $\bar{\omega}$ by $f_{k+1} \mapsto \bar{\mathbf{D}} \cdot f_{k+1} = f^*_{k+1} \bar{\omega}$ .

Now, we are  able to consider the formal invariance of a differential equation.

<u>Proposition</u> .8 Let $\hat{\Pi}^k(W)$ be the subbundle of $J\mathcal{D}^k(W \times W)$ as defined in Section 2.  One has a natural action of $\hat{\Pi}^k(W)$ on the manifold $J\mathcal{D}^k(W)$ such that the following diagram

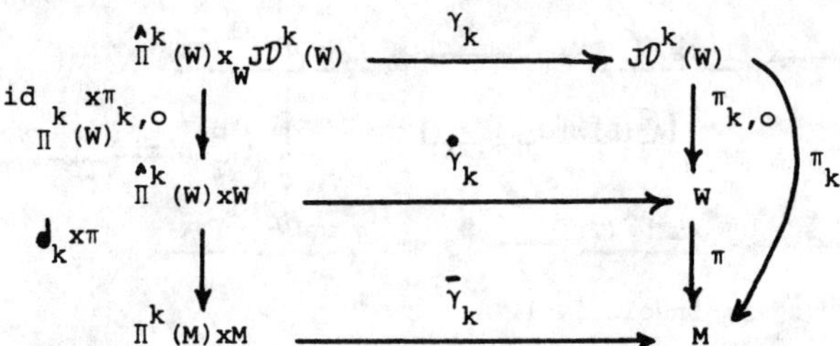

is commutative; where the map $\Phi_k$ is given by $D^k f_W(p) \mapsto D^k f_M(\pi(p))$, for any $(f_W, f_M) \in \text{Hom}(C(W))$.  Furthermore, the action is associative and the unities of $\hat{\Pi}^k(W)$ give the transformation identity on $J\mathcal{D}^k(W)$, i.e. the following diagrams

$$\hat{\Pi}^k(W)\times_W\hat{\Pi}^k(W)\times_W J\mathcal{D}^k(W)\xrightarrow{\ id_{\hat{\Pi}^k(W)}\times\gamma_k\ }\hat{\Pi}^k(W)\times_W J\mathcal{D}^k(W)$$

$$\Big\downarrow\alpha\times id_{J\mathcal{D}^k(W)}\qquad\qquad\Big\downarrow\gamma_k$$

$$\hat{\Pi}^k(W)\times_W J\mathcal{D}^k(W)\xrightarrow{\quad\gamma_k\quad} J\mathcal{D}^k(W)$$

$$J\mathcal{D}^k(W)\xrightarrow{\ (id_k\bullet\pi_{k,o},\,id_{J\mathcal{D}^k(W)})\ }\hat{\Pi}^k(W)\times_W J\mathcal{D}^k(W)$$

$$\Big\downarrow\gamma_k$$

$$J\mathcal{D}^k(W)$$

(associativity)        (identity)

are commutative, being $\alpha$ the composition law in $\hat{\Pi}^k(W)$ and $id_k\equiv D^k(id_W)$.

<u>Proof.</u> In fact the map $\gamma_k:\hat{\Pi}^k(W)\times_W J\mathcal{D}^k(W)\to J\mathcal{D}^k(W)$ defined by:

$$\gamma_k(u\equiv D^k f(p),q\equiv D^k s(\pi(p)))=D^k(f_W\bullet s\bullet f_M^{-1})(f_M(x)),\quad x\equiv\pi(p'),\ s(x)=p,\ f\equiv(f_W,f_M),$$

does not depend on the rapresentative for $u$ anf $q$. In fact, if $u=D^k f'(p)$ and $q=D^k s'(x)$ one has:

$$\gamma_k(D^k f'(p),D^k s'(x))=D^k(f'_W\bullet s\bullet f'^{-1}_M)(f'_M(x))\equiv\mathcal{D}^k(f'_M,f'_W)(D^k s'(x))\equiv\mathcal{D}^k(f'_M,f'_W)(D^k s(x))$$

$$=D^k(f'_W\bullet s\bullet f'^{-1}_M)(f'_M(x))\equiv\mathcal{D}^k(s\bullet f'^{-1}_M,id_W)\bullet D^k f'_W\bullet s\bullet f'^{-1}_M\bullet f'_M(x)$$

$$\equiv\mathcal{D}^k(s\bullet f'^{-1}_M,id_W)(D^k f_W(p)=D^k(f_W\bullet s\bullet f'^{-1}_M)(f_m(x))$$

$$\equiv\mathcal{D}^k(f_M,f_W)\bullet D^k s\bullet f'^{-1}_M\bullet f_M(x)\equiv\mathcal{D}^k(f_M,f_W)(D^k s(x))$$

$$=D^k(f_W\bullet s\bullet f_M^{-1})(f_M(x))$$

$$=\gamma_k(D^k f(p),D^k s(x)).\qquad\square$$

<u>Definition</u> .10  1. The differential equation $E_k\subset J\mathcal{D}^k(W)$ is (formally) <u>invariant</u> by the Lie equation $\hat{R}_k\subset\hat{\Pi}^k(W)$ if the canonical map $\gamma_k:\hat{\Pi}^k(W)\times_W J\mathcal{D}^k(W)\to J\mathcal{D}^k(W)$ induces the following commutative diagrams:

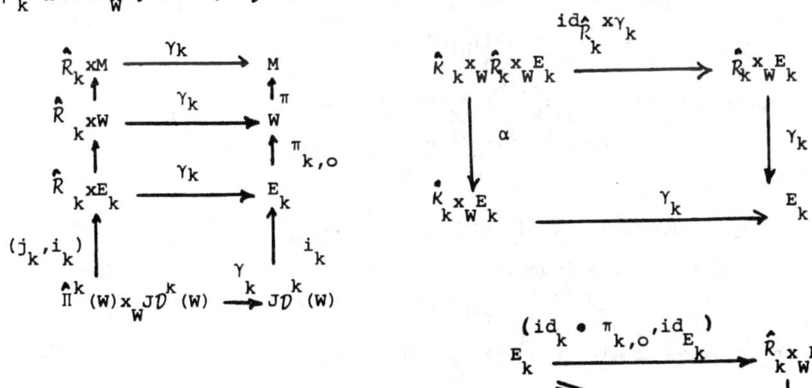

where $j_k$ is the canonical embedding $\hat{R}_k\longrightarrow\hat{\Pi}^k(W)$.

2. The <u>biggest</u> Lie groupoid $\overset{\wedge}{R}_k \subseteq \overset{\wedge k}{\Pi}(W)$ of invariance of $E_k \subseteq JD^k(W)$ is:

$$\overset{\wedge}{R}_k = \{\ g_k \in \overset{\wedge k}{\Pi}(W)\ |\ \gamma_k(g_k, E_k) = E_k\ \}\ .$$

3. Let $\overset{\bullet k}{\Pi}(W)$ be the subgroupoid of $\overset{\wedge k}{\Pi}(W)$ of k-derivatives of local fiber bundle diffeomorphisms $f \equiv (f_W, f_M)$ of W such that $f_M = \mathrm{id}_M$: $\overset{\bullet k}{\Pi}(W) = \overset{\wedge k}{\Pi}(W) \cap JD^k(W x_M W)$. We call $\overset{\bullet k}{\Pi}(W)$ the <u>gauge-subgroup</u> of $\overset{\wedge k}{\Pi}(W)$. If $E_k \subseteq JD^k(W)$ is invariant by a Lie equation $\overset{\wedge}{R}_k \subseteq \overset{\wedge k}{\Pi}(W)$ set $\overset{\bullet}{R}_k \equiv \overset{\wedge}{R}_k \cap \overset{\bullet k}{\Pi}(W)$. We call $\overset{\bullet}{R}_k$ the <u>gauge-sub-groupoid</u> of $\overset{\wedge}{R}_k$.

4. Let $S = (W, B; \mathbb{B})$ be a reduced super-bundle of geometric objects of order k over M. Then, we can identify a subgroupoid $\amalg^k(W)$ of $\overset{\wedge k}{\Pi}(W)$:

$$\amalg^k(W) \equiv \left\{ D^k f(p) \in \overset{\wedge k}{\Pi}(W)\ |\ f_W = \mathbb{B}(f_M) \right\}\ .$$

We call $\amalg^k(W)$ the <u>dynamic subgroupoid</u> of $\overset{\wedge k}{\Pi}(W)$. If $E_k \subseteq JD^k(W)$ is invariant by a Lie equation $\overset{\wedge}{R}_k \subseteq \overset{\wedge k}{\Pi}(W)$, set $\overset{\square}{R}_k \equiv \overset{\wedge}{R}_k \cap \amalg^k(W)$. We call $\overset{\square}{R}_k$ the <u>dynamic subgroupoid</u> of $\overset{\wedge}{R}_k$.

<u>Proposition .9</u> If $E_k \subseteq JD^k(W)$ is a k-order differential equation on the fiber bundle $\pi: W \to M$, such that it is invariant by the Lie equation $\overset{\wedge}{R}_k \subseteq \overset{\wedge k}{\Pi}(W)$, then if $E_{k+q}$ and $\overset{\wedge}{R}_{k+q}$ are both systems of order k+q for some $q \geq 0$, $E_{k+q}$ is an invariant equation by the Lie equation $\overset{\wedge}{R}_{k+q}$.

<u>Remark.</u> Let $\{x^i, y^j_{i_1 \ldots i_\alpha}\}_{\ 0 \leqslant \alpha \leqslant k}$ be local coordinates on $JD^k(W)$. Then, in local coordinates the map $\gamma_k: \overset{\wedge k}{\Pi}(W) x_W JD^k(W) \to JD^k(W)$ looks like:

$$x^i \circ \gamma_k(D^k f(p), D^k s(x)) = x^i \circ f_M(\pi(p)) \equiv f^i_M(\pi(p)) \equiv \bar{x}^i(\pi(p)),\ D^k f(p) \in \overset{\wedge k}{\Pi}(W),\ p \in W,$$

(♥)
$$s(x) = p;$$

$$y^j \circ \gamma_k(D^k f(p), D^k s(x)) = y^j \circ f_W(p) \equiv f^j_W(p) \equiv \bar{y}^j(p);$$

$$y^j_i \circ \gamma_k(D^k f(p), D^k s(x)) = (\partial x_i . f^{-1}_M{}^k)(f_M(\pi(p))) y^\beta_k(p) (\partial y_\beta . f^j_W)(p)$$
$$= (\partial x_i . \bar{x}^k)^{-1}(f_M(\pi(p))) y^\beta_k(p) (\partial y_\beta . \bar{y}^j)(p).$$

......

So, if the local expression of $E_k$ is $F^\tau(x^i, y^j, \ldots, y^j_{i_1 \ldots i_k}) = o$, we could obtain the biggest Lie pseudogroup of invariance of $E_k$ by imposing the equality $F^\tau(x^i, y^j, \ldots, y^j_{i_1 \ldots i_k}) = O = F^\tau(\bar{x}^i, \bar{y}^j, \ldots, \bar{y}^j_{i_1 \ldots i_k})$ and after eliminating the coordinates non-over-lined by means of the expressions (♥) we get the local form for $R_k$.

__Example.__ Let $\pi: W = \mathbb{R} \times \mathbb{R}^2 \to \mathbb{R}$. Let $\{x, y^1, y^2, y^1_x, y^2_x\}$ be local coordinates on $J\mathcal{D}(W)$. Let us consider the equation $E_1 \subset J\mathcal{D}(W): \{\ F \equiv y^2 y^1_x - \omega(x) = 0\ \}$.

Working out the maximum Lie group of invariance we obtain:

$$\bar{y}^2\left[\ y^1_x(\partial x.\bar{x})^{-1}(\partial y_1.\bar{y}^1) + y^2_x(\partial x.\bar{x})^{-1}(\partial y_2.\bar{y}^1)\right] - \omega(\bar{x}) = 0 = y^2 y^1_x - \omega(x).$$

Then we can write:

$$y^1_x\left[\ \bar{y}^2(\partial x.\bar{x})^{-1}(\partial y_1.\bar{y}^1) - y^2\right] + \bar{y}^2 y^2_x(\partial x.\bar{x})^{-1}(\partial y_2.\bar{y}^1) + \omega(x) - \omega(\bar{x}) = 0.$$

So we get:

$$\hat{R}_1 \begin{cases} \bar{y}^2(\partial x.\bar{x})^{-1}(\partial y_1.\bar{y}^1) - y^2 = 0 \\[4pt] (\partial x.\bar{x})^{-1}(\partial y_2.\bar{y}^1)\bar{y}^2 = 0 \\[4pt] \omega(x) - \omega(\bar{x}) = 0 \end{cases} \Rightarrow\ G \begin{cases} \Rightarrow\quad \bar{y}^2 = \dfrac{y^2(\partial\bar{x}.x)}{(\partial y_1.\bar{y}^1)} \\[10pt] \Uparrow \\[4pt] (\partial y_2.\bar{y}^1) = 0 \Rightarrow \bar{y}^1 = \bar{y}^1(x, y^1) \\[6pt] \Rightarrow\quad \bar{x} = f(x) \end{cases}$$

The gauge-subgroupoid of $R_1$ is:

$$\dot{R}_1 \begin{cases} \bar{y}^2(\partial y_1.\bar{y}^1) - y^2 = 0 \\[4pt] (\partial y_2.\bar{y}^1)\bar{y}^2 = 0 \end{cases} \Rightarrow\ \dot{G} \begin{cases} \bar{y}^1 = \bar{y}^1(y^1) \\[6pt] \bar{y}^2 = y^2/(\partial y_1.\bar{y}^1). \end{cases}$$

Now, xe can consider W as a natural bundle having the following functor $\mathbb{B}$:

(1) $\mathbb{B}(U) = U \times \mathbb{R}^2$, for any U open in $\mathbb{R}$ and ; (2) the map $\mathbb{B}(f_{\mathbb{R}}): \mathbb{B}(U) \to \mathbb{B}(\bar{U})$ is given by $\mathbb{B}(f_{\mathbb{R}}) = (f_{\mathbb{R}}, id_{\mathbb{R}^2})$, for any local diffeomorphism $f_{\mathbb{R}}: U \to \bar{U}$ of $\mathbb{R}$ So, we can canonically identify a dynamic sub-groupoid of $R_1$; more precisely one has:

$$y^2 y^1_x(\partial x.\bar{x})^{-1} - \omega(\bar{x}) - y^2 y^1_x + \omega(x) = 0 \quad\Rightarrow\quad \overset{\square}{R}_1 \begin{cases} (\partial\bar{x}.x) = 1 \\[4pt] \omega(x) - \omega(\bar{x}) = 0. \end{cases}$$

If $\omega(x)$ is neither invariant for translations no for a particular translation must be $\overset{\square}{G} = \{id_R\}$. If $\omega(x)$ is invariant for translations then $\overset{\square}{G}$ is the one-parameter group of translations: $x = \bar{x} + c$, $\forall\, c \in \mathbb{R}$. Finally, if $\omega(x)$ is invariant for a particular value of $c \neq 0$, then we get $G = \{id_{\mathbb{R}}, \bar{x} = x - c\ \}$, $c \in \mathbb{R}$ fixed.

__Remark.__ If $E_k \subset J\mathcal{D}^k(W)$ is a k-order differential equation on a fiber bundle $\pi: W \to M$, we can consider the pseudogroup $G$ of local diffeomorphisms $f \in Hom(C(W))$ such that $J\mathcal{D}^k(f)\big|_{E_k} \in Hom(C(E_k))$. Really, $G$ is the set of solutions of a Lie equation $\hat{R}_k \subset \hat{\Pi}^k(W)$. (If $\hat{R}_k$ is not completely integrable one has: $I^k(G) \subset \hat{R}_k$). The set $\Theta$ of solutions of the linearized equation $\hat{R}_k \subset \hat{J\mathcal{D}}^k(TW)$ is the corresponding infinitesimal pseudogroup of invariance of $E_k$. Then, $\blacksquare\hat{R}_k$ results into an involutive distribution on $\hat{R}_k$. We have the following:

**Theorem .5** If $E_k \subset J\mathcal{D}^k(W)$ is invariant by a Lie equation $\hat{R}_k \subset \hat{\Pi}^k(W)$ then on $E_k$ there exists an involutive distribution $\boxtimes R_k$ generated by all vector fields $X \in \hat{C}^\infty(TW)$ such that $\overset{(k)}{X}$ is tangent to $E_k$.

**Proof.** Let us consider the following commutative diagram:

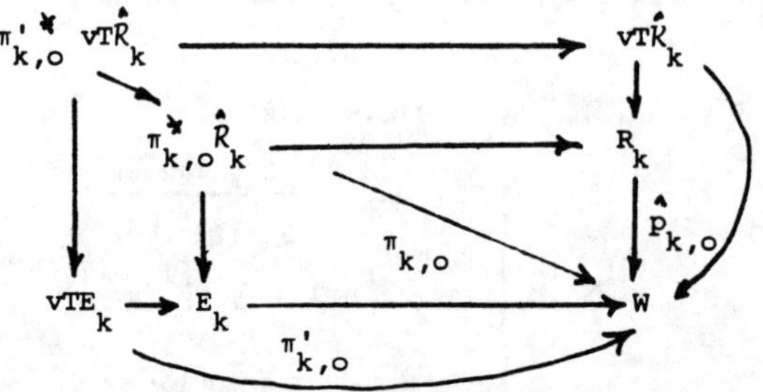

Note that the action map $\gamma_k$ can be written also as $\ell_k : \pi^*_{k,o}\hat{R}_k \longrightarrow E_k$. So, we obtain the following commutative diagram:

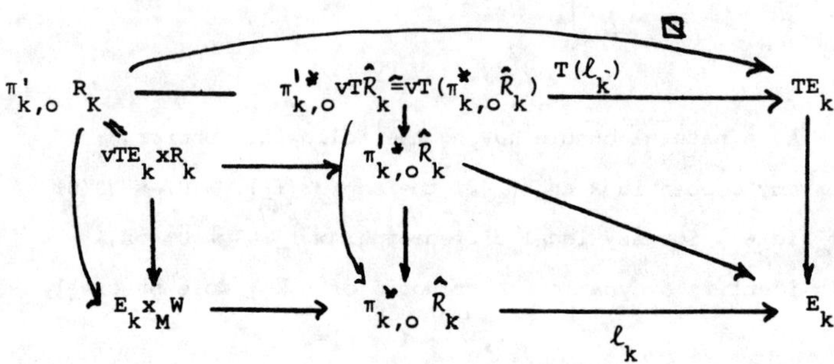

which defines also the map $\boxtimes$. Then, the image of $\pi'^*_{k,o} R_k$ into $TE_k$ by means of $\boxtimes$ defines the cited involutive distribution on $E_k$ that we simply write $\boxtimes R_k$. We can also easily see that $\boxtimes R_k$ is just generated by the all vector fields $X \in \hat{C}^\infty(TW)$ such that $\overset{(k)}{X}$ is tangent to $E_k$. $\square$

**Corollary .2** If $E_k \subset J\mathcal{D}^k(W)$ is invariant by $\hat{R}_k \subset \hat{\Pi}^k(W)$ then $X \in \hat{C}^\infty(TW)$ belongs to $\boxtimes R_k$ iff $\overset{(k)}{X}.F^i_{j_1 \ldots j_\alpha} = 0$ ($\emptyset$) or equivalently

$$(\partial x_\alpha.F^i)X^\alpha + \sum_{o \leqslant \alpha \leqslant k} (\partial y^\beta_{j_1 \ldots j_\alpha}.F^i)X^\beta_{j_1 \ldots j_\alpha}) = 0$$

where $F^i$, $i=(1,\ldots,s)$ are numerical functions on $J\mathcal{D}^k(W)$ locally characterizing $E_k$ being $\{ x^i, y^\beta_{j_1 \ldots j_\alpha} \}_{o \leqslant \alpha \leqslant k}$ local coordinates on $J\mathcal{D}^k(W)$ and $\overset{(k)}{X} = X^\alpha \partial x_\alpha +$

$\sum_{o \leqslant \alpha \leqslant k} X^\beta_{j_1 \ldots j_\alpha} \partial y^{j_1 \ldots j_\alpha}_\beta$ is the k-prolongation of $X = X^\alpha \partial x_\alpha + Y^\beta \partial y_\beta$; $Y^\beta = X^\beta_{j_1 \ldots j_\alpha}$ for $\alpha = 0$.

If $E_k$ is _irreducible_, that is any real valued function f vanishing on $E_k$ must be of the form $f = \sum_i \lambda_i F^i$, where $\lambda_i$ are smooth real valued functions on $\mathcal{JD}^k(W)$, then the computation of ($\phi$) becomes $\overset{(k)}{X}.F^s = \sum_i \lambda_i^s F^i$.

<u>Corollary</u> .2 Let $E_k \subset \mathcal{JD}^k(W)$ be invariant by $\hat{R}_k \subset \hat{\Pi}^k(W)$, then if $E_{k+q}$ and $R_{k+q}$ are both systems of order k+q, $q \geq 0$, then on $E_{k+q}$ there exists an involutive distribution $\boxempty R_{k+q}$ generated by all vector fields $X \in \hat{C}^\infty(TW)$ such that $\overset{(k+q)}{X}$ are tangent to $E_{k+q}$.

<u>Proposition</u> .10 Let $E_k \subset \mathcal{JD}^k(W)$ be a differential equation invariant by a Lie groupoid $\hat{R}_k \subset \hat{\Pi}^k(W)$, then the projection $\pi_{k,o}:E_k \longrightarrow W$ is an epimorphism iff the projection $\gamma_k:\hat{R}_k \underset{W}{\times} E_k \longrightarrow E_k$ is an epimorphism.

<u>Proof.</u> This can be seen by the following commutative diagram:

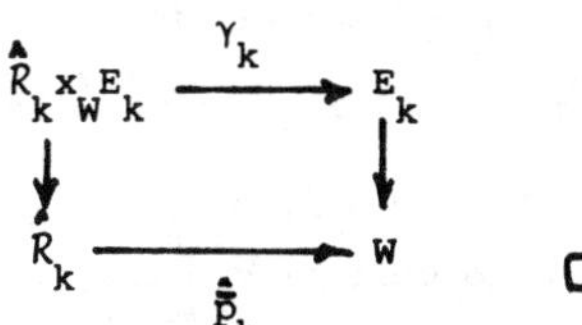

<u>Proposition</u> .11 If a PDE $E_k \subset \mathcal{JD}^k(W)$ is invariant by a Lie equation $\hat{R}_k$, then $E_k$ is transitive when $\hat{R}_k$ is transitive.

<u>Remark.</u> If $\pi:W \rightarrow M$ is a fiber bundle, we have a natural map $\delta_k:\hat{\mathcal{JD}}^k(W,W) \longrightarrow \mathcal{JD}^k(M,M)$ sanding $D^k f_W(p) \longmapsto D^k f_M(\pi(p))$, being $f_W$ a fiber-bundle morphism $W \longrightarrow W$ over a morphism $f_M:M \rightarrow M$. So, if we have a Lie equation $\hat{R}_k \subset \hat{\Pi}^k(W)$ we have the following commutative diagram

$$
\begin{array}{ccccccc}
0 & \longrightarrow & \hat{R}_k & \longrightarrow & \hat{\Pi}^k(W) & \longrightarrow & \hat{\mathcal{JD}}^k(W,W) \\
 & & \downarrow & & \downarrow & & \downarrow \delta_k \\
0 & \longrightarrow & R_k & \longrightarrow & \Pi^k(M) & \longrightarrow & \mathcal{JD}^k(M,M) \\
 & & \downarrow & & & & \\
 & & 0 & & & &
\end{array}
$$

being $R_k \equiv \delta_k(\hat{R}_k) \subset \Pi^k(M)$. (The first vertical line is exact). If $\hat{R}_k$ is a transitive Lie equation so is $R_k$. If $\hat{R}_k$ is a formally inte-grable Lie equation so is $R_k$.

Then, by using former considerations we can build a reduced super-bundle of geometric objects $\pi_G:\hat{G} \rightarrow W$ over W, a section $\hat{s}$ of $\hat{G}$ and an epimorphism $\hat{\Phi}: \hat{\Pi}^k(W) \longrightarrow \hat{G}$ such that $\hat{R}_k = \ker_{\hat{s}}\hat{\Phi}$. So, we have the following commutative diagram:

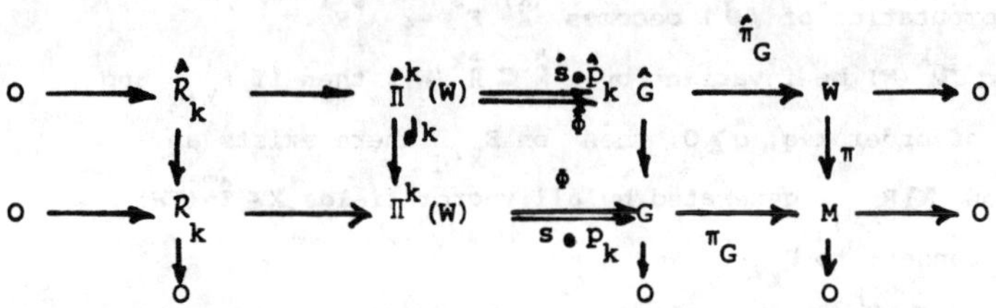

Then, $\hat{R}_k$ is formally integrable iff its defining section $\hat{s}$ verifies a differential equation $\hat{B}_1(\hat{s}|\hat{c})\equiv\ker_{\hat{c}}\hat{I}\subset J\mathcal{D}(\hat{G})$, being $\hat{s}$ a suitable section $\hat{G}\to\hat{G}_1\equiv\Lambda^o_r W\otimes vT\hat{G}/\delta(\Lambda^o_{r-1}W\otimes\hat{N}_1)$, where $\hat{N}_1\equiv\mathrm{im}(\sigma_1(vT(\hat{\Phi}^{(1)})))$ in the following exact sequence of vector bundles over W associated to $\hat{M}_k\equiv$symbol of $\hat{R}_k$:

$0\longrightarrow\hat{M}_{k+1}\longrightarrow S^o_{k+1}W\otimes TW\overset{\sigma_1(-)}{\longrightarrow}S^o_1W\otimes s^*vT\hat{G}$ and $\hat{I}$ is an affine epimorphism $J\mathcal{D}(\hat{G})\to\hat{G}_1$ over $\hat{G}$ ; (see Theorem 4.3).

Furthermore, $\hat{B}_1(\hat{s}|\hat{c})$ is analytic and formally integrable iff the section $\hat{c}$ is a solution of a differential equation $\hat{J}(\hat{c})\subset J\mathcal{D}(\hat{G}_1)$, ($\equiv$ <u>jacobi conditions</u>).

So, we get the following commutative diagram:

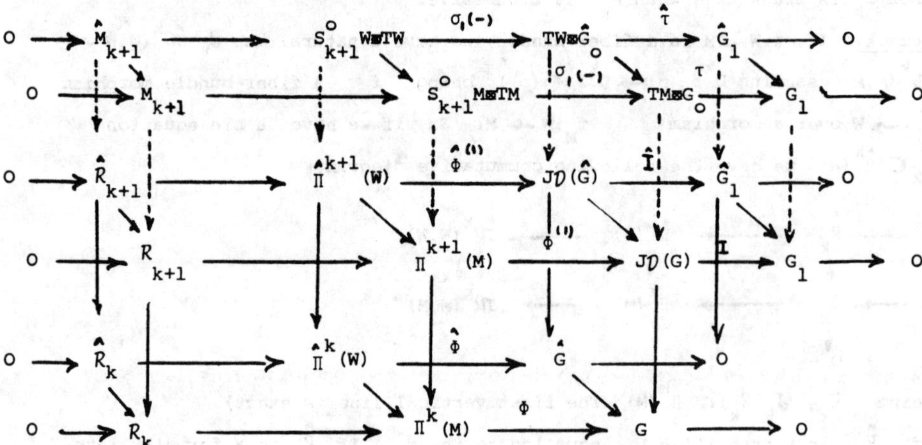

<u>Remark.</u> Let $\hat{R}_k\subset\hat{\Pi}^k(W)$ be a Lie equation of order k on a fiber bundle $\pi:W\to M$. Let $u\in J\mathcal{D}^k(W)$. The <u>orbit</u> of $\hat{R}_k$ trough u is the image $O_u$ of the application $\gamma_{k,u}:\hat{R}_k(p,W)\to J\mathcal{D}^k(W)$, being $p\equiv\pi_{k,o}(u)\in W$ on $\hat{R}_k(p,W)\equiv p_k^{-1}(p)\subset\hat{R}_k$. One says that $\hat{R}_k$ acts <u>regularly</u> on $J\mathcal{D}^k(W)$ if every orbit $O$ has the same dimension and if for each $a\in O$, and for each neighbourhood U containing a, one has that $O'\cap U$ is pathwise connected, where $O'$ is any orbit having non empty intersection with U. Furthermore, the saturation of a subset S of $J\mathcal{D}^k(W)$ is the union of all the orbits passing trough S. On the quotient space $J\mathcal{D}^k(W)/\hat{R}_k$ there is a natural

topology given by the images of saturated open subsets of $JD^k(W)$, then $JD^k(W)/\hat{R}_k$ has a natural structure of smooth manifold such that the projection $\pi^k/:JD^k(W) \to JD^k(W)/\hat{R}_k$ is a smooth map. The kernel of $T(\pi^k/)$ is just the involutive distribution $\mathbf{N}R_k$. Note that the orbit $O_u$ coincides with the maximal integral manifold at u of this involutive distribution.

A section $s_k$ of $\pi_k:JD^k(W) \to M$ is $\hat{R}_k$-invariant iff its image $s_k(M) \subset JD^k(W)$ is a $\hat{R}_k$-invariant submanifold (of dimension n) of $JD^k(W)$. A necessary condition for $JD^k(W)$ to admit such sections is that the dimension $\ell$ of the orbit $O_u$, $\forall\, u \in S_u(M)$ is not higher than $n : \ell \leq n$. One has the following fiber bundle epimorphism:

$$
\begin{array}{ccc}
JD^k(W) & \xrightarrow{\ \pi^k/\ } & JD^k(W)/\hat{R}_k \\[2pt]
{\scriptstyle \pi_k}\downarrow & & \downarrow{\scriptstyle \pi_k/} \\[2pt]
M & \xrightarrow[\ \pi_k/\ ]{} & M/\hat{R}_k
\end{array}
$$

If $C^\infty(JD^k(W))^{\hat{R}_k}$ denotes the space of $\hat{R}_k$-invariant sections of $\pi_k$, one has the following one to one map $C^\infty(JD^k(W))^{\hat{R}_k} \quad C^\infty(JD^k(W)/\hat{R}_k)$ given by

$$s_k \longmapsto s_k/ , \quad s_k/([x]_{\hat{R}_k}) = \pi^k/(s_k(x)) \equiv [s_k(x)]_{\hat{R}_k} .$$

One can easily see that the correspondence does not depend on the representative used.

Thus, the following diagram:

$$
\begin{array}{ccc}
JD^k(W) & \xrightarrow{\ \pi^k/\ } & JD^k(W)/\hat{R}_k \\[2pt]
{\scriptstyle s_k}\uparrow & & \uparrow{\scriptstyle s_k/} \\[2pt]
M & \xrightarrow[\ \pi_k/\ ]{} & M/\hat{R}_k
\end{array}
$$

is commutative for any $\hat{R}_k$-invariant section $s_k$ of $\pi_k$.

As the fibered action of $\hat{R}_k$ on $JD^k(W)$ gives an action of $\hat{R}_k$ on W, we can define $W/\hat{R}_k$ and a canonical epimorphism $\pi/:W \to W/\hat{R}_k$.

In this way a section s of $\pi$ is $\hat{R}_k$-invariant iff its image $s(M) \in W$ is $\hat{R}_k$-invariant. In particular $s_k = D^k s$ is a $\hat{R}_k$-invariant section of $\pi_k$ iff s is a $\hat{R}_k$-invariant section of $\pi$. Set:

$$JD^k(W)^{\hat{R}_k} \equiv \{\, u \in JD^k(W) \mid \exists\ \hat{R}_k\text{-invariant section } s \in C^\infty(W)^{\hat{R}_k} ;\ D^k s(\pi_k(u))=u \,\} .$$

There is a natural fiber bundle morphism over $\pi/$, $\ {}^k\pi/:JD^k(W)^{\hat{R}_k} \longrightarrow JD^k(W/\hat{R}_k)$ given by $u = D^k s(x) \longmapsto (D^k s/)([x]_{\hat{R}_k})$, [(+)] namely the following diagram

_______________

(+) $s/$ is the section of $/\pi$ canonically associated to a $\hat{R}_k$-invariant section s of $\pi$.

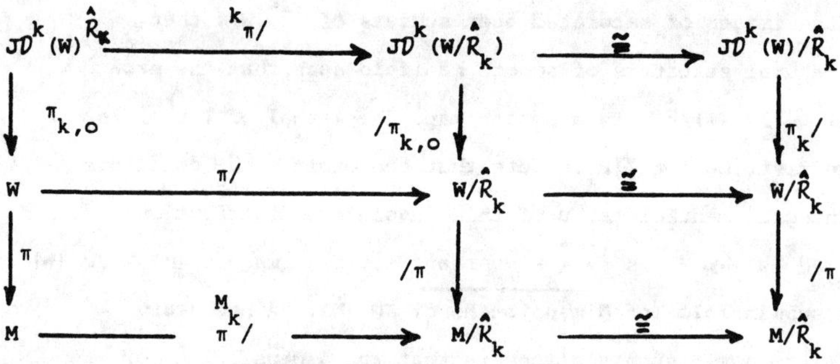

$$
\begin{array}{ccccc}
J\mathcal{D}^k(W)^{\hat{R}_*} & \xrightarrow{\ ^k\pi/\ } & J\mathcal{D}^k(W/\hat{R}_k) & \xrightarrow{\ \cong\ } & J\mathcal{D}^k(W)/\hat{R}_k \\
\downarrow{\scriptstyle \pi_{k,o}} & & \downarrow{\scriptstyle /\pi_{k,o}} & & \downarrow{\scriptstyle \pi_k/} \\
W & \xrightarrow{\ \pi/\ } & W/\hat{R}_k & \xrightarrow{\ \cong\ } & W/\hat{R}_k \\
\downarrow{\scriptstyle \pi} & & \downarrow{\scriptstyle /\pi} & & \downarrow{\scriptstyle /\pi} \\
M & \xrightarrow{\ ^{M_k}\pi/\ } & M/\hat{R}_k & \xrightarrow{\ \cong\ } & M/\hat{R}_k
\end{array}
$$

is commutative. $^k\pi/$ is a diffeomorphism on fibers .

So, we have the following

<u>Theorem</u> .6   Let $E_k \subset J\mathcal{D}^k(W)$ be a k-order differential equation on $\pi:W \to M$
Let $\hat{R}_k(E_k) \equiv E_k \cap J\mathcal{D}^k(W)^{\hat{R}}$ be the system corresponding to $E_k$ for $\hat{R}_k$-invariant
solutions of $E_k$. If $\hat{R}(E_k)$ is invariant under $\hat{R}_k$ then a section s
of $\pi$ is a $k$-invariant solution to $E_k$ iff. s/ is a <u>solution</u> of the <u>reduced</u>
differential equation : $(E_k)^{\hat{R}_k} \equiv {}^k\pi/((\hat{R}^k(E_k)))$.
In particular , if $E_k$ is $\hat{R}_k$-invariant, then $(E_k)^{\hat{R}_k}$ is so.

# 5. - THE EULER-LAGRANGE OPERATOR FOR LAGRANGIANS OF ANY ORDER

Let $M$ be an oriented $C^\infty$ differentiable manifold with dimension n. Let $\eta$ be the volume form on M.

Let us give the following fundamental

**Definition .1** A <u>dynamic conservation law</u> of a continuum system, characterized by a differential equation $E_k \subset J\mathfrak{J}^k(C)$ on a fiber bundle $\pi_C : C \to M$ is any differential $(n-1)$-form $B : J\mathfrak{J}^h(C) \to \Lambda^0_{n-1} J\mathfrak{J}^h(C)$, $k \leq h$, such that there are solutions $c : M \to C$ of $E_k$ for which $d(D^h c)^* B = 0$. Then, we say that the dynamic configuration c <u>leads</u> the conserved entity $\beta \equiv (D^h c)^* B : M \to \Lambda^0_{n-1} M$.

<u>Examples</u> We shall prove later that all the conservation laws of Noetherian type for Lagrangian equations are dynamic ones. Further, some conservation laws related to the socalled solitons, that are particular solutions of field equations, that arise when the theory exhibites spontaneous symmetry breakdown, are observed; these are usually called <u>topological conservation laws</u> $\begin{bmatrix} 1:q,8 \end{bmatrix}$ ). In order to develop a suitable intrinsic geometric framework to build dynamic conservation laws, we shall consider an intrinsic variational calculus for Lagrangians of any order.

The procedure developed in $[1:n]$ to give an intrinsic expression of the Euler-Lagrange operator for Lagrangians of first order is fit to be extended to Lagrangians of any order. In this section we shall prove that this can be done. Recall $[1,0]$ that to a numerical function f on $J\mathfrak{J}^k(\mathbb{C})$ one can associate the map

$$\overset{(h',h)}{\underset{s}{\delta}} f \equiv \overset{h'}{d}f - \overset{h}{d}f : J\mathfrak{J}^k(\mathbb{C}) \to vT^{\overset{(h',h)}{*}} J\mathfrak{J}^k(\mathbb{C}) \equiv T^{\overset{h'}{*}} J\mathfrak{J}^k(\mathbb{C}) \underset{J\mathfrak{J}^k(\mathbb{C})}{\oplus} T^{\overset{h}{*}} J\mathfrak{J}^k(\mathbb{C}) , \quad -1 \leq h' \leq h \leq k,$$

where $\overset{s}{d}f$, $-1 \leq s \leq k$, is the vertical differential of f with respect to the fibered structure $\pi_{k,s} : J\mathfrak{J}^k(\mathbb{C}) \to J\mathfrak{J}^s(\mathbb{C})$, $s \leq k$. We set $\pi_{k,-1} = \pi_k : J\mathfrak{J}^k(\mathbb{C}) \to M$ and $\overset{-1}{d}f = df$. One has also $\overset{k}{d}f = 0$. Let $\{x^i, y^j_{i_1 \dots i_\alpha}\}$ , $1 \leq i \leq n$, $1 \leq j \leq m$, $0 \leq \alpha \leq k$, $1 \leq i_1, \dots, i_\alpha \leq n$ , be a coordinate system on $J\mathfrak{J}^k(\mathbb{C})$. Then, the local expression of $\overset{(h',h)}{\delta} f$ is:

$$\overset{(h',h)}{\delta}f = \sum_{\substack{1\leq j\leq m \\ 1\leq i_1,\dots,i_\alpha\leq n \\ h'+1\leq \alpha\leq h}} (\partial y_j^{i_1\dots i_\alpha}.f)\, dy^j_{i_1\dots i_\alpha}\;.$$

Set $\overset{s}{\delta}f \equiv \overset{(s-1,s)}{\delta}f : J\mathfrak{d}^k(\mathbb{C}) \to vT^{*(s-1,s)}J\mathfrak{d}^k(\mathbb{C})$. The local expression of $\overset{s}{\delta}f$ is of course:

$$\overset{s}{\delta}f = \sum_{\substack{1\leq j\leq m \\ 1\leq i_1,\dots,i_\alpha\leq n}} (\partial y_j^{i_1\dots i_s}.f)\, dy^j_{i_1\dots i_s}$$

In particular for $s=k$ one has $\overset{k}{\delta}f : J\mathfrak{d}^k(\mathbb{C}) \to O\oplus \overset{k-1}{\check{T}}{}^* J\mathfrak{d}^k(\mathbb{C}) \cong \pi^*_{k,o}(S^k_o M\boxtimes vT^*\mathbb{C})$.

<u>Proposition</u> .1 Let $f:J\mathfrak{d}^k(\mathbb{C}) \to \mathbb{R}$ be a numerical function of order $k$ on $\mathbb{C}$. Then there exists a unique map $\overset{s}{\check{\delta}}f : J\mathfrak{d}^k(\mathbb{C}) \to \pi^*_{k,o}\overset{s-1}{\check{T}}{}^* J\mathfrak{d}^s(\mathbb{C})$, $0\leq s<k$, such that the following diagram

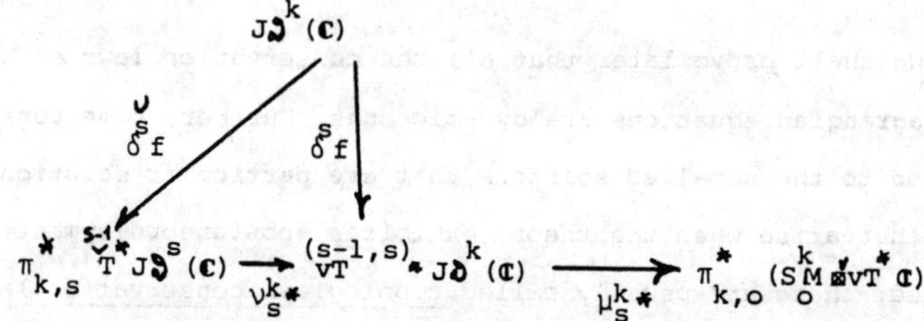

is commutative, where $\nu^k_{s*}$ and $\mu^k_{s*}$ are the canonical vector bundle homomorphisms over $J\mathfrak{d}^k(\mathbb{C})$ canonically induced by means of the inclusions $\overset{k-1}{\check{T}} J\mathfrak{d}^*(\mathbb{C}) \hookrightarrow \overset{b}{T}J\mathfrak{d}^k(\mathbb{C})$, $-1\leq h\leq k-1$ and the isomorphism $\pi^*_{k,o}(S^o_k M\boxtimes vT\mathbb{C}) \cong \overset{k-1}{\check{T}} J\mathfrak{d}^k(\mathbb{C})$.

<u>Proof.</u> Let us observe that the horizontal line in the above diagram is an exact

sequence and that $\mu_s^{k*} \circ \overset{s}{\delta} f = 0$ for $0 \leq s \leq k-1$. Let us put $A = \mathrm{Im}(\nu_s^{k*}) = \ker(\mu_s^{k*})$. Since $\mu_s^{k*} \circ \overset{s}{\delta} f = 0$, it follows that $\underline{\overset{s}{\delta}} f$ maps $J\mathfrak{D}^k(\mathbb{C})$ into $A$ and thereby give rise to a commutative diagram

$$
\begin{array}{ccc}
 & J\mathfrak{D}^k(\mathbb{C}) & \\
 & \overset{g}{\swarrow} \quad \downarrow \overset{s}{\delta} f & \\
\pi_{k,s}^* \overset{s-1}{\vee} T^* J\mathfrak{D}^s(\mathbb{C}) \xrightarrow{\ \zeta\ } A \xleftarrow{\ } \overset{(s-1,s)}{\vee} T \, J\mathfrak{D}^k(\mathbb{C}) & .
\end{array}
$$

But $\zeta$ is an isomorphism. Hence, the map $\zeta^{-1} \circ g$ has the required properties. $\square$

<u>Corollary 1</u>  In a coordinate system the local expression of $\overset{\smallsmile}{\overset{s}{\delta}} f$ coincides with that of $\overset{s}{\delta} f$ .

<u>Definition 2.1</u>  For any $o \leq s \leq k$ we set:

$$
\overset{s}{\underset{\bullet}{\delta}} f \equiv
\begin{cases}
\overset{\smallsmile}{\overset{s}{\delta}} f : J\mathfrak{D}^k(\mathbb{C}) \longrightarrow \pi_{k,o}^* \overset{s}{S} M \boxtimes vT^* \mathbb{C} & , \text{ if } o \leq s < k; \\[2ex]
\overset{k}{\delta} f & , \text{ if } s=k.
\end{cases}
$$

<u>Proposition 2.2</u>  Let $f : J\mathfrak{D}^k(\mathbb{C}) \to \mathbb{R}$ be a numerical function on $J\mathfrak{D}^k(\mathbb{C})$. There exists a canonical (with respect to the structure $(J\mathfrak{D}^k(\mathbb{C}), f)$) fibered morphism $\mathfrak{E}(f) : J\mathfrak{D}^{2k}(\mathbb{C}) \to vT^* \mathbb{C}$ , such that the local expression of the corresponding differential operator (<u>Euler-Lagrange operator</u>)  $\mathfrak{E}(f). : C^\infty(\mathbb{C}) \to C^\infty(vT^* \mathbb{C})$  is given by

$$
(1) \qquad \varepsilon(f).c = \sum_{o \leq \alpha \leq k} (-1)^\alpha \{ \partial x_{i_1} \dots \partial x_{i_\alpha} . ((\partial y_j^{i_1 \dots i_\alpha} . f) \circ D^k c) \} \, dy^j .
$$

<u>Proof.</u>  For any $o \leq s \leq k$  set

$$
\overset{s}{\Theta} f \equiv \bar{\pi} \circ \overset{s}{\underset{\bullet}{\delta}} f : J\mathfrak{D}^k(\mathbb{C}) \to \overset{s}{\underset{o}{S}} M \overset{\smallsmile}{\boxtimes} vT^* \mathbb{C},
$$

where $\bar{\pi}$ is the canonical projection $\pi_{k,o}^* (\overset{s}{\underset{o}{S}} M \overset{\smallsmile}{\boxtimes} vT^* \mathbb{C}) \to \overset{s}{\underset{o}{S}} M \overset{\smallsmile}{\boxtimes} vT^* \mathbb{C}$.

The local expression of $\overset{s}{\Theta} f$ is as follows:

$$
(2) \qquad \overset{s}{\Theta} f = \sum_{1 \leq j \leq m} (\partial y_j^{i_1 \dots i_s} . f) \, \partial x_{i_1} \Theta \dots \Theta \, \partial x_{i_s} \boxtimes dy^j .
$$

Let $\overset{s}{\pi}$ be the canonical fibered morphism $\overset{s}{\underset{o}{S}} M \to TM \boxtimes \overset{s-1}{\underset{o}{S}} M$ given by composition

$$
\begin{array}{ccc}
\overset{s}{\underset{o}{S}} M & \xrightarrow{\ \overset{s}{\pi}\ } & TM \boxtimes \overset{s-1}{\underset{o}{S}} M \\
\downarrow & & \uparrow \\
\overset{s}{\underset{o}{T}} M & \xrightarrow{\qquad} & TM \boxtimes \overset{s-1}{\underset{o}{T}} M
\end{array}
$$

322

Let $\overset{s}{a}_\eta$ be the canonical fibered morphism $\overset{o}{\Lambda}_n M \hat{\otimes} \overset{s}{S} M \check{\otimes} vT^* \mathbb{C} \to \overset{s}{S} M \check{\otimes} vT^* \mathbb{C}$, $1 \le s$.
Let $\overset{s}{\boxdot}$ be the differential operator $\overset{s}{\boxdot} : C^\infty(\overset{s}{S} M \check{\otimes} vT^* \mathbb{C}) \to C^\infty(\overset{s-1}{S} M \check{\otimes} vT^* \mathbb{C})$,
given by

$$\overset{s}{\boxdot}.\sigma = \overset{s}{a}_\eta \bullet d((\overset{s}{\pi} \bullet \sigma) \lrcorner\, \eta) \qquad , \text{ for } s \ge 1$$

and

$$\overset{o}{\boxdot}.\sigma = \sigma .$$

Then the morphism $\varepsilon(f)$ is the one associated to the following differential

operator (<u>Euler-Lagrange operator</u>)

(3)

$$\boxed{\begin{array}{l} \varepsilon(f).: C^\infty(\mathbb{C}) \to C^\infty(vT^* \mathbb{C}) \\[2ex] \varepsilon(f).c = \sum_{o \,\le \alpha \le k} (-1)^\alpha (\, \overset{①}{\boxdot}. \,\ldots\ldots\, \boxdot. \overset{\alpha-1}{\boxdot}. \overset{\alpha}{\boxdot}.(\, \overset{\alpha}{\partial} f \bullet D^k c)). \end{array}}$$

where ① denotes that the term $\overset{1}{\boxdot}$ is present if and only if $\alpha > 0$.

It is easy to see that the local expression of $\overset{1}{\boxdot}. \,\ldots\, \overset{\alpha}{\boxdot}. (\overset{\alpha}{\partial} f \bullet D^k c))$ is as
follows;: $(\partial x_{i_1} . \,\ldots\, \partial x_{i_\alpha} . ((\partial y_j^{i_1 \ldots i_\alpha} . f)) dy^j$. So the proof is complete. $\square$


Let us now give a useful definition which extends the concept of linear differential operator to fiber bundles which are not necessarily vector Fiber
bundles.

<u>Definition</u> .2  Let $\pi_K : \mathbb{K} \to M$ be an affine fiber bundle over M with as-
sociated vector fiber bundle $\pi_K : \overline{\mathbb{K}} \to M$ . A <u>linear differential operator</u>
of order k on $\pi_\mathbb{C}$ is a differential operator $K.: C^\infty(\mathbb{C}) \to C^\infty(\mathbb{K})$ of order k
such that the associated fibered morphism $K: J\overset{k}{\partial}(\mathbb{C}) \to \mathbb{K}$ over M induces an
affine fibered bundle morphism $K': J\overset{k}{\partial}(\mathbb{C}) \to \pi_{k-1}^* \mathbb{K}$ over $J\overset{k-1}{\partial}(\mathbb{C})$. (By abuse
of notation we shall write K at the place of K'). Further, if $\mathbb{C}$ and $\mathbb{K}$ are
vector fibered bundles over M a <u>strictly linear differential operator</u>
$K.: C^\infty(\mathbb{C}) \to C^\infty(\mathbb{K})$ of order k is such that K is a vector fibered homomorphism
over M.

__Proposition .4__   1. If $\{ x^k, y^j_{i_1 \cdots i_\alpha} \}_{0 \leq \alpha \leq k}$ is a coordinate system on $J\partial^k(\mathbb{C})$ and $\{x^\alpha, z^i\}$ is a coordinate system on $\mathbb{K}$, the local expression of a linear differential operator $K. : C^\infty(\mathbb{C}) \to C^\infty(\mathbb{K})$ of order $k$ is as follows

$$ K^j \equiv z^j \bullet K = A^j + \sum_{\substack{0 \leq \beta \leq m \\ 0 \leq i_1, \ldots, i_k \leq n}} y^\beta_{i_1 \cdots i_k} B_\beta^{j i_1 \cdots i_k} $$

where $A^j, B_\beta^{j i_1 \cdots i_k} : J\partial^{k-1}(\mathbb{C}) \to \mathbb{R}$ are numerical functions on $J\partial^{k-1}(\mathbb{C})$.

2. The symbol $\sigma_k(K)$ of $K$ is the fibered homomorphism

$$ \sigma_k(K) \equiv vT(K) \bullet \mu : \pi^*_{k,o}(S^o_k M \otimes vT\mathbb{C}) \to \bar{\mathbb{K}} \cong \mathbb{K}^* v\mathbb{TK} \qquad \text{over } J\partial^k(\mathbb{C}). \quad \sigma_k(K) \text{ is locally} $$

given by

$$ \sigma_k(K)(\upsilon) = B_\beta^{j i_1 \cdots i_k} x^\beta_{i_1 \cdots i_k} \partial z_j \bullet K \; , \qquad \forall \upsilon \in \pi^*_{k,o}(S^o_K M \otimes \bar{\pi}\mathbb{C}) $$

3. A strictly linear differential operator is a linear differential operator too.

4. $\varepsilon(f)$ is a 2k-order linear differential operator.


__Definition .3__   An __extremal__ for $f : J\partial^k(\mathbb{C}) \to \mathbb{R}$ is a solution of the __Euler-Lagrange equation__ $\mathbb{E}_{2k}(f)$ given by $\mathbb{E}_{2k}(f) = \ker \varepsilon(f)$. The set of extremals of $f$ will be denoted by $\mathrm{Ext}(f) = S(\mathbb{E}_{2k}(f))$.

324

<u>R E M A R K</u>

In order to make the paper self-contained, let us deduce in a geometrical way
the Euler-Lagrange equation from a variational problem.

We shall make some preliminary considerations of the differential geometry of
infinite jet-derivative spaces (see also refs. $\left[1;i,o\right]$ ).

Set $J\mathcal{D}^\infty(\mathbb{C})\equiv\varprojlim J\mathcal{D}^k(\mathbb{C})$. One has the following canonical projections: $\pi_{\infty,k}:J\mathcal{D}^\infty(\mathbb{C})\to J\mathcal{D}^k(\mathbb{C})$
$\pi_\infty:J\mathcal{D}^\infty(\mathbb{C})\to M$. A function $f:J\mathcal{D}^\infty(\mathbb{C})\to\mathbb{R}$ is said to be $C^\infty$ if for each $u\in J\mathcal{D}^\infty(\mathbb{C})$
there are a $k\in\mathbb{N}$ and a neighborhood $U_k$ of $\pi_{\infty,k}(u)$ in $J\mathcal{D}^k(\mathbb{C})$ and a $C^\infty$ function
$f_k:U_k\to\mathbb{R}$ such that $f|\pi_{\infty,k}^{-1}(U_k)=f_k\circ\pi_{\infty,k}$ . The vector space of $C^\infty$ functions
on $J\mathcal{D}^\infty(\mathbb{C})$ is denoted by $A(\mathbb{C})$. $J\mathcal{D}^\infty(\mathbb{C})$ has a natural structure of paracompact manifold. A vector field on $J\mathcal{D}^\infty(\mathbb{C})$ is a linear derivation on $A(\mathbb{C})$. The tangent space
$T_u J\mathcal{D}^\infty(\mathbb{C})$ at the point $u\in J\mathcal{D}^\infty(\mathbb{C})$ is the space of linear derivations from $A(\mathbb{C})$ to
$\mathbb{R}$ in $u$. A vector field $X$ on $J\mathcal{D}^\infty(\mathbb{C})$ determines and is determined by a sequence
of mappings $X_i:J\mathcal{D}^\infty(\mathbb{C})\to TJ\mathcal{D}^i(\mathbb{C})$, $i=0,1,2,\ldots$, $X_i(u)=T(\pi_{\infty,i})(X(u))$ such that:
(1) $X_i(u)\in T_{\pi_{\infty,i}(u)}J\mathcal{D}^i(\mathbb{C})$; (2) $T(\pi_{i,i-1})(X_i(u))=X_{i-1}(u)$; (3) For each smooth
function $f:J\mathcal{D}^i(\mathbb{C})\to\mathbb{R}$, the function $J\mathcal{D}^\infty(\mathbb{C})\ni u\mapsto X_i(u)(f)$ is
smooth on $J\mathcal{D}^\infty(\mathbb{C})$.

Therefore, we set $X\equiv(X_0,X_1,\ldots)$. For each $u\in J\mathcal{D}^\infty(\mathbb{C})$, $T_u J\mathcal{D}^\infty(\mathbb{C})$ is canonically isomorphic to the inverse limit of $\{T_{\pi_{\infty,k}(u)}J\mathcal{D}^k(\mathbb{C}), T(\pi_{k,e})\}$ $\quad k\geq e;\ e,k\in\mathbb{N}$
Examples of vector fields on $J\mathcal{D}^\infty(\mathbb{C})$ can be obtained by means of vector fields
given on $M$ and $\mathbb{C}$. In the following table we resume some of the most important
ones.

<u>Tab. 1 - Vector fields on $J\mathcal{D}^\infty(\mathbb{C})$ induced by vector fields on $M$ and $\mathbb{C}$.</u>

| name | definition |
| --- | --- |
| horizontal v.f. $X$ : | $(X.f)(u)=(Y.(f\circ D^\infty c))(\pi_\infty(u))$, $\forall\ f\in A(\mathbb{C})$, $Y:M\to TM$ given. (+) |
| integrable v.f. $X$ : | $\bar{X}\equiv(X,\overset{1}{X},\overset{2}{X},\ldots)$; $(\bar{X}.f)(u)=(\overset{k}{X}.\tilde{f})(\pi_{\infty,k}(u))$, $f=\hat{f}\circ\pi_{\infty,k}$, |
| | $\overset{k}{X}\equiv k$-th prolongation of $(X,Y)=\pi_{\mathbb{C}}$-related vector fields (++) |
| vertical integrable v.f. $X$ : | integrable vector field with $Y=0$. |

---

(+) $c$=local cross section of $\pi_{\mathbb{C}}$ defined in a neighborhood of $\pi_\infty(u)$ such that
$\quad D^\infty c(\pi_\infty(u))=u$.

(++) For the definition of $i$-th prolongation of a vector field see next section 4.

We say that a vector field $X$ on $J\mathcal{D}^\infty(\mathbb{C})$ is a <u>vertical</u> field if for each $f:M\to\mathbb{R}$
$X.(f\circ\pi_\infty)=0$. Let $vT_u J\mathcal{D}^\infty(\mathbb{C})$ denote the space of vertical vectors at $u\in J\mathcal{D}^\infty(\mathbb{C})$. Then

one has the canonical splitting of $T_u J\mathfrak{H}^\infty(\mathbb{C})$ as $H_u(\mathbb{C}) \oplus vT_u J\mathfrak{H}^\infty(\mathbb{C})$ where $H_u(\mathbb{C})$ is the space of horizontal vectors at u. The existence of this canonical split-ting is a property of $J\mathfrak{H}^\infty(\mathbb{C})$ which does not exist for the space $J\mathfrak{H}^k(\mathbb{C})$, $k \neq \infty$.

A $\underline{\text{k-form}}$ $\omega$ on $J\mathfrak{H}^\infty(\mathbb{C})$ is a multilinear alternating map, assigning to each k-tuple of vector fields $X_1, \ldots X_k$ on $J\mathfrak{H}^\infty(\mathbb{C})$ a smooth function $\omega(X_1, \ldots X_k)$ on $J\mathfrak{H}^\infty(\mathbb{C})$ in such a way that $\omega(X_1, \ldots X_k)(u)$ is completely determined by $\omega$ and the values $X_i(u)$, $i=1,\ldots,k$. We denote the corresponding alternating k-linear map from $T_u J\mathfrak{H}^\infty(\mathbb{C})$ to R by $\omega(u)$. Then it can be seen that a k-form on $J\mathfrak{H}^\infty(\mathbb{C})$ determines and is determined by a sequence $\{ U_i, \omega_i \}_{i \in \mathbb{N}}$ where $U_i$ is an open set of $J\mathfrak{H}^i(\mathbb{C})$ and $\omega_i$ is a k-form on $U_i$ such that : (1) $\pi_{\infty,i}^{-1}(U_i) \supset \pi_{\infty,i-1}^{-1}(U_{i-1})$,

(2) $\bigcup_{i \in \mathbb{N}} \pi_{\infty,i}^{-1}(U_i) = J\mathfrak{H}^\infty(\mathbb{C})$; (3) $\pi_{i,i-1}^* \omega_{i-1} = \omega_i | \pi_{i,i-1}^{-1}(U_i)$;

(4) for each $u \in \pi_{\infty,i}^{-1}(U_i)$ $\omega(u)(X_1(u),\ldots X_k(u)) = \omega_i(\pi_{\infty,i}(u))(T(\pi_{\infty,i})(X_1(u)),$
$$\ldots, T(\pi_{\infty,i})(X_k(u))).$$

Set $\mathcal{H}_{l,k}(\mathbb{C}) = $ space of (l+k)-forms on $J\mathfrak{H}^\infty(\mathbb{C})$ such that $\omega(X_1, \ldots X_{l+k}) = 0$

if among $X_1, \ldots X_{k+1}$ there are more than l horizontal v.f.

or more than k vertical v.f.

Given a $\omega \in \mathcal{H}_{l,k}(\mathbb{C})$, fixed a local section $c:M \supset U \rightarrow \mathbb{C}$ and a k-tuple $X_1, \ldots X_k$ of vertical symmetries of $\pi_{\mathbb{C}}$ one has an l-form on U $E_\omega(c;X_1, \ldots X_k)$ given by $E_\omega(c;X_1, \ldots X_k) = D^\infty c^* \omega(\bar{X}_1, \ldots \bar{X}_k)$, where $\bar{X}_i$, $i=1,\ldots k$ are the vertical integrable vector fields corresponding to $X_i$.

The operators $\wedge$, $i_X$, d and $\mathcal{L}_X$ can be formally extended to differential forms on $J\mathfrak{H}^\infty(\mathbb{C})$. An important property of the exterior differential is the following.

If $\omega \in \mathcal{H}_{l,k}(\mathbb{C})$ then $d\omega \in \mathcal{H}_{l+1,k}(\mathbb{C}) \oplus \mathcal{H}_{l,k+1}(\mathbb{C})$. So we can define the operators $\overset{o}{d}: \mathcal{H}_{l,k}(\mathbb{C}) \rightarrow \mathcal{H}_{l+1,k}(\mathbb{C})$ and $\overset{v}{d}: \mathcal{H}_{l,k}(\mathbb{C}) \rightarrow \mathcal{H}_{l,k+1}(\mathbb{C})$ such that $d = \overset{o}{d} + \overset{v}{d}$. $\overset{v}{d}$ is called the $\underline{\text{vertical exterior derivative}}$ and $\overset{o}{d}$ the $\underline{\text{horizontal exterior deri-}}$ $\underline{\text{vative}}$. Note that being $d \circ d = 0$ we also get $\overset{v}{d} \circ \overset{v}{d} = 0$, $\overset{o}{d} \circ \overset{o}{d} = 0$ and $\overset{v}{d} \circ \overset{o}{d} + \overset{o}{d} \circ \overset{v}{d} = 0$.

One has the following complexes of cochains:

and a similar one by replacing each $\mathcal{H}_{l,k}(\mathbb{C})$ by its sheaf $\tilde{\mathcal{H}}_{l,k}(\mathbb{C})$ of germs of sections. One has the following properties:

1) In the diagram of the sheaf of germs of sections the rows and the columns are exact ($\check{d}$-exactness and $\overset{\bullet}{d}$-exactness);

2) The image of $\overset{\bullet}{d}:\tilde{\mathcal{H}}_{n-1,k}(\mathbb{C}) \to \tilde{\mathcal{H}}_{n,k}(\mathbb{C})$ is characterized as follows: $\omega \in \tilde{\mathcal{H}}_{n,k}(\mathbb{C})_u$, $k>0$, is in the image of $\overset{\bullet}{d}$ if and only if there are representative $\hat{\omega}$ of $\omega$ in $\mathcal{H}_{n,k}(\mathbb{C})$ , and a neighborhood $U$ of $u$ in $J\mathcal{D}^{\infty}(\mathbb{C})$ such that for each local section $c$ , defined on a neighborhood of $\pi_{\infty}(u)$ and each $V \subset \bar{V} \subset D^{\infty}c^{-1}(U)$, $V$ a bounded oriented open subset of $M$ $\int_V E_{\hat{\omega}}(c;X_1,\ldots X_k)$ depends only on the germs of the vertical symmetries $X_1,\ldots X_k$ along $c(\partial V)$, but not on $X_1,\ldots X_k$ away from $c(\partial V)$.

3) Set $\tilde{\mathcal{G}}^i(\mathbb{C}) \equiv \tilde{\mathcal{H}}_{n,i}(\mathbb{C})/\overset{\bullet}{d}(\check{\mathcal{H}}_{n-1,i}(\mathbb{C}))$. One has the canonical map $\mathbb{D}:\mathcal{H}_{n,i}(\mathbb{C}) \to \tilde{\mathcal{G}}^i(\mathbb{C})$;

4) $\check{d}$ is defined on $\tilde{\mathcal{G}}^i(\mathbb{C})$ because $\overset{\bullet}{d}(\check{d}(\tilde{\mathcal{H}}_{n-1,i}(\mathbb{C}))) \subset \overset{\bullet}{d}(\tilde{\mathcal{H}}_{n-1,i+1}(\mathbb{C}))$. So we get the $\check{d}$-exactness on $\tilde{\mathcal{G}}^i(\mathbb{C})$: if $\omega \in (\tilde{\mathcal{G}}^i(\mathbb{C}))_u$, $i>0$, with $\check{d}\omega = 0$, then there is an $\omega \in \tilde{\mathcal{G}}^{i-1}(\mathbb{C})$ such that $\check{d}\eta = \omega$ ; and $\check{d}:\tilde{\mathcal{G}}^0(\mathbb{C}) \to \tilde{\mathcal{G}}^1(\mathbb{C})$ is surjective.

5) <u>$\overset{\bullet}{d}$-cohomology</u>

$$\frac{\ker(\overset{\bullet}{d}:\mathcal{H}_{k,i}(\mathbb{C}) \to \mathcal{H}_{k+1,i}(\mathbb{C}))}{\mathrm{Im}(\overset{\bullet}{d}:\mathcal{H}_{k-1,i}(\mathbb{C}) \to \mathcal{H}_{k,i}(\mathbb{C}))} \cong \begin{cases} 0 & \text{if } i \neq 0,\ 0 < k < n \\ H^k(\mathbb{C};\mathbb{R}) & \text{if } i = 0,\ 0 < k < n; \end{cases}$$

$\overset{\bullet}{d}:\mathcal{H}_{0,i}(\mathbb{C}) \to \mathcal{H}_{1,i}(\mathbb{C})$ is injective provided $i>0$ and

$$\frac{\ker(\mathbb{D}:\mathcal{H}_{n,i}(\mathbb{C}) \to \tilde{\mathcal{G}}^i(\mathbb{C}))}{\mathrm{Im}(\overset{\bullet}{d}:\mathcal{H}_{n-1,i}(\mathbb{C}) \to \mathcal{H}_{n,i}(\mathbb{C}))} \cong \begin{cases} 0 & \text{if } i \neq 0 \\ H^n(\mathbb{C};\mathbb{R}) & \text{if } i = 0. \end{cases}$$

6) $\tilde{\mathcal{G}}$-cohomology.

$$\frac{\ker(\check{d}:\tilde{\mathcal{G}}^i(\mathbb{C}) \to \tilde{\mathcal{G}}^{i+1}(\mathbb{C}))}{\mathrm{Im}(\mathbb{D}\circ\check{d}:\mathcal{H}_{n,i-1}(\mathbb{C}) \to \tilde{\mathcal{G}}^i(\mathbb{C}))} \cong H^{n+i}(\mathbb{C};\mathbb{R}).$$

A variational problem can be identified with a Lagrangean $L$ belonging to $\mathcal{H}_{n,0}(\mathbb{C})$. Such a $L$ assigns to each (local) section $c$ an $n$-form $E_L(c)$ on $M$ . A form $\Xi \in \mathcal{H}_{n,1}(\mathbb{C})$ is called a <u>source-form</u> if for each section $c$, $p \in \mathrm{Domain}(c)$, and vertical symmetry $X$ , $E_\Xi(c;X)(p)$ depends only on $\Xi$ , $c$, $p$ and $X(c(p))$, but not on the higher derivatives of $X$ in $c(p)$. One can see that each $\omega \in \mathcal{H}_{n,1}(\mathbb{C})$ can be written uniquely as $\omega = \bar{\omega} + \Pi\omega$ , with $\Pi\omega$ a source form and $\bar{\omega} \in \mathrm{Im}(\overset{\bullet}{d})$. In particular if $X = X^i \partial y_i$

and
$$E_\omega(c;X)(p) = \sum_{\substack{1 \le i \le m \\ 1 \le j_1 \le \cdots \le j_k \le n}}^{\ast} \Omega(c)_{ij_1\ldots j_k}(p)(\partial x_{j_1}\ldots\partial x_{j_k}.X^i(p,c(p))dx^1\wedge\ldots\wedge dx^n,$$

where $\ast$ means that the summation is locally finite, we get

$$(\bullet)\qquad E_{\Pi\omega}(c;X)(p) = \sum_{\substack{1 \le i \le m \\ 1 \le j_1 \le \cdots \le j_k \le n}}^{\ast} (-1)^k(\partial x_{j_1}\ldots\partial x_{j_k}.(\Omega(c)_{ij_1\ldots j_k}(p))X^i(p,c(p))dx^1\wedge\ldots\wedge dx^n$$

Now let $A \subset M$ be a bounded open oriented subset of M, and $K \subset A$ some compact subset. Let c be a section of $\pi_\alpha$ defined on A and $\tilde{c}$ be any deformation of c subjected to the conditions $\tilde{c}_t(p) = \tilde{c}_o(p) = c(p)$ whenever $p \in A-K$. Then for each Lagrangean L set $I(A,c)(t) \equiv \int_A E_L(c_t)$. Then one has:

$$dI(A,c)(0) = \int_A E_{dL}(\tilde{c}_o;\partial\tilde{c}).$$

Note that the above equation remains true of course if we add to dL an element which belongs to $\text{Im}(\overset{o}{d}: \mathcal{X}_{n-1,1}(\mathbb{C}) \to \mathcal{X}_{n,1}(\mathbb{C}))$. In particular we can write

$$dI(A,c)(0) = \int_A E_{dL+\bar\omega}(\tilde{c}_o;\partial\tilde{c}) = \int_A E_{\Pi dL}(\tilde{c}_o;\partial\tilde{c}),$$

where $\Pi dL$ is the unique source form such that $dL - \Pi dL$ belongs to $\text{Im}(\overset{o}{d})$.

The relation between $E_{\Pi dL}$ and the Euler-Lagrange operator for a k-order Lagrangian as given in Proposition 5.2 is now given by the following

<u>Proposition ,5</u>  Let $f : J\mathcal{Y}^k(\mathbb{C}) \to \mathbb{R}$ be a k-order Lagrangian on $\mathbb{C}$. Then $\varepsilon(f)$ identifies a (n+1)-differential form $\varepsilon[f]$ on $J\mathcal{Y}^\infty(\mathbb{C})$ which coincides with $E_{\Pi d\check\Omega}[f]$ being $\check\Omega[f]$ the lifting on $J\mathcal{Y}^\infty(\mathbb{C})$ of the n-form $\Omega[f] = f.\pi_k^\ast\eta$ on $J\mathcal{Y}^k(\mathbb{C})$.

<u>Proof.</u>  In fact the map $\varepsilon(f):J\mathcal{Y}^{2k}(\mathbb{C}) \to vT^\ast\mathbb{C}$ can be lifted on a map $\check\varepsilon(f) : J\mathcal{Y}^\infty(\mathbb{C}) \to T^\ast J\mathcal{Y}^\infty(\mathbb{C})$ given by $\check\varepsilon(f)(u)(X_u) = \varepsilon(f)(\pi_{\infty,2k}(u))(vT(\pi_{\infty,o})(\check X_u))$ where $\check X_u$ is the vertical componenet of $X_u \in T_u J\mathcal{Y}^\infty(\mathbb{C})$. Then the (n+1)-form $\varepsilon[f]$ on $J\mathcal{Y}^\infty(\mathbb{C})$ is given by

$$\boxed{\varepsilon[f] = \check\varepsilon(f)\wedge\pi_\infty^\ast\eta}$$

where $\pi_\infty^\ast\eta$ is the lifting on $J\mathcal{Y}^\infty(\mathbb{C})$ of the volume form on M.
Now the coincidence of $\varepsilon[f]$ with $E_{\Pi d\check\Omega}[f]$ is seen by considering the rispective local expressions of these forms ( see eqs. (3) and $(\bullet)$). $\qquad\square$

<u>Examples</u>    Let us give the explicity local expression of the Euler-Lagrange form for k=0,1,2,3.  Let $\{x^k, y^j_{i_1 \ldots i_\alpha}\}_{0 \le \alpha \le 2k}$ be local coordinates on $J\mathfrak{H}^{2k}(\mathbb{C})$.

$\boxed{k=0}$ $\quad \varepsilon[f] = d\check{\Omega}[f] = (\partial y_j . f) dy^j \wedge dx^1 \wedge \ldots \wedge dx^n \; ; \; \Omega[f] = f . \pi^\times_{\mathbb{C}} \eta$ .

$\boxed{k=1}$ $\quad \varepsilon[f] = \{A_j - y^\alpha_{ks}(\partial y^s_\alpha (\partial y^k_j . f))\} \; dy^j \wedge dx^1 \wedge \ldots \wedge dx^n$ , where $A_j$, $j=1,\ldots,m$, are numerical functions on $J\mathfrak{H}(\mathbb{C})$ given by

$$A_j = (\partial y_j . f) - (\partial x_k . (\partial y^k_j . f)) - y^\alpha_k (\partial y_\alpha (\partial y^k_j . f)) \; .$$

$\boxed{k=2}$ $\quad \varepsilon[f] = \{A_j + y^\gamma_{ikrs}(\partial y^{rs}_\gamma . (\partial y^{ik}_j . f))\} \; dy^j \wedge dx^1 \wedge \ldots \wedge dx^n$ , where $A_j$, $j=1,\ldots m$, are numerical functions on $J\mathfrak{H}^3(\mathbb{C})$ given by:

$$A_j = (\partial y_j . f) - (\partial x_r (\partial y^r_j . f)) + (\partial x_s \partial x_r (\partial y^{rs}_j . f)) +$$

$$\sum_{0 \le \alpha \le 2} \left[ -(\partial y^{i_1 \ldots i_\alpha}_i . (\partial y^r_j . f)) + 2(\partial y^{i_1 \ldots i_\alpha}_i . (\partial x_s (\partial y^{rs}_j . f))) \right] y^i_{i_1 \ldots i_\alpha r}$$

$$+ \sum_{0 \le \alpha, \beta \le 2} \left[ \partial y^{e_1 \ldots e_\beta}_e . (\partial y^{i_1 \ldots i_\alpha}_i . (\partial y^{rs}_j . f))) \right] y^i_{i_1 \ldots i_\alpha r} y^e_{e_1 \ldots e_\beta s}$$

$$+ \sum_{0 \le \alpha \le 1} (\partial y^{i_1 \ldots i_\alpha}_i . (\partial y^{rs}_j . f)) y^i_{i_1 \ldots i_\alpha rs} \; .$$

$\boxed{k=3}$ $\quad \varepsilon[f] = \{A_j - y^i_{uvzrst}(\partial y^{uvz}_i . (\partial y^{rst}_j . f))\} \; dy^j \wedge dx^1 \wedge \ldots \wedge dx^n$ , where $A_j$, $j=1,\ldots,m$, are numerical functions on $J\mathfrak{H}^5(\mathbb{C})$ given by

$$
\begin{aligned}
A_j = \;& \sum_{0 \le \alpha \le 3} (-1)^{\alpha}\, \partial x_{i_1} \dots \partial x_{i_\alpha}\, (\partial y_j^{\,i_1 \dots i_\alpha}. f)) \\[6pt]
&+ \sum_{0 \le \alpha \le 3} \{\, y^i_{i_1 \dots i_\alpha r}\,[\,-(\partial y_i^{\,i_1 \dots i_\alpha}.\;(\partial y_j^{\,r}.f)) + 2(\partial y_i^{\,i_1 \dots i_\alpha}(\partial x_s(\partial y_j^{\,rs}.f))) \\[4pt]
&\qquad\qquad -3(\partial y_i^{\,i_1 \dots i_\alpha}.(\partial x_s.\partial x_t.(\partial y_j^{\,rst}.f)))\,] \\[4pt]
&+ y^i_{i_1 \dots i_\alpha rs}\,[\,(\partial y_i^{\,i_1 \dots i_\alpha}.(\partial y_j^{\,rs}.f)) - 3(\partial y_i^{\,i_1 \dots i_\alpha}(\partial x_t(\partial y_j^{\,rst}.f)))\,] \\[4pt]
&- \sum_{0 \le \alpha \le 2} ((\partial y_i^{\,i_1 \dots i_\alpha}\,\partial y_j^{\,rst}.f))\, y^i_{i_1 \dots i_\alpha rst} \\[4pt]
&+ \sum_{0 \le \alpha, \beta \le 3} \{\, y^i_{i_1 \dots i_\alpha r}\, y^e_{e_1 \dots e_\beta s}\,[\,\partial y_e^{\,e_1 \dots e_\beta}(\partial y_i^{\,i_1 \dots i_\alpha}.(\partial y_j^{\,rs}.f))) \\[4pt]
&\qquad -3(\partial y_e^{\,e_1 \dots e_\beta}(\partial y_i^{\,i_1 \dots i_\alpha}.(\partial x_t(\partial y_j^{\,rst}.f))))\,] - 3 y^i_{i_1 \dots i_\alpha rt}\, y^e_{e_1 \dots e_\beta s} \\[4pt]
&\qquad\qquad \partial y_e^{\,e_1 \dots e_\beta}.(\partial y_i^{\,i_1 \dots i_\alpha}\,\partial y_j^{\,rst}.f)))\,\} \\[6pt]
&- \sum_{0 \le \alpha, \beta \le 3} y^i_{i_1 \dots i_\alpha r}\, y^e_{e_1 \dots e_\beta s}\, y^p_{p_1 \dots p_\gamma t}\, \partial y_p^{\,p_1 \dots p_\gamma}.(\partial y_e^{\,e_1 \dots e_\beta}. \\[6pt]
&\qquad\qquad (\partial y_i^{\,i_1 \dots i_\alpha}.(\partial y_j^{\,rst}.f))).
\end{aligned}
$$

## 6. - CARTAN FORM AND NOETHER CONSERVATION LAWS FOR LAGRANGIANS OF ANY ORDER

Dynamical conservation laws occur in connection with the existence of differential invariants on the state space $J\mathfrak{J}^k(\mathbb{C})$ of a continuum system. In fact under suitable conditions, these differential invariants can be considered as conserved Lagrangians of order k and so they generate conservation laws. Therefore, in this section, a rigorous geometric characterization of Noether conservation laws for Lagrangians of any order is given.

Let $\pi : E \longrightarrow M$ be a fiber bundle and let X be a vector field $\pi$-related with a vector field $\hat{X}$ on M. Set $\widetilde{X} = \partial\widetilde{\phi}$ and $X = \partial\phi$, i.e. $\widehat{\phi}$ is the 1-parameter group of transformations of M generated by $\widetilde{X}$. The q-th prolongation $\overset{q}{X}$ of X is $\overset{q}{X} = \partial\phi^{(q)}$, with $\phi^{(q)}_t = J\mathfrak{J}^q(\phi_t)$, $\forall t \in \mathbb{R}$, where $J\mathfrak{J}^q$ is the q-th jet-derivative functor .(see ref. $\begin{bmatrix}1\,2\end{bmatrix}$). The following diagrams

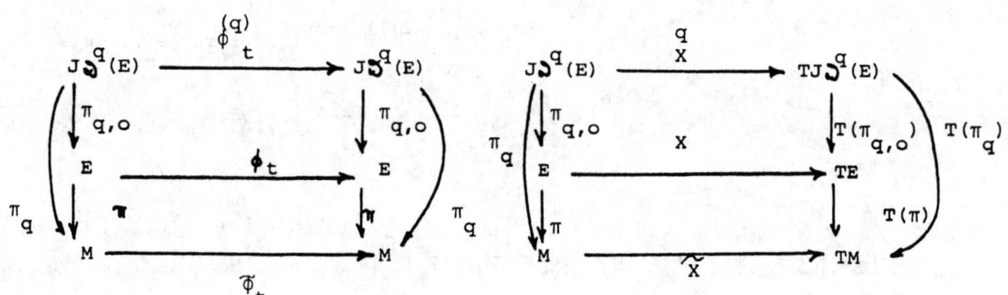

are commutative. Let $\{x^\alpha, y^j_{i_1\ldots i_k}\}$ $o \leq k \leq q$ be a local coordinate system on $J\mathfrak{J}^q(E)$. Let $X = \widetilde{X}^\alpha \partial x_\alpha + X^\beta \partial y_\beta$ be the local expression of X. Note that in general $(\partial y_\alpha . \overset{k}{\widetilde{X}}_q) = o$ and $(\partial x_\alpha . X^\beta) \neq o$. Then the local expression of X is as follows:

$$\overset{q}{X} = \widetilde{X}^k \partial x_k + \sum_{o \leq \alpha \leq q} X^j_{i_1\ldots i_\alpha} \partial y_j^{i_1\ldots i_\alpha}$$

where $X^j_{i_1\ldots i_\alpha}$ are numerical functions on $J\mathfrak{J}^\alpha(E)$ given by

$$x^j_{i_1 \cdots i_\alpha} = \Box_{i_1} \cdots \Box_{i_\alpha} . x^j - \sum_{\substack{\sigma \in B_\alpha \\ 1 \leq p \leq \alpha}} (\partial x_{\sigma(i_p)} \cdots \partial x_{\sigma(i_1)} . \tilde{x}^k) y^j_{k\sigma(i_{p+1}) \cdots \sigma(i_\alpha)}$$

with $\quad \Box_i . f \equiv (\partial x_i . f) + \sum_{\substack{0 \leq s \leq q \\ 1 \leq i_1 \leq \cdots \leq i_s \leq n}} (\partial y_j^{i_1 \cdots i_s} . f) y^j_{i i_1 \cdots i_s} \quad$, for any fun-

ction $f: J\mathfrak{D}^q(C) \longrightarrow \mathbb{R}$ and $B_\alpha$ =set of cyclic permutations of $\alpha$ objects. Set $x^j_{i_1 \cdots i_q} = x^j$, for q=0.

Important cases are the ones with q=1,2,3.

$\boxed{q=1}$
$$\overset{1}{x} = x + \left[-y^j_\sigma(\partial x_\alpha . \tilde{x}^\sigma) + y^\beta_\alpha(\partial y_\beta . x^j) + (\partial x_\alpha . x^j)\right] \partial y^\alpha_j \quad ;$$

$\boxed{q=2}$
$$\overset{2}{x} = \overset{1}{x} + \left[-y^\beta_{j_1\sigma}(\partial x_{j_2} . \tilde{x}^\sigma) - y^\beta_{\sigma j_2}(\partial x_{j_1} . \tilde{x}^\sigma) - (\partial x_{j_1} . \partial x_{j_2} . \tilde{x}^\sigma) y^\beta_\sigma + y^\sigma_{j_1 j_2}(\partial y_\sigma . x^\beta) + \right.$$
$$+ (\partial x_{j_1} . \partial y_\sigma . x^\beta) y^\sigma_{j_2} + (\partial x_{j_1} . \partial x_{j_2} . x^\beta) + (\partial x_{j_2} . \partial y_\sigma . x^\beta) y^\sigma_{j_1} +$$
$$\left. + (\partial y_k(\partial y_j . x^\beta)) y^k_{j_1} y^j_{j_2} \right] \partial y^{j_1 j_2}_\beta \quad ;$$

$\boxed{q=3}$
$$\overset{3}{x} = \overset{2}{x} + \left[ -y^\beta_{j_3 j_2 \sigma}(\partial x_{j_1} . \tilde{x}^\sigma) - y^\beta_{j_1 j_2 \sigma}(\partial x_{j_2} . \tilde{x}^\sigma) - y^\beta_{j_1 j_2 \sigma}(\partial x_{j_3} . \tilde{x}^\sigma) + \right.$$
$$-y^\beta_{j_3 \sigma}(\partial x_{j_2} \partial x_{j_1} . \tilde{x}^\sigma) - y^\beta_{j_1 \sigma}(\partial x_{j_3} \partial x_{j_2} . \tilde{x}^\sigma) - y^\beta_{j_2 \sigma}(\partial x_{j_1} \partial x_{j_3} . \tilde{x}^\sigma) +$$
$$-y^\beta_\sigma(\partial x_{j_3} \partial x_{j_2} \partial x_{j_1} . \tilde{x}^\sigma) + y^\sigma_{j_3 j_2 j_1}(\partial y_\sigma . x^\beta) + y^\sigma_{j_2 j_1}(\partial x_{j_3} \partial y_\sigma . x^\beta) +$$
$$+ y^\sigma_{j_3 j_1}(\partial x_{j_2} \partial y_\sigma . x^\beta) + y^\sigma_{j_2 j_3}(\partial x_{j_1} \partial y_\sigma . x^\beta) + y^\sigma_{j_2}(\partial x_{j_1} \partial x_{j_2} \partial y_\sigma . x^\beta) +$$
$$+ y^\sigma_{j_3}(\partial x_{j_2} \partial x_{j_1} \partial y_\sigma . x^\beta) + y^\sigma_{j_1}(\partial x_{j_2} \partial x_{j_3} \partial y_\sigma . x^\beta) + (\partial y_\omega \partial y_\gamma \partial x_{j_1} . x^\beta) y^\omega_{j_2} y^\gamma_{j_3}$$
$$+ (\partial y_\omega \partial y_\gamma \partial x_{j_2} . x^\beta) y^\omega_{j_1} y^\gamma_{j_3} + (\partial y_\omega \partial y_\gamma \partial x_{j_3} . x^\beta) y^\omega_{j_2} y^\gamma_{j_1} + (\partial y_\omega \partial y_k \partial y_\gamma . x^\beta) y^\omega_{j_1} y^k_{j_2} y^\gamma_{j_3}$$
$$+ (\partial y_k \partial y_\omega . x^\beta) y^k_{j_1 j_2} y^\omega_{j_3} + (\partial y_k \partial y_\omega . x^\beta) y^k_{j_1} y^\omega_{j_3 j_2} + (\partial y_\omega \partial y_k . x^\beta) y^\omega_{j_2} y^k_{j_3 j_1}$$
$$\left. + (\partial x_{j_1} \partial x_{j_2} \partial x_{j_3} . x^\beta) \right] \partial y^{j_1 j_2 j_3}_\beta \quad .$$

The way to generalize to Lagrangians of any order the concept of Cartan form given intrinsically by Goldschmidt and Sternberg [1;9] for Lagrangians of first order, is to follow an iterative procedure. Give, first, some preliminary definitions and propositions.

Let $\pi:E \to X$, $\mathbf{1}:X \to Y$ be fiber bundles over X and Y respectively, $\dim X = n$. Set $\mathbf{1} \cdot \pi \equiv$

<u>Definition .1</u>  Let $\tau$ be the map $\tau: C^\infty(\Lambda_p^o J\mathcal{S}^k(E)) \to C^\infty(\Lambda_p^{\bullet o} J\mathcal{S}^{k+1}(E))$ defined by the universal property $(D^{k+1}\gamma)^* \tau\omega = (D^k\gamma)^*\omega$, $\forall \gamma \in C^\infty(E)$.

Set
$$\Lambda_{n+h,h}^o(\pi) \equiv \{\omega \in C^\infty(\Lambda_{n+h}^o E) \mid \omega(X_1,\ldots,X_n) \in C^\infty(\dot\Lambda_n^o(\pi)) \mid \forall X_i \in C^\infty(vTE)\} \;;$$

$$\Lambda_{n+h,h}^o(\phi;\pi) \equiv \Lambda_{n+h,h}^o(\phi) \cap \dot\Lambda_{n+h}^o(\pi).$$

Let $\overset{+}{\tau}$ be the map $\overset{+}{\tau}: C^\infty(\Lambda_{n+h}^o(\pi_\infty)) \to C^\infty(\Lambda_{n+h,h}^o(\pi_\infty))$ given by $\overset{+}{\tau}\omega(X_1,\ldots X_h) = \tau(\omega(X_1,\ldots,X_h))$, $\forall \omega \in C^\infty(\Lambda_{n+h}^o(\pi_\infty))$, $\forall X_i \in C^\infty(vTJ\mathcal{S}^\infty(E))$. In particular if $\omega \in C^\infty(\Lambda_{n+h}^o J\mathcal{S}^k_{\ })$ then $\overset{+}{\tau}\omega \in C^\infty(\Lambda_{n+h,h}^o(\pi_{k+1};\pi_{k+1,k}))$.

<u>Proposition .1</u>  There exists a canonical vector fiber bundle homomorphism

$\omega_k^s: TJ\mathcal{S}^k(E) \to \pi_{k,s}^* vTJ\mathcal{S}^s(E)$, over $J\mathcal{S}^k(E)$, $0 \leq s \leq k-1$.

<u>Proof.</u>  In fact one has

$$\omega_k^s : \xi_q \in T_q J\mathcal{S}^k(E) \mapsto \omega_k^s(\xi_q) \equiv T(\pi_{k,s})(\xi_q) - T(D^s s \cdot \pi_k)(\xi_q)$$

with s a section of E over $x = \pi_k(q)$ such that $D^k s(x) = q$. This map does not depend on the section s used. $\square$

<u>Remark</u>  By means of the homomorphism $\omega_k^s$ we can associate to the Euler-Lagrange operator $\varepsilon(f):J\mathcal{S}^{2k}(\mathbb{C}) \to vT^*\mathbb{C}$ a differential 1-form $\overset{2k}{\varepsilon}(f)$ on $J\mathcal{S}^{2k}(\mathbb{C})$: $\overset{2k}{\varepsilon}(f): J\mathcal{S}^{2k}(\mathbb{C}) \to T^* J\mathcal{S}^{2k}(\mathbb{C})$ given by $\overset{2k}{\varepsilon}(f)(u)(X_u) = \varepsilon(f)(u)(\omega_{2k}^o(X_u))$, $\forall u \in J\mathcal{S}^{2k}(\mathbb{C})$, $X_u \in T_u J\mathcal{S}^{2k}(\mathbb{C})$. So, we can also consider the $(n+1)$-form $\overset{2k}{\varepsilon}[f]$ on $J\mathcal{S}^{2k}(\mathbb{C})$ given by $\overset{2k}{\varepsilon}[f] \equiv \overset{2k}{\varepsilon}(f) \wedge \pi_{2k}^* \eta$. Of course the form $\overset{2k}{\varepsilon}[f]$ lifted on $J\mathcal{S}^\infty(\mathbb{C})$ coincides with the form $\varepsilon[f]$ and locally has the same expression of this last. For sake of semplicity we shall denote $\overset{2k}{\varepsilon}[f]$ by $\varepsilon[f]$ .

<u>Proposition .2</u>  Let $f:J\mathcal{S}(\mathbb{C}) \to \mathbb{R}$ be a first order Lagrangian. There exists a canonical n-form $\overset{1}{\bullet}[f] \in C^\infty(\dot\Lambda_n^o(\pi_{1,o}))$.

<u>Proof.</u> The form $\overset{1}{\bullet}[f]$ is given by composition

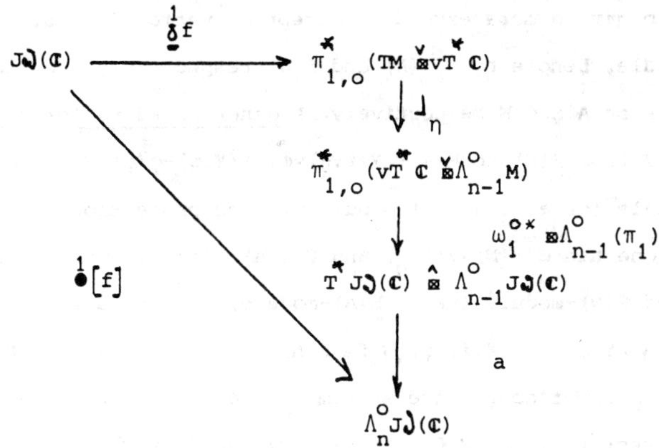

where $\lrcorner_\eta$ is the contraction for the volume form $\eta$ and $a$ is the antisimmetrization operator. The local expression of $\overset{1}{\bullet}[f]$ is as follows

$$\overset{1}{\bullet}[f] = (-1)^{\beta+1}(\partial y^{\beta}_j.f)\, dy^j \wedge dx^1 \wedge \ldots \wedge \widehat{dx^{\beta}} \wedge \ldots dx^n - (\partial y^{\beta}_j.f) y^j_{\beta}\, dx^1 \wedge \ldots \wedge dx^n$$

that shows that $\overset{1}{\bullet}[f]$ belongs to $C^\infty(\overset{\bullet}{\Lambda}{}^o_n(\pi_{1,o}))$.

Definition .2    The <u>Cartan form</u> corresponding to a first order Lagrangian is the n-form $\sigma[f] \in C^\infty(\Lambda^o_n(J\mathfrak{I}(\mathbb{C})))$ given by

$$\boxed{\sigma[f] \equiv \pi^{*}_{2,1}\,\Omega[f] + \overset{1}{\bullet}[f]}$$

<u>Note</u>.    The local expression of $\sigma[f]$ is as follows

$$\sigma[f] = f\, dx^1 \wedge \ldots \wedge dx^n - (\partial y^{\beta}_j.f) y^j_{\beta}\, dx^1 \wedge \ldots \wedge dx^n + (-1)^{\beta+1}(\partial y^{\beta}_j.f)\, dy^j \wedge dx^1 \wedge \ldots \wedge \widehat{dx^{\beta}} \wedge \ldots \wedge dx^n$$

This shows that really $\sigma[f]$ belongs to $C^\infty(\Lambda^o_n(\pi_1;\pi_{1,o}))$.    Remark that to $\sigma[f]$ we can associate a n-form on $J\overset{\infty}{\mathfrak{I}}(\mathbb{C})$, also denoted by $\sigma[f]$ .

Proposition .3    Let $f: J\overset{k}{\mathfrak{I}}(\mathbb{C}) \to \mathbb{R}$ be a k-order Lagrangian. It identifies a (k-1)-order lagrangian on $J\mathfrak{I}(\mathbb{C})$ $\overset{(k-1)}{f}: J\overset{k-1}{\mathfrak{I}}(J\mathfrak{I}(\mathbb{C})) \to \mathbb{R}$.

Proof.    As the canonical map $\beta_{k-1,1}: J\overset{k}{\mathfrak{I}}(\mathbb{C}) \to J\overset{k-1}{\mathfrak{I}}(J\mathfrak{I}(\mathbb{C}))$ is an immersion in $J\overset{k-1}{\mathfrak{I}}(J\mathfrak{I}(\mathbb{C}))$, there exists a tubular neighborhood $[6]$ .

$$\{ \pi_{(k)}:U_{(k-1)} \to \beta_{k-1,1}(J\overset{k}{\mathfrak{I}}(\mathbb{C})) ; h:U_{(k-1)} \to J\overset{k-1}{\mathfrak{I}}(J\mathfrak{I}(\mathbb{C})) \}$$

of the image $\beta_{k-1,1}(J\overset{k}{\mathfrak{I}}(\mathbb{C}))$ with $\pi_{(k)}$ vector bundle and h embedding.    So, we can translate f from $J\overset{k}{\mathfrak{I}}(\mathbb{C})$ on $\beta_{k-1,1}(J\overset{k}{\mathfrak{I}}(\mathbb{C}))$ and them lift it on $U_{(k-1)}$. Finally, we can extend this map smoothly over $J\overset{k-1}{\mathfrak{I}}(J\mathfrak{I}(\mathbb{C}))$. $\square$

Remark      It is useful to give a more extended concept of vector fields.

● Let $\phi:A \to N$ be a fiber bundle. Denote by $F(A)$ and $F(N)$ respectively the rings of the $C^\infty$ numerical functions on $A$ and $N$ respectively. A <u>generalized vector field</u> of $\phi$ is a linear derivation $Y$ from $F(N)$ to $F(A)$: $Y(f.g) = \phi(f)(Y.g) + \phi(g)(Y.f)$, $\forall f,g \in$ These fields form a $F(A)$-module (or also a $F(N)$-module via the projection $\phi$), denoted by $V(\phi)$. Of course one has $C^\infty(TN) = V(\mathrm{id}_N)$ and $C^\infty(TA) = V(\mathrm{id}_A)$. One has also the canonical monomorphism of $F(N)$-module (resp. $F(A)$-module), $C^\infty(TN) \to V(\phi)$ (resp. $C^\infty(TA) \to V(\phi)$) given by $X \mapsto Y$, $Y.f = (X.f) \circ \phi$, $\forall f \in F(N)$, (resp. $X \mapsto Y$, $Y.f = X.(f \circ \phi)$ $\forall f \in F(N)$). If $\{x^i, y^j\}$ is a fiber coordinate system on $\phi:A \to N$, then the local expression of a generalized vector field $Y \in V(\phi)$ is $Y = Y^i \partial x_i$, with $Y^i:U \subset A \to |R$. So, any generalized vector field on $\phi$ can be locally identified with a vector fie on $A$.

● Let $\pi:E \to N$ be a fiber bundle. Let $\overset{\phi}{\pi}_{k+1,k}: J\mathfrak{D}^{k+1}(E) \to J\mathfrak{D}^k(E)$ be the canonical map. Then, we have a homomorphism of $F(A)$-modules $\Pi^k(\phi,\pi):V(\phi) \to V(\overset{\phi}{\pi}_{k+1,k})$ given by $Y \mapsto {}^kY$, where ${}^kY$ is given by $({}^kY.f)(D^{k+1}\gamma(\phi(a))) = (Y.(f\,D^k\gamma))(a)$, $\forall f \in F(J\mathfrak{D}^k(E)$ $\forall a \in A$.

● In particular for $\phi \equiv \pi_s:J\mathfrak{D}^s(E) \to N$ we have the map $\Pi^k_s(\pi) \equiv \Pi^k(\pi_s,\pi):V(\pi_s) \to V(\pi_{[s,k]}$ $[s,k] \equiv \max(s,k+1)$, $Y \mapsto {}^k\overline{Y} = \Pi^k_s(\pi)(Y)$; ${}^k\overline{Y} \in V(\pi_{k+1,k})$ if $k+1 \geq s$ and ${}^k\overline{Y} \in V(\pi_s$ if $k+1 \leq s$.     Further, one has the following properties:

(a)  $h \in F(J\mathfrak{D}^s(E))$, $Y \in V(\pi_s)$ $\Rightarrow$ ${}^k\overline{hY} = h\,{}^k\overline{Y}$, $k+1 \leq s$; ${}^k\overline{hY} = h \circ \pi_{k+1,s}\,{}^k\overline{Y}$, $k+1 \geq s$;

(b)  $Y \in V(\pi_s)$ $\Rightarrow$ ${}^k\overline{\pi^*_{r,s}Y} = \begin{cases} \pi^*_{k,s}\,{}^k\overline{Y} & , \ s \geq k+1 \\[2ex] \pi^*_{r,k+1}\,{}^k\overline{Y} & , \ r \geq k+1 \geq s \\[2ex] {}^k\overline{Y} & , \ k+1 \geq r \end{cases}$

(c)  $Y \in V(\pi_s)$ $\Rightarrow$ ${}^k\overline{Y}.\pi^*_{k,e} = \begin{cases} {}^e\overline{Y} & , \ s \geq k+1 \\[2ex] \pi^*_{k+1,e}\,{}^e\overline{Y} & , \ k+1 \geq s \geq e+1 \\[2ex] \pi^*_{k+1,e+1}\,{}^e\overline{Y} & , \ e+1 \geq s \end{cases}$

(d)  For any $Y \in V(\pi_s)$, ${}^k\overline{Y}$ is characterizd by the following universal property

$$(D^{[s,k+1]}\gamma)^{\dagger} \; {}^{k}\bar{Y} = (D^{s}\gamma)^{\dagger}.Y\,(D^{k}\gamma)^{\dagger} \qquad\qquad \gamma \; C^{\infty}(\pi);$$

(e)  The local expression of ${}^{k}\bar{Y}$ is ${}^{k}\bar{Y} = Y^{j}\circ\pi_{[s,\,k+1],s}\left(\partial x_{j} + \sum_{0\le\alpha\le k} y^{a}_{i_{1}\ldots i_{\alpha}}\,\partial\,y^{i_{1}\ldots i_{\alpha}}_{a}\right)$

if $Y = Y^{j}\,\partial x_{j}$ , $Y^{j}:J\mathfrak{J}^{s}(E)\to \rm I\!R$ , is the local expression of $Y$ and $\{x^{i}, y^{a}_{i_{1}\ldots i_{\alpha}}\}$ is a local coordinate system on $J\mathfrak{J}^{p}(E)$.

We have a homomorphism $\bar{\Pi}_{s}(\pi)$ of $F(J\mathfrak{J}^{s}(E))$-modules $\bar{\Pi}_{s}(\pi):V(\pi_{s})\to C^{\infty}(TJ\mathfrak{J}^{\infty}(E))$, $0\le s\le\infty$ given by $\bar{\Pi}_{s}(\pi):\; Y\mapsto\bar{Y}\equiv(Y,{}^{0}\bar{Y},{}^{1}\bar{Y},\ldots)$.  More precisely one has

$$\bar{Y}.f = \begin{cases} ({}^{k}\bar{Y}.\hat{f})\bullet\pi_{\infty,k+1} & \text{if } k+1\ge s \\[2em] ({}^{k}\bar{Y}.\tilde{f})\bullet\pi_{\infty,s} & \text{if } k+1\le s \end{cases}$$

for any $f\in A\!\!\setminus(E)$, with $\tilde{f}$ the local function defined on a neighborhood of $\pi_{\infty,k}(u)$ on $J\mathfrak{J}^{k}(E)$ , $k\in\rm I\!N$ , such that $f(u)=(\tilde{f}\circ\pi_{\infty,k})(u)$ , $u\in J\mathfrak{J}^{\infty}(E)$.

The set of horizontal vector fields on $J\mathfrak{J}^{\infty}(E)$ can be seen as a $F(N)$-sub-module of $C^{\infty}(TJ\mathfrak{J}^{\infty}(E))$ given as the image of the following homomorphism of $F(N)$-modules $\daleth:C^{\infty}(TN)\to C^{\infty}(TJ\mathfrak{J}^{\infty}(E))$ given by composition

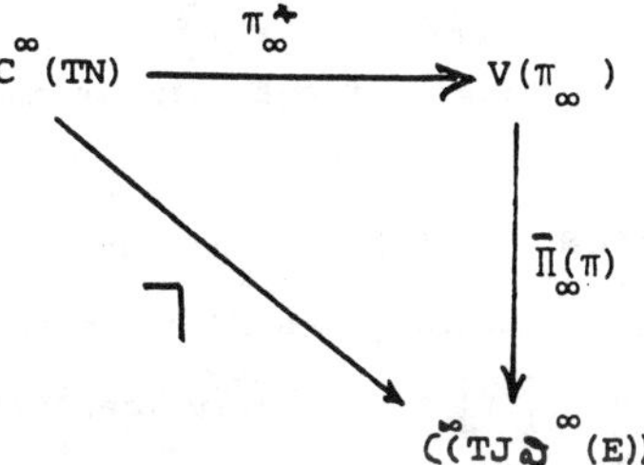


$\daleth$ is just the canonical connection on $J\mathfrak{J}^{\infty}(E)$ and one has the canonical splitting $C^{\infty}(TJ\mathfrak{J}^{\infty}(E)) \cong C^{\infty}(vTJ\mathfrak{J}^{\infty}(E)) \oplus \overline{V(\pi_{\infty})}$, where $\overline{V(\pi_{\infty})}\equiv\bar{\Pi}_{\infty}(\pi)(V(\pi_{\infty}))$.  $\daleth$ is an homomorphism of Lie algebras. $\overline{V(\pi_{\infty})}$ is a Lie algebra.  Further, one can see that for any $X\in C^{\infty}(TN)$, $\daleth(X)$ identifies a set of generalized vector fields ${}^{k}\bar{X}\in V(\pi_{k+1,k})$ , $k\in\rm I\!N$.  The local expression of ${}^{k}\bar{X}$ is ${}^{k}\bar{X}=X^{j}\left(\partial x_{j} + \sum_{0\le\alpha\le k} y^{a}_{i_{1}\ldots i_{\alpha}\,j}\,\partial\,y^{i_{1}\ldots i_{\alpha}}_{a}\right)$ if $X=X^{j}\,\partial x_{j}$ , $X^{j}:N\to\rm I\!R$ , and $\{x^{i}, y^{a}_{i_{1}\ldots i_{\alpha}}\}$ is a coordinate system on $J\mathfrak{J}^{p}(E)$.

We shall utilize the following

<u>Theorem</u>  ·1   $\forall\,\beta\in C^{\infty}(\Lambda^{o}_{n+1,1}(\pi_{k};\pi_{k,1}))$ there exists a pair of forms

$\hat{\sigma}\beta \in C^{\infty}(\Lambda^{o}_{n,1}(\pi_{k};\pi_{k,o}))$ , $\hat{\varepsilon}\beta \in C^{\infty}(\Lambda^{o}_{n+1,1}(\pi_{k+1};\pi_{k+1,o}))$ such that the following decomposition

($\maltese$) $\qquad \tau(\bar{X}\lrcorner\beta) = \tau d(\bar{X}\lrcorner\hat{\sigma}\beta) + \tau(X\lrcorner\hat{\varepsilon}\beta)$ $\qquad$ (+)

holds for any $X \in V(\pi_{\infty,o})$.

<u>Proof.</u> See ref. $[1:i]$. $\square$

We are now able to give the following fundamental

<u>Definition</u> .3 Let $f:J^{k}_{\mathfrak{O}}(\mathbb{C}) \to \mathbb{R}$ be a k-order Lagrangian. The <u>Cartan form</u> of f is the differential n-form $\sigma[f] \in C^{\infty}(\Lambda^{o}_{n,1}(\pi_{2k-1};\pi_{2k-1,k-1}))$ given by the following induction procedure: $\boxed{\sigma[f] = \alpha[f] + \hat{\sigma}\beta[f]}$ , where:

(a) $\alpha[f] = \pi^{*}_{2k-1,2k-2}(\beta^{*}_{2(k-1)-1,1} \ \sigma[\overset{(k-1)}{f}])$,

$\sigma[\overset{(k-1)}{f}] \in C^{\infty}(\Lambda^{o}_{n,1}((\pi_{1})_{2(k-1)-1};(\pi_{1})_{2(k-1)-1,(k-1)-1})) = $ Cartan form of the (k-1)-order Lagrangian $\overset{(k-1)}{f}:J^{k-1}_{\mathfrak{O}}(J_{\mathfrak{O}}(\mathbb{C})) \to \mathbb{R}$ induced by means of Proposition 4.3;

(b) $\beta[f] = \beta^{*}_{2(k-1),1}\ \varepsilon[\overset{(k-1)}{f}]$,

$\varepsilon[\overset{(k-1)}{f}] \in C^{\infty}(\Lambda^{o}_{n+1,1}((\pi_{1})_{2(k-1)};(\pi_{1})_{2(k-1),1})) = $ Euler-Lagrange form of the (k-1)-order Lagrangian $\overset{(k-1)}{f}$.

Note that as the Cartan form is explicetely defined for k=1 (Definition 6.2 and the Euler-Lagrange form has been previously given ( for Lagrangians of any order) the above definition is correct. Note further that $\sigma[f]$ canonically identifies a differential form on $J^{\infty}_{\mathfrak{O}}(\mathbb{C})$ also denoted by $\sigma[f]$.

<u>Proposition</u> .4 Let $f:J^{k}_{\mathfrak{O}}(\mathbb{C}) \to \mathbb{R}$ be a k-order Lagrangian . Let $\{x^{i},y^{j}_{i_{1}\dots i_{\alpha}}\}_{0 \leq \alpha \leq 2k-1}$ be a coordinate system on $J^{2k-1}_{\mathfrak{O}}(\mathbb{C})$. Then the local expression of $\sigma[f]$ is as follows:

$$\boxed{\sigma[f] = A\,dx^{1}\wedge\cdots\wedge dx^{n} + \sum_{0 \leq \alpha \leq k-1} A_{j}^{\beta i_{1}\dots i_{\alpha}}\,dy^{j}_{i_{1}\dots i_{\alpha}}\wedge dx^{1}\wedge\cdots\wedge \widehat{dx^{\beta}}\wedge\cdots\wedge dx^{n}}$$

where A and $A_{j}^{\beta i_{1}\dots i_{\alpha}}$ are numerical functions on $J^{2k-1}_{\mathfrak{O}}(\mathbb{C})$ given by

---

(+) Note that if $\omega \in C^{\infty}(\overset{\circ}{\Lambda}_{p}(\pi))$, p>0, the meaning of $X\lrcorner\omega$ for any $X \in V(\pi)$ is correctl given by putting $X\lrcorner\omega = Y\lrcorner\omega$ with Y any vector field on E such that $X=Y\bullet\pi^{*}$. This clarifies the meaning of the form $X\lrcorner\hat{\varepsilon}\beta$ in ($\maltese$).

$$A = f - \sum_{0 \leq \alpha \leq k-1} (-1)^{\beta+1} A_j^{\beta i_1 \ldots i_\alpha} y_{i_1 \ldots i_\alpha \beta}^j$$

$$A_j^{\beta i_1 \ldots i_\alpha} = (-1)^{\beta+1} \left[ (\partial y_j^{\beta i_1 \ldots i_\alpha} \cdot f) - (-1)^{\gamma+1} (\partial x_\gamma . A_j^{\gamma \beta \, i_1 \ldots i_\alpha}) - (-1)^{\gamma+1} \sum_{\substack{\omega \geq 0 \\ 0 \leq \alpha \leq k-2}} (\partial y_e^{e_1 \ldots e_\omega} . A_j^{\gamma \beta \, i_1 \ldots i_\alpha}) y_{e_1 \ldots e_\omega \gamma}^e \right]$$

$$A_j^{\beta \, i_1 \ldots i_{k-1}} = (-1)^{\beta+1} (\partial y_j^{\beta \, i_1 \ldots i_{k-1}} . f) \quad .$$

<u>Proposition .5</u>    The relation between the Euler-Lagrange form and the Cartan form is given by the following equation:  $\boxed{\varepsilon[f] = \tau^+ d \, \sigma[f]}$

<u>Proposition .6</u>    Let $f: J^k_{\mathfrak{d}}(\mathbb{C}) \to \mathbb{R}$ be a k-order Lagrangian . Then the differential forms $\Omega[f], \varepsilon[f], \sigma[f]$ are related by the following formula (<u>first variation formula</u>).

$$(4) \quad \boxed{\pi^*_{\infty \, ,k+1} \tau \underset{X}{\mathcal{L}_k} \Omega[f] = \tau(\bar{X} \lrcorner \varepsilon[f]) + \tau d(\bar{X} \lrcorner \sigma[f]).}$$

with $\bar{X} \equiv (X, \overset{1}{X}, \overset{2}{X}, \ldots)$ being X an infinitesimal symmetry of $\pi_{\mathbb{C}}$.

<u>Proof</u>.  The proof can be obtained by local computations.  $\square$

▲ <u>Theorem .2</u>   Let $\Omega[f] \equiv f . \pi^*_k \eta$ be the Lagrange density associated to the Lagrangian $f: J^k_{\mathfrak{d}}(\mathbb{C}) \to \mathbb{R}$ . Let X be an infinitesimal symmetry of $\pi_{\mathbb{C}}$ such that $D^k c \underset{X}{\overset{k}{\mathcal{L}_k}} \Omega[f] = 0$, for any extremal c of f. Then the form

$$(5) \quad \beta = \tau(\bar{X} \lrcorner \sigma[f]) \in C^\infty(\Lambda^0_{n-1} J^\infty_{\mathfrak{d}}(\mathbb{C})) \quad , \quad \bar{X} = (X, \overset{1}{X}, \overset{2}{X}, \ldots) ,$$

over $J^\infty_{\mathfrak{d}}(\mathbb{C})$ is a <u>conservation law</u> , that is

$$(6) \quad \boxed{d \{ (D^\infty c)\tau^* (\bar{X} \lrcorner \sigma[f]) \} = 0,}$$

for every extremal c of f.

<u>Proof</u>.   The proof discends directly from (4).  $\square$

338

**Proposition .7**  Let $\Omega[f] = {\textstyle\xi}\pi_k^*\eta$  be the Lagrangian density associated to the Lagrange function $f: J\mathfrak{D}^k_{\mathbb{C}}(\mathbb{C}) \to \mathbb{R}$ . Let X be an infinitesimal symmetry of $\pi_{\mathbb{C}}$ such that there exists $\mu \in C^\infty(\Lambda^o_{n-1} J\mathfrak{D}^h(\mathbb{C}))$ , $o \leq h \leq \infty$ such that $(D^k c)^* \mathcal{L}^k_X \Omega[f] = d(D^h c)^* \mu$ , for any extremal c of f.  Then

$$\alpha = \left[ \tau \left( \bar{X} \; \lrcorner \; \sigma[f] \right) - \pi^*_{\infty,h} \mu \right] : J\mathfrak{D}^\infty(\mathbb{C}) \to \Lambda^o_{n-1} J\mathfrak{D}^\infty(\mathbb{C})$$

is a conservation law.

**Proof.**  We must prove that $d\{(D^\infty c)^* \alpha\} = o$  for every extremal c of f. In fact from (4) we get

$$d((D^\infty c)^* \mu) = d((D^\infty c)^* \mathcal{L}^\infty_X \Omega[f]) = (D^\infty c)^* \tau \, d(\bar{X} \; \lrcorner \; \sigma[f]) + (D^\infty c)^* \tau \, (\bar{X} \; \lrcorner \; \varepsilon[f])$$

$$= d\left[ (D^\infty c)^* (\tau(\bar{X} \; \lrcorner \; \sigma[f] ))) \right] \; . \qquad \square$$

**Theorem .3**  Let $\beta \in J\mathfrak{D}^\infty(\mathbb{C}) \to \Lambda^o_{n-1} J\mathfrak{D}^\infty(\mathbb{C})$ be  a conservation law for $f : J\mathfrak{D}^k(\mathbb{C}) \to \mathbb{R}$ , that is $d((L^\infty c)^* \beta) = o$,  for any extremal c of f. Let A, B be two (n-1)-cocycles on M that cobound, i.e. are the boundaries of a n-dimensional region C of M: $A - B = \partial C$ .  Then one has:

$$(7) \qquad \int_A (D^\infty c)^* \beta \; = \; \int_B (D^\infty c)^* \beta$$

**Proof.** $\int_A (D^\infty c)^* \beta - \int_A (D^\infty c)^* \beta = \int_{A-B} (D^\infty c)^* \beta = \int_{\partial C} (D^\infty c)^* \beta = \int_C d(D^\infty c)^* \beta = o$ $\qquad \square$

The physical interpretation of Theorem   is given by the following

**Corollary .1**   Let M be a 4-dimensional space-time.  Let specialize A and B of above thoerem with two 3-dimensional space-like regions individued by a framing on M. Let us define the <u>charge map</u>

$$\boxed{\; Q: \mathbb{R} \to \mathbb{R} \; , \qquad Q(t) = \int_{A_t} (D^\infty c)^* \beta \;}$$

being $A_t$ a 3-dimensional space-like region $\psi$-related ($\psi$ = frame) to A and B. Then one has

$$(8) \qquad\qquad\qquad\qquad dQ = o.$$

In other words Q is a constant map.  Thus, to a conservation law it is associated a conserved "charge".

# 7. - SYMMETRY PROPERTIES AND DYNAMIC CONSERVATION LAWS

The symmetry group $G$ of a continuum system $CS=(\mathcal{C},E_k)$ , is the group of fiber bundle transformations of $\pi_k:J\mathcal{S}^k(\mathbb{C}) \to M$ and $\pi_{k,o}:J\mathcal{S}^k(\mathbb{C}) \to \mathbb{C}$, such that preserve $E_k$ and the structure of $M$; in particular preserve the volume form .

**Proposition .1** A symmetry $\psi$ of $CS$ is such that $\psi = J\mathcal{S}^k(\phi)$ for some automorphism $\phi$ of $\pi_\mathbb{C}$.

**Proof.** In fact $J\mathcal{S}^k(\mathbb{C})$ has a natural structure of super-bundle of geometric objects (see refs. $[11,12,15]$) over $\pi_\mathbb{C}$, so the transformations of $J\mathcal{S}^k(\mathbb{C})$ must be the natural functorial extension of $\pi_\mathbb{C}$ ones by means of the functor $J\mathcal{S}^k$.
$\square$

Before to state the fundamental theorem for dynamic conservation laws, we shall consider the following

**Lemma .1** Let $X:J\mathcal{S}^k(\mathbb{C}) \to TJ\mathcal{S}^k(\mathbb{C})$ be a vector field on $J\mathcal{S}^k(\mathbb{C})$ $\pi_k$-related with a vector field $\check{X}:M \to TM$ on $M$, i.e. the following diagram

$$
\begin{array}{ccc}
J\mathcal{S}^k(\mathbb{C}) & \xrightarrow{\ X\ } & TJ\mathcal{S}^k(\mathbb{C}) \\
\pi_k \downarrow & & \downarrow T(\pi_k) \\
M & \xrightarrow{\ \tilde{X}\ } & TM
\end{array}
$$

is commutative. Let $\partial\psi = X$ and $\partial\tilde{\phi} = \hat{X}$ and suppose that

$$(9) \qquad \tilde{\phi}_t^* \eta = \eta$$

Let $f:J\mathcal{S}^k(\mathbb{C}) \to \mathbb{R}$ be a numerical function on $J\mathcal{S}^k(\mathbb{C})$ and set $\Omega[f] = f.\pi_k^* \eta$
The following propositions are equivalent:

(a) $X.f = o$.

(b) $f\circ\psi_t = f$ , $\forall\ t\in\mathbb{R}$;

(c) $\mathcal{L}_X f = o$ ;

(d) $\psi_t^* \, \Omega[f] = \Omega[f]$ ;

(e) $\mathcal{L}_X \, \Omega[f] = o$.

<u>Theorem</u> .1 (<u>Dynamic conservation laws</u>)   Let $H \subset G$ be a r-dimensional Lie sub-group of the group of symmetry G of CS $\equiv \mathscr{C}(E_k)$. Let $E_H$ be the involutive r-distribution generated by the infinitesimal generators of the restricted action of H on $E_k$.  Let $\{\phi^\alpha\}_{\alpha \in (1, \cdots, s)}$ , $s \leq (q-r)$, $q = \dim E_k$ , be a fundamental set of differential invariants    of $\mathbb{E}_H$ . Then, for any function $\phi^\alpha$ there are r-independent conservation laws (<u>dynamic conservation laws</u>). More precisely one has:

$$c \in \mathrm{Ext}(\phi^\alpha) \cap \mathscr{S}(E_k) \neq \emptyset \quad \Rightarrow \quad d\left\{ (D^{\cdot}{}^{*}c)\, \beta^\alpha_{(e)} \right\} = 0,$$

with

$$\beta^\alpha_{(e)} \equiv \tau(\bar{X}_{(e)} \lrcorner \, \sigma[\phi^\alpha])$$

being $\{X_{(1)}, \cdots , X_{(r)}\}$ a basis of $\mathbb{E}_H$.

<u>Proof</u>. As each $\phi^\alpha$ is a differential invariant of $\mathbb{E}_H$, one has $X.\phi^\alpha = 0$, for any infinitesimal generator X of H.  So by using Lemma 7.1 one has

$$(D^k c)^{*} \mathscr{L}_X \, \Omega[\phi^\alpha] = 0 \quad \text{with} \quad \Omega[\phi^\alpha] \equiv \phi^\alpha . \pi_k^{*}\eta \quad \text{and} \quad c \in C^\infty(\mathbb{C}).$$ So one has

$$(10) \qquad d\left\{ (D^\infty {\cdot} c)^{*} \beta^\alpha \right\} = -(D^\infty {\cdot} c)^{*} \, j^\alpha$$

with

$$\beta^\alpha \equiv \tau(\bar{X} \lrcorner \, \sigma[\phi^\alpha])$$

and

$$j^\alpha \equiv (\bar{X} \lrcorner \, \epsilon[\phi^\alpha]).$$

Then if $c \in \mathrm{Ext}(\phi^\alpha) \cap S(E_k) \neq \emptyset$, since $(D^\infty {\cdot} c)^{*} j^\alpha = 0$ eq.(10) gives $d\{ (D^\infty c)^{*} \beta^\alpha \} = 0$.  $\square$

<u>Corollary</u> .1    Let $E_k$ be completely integrable (+) . Let $H \subset G$ be a r-dimensional Lie group.  Then to any differential invariant $\phi^\alpha$ of $\mathbb{E}_H$, there corrisponds r independent dynamic conservation laws if and only if

$$(11) \qquad E_{k+k} \cap \mathbb{E}_{2k}(\phi^\alpha) \neq \emptyset.$$

---

(+) Recall that a k-order differential equation $E_k \subset J\mathscr{D}^k(\mathbb{C})$ is <u>completely</u> <u>integrable</u>  if for any $q \in E_k$ there exists a solution c of $E_k$ such that $D^k c(x) = q$ with $x = \pi_k(q)$.  The completely integrability implies the surjectivity of the map $p_{k+h,k} : E_{k+h} \to E_k$, $h \geq 0$, which is the restriction of $\pi_{k+h,k}$ to $E_{k+h}$.

<u>Proof</u>. Since $E_k$ is completely integrable all solutions of $E_k$ will be also solutions of $E_{k+k}$. So if condition (11) is verified one has also that $\text{Ext}(\phi^\alpha) \cap S(E_k) \neq \emptyset$ . $\square$

<u>Corollary</u> .2    To any dynamical conservation law there corresponds a conserved charge.

The local characterization of the symmetry group of a continuum system is given by means of the following.

<u>Proposition</u>.2    $X: \mathbb{C} \to T\mathbb{C}$ is an infinitesimal symmetry of $CS = (\mathscr{C}, E_k)$ if and only if

$$\overset{k}{X}.F^i = o;$$

or

$$(\partial x_\alpha .F^i)\overset{\alpha}{X} + (\partial y_\alpha .F^i)x^\alpha + \sum_{1 \leq \alpha \leq k} (\partial y_\beta^{j_1 \ldots j_\alpha} .F^i)x^\beta_{j_1 \ldots j_\alpha} = o$$

where $F^i$, $j=(1,\ldots,s)$, are numerical functions on $J\mathring{\mathscr{D}}^k(\mathbb{C})$ locally characterizing $E_k$ and $\overset{k}{X} = $ k-th prolongation of X.

<u>Lemma</u>.2    Let X be an infinitesimal symmetry of $CS=(\mathscr{C}, E_k)$. Let us consider the h-th prolongation $E_{k+h}$ of $E_k$. Then $\overset{k+h}{X}$ is tangent to $E_{k+h}$.

<u>Theorem</u>.2    Let $CS = (\mathscr{C}, E_k)$ be a continuum system such that $E_k$ is completely integrable. Then to any infinitesimal symmetry $X: \mathbb{C} \to T\mathbb{C}$ of CS one can associates conservation laws related to the equations $E_{k+h}$ , $h \geq o$ .

<u>Proof</u>.  In fact for Lemma .2 $\overset{k+h}{X}$ is tangent to $E_{k+h}$, $h \geq 0$ and so one can associates some differential invariants $\phi$ , $\overset{k+h}{X}.\phi = o$, such that if $\text{Ext}(\phi) \cap S(E_{k+h}) = \text{Ext}(\phi) \cap S(E_k) \neq \emptyset$ they generate some effective dynamical conservation laws. $\square$

342

The relation between dynamic conservation laws and Noether ones is given by the following non trivial theorem.

<u>Theorem.3</u>    A Noether conservation law is a dynamic one.

<u>Proof.</u>  Let the dynamical equation $E_{2h} \subset J\mathfrak{D}^{2h}(\mathbb{C})$ of CS be expressed by means of a h-order Lagrangian $f:J\mathfrak{D}^h(\mathbb{C}) \to \mathbb{R}$, $\mathbb{E}_{2h}(f) = \ker_\varepsilon(f)$. Let X and $\widetilde{X}$ $\pi_\mathbb{C}$-related vector fields on $\mathbb{C}$ and M respectively such that $\overset{h}{X} = J\mathfrak{D}^h(\mathbb{C}) \to TJ\mathfrak{D}^h(\mathbb{C})$ is an infinitesimal Noether symmetry for f, that is $\overset{h}{X}.f=o$ and $\phi_t^* \eta=\eta$ , with $\widetilde{X}=\partial\widetilde{\phi}$. Let us put $\overset{\vee}{f} = f\cdot\pi_{2h,h}:J\mathfrak{D}^{2h}(\mathbb{C}) \to \mathbb{R}$ .  Then

(12)     $\overset{2h}{X}.\overset{\vee}{f} = o.$

In fact, set $X=\partial\psi$ , $\overset{h}{X} =\partial\psi^{(h)}$

$o= X.f \;\Rightarrow\; f\cdot\psi_t^{(h)} =f \;\Rightarrow\; f\circ\psi_t^{(h)} \circ \pi_{2h,h} = f\circ\pi_{2h,h} \;\Rightarrow\; f\circ\pi_{2h,h}\circ \overset{(2h)}{\psi}_t= f\circ\pi_{2h,h}$

$$\Rightarrow\; \overset{\vee}{f}\circ\overset{(2h)}{\psi}_t = \overset{\vee}{f} \;\Rightarrow\; \overset{2h}{X}.\overset{\vee}{f} = o.$$

Thus for the 2h-order Lagrangian $\overset{\vee}{f}$, $\overset{2h}{X}$ is an infinitesimal Noether symmetry. Let us now prove that $\overset{2h}{X}|E_{2h}$ is a vector field tangent to $E_{2h}$ .  For this it is sufficent   to prove that if s is a solution of $E_{2h}$ one has that $\tilde{s} =\psi_t\circ s\circ\widetilde{\phi}_t^{-1}$ is a solution of $E_{2h}$ too.  In fact if this hypothesis is verified one has

$$D^{2h}s(x)\in E_{2h} \;\Rightarrow\; \overset{(2h)}{\psi}_t(D^{2h}s(x)) = J\mathfrak{D}^{2h}(\psi_t)(D^{2h}s(x))=(D^{2h}\tilde{s})(\widetilde{\phi}_t(x)) \in E_{2h} \;,$$

so $\overset{(2h)}{\psi}_t$ is a 1-parameter group of transformations of $E_{2h}$.   Now by using Lemma 5.1  one has that the condition $\overset{h}{X}.f = o$ is equivalent to $f\circ\psi_t^{(h)} = f$.  So we get for the corresponding action integral:

$$I_A\bigl[s\bigr] = \int_A (f\cdot D^h s)\eta = \int_A (f\cdot\psi_t^{(h)}\cdot D^h s )\cdot\eta$$

Therefore, $\psi_t\circ s$ is an extremal for f if s thus, i.e. $D^{2h}(\psi_t\cdot s)(x) \in (E_{2h})_x$ This implies also that $\psi_t\circ s\circ\widetilde{\phi}_t^{-1}$ is a solution of $E_{2h}$ since $D^{2h}(\psi_t\circ s\circ\widetilde{\phi}_t^{-1})(x)=$ $=D^{2h}(\psi_t\circ s)\widetilde{\phi}_t^{-1}(x) \in (E_{2h})_{\widetilde{\phi}_t^{-1}(x)}$ .

From the above considerations we prove that $\overset{2h}{X}$ belongs to the set of infinitesimal symmetries of CS and therefore from eq. (12) $\overset{\vee}{f}$ is a differential invariant of an involutive distribution $\mathbb{E}$ generated by $\overset{\smile}{X}{}^{2h}$. As a final step we must prove the following equations:

A) $\operatorname{Ext}(\overset{\smile}{f}) = S(E_{2h}(f))$;

B) $\quad \cdot \left[\tau(\bar{X} \quad \lrcorner\ \sigma[f]\ )\ \right] = \tau(\bar{X} \quad \lrcorner\ \sigma[\overset{\smile}{f}])\quad .$

We shall use the following

__Lemma__ .3  Let $f$ be a $k$-order Lagrangian and let $\overset{\smile}{f}:J\overset{\smile}{\mathfrak{J}}{}^{k+h}(\mathbb{C}) \to \mathbb{R}$ be the natural lifting of $f$ on $J\overset{\smile}{\mathfrak{J}}{}^{k+h}(\mathbb{C})$, that is $\overset{\smile}{f} = f \circ \pi_{k+h,k}$. Then one has:

I)  $\operatorname{Ext}(f) = \operatorname{Ext}(\overset{\smile}{f})$;

II)  $\varepsilon[\overset{\smile}{f}] \quad = \quad \varepsilon[f]$ ;

III)  Let us put $\Omega[f] = f\,\pi_k^{\ast}\,\eta$ , $\Omega[\overset{\smile}{f}] = \overset{\smile}{f}.\,\pi_{k+h}^{\ast}\eta$ .  Then one has

$$\Omega[\overset{\smile}{f}] \;=\; \pi_{k+h,k}^{\ast}\ \Omega[\,f]$$

IV)  $$\sigma[f] = \sigma[\overset{\smile}{f}]\quad .$$

__Proof of Lemma__ .3  I)  Note that the $2(k+h)$-order differential operator $\varepsilon(\overset{\smile}{f}):J\overset{\smile}{\mathfrak{J}}{}^{2(k+h)}(\mathbb{C}) \to vT^{\ast}\mathbb{C}$  is reducible to $\varepsilon(f)$, i.e.

$$\varepsilon(\overset{\smile}{f}) = \varepsilon(f) \circ \pi_{2(k+h),2k}$$

Therfore, $\pi_{2(k+h),2k}^{-1}\ (E_{2k}) = E_{2(k+h)}$ .

II)  It is a direct consequence of  I) and Proposition 5.5

III)  $\pi_{k+h,k}^{\ast}\ \Omega[f] \;=\; \Lambda_n^{\circ}(\pi_{k+h,k}) \circ \Omega[f] \circ \pi_{k+h,k}$

$$= \Lambda_n^{\circ}(\ \pi_{k+h,k}) \circ (f\pi_k^{\ast}\eta\ )\pi_{k+h,k}$$

$$= f \circ \pi_{h+h,k} \cdot \Lambda_n^{\circ}(\ \pi_{k+h,k})\ \pi_k^{\ast}\ \eta \circ \pi_{k+h,k}$$

$$= \overset{\smile}{f}.\,\pi_{k+h}^{\ast}\ \eta = \Omega[\overset{\smile}{f}]\quad .$$

IV)  It is a direct consequence of the above points I,II,III, first formula of variation  and of the following lemmas

__Lemma .4__  Let $\alpha$ be a p-form on $J_{\mathfrak{H}}^k(\mathbb{C})$ and X a vector field on $J_{\mathfrak{H}}^k(\mathbb{C})$. One has

$$\overset{h}{X} \lrcorner \ \pi^*_{k+h,k} \, \alpha = \pi^*_{k+h,k} (X \lrcorner \alpha )$$

__Lemma .5__  Let $\alpha$ be a p-form on $J_{\mathfrak{H}}^k(C)$. One has

$$\tau(\pi^*_{k+h,k} \alpha ) = \pi^*_{k+h+1,k+1} (\tau\alpha ).$$

To conclude the proof of Theorem .3 we shall simply observe that points A) and B) are justified by above Lemmas.

$$\square$$

Let us conclude this section with some important examples of differential equations which admit dynamic conservation laws.

We shall also see that some of discovered dynamic conservation laws are trivial; in fact they are the horizontal part of the differential of some functions.  Since these entities are conserved for any equation, they cannot give any non-trivial information about the differential equations considered. This aspect must be always considered when one works with dynamic conservation laws.

### EXAMPLES

__A - KORTEWEG DE VRIES EQUATION__      The configuration bundle is $\mathcal{C} \equiv \{\pi_{\mathbb{C}} : \mathbb{C} \equiv \mathbb{R}^2 \times \mathbb{R} \to \mathbb{R}^2\}$ with coordinates $(x,t,u)$. As on the manifold $\mathbb{R}^2$ we do not impose any structure, the continuum system considered is given by $CS \equiv (\mathcal{C}, (KdV))$, where $(KdV) \subset J_{\mathfrak{H}}^3(\mathbb{C})$, is the equation characterized by the function $F = u_{xxx} + 6uu_x - u_t : J_{\mathfrak{H}}^3(\mathbb{C}) \to \mathbb{R}$.  Then the differential equation which identifies the algebra of infinitesimal symmetries of $CS$ is given by

$$\overset{3}{X}.F = \lambda F \ , \ \text{where } X = \bar{X}^t \partial t + \bar{X}^x \partial x + X \partial u$$

and $\lambda$ is a numerical function on $J_{\mathfrak{H}}^3(C)$. One gets:

$$(13) \quad 6Xu_x +$$
$$6u\left[ u_x(-\bar{X}^x_x + X_u) - u_t\bar{X}^t_x + X_x\right] +$$

345

$$-\left[-u_x \bar{X}^x_t + u_t(X_u - \bar{X}^t_t) + X_t\right]+$$
$$+\left[u_x(3X_{xxu} - \bar{X}^x_{xxx}) - u_t \bar{X}^t_{xxx} + u_{xx}(-3\bar{X}^x_{xx} + 3X_{xu} + 3u_x u_{xx} X_{uu}) - 3\bar{X}^t_{xx} u_{xt} + 3u_x^2 X_{uux} +\right.$$
$$\left. + u_x^3 X_{uuu} + u_{xxx}(X_u - 3\bar{X}^x_x) - 3\bar{X}^t_x u_{xxt} + X_{xxx}\right]$$

$$= \lambda(u_{xxx} + 6uu_x - u_t).$$

As the third order derivatives $u_{xxx}$, $u_{xxt}$ occur only linearly in the first term, $\lambda$ can only depend on $(x,t,u,u_x,u_t,u_{xx},u_{xt},u_{tt},u_{xtt},u_{ttt})$. The coefficient of $u_{xxx}$ in (13) is $X_u - 3\bar{X}^x_x = \lambda$, implying that $\lambda$ is a function of $(x,t,u)$ alone. The coefficient of $u_{xxt}$ is $-3\bar{X}^t_x = 0$, this implies that $\bar{X}^t$ is a function of $t$ alone. The coefficient of $u_{xx}$ is $-3\bar{X}^x_{xx} + 3X_{xu} + 3X_{uu} u_x = 0$. Thus we get $X_{uu} = 0$, $\bar{X}^x_{xx} = X_{xu}$. From the first equation we get $X = \alpha(x,t) + u\,\beta(x,t)$. Therefore one has also $\bar{X}^x_{xx} = \beta_x$. The coefficient of $u_x$ is $6X + 6u(-\bar{X}^x_x + X_u) + \bar{X}^x_t + 3X_{xxu} - \bar{X}^x_{xxx} = 6u\lambda$. So we try $6u(\beta + 2\bar{X}^x_x) + 6\alpha + \bar{X}^x_t + 2\beta_{xx} = 0$, i.e. $\beta = -2\bar{X}^x_x$, $6\alpha = -\bar{X}^x_t - 2\beta_{xx}$. The coefficient of $u_t$ is $-(X_u - \bar{X}^t_t) = -\lambda$, i.e. $\lambda = \lambda(x,t)$ and $\bar{X}^t_t = 3\bar{X}^x_x \Rightarrow \bar{X}^x_{xx} = 0 = \beta_x \Rightarrow \beta = \beta(t)$,

$$(14) \qquad 6\alpha = -\bar{X}^x_t \quad, \quad -2\bar{X}^x_x = \beta(t).$$

Finally, the terms in (1) not involving any derivative of $u$ are $6uX_x - X_t + X_{xxx} = 0$. From which we get $u(6\alpha_x - \beta_t) - \alpha_t + \alpha_{xxx} = 0$, i.e. $(\bullet)\ 6\alpha_x = \beta_t$, $\alpha_t = \alpha_{xxx} \Rightarrow$ $\alpha_{xxx} = 0 \ne \alpha_t = 0$, $\alpha = \alpha(x)$. . From eq. (14) we get $6\alpha_x = \tfrac{1}{2}\beta_t$. By comparison of this equation with $(\bullet)$ we conclude that $\alpha_x = \beta_t = 0$. So we have $\alpha = c_1$, $\beta = c_2$ and $X = c_1 + c_2 u$. Now from the first eq. (14) we get $\bar{X}^x = -6c_1 t + \delta(x)$ and from second eq. (14) we get $\delta(x) = -\tfrac{1}{2}c_2 x + c_3$. So we have $\bar{X}^x = -6c_1 t - \tfrac{1}{2}c_2 x + c_3$. Finally, from eq. $\bar{X}^t_t = -\tfrac{3}{2}c_2$ we get $\bar{X}^t = -\tfrac{3}{2}c_2 t + c_4$, where $c_i$, $1 \le i \le 4$ are arbitrary real constants. Therefore, the infinitesimal symmetry algebra of the KdV continuum system is four dimensional with the following basis:

$$X_1 = -6t\,\partial x + \partial u$$
$$X_2 = -\tfrac{1}{2}x\,\partial x - \tfrac{3}{2}t\,\partial t + u\partial u$$
$$X_3 = \partial x$$
$$X_4 = \partial t .$$

The corresponding constant of structure of the symmetry group of $CS \equiv (\Gamma, (KdV))$ are the following: $C^k_{ii} = 0$, $C^k_{ij} = -C^k_{ji}$, $C^1_{12} = 1$, $C^3_{14} = 6$, $C^3_{23} = \tfrac{1}{2}$, $C^4_{24} = \tfrac{3}{2}$, the remaining are zero.

(KdV) is not a Lagrangian equation. Now, the Euler-Lagrange equation associated to the function $F \equiv u_{xxx} + 6uu_x - u_t$ is just $J \mathcal{S}^6(\mathbb{C})$ and so one has $\mathrm{Ext}(F) \cap \mathcal{S}(W)) = \mathcal{S}(W))$. The Cartan form $\sigma[F]$ associated to $F$ is given by

$$\sigma[F] = 6udu \wedge dt + du \wedge dx + du_{xx} \wedge dt.$$

The corresponding conserved entities and conserved charges are the following:

$$\beta_1 \equiv \tau(\overset{\infty}{X}_1 \lrcorner \sigma[F]) = (6tu_x + 1)dx + 6(u + tu_t)dt \quad ; \quad Q_1 = \int (6tu_x + 1)dx;$$

$$\beta_2 \equiv \tau(\overset{\infty}{X}_2 \lrcorner \sigma[F]) = (\tfrac{1}{2}xu_x + 9tuu_x + \tfrac{3}{2}tu_{xxx} + u)dx + (\tfrac{1}{2}xu_t + 9tuu_t + \tfrac{3}{2}tu_{xxt} + 6u^2 + 2u_{xx})dt$$

$$Q_2 = \int (\tfrac{1}{2}xu_x + 9tuu_x + \tfrac{3}{2}tu_{xxx} + u)dx$$

$$\beta_3 \equiv \tau(\overset{\infty}{X}_3 \lrcorner \sigma[F]) = u_x dx + u_t dt; \qquad Q_3 = \int u_x dx;$$

$$\beta_4 \equiv \tau(\overset{\infty}{X}_4 \lrcorner \sigma[F]) = -(6uu_x + u_{xxx})dx - (6uu_t + u_{xxt})dt \quad ; \quad Q_4 = -\int (6uu_x + u_{xxx})dx.$$

Note that among these four "entities" only the 1-form $\beta_2$ gives rise to a non trivial conservation law. The remaining three "conserved entities" give rise to trivial conservation laws; in fact, each one of these quantities is the horizontal part of the differential of function and, therefore, such quantities are conserved independently on the KdV-equation.

<u>B - WAVE EQUATION</u>     The system is given by CS $\equiv$ ( $\mathcal{C}$, (W)), where $\mathcal{C}$ is as in the above cases and (W)$\subset J\mathfrak{J}^2(\mathbb{C})$ is a 2-order differential equation on $\mathbb{C}$ characterized by the following function $F\equiv u_{tt}-u_{xx}:J\mathfrak{J}^2(\mathbb{C})\to \mathbb{R}$. The differential equation (11) is now given by

$$(16)\ [u_{tt}(-2\bar{X}^t_t+X_u)+u_{tx}(-2\bar{X}^x_t)+u_t(-\bar{X}^t_{tt}+2X_{ut})+$$

$$+u_x(-\bar{X}^x_{tt})+X_{tt}+u^2_t X_{uu}]+$$

$$-[u_{xt}(-2\bar{X}^t_x) + u_{xx}(-2\bar{X}^x_x+X_u) + u_t(-\bar{X}^t_{xx}) + X_{xx} + u_x(-\bar{X}^x_{xx}+2X_{ux}) + u^2_x X_{uu}] = \lambda(u_{tt}-u_{xx}).$$

By using a proceeding similar to the above ones, one gets that the infinitesimal symmetry algebra of CS is the space of all vector fields $X=\bar{X}^t\partial t+\bar{X}^x\partial x+X\partial u$ on $\mathbb{C}$ with

$$\bar{X}^t = f_1(x+t) - g_1(x-t) + c_1$$

$$\bar{X}^x = f_1(x+t) + g_1(x-t)$$

$$X = f_2(x+t) +g_2(x-t) + c_2 u,$$

where $f_i, g_i$, $1\leq i\leq 2$ are arbitrary numerical functions of one variable and $c_i$, $1\leq i\leq 2$, are arbitrary constants.

Wave system          is       a Lagrangian system with Lagrangian function $f\equiv\frac{1}{2}(u^2_x-u^2_t):J\mathfrak{J}(\mathbb{C})\to \mathbb{R}$. An infinitesimal generator of the symmetry group of this system is characterized by the following vector field on $\mathbb{C}$:

$$X= (f_1-g_1+c_1)\partial t+ (f_1+g_1)\partial x + (f_2+g_2+c_2 u)\partial u ,$$

The Euler-Lagrange equation $\mathcal{E}(F)$ associated to the function $F\equiv u_{tt}-u_{xx}$ is just $J\mathfrak{J}^4(\mathbb{C})$ so we also have $\mathrm{Ext}(F)\cap S((W))= S((W))$. The Cartan form $\sigma[F]$ associated to F is given by $\sigma[F] = -du_x\wedge dt - du_t\wedge dx$. So we get the conserved 1-form

$$\beta \equiv \tau(\overset{\infty}{X}\lrcorner\sigma[F])= [\,(u_{tt}+u_{xt})f_1+(u_{tt}-u_{xt})g_1+(u_x+u_t)\dot{f}_1+(u_x-u_t)\dot{g}_1-\dot{f}_2-\dot{g}_2-u_x c_2+u_{xt}c_1\,]\,dt+$$

$$[\,(u_{xx}+u_{tx})f_1+(-u_{xx}+u_{tx})g_1+(u_x+u_t)\dot{f}_1+(-u_x+u_t)\dot{g}_1-\dot{f}_2+\dot{g}_2+c_1 u_{xx}-u_t c_2\,]\,dx.$$

Therefore, the corresponding conserved charge is

$$Q = \int[f_1(u_{xx}+u_{tx}) + g_1(-u_{xx}+u_{tx}) + \dot{f}_1(u_x+u_t) + \dot{g}_1(u_t-u_x) - \dot{f}_2 + \dot{g}_2 + c_1 u_{xx} - u_t c_2]\,dx .$$

In this case we have infinite independent conserved charges.

(Note that the point over $f_1$ and $g_1$ denotes derivation).

348

## C - EULER EQUATION

A perfect incompressible fluid can be considered geometrically defined by
means of the following structure (see refs. $[9,13,14]$)  : $\text{PIF} \equiv (\mathcal{C}, (E); \psi)$,
where : 1) $\mathcal{C} = \{\pi: \mathbb{C} \equiv J\vartheta(\varsigma) \times_M \Pi \to M |$  is a fiber bundle over the affine
Galilean 4-dimensional space-time M; with $J\vartheta(\varsigma) \equiv$ <u>fiber bundle of velocity</u>,
$\Pi \equiv M \times \mathbb{R} =$<u>fiber bundle of pressure</u>; 2) $(E) \equiv \ker \mathbb{D} = (\ker \mathcal{Z}, \ker_\psi \vartheta) \subset J\vartheta(\mathbb{C})$,
<u>Euler equation</u> , with $\dot{\psi} \equiv$<u>velocity of the frame</u> $\psi$ ,
$\mathbb{D} \equiv (\mathcal{Z}, \vartheta): J\vartheta(\mathbb{C}) \to \mathbb{K} \equiv \overset{o}{T}M \times_M J\vartheta(\varsigma) \cong \overset{o}{T}M \times_M vTM$ (this isomorphism is canonical
in the structure PIF), is the <u>Euler operator</u> given by $\mathcal{Z}.c \equiv \mathcal{Z}.(v,p) = \mathrm{div}(\rho v)$,
$\vartheta.c \equiv \vartheta.(v,p) = \mathrm{div}(\rho v \boxtimes v + pg) - \rho B$, (the divergence is performed with respect
to the canonical connection of M), v=<u>velocity field</u>, p=<u>pressure field</u>, $\rho$ =<u>fluid
density</u> (constant), B=<u>density of the body force</u>, (in all this paper we shall
assume that B is a trivial constitutive map, that is it depends only on the
events $x \in M$);   3) $\psi$ is an inertial frame.

By using the canonical isomorphism $J\vartheta(\varsigma) \cong vTM$, we can also identify the Euler
operator $\mathbb{D}$ with a fiber bundle morphism over M  $\mathbb{D}: J\vartheta(\mathbb{C}) \to vTM$. Further,
it is useful to regard the Euler equation from some different points of views.
In fact, by using some natural isomorphisms we also have the following commutative diagram:

$$
\begin{array}{ccccc}
\ker \mathbb{D} \equiv (E) & \hookrightarrow & J\vartheta(J\vartheta(\varsigma) \times_M \Pi) & \xrightarrow{\;\;\mathbb{D}\;\;} & J\vartheta(\varsigma) \cong vTM \\[4pt]
\| & & \| & & | \\[4pt]
\ker(\mathbb{D}_\wedge + \blacklozenge_\psi) \equiv (E)_\wedge & \hookrightarrow & J\vartheta(vTM \times_M \Pi) & \xrightarrow{\;\mathbb{D}_\wedge + \blacklozenge_\psi\;} & J\vartheta(\varsigma) \cong vTM \\[4pt]
\| & & \| \; j & & \downarrow\, 'g \\[4pt]
\ker \underline{\mathbb{D}} \equiv (\underline{E}) & \hookrightarrow & J\vartheta(v\overset{*}{T}M \times_M \Pi) & \xrightarrow{\;\;\underline{\mathbb{D}}\;\;} & v\overset{*}{T}M
\end{array}
$$

where $\mathbb{D}_\wedge(v_{\wedge\psi}, p) = \mathrm{div}(\rho v_{\wedge\psi} \boxtimes v_{\wedge\psi} + pg) - \rho B$ ; $\blacklozenge_\psi(v_{\wedge\psi}, p) = \mathrm{div}\,\rho(v_{\wedge\psi} \boxtimes \dot\psi + \dot\psi \boxtimes v_{\wedge\psi} + \dot\psi \boxtimes \dot\psi)$ ;
$\underline{\mathbb{D}} \equiv 'g \circ (\mathbb{D}_\wedge + \blacklozenge_\psi) \circ j^{-1}$.

Of course, we can identify (E) with $(E)_\wedge$ and $(\underline{E})$ .

By using the formal theory of partial differential equations, we get the
following:

<u>Proposition **A**</u>   1. $\mathbb{D}$  is an epimorphism.

2. (E) is a first order involutive formally integrable differential equation.

3. (<u>Existence of analytic solutions</u>). For any $q \in (E)_{+h}$ ( <u>generalized initial
condition</u> ), there exists an analytic solution c of (E) over a neighborhood

of $x = \pi_{1+h}(q) \in M$, such that $D^{1+h} c(x) = q$; ($\pi_s$ is the canonical projection $J\mathfrak{J}^s(\mathbb{C}) \to M$).

Let us now investigate about the conservation laws associated to (E). As the Euler equation is not the Euler-Lagrange equation of some a Lagrangian built on the configuration space $\mathbb{C}$, we can not use the Noether's theorem.

However, we have developed a general method to obtain dynamic conservation laws directly by studing the symmetry properties of the dynamic equation.

Now, the infinitesimal generators of the symmetry group of (E) are vector fields $X : \mathbb{C} \to T\mathbb{C}$ on $\mathbb{C}$ $\pi$-related with vector fields $X$ on $M$ which belong to the infinitesimal algebra of the Galilei group and such that the first prolongation $\overset{1}{X} : J\mathfrak{J}(\mathbb{C}) \to TJ\mathfrak{J}(\mathbb{C})$ of $X$ is tangent to the submanifold (E) of $J\mathfrak{J}(\mathbb{C})$. Let $\{x^\alpha, \overset{i}{x}, \lambda\}$ be a coordinate system on $\mathbb{C}$. Then, one can prove the following theorem. (See ref. [14]).

<u>Theorem A</u>. The infinitesimal symmetry algebra of a perfect incompressible fluid is generated by vector fields $X = \overline{X}^\alpha \partial x_\alpha + X^i \partial \overset{.}{x}_i + Y \partial \lambda : \mathbb{C} \to T\mathbb{C}$ such that $\overline{X} = \overline{X}^\alpha \partial x_\alpha$ is a Galilean infinitesimal symmetry and $X^i$ and $Y$ are numerical functions on $\mathbb{C}$ given by $X^i = \sum_{1 \leq r \leq 3} (\partial x_r . \overline{X}^i) \overset{.}{x}^r$, $Y = \rho [ \sum_{i,s} g_{is} (\partial x_o . \overline{X}^i) \overset{.}{x}^s ] + \ell(x^\alpha)$, with $\ell : M \to \mathbb{R}$ numerical function on $M$ such that $(\partial x_s . \ell) = -\sum_i g_{js} (\mathcal{L}_X B)^1$,

Then, the general theorem for dynamic conservation laws of (E) is given by the following

<u>Theorem B</u>. Let $H$ be a 1-dimensional Lie sub-group of the symmetry group of (E). Let $X$ be the corresponding vector field on $\mathbb{C}$. Let $\{f^\alpha\}$ be a fundamental set of differential invariants of the involutive distribution generated by $\overset{1}{X}$. Then, for any function $f^\alpha : J\mathfrak{J}(\mathbb{C}) \to \mathbb{R}$, there is a conservation law (<u>dynamic conservation law</u>). More precisely, one has: $c \in \text{Ext}(f^\alpha) \cap \mathfrak{J}(E) \neq \phi \implies d((Dc)^* \beta^{(\alpha)}) = 0$, where $\text{Ext}(f^\alpha)$ =set of extremals of the function $f^\alpha$, $\mathfrak{J}(E)$ =set of solutions of the Euler equation, and $\beta^{(\alpha)}$ is a differential 3-form on $J\mathfrak{J}(\mathbb{C})$ given by $\beta^{(\alpha)} = \tau(\overset{1}{X} \lrcorner \sigma(f^\alpha))$, with $\sigma(f^\alpha)$ =Cartan form associated to $f^\alpha$, and $\tau$ is the map $\tau : \Gamma(\Lambda_p^\circ J\mathfrak{J}(\mathbb{C})) \to \Gamma(\Lambda_p^\circ J\mathfrak{J}^2(\mathbb{C}))$, ($\Lambda_p^\circ \mathfrak{J}(\mathbb{C})$ =space of horizontal p-forms on the fiber bundle $\pi_2 : J\mathfrak{J}^2(\mathbb{C}) \to M$), defined by the universal property: $(D^2 \gamma)^* \tau \omega = (D\gamma)^* \omega$, $\forall \gamma \in \Gamma^\infty(\mathbb{C})$.

350

Proof. In fact as $\overset{\shortmid}{X}\cdot\overset{\shortmid}{f}=0$ , we can consider $\overset{\alpha}{f}$ as a first order Lagrangian invariant under the group H. So , we can apply to $\overset{\alpha}{f}$ the Noether's theorem. The only restriction is that the configurations c for which we can observe these conservation laws, must be at the same time extremals for $\overset{\alpha}{f}$ and, of course, dynamic configurations for (E), that are solutions of (E). $\square$

Corollary A   Let $f$ be a differential invariant of X ($\equiv$infinitesimal symmetry of (E)). Let $\mathbb{E}(f)$ be the Euler-Lagrange equation of $f$ . Then, we can associate to f  a dynamic conservation law if and only if ($\bullet$) (E)$_{+1} \cap \mathbb{E}(f) \neq \emptyset$ .

Proof. Since (E) is completely integrable all solutions of (E) will be also solutions of (E)$_{+1}$ . So, if condition ($\bullet$) is verified one has that $\mathrm{Ext}(f) \cap \mathfrak{Y}(E) \neq \emptyset$ . $\square$


Now, on the Galilean space-time there is a canonical volume form $\eta$ which allows to globally orient M.   More precisely,   $\eta = \sigma \wedge \underline{\eta}$ , where $\sigma$ is the canonical differential 1-form on M and $\underline{\eta}$ is the canonical vertical differntial 3-form on M associated to the space-like metric field g.   In adapted coordinates $\{x^{\alpha}\} \equiv \{x^{0},x^{1},x^{2},x^{3}\}$   we get   $\sigma = dx^{0}$, $g = g_{ij}\,dx^{i} \otimes dx^{j}$, $\eta = \sqrt{\det(g_{ij})}\,dx^{1} \wedge dx^{2} \wedge dx^{3}$   and   $\eta = \sqrt{\det(g_{ij})}\,dx^{0} \wedge dx^{1} \wedge dx^{2} \wedge dx^{3}$ . Set   $h \equiv \sqrt{\det(g_{ij})} : M \to \mathbb{R}$.
Then, the Euler-Lagrange equation and the Cartan form for any first order lagrangian $f : J\mathfrak{D}(W) \to \mathbb{R}$ , where   $\pi : W \to M$ is a fiber bundle over M is given by the following formulas in adapted coordinates $\{x^{\alpha},y^{i},y^{j}_{\alpha},y^{j}_{\alpha\beta}\}$
$$0 \leq \alpha,\beta \leq 3$$
$$1 \leq j \leq m$$
(m$\equiv$dimension of the fiber of W) on $J\mathfrak{D}^{2}(W)$ . (For a proof see the corresponding intrinsic formulas given in ref.[12]

(17)(Euler-Lagrange Equation)   $\mathcal{E}[f] \subset J\mathfrak{D}^{2}(W)$

$$h\left[ (\partial y_{j}.f) - (\partial x_{\alpha}.(\partial y^{\alpha}_{j}.f)) - y^{q}_{\alpha}(\partial y_{q}.(\partial y^{\alpha}_{j}.f)) - y^{\alpha}_{\rho\alpha}.(\partial y^{\alpha}_{q}.(\partial y^{\beta}_{j}.f)) \right] - (\partial y^{\alpha}_{j}.f)(\partial x_{\alpha}.h) = 0$$

$$1 \leq j \leq m , \qquad 0 \leq \alpha,\beta \leq 3 .$$


(18)(Cartan form)   $\sigma[f] : J\mathfrak{D}(W) \to \overset{o}{\Lambda^{o}_{4}} J\mathfrak{D}(W)$

$$\sigma[f] = h\Big\{ f\,dx^{0} \wedge dx^{1} \wedge dx^{2} \wedge dx^{3} + \sum_{\substack{0 \leq \beta \leq 3 \\ 1 \leq j \leq m}} (\partial y^{\beta}_{j}.f)\,dy^{j} \wedge dx^{0} \wedge dx^{1} \wedge \ldots \widehat{dx^{\beta}} \ldots \wedge dx^{3} +$$

$$- \sum_{\substack{0 \leq \beta \leq 3 \\ 1 \leq j \leq m}} (\partial y^{\beta}_{j}.f)\,y^{j}_{\beta}\,dx^{0} \wedge dx^{1} \wedge dx^{2} \wedge dx^{3} \Big\}$$

Let H=<u>T</u> =<u>additive group of the time translations</u>.

H≡<u>T</u>    $\phi$: <u>T</u> x($\mathbb{C}$≡MxIx $\mathbb{R}$) → $\mathbb{C}$≡MxIx $\mathbb{R}$  ; ($\mu$,x,v,p) ↦ ($\psi$(x,$\mu$),v,p).

Let X:$\mathbb{C}$ → T$\mathbb{C}$ be the infinitesimal generator of $\phi$ ; in a frame coordinate system $\{x^\alpha,\dot{x}^i.,\lambda\}$ on $\mathbb{C}$ we can write $X=\partial x_0$ ; the corresponding vector field $\bar{X}$ on M looks like $\bar{X}=\partial x_0$ . Assuming $\mathcal{L}_{\bar{X}}B=0$ we get that the Euler equation admits the following infinitesimal symmetry $X=\partial x_0$ . We have that

$$F_1\equiv G^j_{jk}\dot{x}^{k\cdot}+\dot{x}^i_i \;\; ; \;\; F_2^{(s)}\equiv\rho[G^s_{ik}\dot{x}^i\dot{x}^k+\dot{x}^s_0+\dot{x}^i\dot{x}^s_i] +\lambda_i g^{is}-\rho B^s \;\;,$$

are the numerical functions on $J\vartheta(\mathbb{C})$ which characterize (E); so we get $\dot{X}.F_1=0$, $\dot{X}.F_2^{(s)}=0$ , ($G^\alpha_{\rho\gamma}$ =numerical functions on M which represent the components of the canonical connection on M and $\{x^\alpha,\dot{x}^i,\lambda,\dot{x}^i_\alpha,\lambda_\alpha\}_{0\leq\alpha\leq3,\,1\leq i\leq3}$ is the coordinate system on $J\vartheta(\mathbb{C})$ induced from the coordinate system $\{x^\alpha,\dot{x}^i,\lambda\}$ on $\mathbb{C}$ . So, we can consider $F_1$ as a differential invariant of $X=\partial x_0$ .

The Cartan form associated to $F_1$ is the following 4-form on $J\vartheta(\mathbb{C})$:

$$\sigma[F_1] \equiv h\Big\{ G^j_{jk}\dot{x}^k\, dx^0\wedge dx^1\wedge dx^2\wedge dx^3 - d\dot{x}^1\wedge dx^0\wedge dx^2\wedge dx^3 + d\dot{x}^2\wedge dx^0\wedge dx^1\wedge dx^3+$$
$$- d\dot{x}^3\wedge dx^0\wedge dx^1\wedge dx^2 \Big\}$$

On the other hand the Euler-Lagrange equation associated to $F_1$ is given by $J\vartheta^2(\mathbb{C})$ as $\mathcal{E}[F_1]=0$ . So that, for any solution c of the Euler equation (E) we get the following conserved entity:

$$\beta \equiv \zeta\,(\partial x_0,\,\lrcorner\sigma[F_1]\,)=h\Big\{[G^j_{jk}\dot{x}^k+\dot{x}^s_s]dx^1\wedge dx^2\wedge dx^3 + \dot{x}^1_0\, dx^0\wedge dx^2\wedge dx^3+$$
$$-\dot{x}^2_0\,dx^0\wedge dx^1\wedge dx^3 +\dot{x}^3_0\,dx^0\wedge dx^1\wedge dx^3 \Big\}.$$

In fact, for any dynamic configuration c≡(v,p), one has $d((Dc)\beta^*)=0$.

The conserved charge (always zero) associated to $\beta$ is

$$Q \equiv \int_A [G^j_{jk}v^k+(\partial x_s.v^s)]\,\underline{\eta} = \int_A (\text{div }v).\underline{\eta}$$

for any vector field v, solution of (E) and space-like 3-chain A.

By using Stokes formula we get that Q= flux$_{\partial A}$ v . So, we try the well known theorem (Lagrange's theorem) of conservation of flux of the velocity through a closed space-like surface as a consequence of the time-translation invariance of the Euler equation.

## PART B : QUANTIZED PDE AND CONSERVATION LAWS

In non-linear field theory we can distinguish three different types of non-linearity: (A) base manifold M=non-affine manifold; (B) configuration bundle C = non-vector fiber bundle; (C) dynamic equation $E_k$ = non-linear differential equation.  These different aspects can be present separately or together.  Now, it is well known [3] that above non-linearity makes unappliable the classical quantum framework, where physical entities are considered as hermitian operators on Hilbert spaces (=<u>Dirac's approach</u>).  For example, also for the more simple situation of scalar fields, the classical proceeding of construction of an Hilbert space by means of the representation of Fock via the identification of a vacuum state fails because this construction for curved space-times is not canonical but it is related to the adapted coordinate system chosen for a physical observer.  Further, the vacuum states have an infinite density of energy that can not be rescaled (renormalized) as one makes in the Minkowski space-time, since in non-gravitational physics the energy as such is not measurable. Now, scope of this chapter is to present a geometric point of view in order to describe charges of a continuum system general enough to include the quantum effects.  Note, that for the relativistic uncertainty principle the only entities that have physical meaning are the global ones (tested by a suitable classe of test functions).  In other words, we can characterize a microphysical system just by means of charges.  Parentetically, the cross-section of a scattering process is just a charge in this sense.

## 8. - QUANTUM CHARGES

Our first scope is to give a precise meaning to the proceeding of measuring
of physical entities defined for a classical continuum system which should
not be necessarily scalar.  This will conduce us to the concept of charge
associated to a physical entity.

Let us first recall some fundamental definitions and results about homology
and cohomology.

A <u>standard p-simplex</u>  $\Delta^p$ in $\mathbb{R}^p$ is given by  $\Delta^p \equiv \{ (a_1,\ldots,a_p) \in \mathbb{R}^p \mid \sum_{i=1}^{p} a_i \leq 1, a_i \geq 0 \}$.
For each $p \geq 0$, we define <u>boundary map</u>    $\partial_i^p : \Delta^p \to \Delta^{p+1}$ , $0 \leq i \leq p+1$, given by

$$p=0: \begin{cases} \partial_o^o(0)=1 \\ \\ \partial_1^o(0)=0 \end{cases} ; \quad p \geq 1 : \begin{cases} \partial_o^p(a_1,\ldots,a_p)=(1-\sum_{i=1}^{p} a_i, a_1,\ldots,a_p) \\ \\ \partial_i^p(a_1,\ldots,a_p)=(a_1,\ldots,a_{i-1},0,a_i,\ldots,a_p), (1 \leq i \leq p+1). \end{cases}$$

A <u>(differentiable) singular p-simplex</u>  u in a differentiable manifold X is
a map $u:\Delta^p \to$ X  which extends to be a differentiable ($C^\infty$) map of a neighborhood of  $\Delta^p$ in $\mathbb{R}^p$ into X.  If $p \geq 1$ , we define <u>itth-face</u>  , $0 \leq i \leq p$, of u
the (p-1)-singular simplex  $\partial_i u \equiv u \circ \partial_i^{p-1} : \Delta^{p-1} \to$ X.  We call <u>support</u> of u the
image $u(\Delta^p) \subset$ X.

A <u>p-chain</u>  A in X (with real coefficients) is a finite linear combination
$A = \sum_i \alpha_i u_i$  of p-simplices $u_i$ in X where the  $\alpha_i$ are real numbers.  The set
$C_p(X)$=free Abelian group generated by the singular p-simplices of X is called
the group of <u>p-chains</u>  of X.  Let  $\partial : C_p(X) \to C_{p-1}(X)$ be the homomorphism
defined by $\partial u = \sum_{i=0}^{p} (-1)^i \partial_i u : \Delta_{p-1} \to$ X; $\partial$  is called the <u>boundary</u>.  As $\partial \bullet \partial = 0$,
one has that $\{C_p, \partial\}$  is a chain complex. (For O-chains one has  $\partial \equiv 0$).  $C_p$
identifies a covariant functor from the category of differentiable manifolds
and differentiable mappings to the category of free Abelian groups.

Let  $\omega$ be a differentiable p-form defined on a  neighborhood of the image
of a p-chain in X.  We define <u>integral of $\omega$ over  A</u> by the following formula:
$\int_A \omega = \sum_i \alpha_i \int_{u_i} \omega$  if $A \equiv \sum_i \alpha_i u_i$  and  $\int_{u_i} \omega = \int_{\Delta^p} u_i^* \omega$ , the last integral
being the standard Riemannian integral.

Let $E^\bullet(X)$ be the real vector space of differential forms on X and $C^\bullet(X)$ the
dual space of $C_\bullet(X)$. So, one has a natural homomorphism  $\kappa^\bullet : E^\bullet(X) \to C^\bullet(X)$  given
by  $\omega \mapsto \kappa^\bullet(\omega)$ ,   $\kappa^\bullet(\omega)(\sigma) = \int_\sigma \omega$ .

The <u>singular homology</u> $H_\bullet(X)$ on $X$ is the homology of the chain complex $\{C_p(X), \partial\}$. One has the following isomorphisms $H^\bullet(X) \cong (H_\bullet(X))^*$, where $H^\bullet(X)$ is the usual (de Rham) cohomology of $X$. So, the map $\kappa^\bullet$ passes to the quotient.

Let us consider, now, the <u>n-th homotopy group</u> $\pi_n(X) \equiv [S^n; X] \equiv$ set of all homotopy classes of continuous functions $f: S^n \to X$; $\pi_n(X)$ is a group for any positive integer $n$ and is Abelian for $n > 1$. Then, one has a relation between $\pi_n(X)$ and $H_n(X)$ given by means of the following canonical homomorphism (<u>Hurewicz map</u>) $\rho: \pi_n(X) \to H_n(X)$, $\rho: [f: S^n \to X] \mapsto \rho[f] \equiv H_n(f)(s)$, where $s = 1$ is the generator of the infinite cyclic group $H_n(S^n) \cong \mathbb{Z}$.

Further, if $X$ is a $(n-1)$-connected space $(n \geq 1)$ (that is $\pi_p(X) = 0$, $0 \leq p \leq (n-1)$), then $\rho: \pi_n(X) \to H_n(X)$ is an isomorphism. (For example, $\pi_r(S^n) \cong H_r(S^n)$, $1 \leq r < n$, $\pi_n(S^n) = H_n(S^n) = \mathbb{Z}$ ).

The dual concept of homotopy is the <u>n-th cohomotopy set</u>: $\pi^n(X) \equiv [X; S^n]$ $\pi^n$ is a controvariant functor from the category of (pointed) topological spaces to the category of (pointed) sets. In particular, for any homotopy classe $[f]$ of map $f: X \to Y$, $\pi^n[f]: \pi^n(Y) \to \pi^n(X)$ is given by $\pi^n[f][g]: [g \bullet f] \in [X; S^m]$. (For example, if $X$ is a compact connected $n$-manifold one has a (non-canonical) isomorphism : (a) $\pi^n(X) \cong \mathbb{Z}$ if $X$ is orientable and $\partial X = \emptyset$; (b) $\pi^n(X) \cong \mathbb{Z}_2$ if $X$ is non-orientable and $\partial X = \emptyset$; (c) $\pi^n(X) = 0$ if $\partial X \neq \emptyset$).

Taking into account that one has the following canonical injective map $[X; Y] \to \mathrm{Hom}(H^n(Y), H^n(X))$, $\forall n \geq 0$, given by $[f] \mapsto [f]_* : [\omega] \mapsto [f_* \omega]$ , where $\omega$ is a closed $n$-form on $Y$, we get some relations between homotopy, cohomology, homotopy and cohomology given by means of the following canonical maps:

(a) $\iota_n: \pi_n(Y) = [S^n; Y] \to \mathrm{Hom}(H^n(Y), H^n(S^n)) \cong \mathrm{Hom}(H^n(Y), \mathbb{R}) \cong H^n(Y)^*$

$$[f] \mapsto [f]_* : [\omega] \mapsto [f^* \omega] \equiv [\lambda \mu] \equiv \lambda \in \mathbb{R}$$

where $\mu$ is the canonical volume form on $S^n$ (+).

---

(+) Recall that on $S^n \subseteq \mathbb{R}^{n+1}$ there exists a canonical volume form $\mu$. In fact, we can induce in a canonical way on $S^n$ a volume form by using the canonical volume form $\omega = dx^1 \wedge \ldots \wedge dx^{n+1}$ of the $(n+1)$-dimensional Euclidean space $\mathbb{R}^{n+1}$. Thus, may be made in the following way: $\mu: S^n \to \Lambda^\circ_n S^n$, $\mu = X \lrcorner \mu \,|\, S^n$, where $X$ is the vector field on $\mathbb{R}^{n+1}$ given by $X: x \mapsto X(x) = \psi_x^{-1}(x) \in T_x \mathbb{R}^{n+1}$, being $\psi_x$ the canonical isomorphism $\psi_x: T_x \mathbb{R}^{n+1} \cong \mathbb{R}^{n+1}$. So, we get $\psi(x^1, \ldots, x^{n+1}) = \sum_{k=1}^{n+1} (-1)^{k+1} x^k \, dx^1 \wedge \ldots \wedge \widehat{dx^k} \wedge \ldots \wedge dx^{n+1}$ .

(b) $j_n : \pi^n(X) \equiv [X; S^n] \to \mathrm{Hom}(H^n(S^n), H^n(X)) \cong \mathrm{Hom}(\mathbb{R}, H^n(X)) \cong H^n(X)$

$$[f] \mapsto [f^* \mu] \in H^n(x).$$

In particular, if X is a $(n-1)$-connected manifold, one has $\pi_n(X) \cong H_n(X)$ and we get the canonical map $\pi^n(X) \to \pi_n(X)^*$ given by $[f] \mapsto ([\gamma] \mapsto \int_{S^{n-1}} (f \circ \gamma)\mu)$, e.g, the following diagram

$$\begin{array}{ccc} \pi^n(X) & \longrightarrow & H^n(X) \\ \downarrow & & \| \wr \\ \pi_n(X) & \overset{\tilde{=}}{\longleftarrow} & H_n(X)^* \end{array}$$

is commutative (if $\dim H^n(X) =$ finite).

The relation between homotopy and fiber spaces is given by the following exact sequence (homotopy sequence of the fiber space $p: X \to B$).

$$\cdots \to \pi_n(F) \to \pi_n(X) \to \pi_n(B) \to \pi_{n-1}(F) \to \pi_{n-1}(X) \to \cdots \to \pi_1(B) \to \pi_0(F)$$
$$\to \pi_0(X) \to \pi_0(B).$$

If the fibration $p: X \to B$ has a cross-section $\lambda: B \to X$, then the homotopy sequence breaks up into a family of splittable short exact sequences

$$(\cdot \cdot) \qquad 0 \to \pi_q(F) \to \pi_q(X) \to \pi_q(B) \to 0 \qquad , q \geq 2 .$$

So, we have the isomorphism: $\pi_q(X) \cong \pi_q(F) \oplus \pi_q(B)$. $\pi_1(X)$ is the semidirect product of $\pi_1(F)$ by $\pi_1(B)$.

If $p: \check{B} \to B$ is a covering map, then $\pi_p(p): \pi_p(\check{B}) \to \pi_p(B)$ is an isomorphism. If F is contractible in X one has that the homotopy sequence $(\cdot \cdot)$ breaks up into a family of splittable short sequences: $0 \to \pi_n(X) \to \pi_n(B) \to \pi_{n-1}(F) \to 0$. So, we have the isomorphism $\pi_n(B) \cong \pi_n(X) \oplus \pi_{n-1}(F)$, $n \geq 2$.

Now, we are able to consider the problem of measuring of physical entities.


Let $K: E_k \subset J\mathfrak{d}^k(C) \to \mathbb{K}$ be a physical entity on a dynamic equation $E_k$ of a continuum system with the configuaration bundle $\pi_C : C \to M$. Let $\Lambda_\bullet \mathbb{K}$, $\Lambda_\bullet J\mathfrak{d}^k(C)$ be the graded bundles of exterior forms on $\mathbb{K}$ and $J\mathfrak{d}^k(C)$ respectively. Then, for any section $u: M \to E_k$ K induces, for functorial property , a vector fiber bundle morphism over K: $\Lambda_\bullet(K.u) : \Lambda_\bullet \mathbb{K} \to \Lambda_\bullet M$. So, by passing to dual bundles and restricting to the sub-space of chains $C_\bullet(M)$ of the space $C^\infty(\Lambda^\bullet M))$ via the canonical monomorphism $i: C_\bullet(M) \to C^\infty(\Lambda_\bullet M)^* \cong C^\infty(\Lambda^\bullet M)$ , we get the linear mapping: $Q(K.u): C_\bullet(M) \to C^\infty(\Lambda_\bullet \mathbb{K})^*$ given by $Q(K.u): A \mapsto Q(K.u)(A)$, $Q(K.u)(A): \beta \mapsto \int_A (K.u)^* \beta$ .

__Definition__ .1    The <u>charge map</u> of $K: E_k \subset J\mathcal{D}^k(C) \to \mathbb{K}$ at the section $u:M \to E_k$ is the map $Q(K.u): C_\bullet(M) \times C^\infty(\Lambda_\bullet \mathbb{K}) \to \mathbb{R}$, given by $Q(K.u): (A, \beta) \mapsto \int_A (K.u)^* \beta$ .

__Proposition__ .1    Let $\{x^\alpha\}$ be a coordinate system on M and $\{\xi^\alpha\} \equiv \{x^\alpha, z^j\}$ a fibered coordinate system on $\mathbb{K}$ .   Then, the evaluation of K.u on a differential p-form $\omega = \omega_{i_1 \ldots i_p} d\xi^{i_1} \wedge \ldots \wedge d\xi^{i_p}$ is made by means of the following p-form on M :

$$(19) \qquad (K.u)^* \omega = A^{i_1 \ldots i_p}_{j_1 \ldots j_p} \, \omega_{i_1 \ldots i_p} \cdot (K.u) \, dx^{j_1} \wedge \ldots \wedge dx^{j_p}$$

where $A^{i_1 \ldots i_p}_{j_1 \ldots j_p}$ is the minor of the jacobian of the map $K.u: M \to \mathbb{K}$ obtained by taking the $i_1 \ldots i_p$ rows and the $j_1 \ldots j_p$ columns.

__Remark.__  Note, that in general, we have not a canonical way to choice the differential forms $\omega$ on $\mathbb{K}$.  However, we can recognize a class of physical entities for which a distinguished class of measures can be finded.  In fact, if $\mathbb{K} \equiv \Lambda_\bullet J\mathcal{D}^k(C)$ (or $\mathbb{K} \equiv \Lambda^o_p J\mathcal{D}^k(C)$, $p>0$) and K is a section of the fiber bundle $\Lambda_\bullet J\mathcal{D}^k(C) \to J\mathcal{D}^k(C)$, one has differential forms $\omega$ on $\mathbb{K}$ such that $K^* \omega = K$. So, we can consider equivalent two differential forms $\omega, \omega'$ on $\mathbb{K}$ if $K^* \omega = K = K^* \omega'$ . This justifies the following

__Definition__ .2   A <u>physical observable</u> is a physical entity $K: J\mathcal{D}^k(C) \to \mathbb{K}$ with $\mathbb{K} = \Lambda_\bullet J\mathcal{D}^k(C)$ (or $\mathbb{K} \equiv \Lambda^o_p J\mathcal{D}^k(C)$, for some $p \in \mathbb{N}$) and $\pi'_K \bullet K = id_{J\mathcal{D}^k(C)}$, where $\pi'_K$ is the projection $\Lambda_\bullet J\mathcal{D}^k(C) \to J\mathcal{D}^k(C)$.

__Theorem__ .1  (Noether's theorem)   For any physical entity $K.u: M \to \mathbb{K}$, at the section $u: M \to E_k$, evaluated on a p-form $\beta: M \to \Lambda^o_p \mathbb{K}$ on $\mathbb{K}$ one has that if A and A' are two p-chains that cobound in M, that is $A = A' + \partial B$, where B is a (p+1)-chain, then if $(K.u)^* \beta \in H^\bullet(M)$, the charge $Q(K.u) = \int_A (K.u)^* \beta$ is conserved: $\int_A (K.u)^* \beta = \int_{A'} (K.u)^* \beta$.  In particular, if $u = D^k c$, for a section $c: M \to C$ we say that the dynamic configuration c <u>leads</u> the conserved entity $(K.u)^* \beta$.

__Remark.__  Closed differential forms $\gamma \equiv (D^k c)^* \beta$ with $\beta: E_k \subset J\mathcal{D}^k(C) \to \Lambda^o_{n-1} J\mathcal{D}^k(C)$ and c a solution of $E_k$, can be built directly by the symmetry properties of the dynamic equations; these are the so called dynamic conservation laws that we have extensively considered in the previous chapter.

The concept of conserved charges may be related to an observer. In fact, even if the differential p-form $\omega: M \to \Lambda^o_p M$ entering in the charge $Q = \int_A \omega$, where A is any p-chain on M, is not closed, can exist physic observers for which Q is conserved.

Before to enter in the details on this point, let us first consider some fundamental results on the meaning of frame.

Let M be a 4-dimensional, global hyperbolic pseudo-Riemannian manifold with signature (+---) (space-time) space and time orientable. The global hyperbolicity assures the existence of Cauchy hypersurfaces and time functions on M. More precisely, we have the following:

<u>Proposition</u> .2  The following propositions are equivalent:

(a)  There exists a (global) Cauchy surface in (M,g).

(b)  There exists a $C^{\infty}$ numerical function   $\tau:M \to \mathbb{R}$ (<u>time-function</u>) such that :(i)  $d\tau$   is every time-like; (ii) $\tau$   increases along every future directed non-space-like curve; (iii) along any inextendible non-space-like curve, $\tau$   takes all values in   $]-\infty,+\infty[$   .

In order to obtain a splitting of M into space and time we shall also introduce a frame.

<u>Definition</u> .3  A <u>frame</u> into (M,g) is a 1-parameter group of transformations of M, $\psi: \mathbb{R} \times M \to M$ such that its velocity  $\dot{\psi}=\partial\psi$  is a time-like vector field on M.

<u>Theorem</u> .2  The choice of a time function $\tau$ and of a frame $\psi$  allows us the identification of M with a product $M=\mathbb{R} \times M_{\psi}$ , where $M_{\psi}$ is a 3-dimensional manifold.

<u>Proof.</u>  Let $M_{\psi}$ be the quotient manifold $M_{\psi}=M/\sim$ , where the equivalence relation $\sim$ is : $p \sim p'$ $\Leftrightarrow p'=\psi(t,p)$ for some $t \in \mathbb{R}$ . Then, $M_{\psi}$ has a natural structure of 3-dimensional manifold. Let us denote by  $[p]_{\psi}$  a point of $M_{\psi}$ identified by means of the point  $p \in M$.  Let us also denote by $\psi/$ the natural projection $M \to M_{\psi}$ . Then, the identification $M \equiv \mathbb{R} \times M_{\psi}$ is given by means of the following one-to-one map: $j_{\psi}:M \to \mathbb{R} \times M_{\psi}$ , $p \mapsto (\tau(p),[p]_{\psi} \equiv \psi/(p))$, and its inverse: $j_{\psi}^{-1} : \mathbb{R} \times M_{\psi} \to M$ , $(t,[p]_{\psi}) \mapsto \psi(t,p)$. $\square$

<u>Definition</u> .4  We call <u>physical frame</u> of (M,g) a couple $\Psi=(\tau,\psi)$ where $\tau$ is a time function of M and $\psi$ is a frame of (M,g) such that $\langle d\tau,\dot{\psi}\rangle=1$. $\tau$ is called the <u>proper time</u> of the physical frame  .

<u>Theorem</u> .3  To any physical frame  $\Psi \equiv (\tau,\psi)$ of (M,g) there corresponds a splitting of M in space  $M_{\psi}$  and time $\mathbb{R}$.

Associated to any physical frame we recognize adapted coordinate systems.

<u>Definition</u> .5  Let  $\Psi=(\tau,\psi)$ be a physical frame of (M,g). A coordinate system

on M __adapted__ to $(\tau,\psi)$ is a coordinate system $\{x^\alpha\}=\{x^0,x^1,x^2,x^3\}$ on M such that $\partial x_0 = \dot\psi$ and $dx^0 = d\tau$ . Then, we say that $t=x^0$ is the __time coordinate__ and $(x^1,x^2,x^3)$ are the __space-coordinates__ of $\{x^\alpha\}$ .

__Definition  .6__   A charge $Q=\int_A \omega$ , $\omega \in C^\infty(\Lambda_p^0 M)$ , A=p-chain on M, is preserved for a frame $\psi$ if $\int_A \omega = \int_{\psi_t(A)} \omega$ , $\forall t \in \mathbb{R}$,

__Proposition  .3__   The charge $Q=\int_A \omega$ is preserved for a frame $\psi$ iff $\mathcal{L}_{\dot\psi}\omega = 0$ or equivalently $\psi_t^* \omega - \omega = 0$ , $\forall t \in \mathbb{R}$, where $\dot\psi \equiv \partial\psi$ .

__Proof.__   Consider the differential p-chain written as $A = \Sigma \lambda_i \sigma^i$ , $\lambda_i \in \mathbb{R}$, $\sigma^i \equiv \{ \Delta_p \rightarrow M \}$ a p-simplex. The corresponding deformated p-chain $\psi_t(A)$ is $\psi_t(A) = \Sigma \lambda_i \psi_t(\sigma^i)$, where $\psi_t(\sigma^i) \equiv \{ \psi_t \circ \sigma^i : \Delta_p \rightarrow M \}$ .

Now, assume $\mathcal{L}_{\dot\psi}\omega = 0$.   Then, we get

$$Q = \int_A \omega = \Sigma_i \lambda_i \int_{\sigma^i} \omega = \Sigma_i \lambda_i \int_{\Delta_p} \sigma^{i*}\omega = \Sigma \lambda^i \int_{\Delta_p} \sigma^{i*}(\psi_t^*\omega) = \Sigma\lambda_i \int_{\Delta_p}(\psi_t \circ \sigma^i)^*\omega = \int_{\psi_t(A)} \omega .$$

Conversely, assume $\int_A \omega = \int_{\psi_t(A)} \omega$ to hold, then we get

$$\Sigma_i \lambda_i \int_{\Delta_p} \sigma^{i*}\omega = \Sigma \lambda_i \int_{\Delta_p} \sigma^{i*}(\psi_t^*\omega) , \text{ hence } \sigma^{i*}(\omega - \psi_t^*\omega) = 0, \forall \sigma^i \Rightarrow \omega - \psi_t^*\omega = 0$$

$\forall t \in \mathbb{R}$.   $\square$

__Proposition  .4__   Let A be a p-chain ($p \leq n = \dim M$) and $\omega$ is a p-differential form on $J\mathfrak{D}^k(C)$. Assume that $\omega$ is preserved for a frame $\psi$ . If the dynamic configuration $c:M \rightarrow C$ is k-invariant for the frame $\psi$ , that is the following diagram

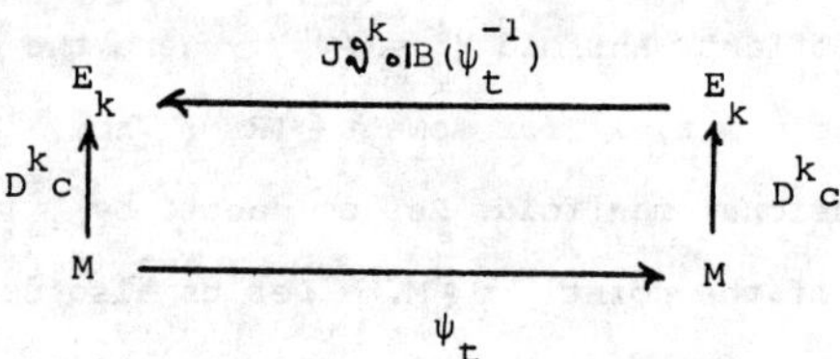


is commutative, (where $B$ is the covariant functor which characterizes the fiber bundle of geometric objects $C \rightarrow M$), or equivalently $J\mathfrak{D}^k \circ B(\psi_t)$ is a transformation of $D^k c(M) \subset E_k$, then we have also that $Q = \int_A (D^k c)^* \omega = \int_{\psi_t(A)} (D^k c)^* \omega$ , $\forall t \in \mathbb{R}$.   So, we can say that the __configuration c leads the charge Q that is preserved with respect to the frame.__

__Proof.__   In fact under above hypotheses we get the following implication:

$$\phi_t^* \omega - \omega = 0 \implies \psi_t^{-1*}\left[(D^k c)^*\omega\right] - (D^k c)^*\omega = 0 ,$$

being $\phi_t \equiv (J\mathfrak{D}^k \circ B)(\psi_t)$ .   $\square$

__Proposition  .5__   If the charge $Q = \int_A (D^k c)^* \omega$ has the p-differential form $\gamma \equiv (D^k c)^* \omega$ closed, then it is conserved for any frame $\psi$ such that the induced transformation $H_p(\psi_t) : H_p(M) \rightarrow H_p(M)$ preserves the homological classe

$[A]$ of A for any $t \in \mathbb{R}$.

Proof. In fact, as $A - \psi_t(A) = \partial B$, where B is the (p+1)-chain generated by the flow $\psi_t$ between A and $\psi_t(A)$ one has that $\int_A \gamma - \int_{\psi_t(A)} \gamma = \partial \int_B \gamma = \int_B d\gamma = 0.$ $\square$

In the following we shall assume that M is a n-dimensional space-time and all physical entities will be assumed differential forms on the state space $J\mathfrak{H}^k(C)$ of a continuum system.

Theorem .4 ( Hodge structure of charges)   Any configuration $c:M \to C$ of a continuum system which leads a charge $Q = \int_A (D^k c)^* \omega$ , where $A \in C_p(M)$ and $\omega \in C^\infty(\Lambda^o_p J\mathfrak{H}^k(C))$, can be decomposed into three parts $Q = Q_\partial + Q_u + Q_c$ , where $Q_\partial$ is the charge on the boundary $\partial A$, $Q_u$ is the unconserved term and $Q_c$ is the conserved one.

Proof.   Let $E^p(M)$ be the real vector space of differential p-forms on M. One has the following direct sum decomposition (Hodge theorem $[6]$):
$E^p(M) = dE^{p-1}(M) \oplus \delta E^{p+1}(M) \oplus \Delta^p(M)$, where $\delta$ is the symbol of codifferential
$\delta \equiv (-1)^p *^{-1} d * : E^p(M) \to E^{p-1}(M)$   and   $\Delta^p(M)$   is the space of harmonic forms:
$\Delta^p(M) \equiv \ker \Delta, \quad \Delta \equiv d \cdot \delta + \delta \cdot d : E^p(M) \to E^p(M)$   . Therefore, we have:

$$Q = Q_\partial + Q_u + Q_c \equiv \underbrace{\int_A d(D^k c)\alpha}_{\int_{\partial A}(D^k c)\,\alpha} + \int_A (D^k c)\delta\beta + \int_A (D^k c)\gamma$$

with $\alpha \in E^{p-1}(M)$, $\beta \in E^{p+1}(M)$ $\gamma \in \Delta^p(M)$. Note, that $\gamma \in (\ker \delta) \cap (\ker d)$.
So, $Q_c$ is a conserved charge. Further, if $A = A' + \partial B$, we get

$$\int_{\partial A}(D^k c)^* \alpha = \int_{\partial A'}(D^k c)^* \alpha + \underbrace{\int_{\partial \partial B}(D^k c)^* \alpha}_{O}$$

So, $Q$   is the charge part of Q which gives the charge on the boundary $\partial A$. $\square$

Examples.   Electric charges and angular momentum in general relativity, (see e.g. ref. $[1:D]$).

Let us, now, define the concept of quantum numbers.

Definition .7   A configuration $c:M \to C$ leads a quantum number $n \in \mathbb{Z}$ if there is a charge $Q = \int_A (D^k c)^* \beta$ , $A \in C_*(M)$, $\beta \in E^*(J\mathfrak{H}^k(C))$, such that $Q = \kappa \cdot n$, $k \in \mathbb{R}$. We say also that Q is quantized .

In order to recognize configurations with quantum numbers , let us first recall some fundamental definitions and results on the degree of a map.

Let X, Y be connected and oriented manifolds of dimension n and $f:X \to Y$ a <u>proper</u> map (that is for any compact $U \subset Y$ one has that $f^{-1}(U) \subset X$ is also compact), then, there exists an integer, $\deg(f) \in \mathbb{Z}$, such that $\int_X f^* \omega = \deg(f) \int_Y \omega$ , for any differentiable n-form $\omega$ with compact support on Y. $\deg(f)$ is called the <u>Brouwer degree</u> of f. In particular, if f is a diffeomorphism one has $\deg(f)=1$. Further, if X and Y are compact n-manifolds, since $H^n(X) \cong \mathbb{R}$, $H^n(Y) \cong \mathbb{R}$, via the integration map, we get the following linear map $\bar{f} : \mathbb{R} \to \mathbb{R}$ given by means of the following commutative diagram:

$$H^n(X) \xleftarrow{\quad f_* \quad} H^n(Y)$$
$$\cong \Big\downarrow \qquad \bar{f}_* \qquad \Big\downarrow \cong$$
$$\mathbb{R} \xleftarrow{\qquad\qquad} \mathbb{R}$$

and one has $\deg(f)=\bar{f}_*(1)$.

Let X, Y be compact n-manifolds without boundary with Y connected:

(a) Homotopic maps $X \to Y$ have the same degree if X, Y are oriented (and the same mod 2 degree otherwise). So, for the oriented case we have a map:

$$\underline{\deg}:[X;Y] \to \mathbb{Z} \quad , \quad \underline{\deg}[f] = \deg(f) \in \mathbb{Z} \; ;$$

otherwise

$$\underline{\deg}:[X;Y] \to \mathbb{Z}_2, \quad \underline{\deg}[f] = \deg_2(f) \in \mathbb{Z}_2.$$

(b) Let $X = \partial W$ , W=compact. Suppose a map $f:X \to Y$ extends to W. Then, $\deg(f)=0$ if W and Y are orientable, and $\deg_2(f)=0$ otherwise.

Let $X \xrightarrow{g} Y \xrightarrow{f} Z$ be a short sequence of compact connected oriented n-manifolds without boundaries and continuous maps. Then, $\deg(f \circ g)=\deg(g).\deg(f)$. The same holds mod 2 if X,Y,Z are not orientable.

Now, we are able to give the following

<u>Theorem</u> .5  Let $\omega \in H^p(J\Lambda^k(C))$, a=differentiable singular p-simplex on M. Let c be a configuration which leads the charge $Q=\int_a (D^k c)^* \omega$ . Then if $\gamma \equiv (D^k c)^* \omega = k.\theta$ , where $\theta$ belongs to the image of the canonical map $j_p \pi^p(M) \to H^p(M)$, p=degree of $\omega$ and $k \in \mathbb{R}$, the configuration c leads a quantum number.

<u>Proof.</u>  Recall, that any p-form $\alpha$ on $S^p$ can be written as $\alpha = b\mu + d\beta$ where $b \in \mathbb{R}$, $\mu$ is the canonical volume form on $S^p$ and $\beta$ is a (p-1)-form on $S^p$. In fact, $H^p(S^p) \cong \mathbb{R}$. Then, we get

$$\int_a (D^k c)^* \omega = k.\int_a \theta = k\int_\Delta f^* \mu \quad = k \deg(f) \int_{S^p} \mu \quad ,$$

where $\Delta \equiv$ support of a and $\bar{f}=f|\Delta$, being $f:M \longrightarrow S^p$ the representation of the cohomology class $[f] \in \pi^p(M)$. So, we get:

$$\int_a (D^k c)^* \omega = \bar{k}. n, \text{ where } \bar{k} \equiv k.\mathrm{vol}(S^p) \text{ and } n=\deg(\bar{f}).$$
$\square$

<u>Corollary</u> .1   If the configuration $c:M \to C$ leads the charge $Q=\int_a (D^k c)^* \alpha$, $\alpha \in E^p(J\mathfrak{J}^k(C))$, and M is (p-1)-connected and $\pi^p(M) \cong H^p(M)$, one has that Q is quantized if a is a differentiable singular p-simplex of M and so c leads a quantum number.

Let us now relate the concept of quantum number to that of "vacuum configuration".

<u>Definition</u> .8   1. We call <u>vacuum configuration</u> of a gauge continuum system $G(M)=(\beta, E_k \subset J\mathfrak{J}^k(C))$ (see ref. [12] ) a dynamic configuration $c_o \equiv (c_o, \sigma_o):M \to C \equiv \underset{\smile}{C} \times C(P)$, such that $\overset{\sigma_o}{\nabla} c_o=0$, ($\nabla^{\sigma_o}$ denotes the absolute differential performed with respect to the connection $\sigma_o$).

2.   Let $CG_{au}$ be the dynamic gauge group of $G(M)$. Recall, [12] that $CG_{au}$ is the set of fiber bundle transformations $f_M, f_C, f_{J\mathfrak{J}^k(C)}$ on M, C and $J\mathfrak{J}^k(C)$ respectively with $f_M=id_M$, $f_C= \mathbb{B}(f_P,id_M)$, $f_{J\mathfrak{J}^k(C)}=J\mathfrak{J}^k(f_C,id_M)$, being $(f_P,id_M) \in G_{au}(P)$=gauge group of P=principal bundle which defines the gauge structure of $G(M)$, $\mathbb{B}$ is the covariant functor of the super-bundle structure $\beta$ of $G(M)$. We call <u>vacuum manifold</u> of $G(M)$: $VM=CG_{au}.c_o$. VM can be considered a homogeneous space of $CG_{au}$, that is $VM=CG_{au}/CG^o_{au}$, where $CG^o_{au}$ =isotropy group of $c_o$.

<u>Proposition</u> .6   The points of VM are solutions c of $E_k$ such that $\nabla c=0$ too.

<u>Proof</u>.   In fact, $c \equiv f_C \bullet c_o$, for some $(f_C,id_M) \in CG_{au}$. So, we get
$$\overset{\sigma}{\nabla} c= \Gamma \bullet Dc=\Gamma \bullet J\mathfrak{J}(f_C,id_M) \bullet Dc_o =\psi \bullet \Gamma_o \bullet Dc_o =\psi \bullet \overset{\sigma_o}{\nabla} c_o =\psi (0)=0,$$
where $\psi \equiv id_{T^*M} \boxtimes vT(f_C)$ and $\Gamma_o, \Gamma$ are the connection maps $J\mathfrak{J}(C) \to T^*M \boxtimes vTC$ induced from the connection fields $\sigma_o$ and $\sigma$ respectively, e.g. the following diagram

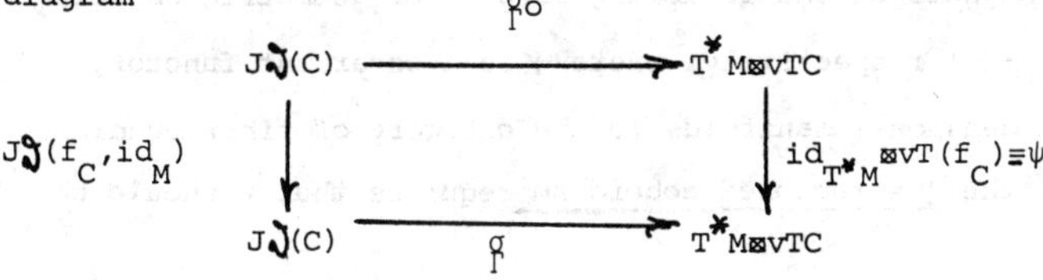


is commutative.
$\square$

Definition .9    A gauge dynamic symmetry associated to a continuum system is <u>spontaneously broken</u> if there is a vacuum manifold VM obtained from a given vacuum configuration $c_o$. In particular, we can distinguish the following cases:

(a) $CG_{au} = CG_{au}^o$ : the dynamic gauge symmetry is <u>exact</u>, the vacuum $c_o$ is unique.

(b) $e \; CG_{au}^o \; CG_{au}$ : the dynamic gauge symmetry is <u>partly spontaneously broken</u>.

(c) $CG_{au} = \{e\}$ : the dynamic gauge symmetry is <u>completely broken</u>.

<u>Remark</u>. Conserved charges associated to solutions $c_o$ belonging to VM are called <u>topological conserved charges of vacuum</u>. In particular, if c is a dynamic solution of $E_k$ such that on the boundary of a (n-1)-chain A coincides with a vacuum solution $c_o \in$ VM: $D^k c|\; \partial A = D^k c_o|\; \partial A$ , then, we get that any conserved charge $Q = \int_A d(D^k c)^* \beta$ , whare $\beta$ is a (n-2)-form on $J^k(C)$ is a topological conserved charge of vacuum. In fact, one has:

$$Q = \int_A d(D^k c)^* \beta \;=\; \int_{\partial A} (D^k c)^* \beta \;=\; \int_{\partial A} (D^k c_o)^* \beta \;=\text{topological conserved charge of vacuum.}$$

Further, if $\int_{\partial A} (D^k c_o)^* \beta$ is quantized, the configuration c leads a conserved quantum number of the vacuum.


# 9. - QUANTUM SITUS

In this section we shall utilize the machinery on the geometric structure of PDE in order to recognize suitable structures that should be able to translate the physical proceeding of quantization.

Introduce, first, some fundamental concepts on the cobordism. (See also refs. [6]).

Definition .1    1. Two (closed) n-dimensional manifolds M, M' are <u>cobordant</u> if there is a (compact) smooth (n+1)-dimensional manifold W with boundary  W (+) such that    $\partial W = M \cup M'$ , the disjoint union of M and M'.

2. If the manifolds are <u>K-structured</u>  : $(M, \kappa)$, $(M', \kappa')$, where $\kappa$  and  $\kappa'$ are fields of geometric objects of the following bundles of geometric objects $(\mathbb{K}, M, \pi_{\mathbb{K}}; K)$ and $(\mathbb{K}', M', \pi_{\mathbb{K}'}; K)$ respectively, where $K$ is a covariant functor, from the category of n-dimensional manifolds to the category of fiber bundles ( see refs. $\begin{bmatrix}11,12\end{bmatrix}$)., the <u>K-structured cobordism</u> requires that W should be

---

(+) The boundary  $\partial X$ of a (topological) manifold X of dimension n, can be characterized as    $\partial X \equiv \{ \; x \in X \;\mid\; H_n(X, X - \{x\}; \mathbb{Z}) = 0 \; \}$ .

structured $(W,\theta)$ , where $\theta$ is a field of geometric objects of a bundle of geometric objects $(\theta \equiv K(W), W, \pi_\theta ; K)$ and $(\partial W, \partial \theta) = (M,\kappa) \dot{\cup} (M',\kappa')$, where $\partial \theta = \theta | \partial W$

3. If $M_n$ is the set of smooth n-dimensional manifolds the $K$-structured cobordism introduces in $M_n$ an equivalence relation; the corresponding set $\overset{n}{\Omega}_K$ of equivalence classes becomes an Abelian group under the operation of disjoint union.

Now, let us be given a continuum system characterized by a differential equation $E_k \subset JD^k(C)$ over a configuration bundle $\pi_C : C \to M$ with basis manifold M of dimension n. Recall $[10,12]$ that a (local) <u>admissible state</u> is a (local) section s of the fiber bundle $\pi_k : JD^k(C) \to M$ such that there exists a (local) configuration c such that $D^k c = s$. A <u>dynamic state</u> is an admissible state s such that $s: M \to E_k$ . All the submanifolds N (<u>Vinogradov submanifolds</u>) of $JD^k(C)$ such that $N = s(M)$ (or $N = s(U)$, $U \subset M$) for some admissible state s are integral manifolds of a distribution $\mathbb{E}$ on $JD^k(C)$. We call $\mathbb{E}$ the <u>k-th Cartan distribution</u> of the continuum system. The distribution $\mathbb{E}$ is characterized as the sub-vector bundle of $TJD^k(C) \to JD^k(C)$ which is the annihilator of the <u>contact module</u> $\Omega^k(C)$ defined by:

$$\Omega^k(C) \equiv \begin{cases} \text{submodule of the module of one forms (over the zero forms) on } JD^k(C) \\ \text{locally spanned by the one forms } \theta^j_{i_1 \cdots i_\beta} \equiv dy^j_{i_1 \cdots i_\beta} - y^j_{i_1 \cdots i_\beta i} dx^i, \\ 0 \leq \beta \leq k-1, \ 1 \leq j \leq m, \ 1 \leq i_1, \ldots i_\beta, i \leq n \end{cases}$$

where $\{x^i, y^j_{i_1 \cdots i_\beta}\}$ is a fibered coordinate system on $JD^k(C)$. Any admissible state is characterized by the equations $s^* \theta^j_{i_1 \cdots i_\beta} = 0$.

Any symmetry of the Cartan distribution $\mathbb{E}$ transforms any Vinogradov submanifold of $JD^k(C)$ into another one. These symmetries are also characterized as <u>contact transformations</u> that are local diffeomrphisms $f: JD^k(C) \to JD^k(C)$, which preserve the contact module $f^* \Omega^k(C) \subset \Omega^k(C)$. Any contact transformation $f: JD^k(C) \to JD^k(C)$ is obtained as the k-th prolongation of a some (local) fiber bundle transformation $f_C : C \to C$, $f = JD^k(f_C)$. A <u>Bäcklund transformation</u> is a contact transformation f restricted to a differential equation $E_k \subset JD^k(C)$. This dermines another differential equation $f(E_k) \subset JD^k(C)$ that is <u>Bäcklund-related</u> to $E_k$.

Further, $N \subset JD^k(C)$ is the image of an admissible state $s = D^k c$ with c a solution of the differential equation $E_k \subset JD^k(C)$ iff N is an integral manifold of the distribution (<u>dynamic distribution</u>) $\mathbb{E}(E_k) \equiv \mathbb{E} \cap TE_k$. Then, $N = D^k c(M)$ is called a **dynamic submanifold** of $E_k$.

**Definition .1** Let $E_k \subset \mathcal{JD}^k(W)$ be a differential equation on the fiber bundle $\pi: W \to M$. The quantization of $E_k$ is the couple $Q(E_k) \equiv (S_p^k(W), C(E_k))$, where:

(1) $S_p^k(W)$ is the **non-linear Spencer equation** on $W$ given by the kernel of of the non-linear Spencer operator on $W$: $S_p^k(W) = \ker(D) \subset \mathcal{JD}(\mathcal{JD}^k(W))$. So, the following sequence is exact: $0 \to S_p^k(W) \to \mathcal{JD}(\mathcal{JD}^k(W)) \xrightarrow{\;D\;} T^* M \otimes \mathcal{JD}^{k-1}(vTW)$;

(2) $C(E_k)$ is the set of Cauchy data hypersurfaces for $S_p^k(W)$ belonging to $E_k$.

**Remark.** Let us, now, shortly consider the formal properties of $S_p^k(W)$. Note that $S_p^k(W)$ is not a formally integrable differential equation. Furthermore, one can see that $S_p^k(W)$ can be identified with the $(k+1)$-order sesquiholonomic prolongation of $W$: $S_p^k(W) \cong \overset{v}{\mathcal{JD}}{}^{k+1}(W)$. Moreover, there exists an exact sequence: $0 \to \mathcal{JD}^{k+1}(W) \to \overset{v}{\mathcal{JD}}{}^{k+1}(W) \to \delta(T^* M \otimes \overset{o}{S}_k M \otimes vTW)$ of affine fiber bundles on $\mathcal{JD}^k(W)$ (via pull-back). So, we have the isomorphism $S_p^k(W) \cong \mathcal{JD}^{k+1}(W) \times_M \delta(T^* M \otimes \overset{o}{S}_k M \otimes vTW)$. The curvature of $S_p^k(W)$ is determined by the following exact sequence: $(S_p^k(W))_{+1} \to S_p^k(W) \to \delta(T^* M \otimes \overset{o}{S}_k M \otimes vTW) \to 0$. So, the PDE $(S_p^k(W))_{+1}^{(1)} \overset{=}{\equiv} \pi_{2,1}((S_p^k(W))_{+1}) = \mathcal{JD}^{k+1}(W) \subset S_p^k(W) \subset \mathcal{JD}(\mathcal{JD}^k(W))$ is formally integrable and has the same solutions of $S_p^k(W)$.

**Definition .2** We call **quantum situs** of $E_k$ the set $\Omega(E_k)$ of **Vinogradov submanifolds** of $\mathcal{JD}^k(W)$ that cobord two Cauchy hypersurfaces data of $C(E_k)$. If $\Omega(E_k) \neq D^k(S(E_k))(M) = \Omega_c(E_k) =$ set of Vinogradov submanifolds of $E_k$, $E_k$ is said to be **quantizable**. $\Omega_c(E_k)$ is called the **classic limit** of $\Omega(E_k)$.

**Definition .3** The elements of a quantum situs are called **quantum cobords**. If $\sigma \in \Omega(E_k)$ belongs also to $\Omega_c(E_k)$ $\sigma$ is called **dynamic cobord** for $E_k$.

**Definition .4** A **scattering process** is the assignment of two Cauchy data hypersurfaces $(c_{in}, c_{out}) \in C(E_k) \times C(E_k)$. The scattering is **classic** if there exist only dynamic cobords cobording $c_{in}$ and $c_{out}$. A **quantum scattering** is a scattering $c_{in} \to c_{out}$ such that there are quantum cobords cobording $c_{in}$ and $c_{out}$.

**Defintion .5** For a quantum cobord $W$, $\partial W = D^k c'(A') \cup D^k c''(A'')$, we say that it has a **bordism-deformation** if there is another quantum cobord $W'$ such that $\partial W' = D^k c'(A') \cup D^k c''(A'')$. The set of bordism-deformations of a quantum cobord $W$ is called **statistical set** of a scattering process $(A',c')$ $(A'',c'')$ and we denote it by $\Omega(A',c';A'',c'')$ or simply by $\Omega$ if confusion does not arise.

365

if a statistical set $\Omega$ of $E_k \subset J\mathcal{D}^k(C)$ contains a dynamic Vinogradov sub-manifold $D^k c(M) \subset E_k$, we say that $\Omega$ is a __quantum fluctuation__ of $D^k c(M)$.

__Proposition .1__    1.  The statistical set   $\Omega$ of a scattering process for $E_k$ is identified with the set of solutions of $S_p^k(W)$ with boundary conditions $(c_{in}, c_{out})$.

2.  The quantum situs  $\Omega(E_k)$ of $E_k$ is identified with the set of solutions of $S_p^k(W)$ with any boundary condition $(c_{in}, c_{out}) \in C(E_k) \times C(E_k)$.


__Remark.__    As the relativistic uncertainty principle prohibits to represent quantum physical observables as punctual fields, the only right description results into a global one by means of a charge associated to physical enti-ties.  So, for any classical entity  $K: J\mathcal{D}^k(C) \to \mathbb{K}$ (=differential operator on the configuration bundle  $\pi: C \to M$, being  $\pi_{\mathbb{K}}: \mathbb{K} \to M$ a fiber bundle over M), we can associate a map ,(__charge map__), $Q: C^\infty(C) \times C_\bullet(M) \to C^\infty(\Lambda \mathbb{K})$, where $C_\bullet(M)$ is the space of chains on M and $C^\infty(\Lambda\mathbb{K})$ is the space of differential forms on K, given by  $Q(c, A): \beta \mapsto \int_A (K \bullet D^k c)^* \beta$ .    Really, a charge $Q \equiv \int_A (K \bullet D^k c)^* \beta$ identifies a numerical function  $f_Q: \Omega \to \mathbb{R}$ on the statistical set  $\Omega$ of a scattering process $c_{in} \to c_{out}$, given by  $Q: \sigma \equiv D^k c(M) \mapsto \int_A (K \bullet D^k c)^* \beta$ . So, on quantum level a proceeding of measure of a physical entity it is identified by means of a numerical function on  $\Omega$ .    Therefore, if  $\mu$ is a measure on  $\Omega$  , we can obtain the __quantum mean value__ of a physical entity K, by considering the Feynman-like integral $\langle\!\langle f_Q \rangle\!\rangle = \int_\Omega f_Q \, d\mu / \int_\Omega d\mu$ , of the numerical function $f_Q$ associated to K.


__Definition  .6__    A physical entity  $B: J\mathcal{D}^k(C) \to \Lambda^o_{n-1} J\mathcal{D}^k(C)$   is a __quantum dynamic conserved entity__ in a scattering process  for  a  differential equation $E_k \subset J\mathcal{D}^k(C)$ if $d(s^* B) = 0$, $\forall s \in \Omega$ except for some s belonging to a subset   $\Omega_o \subset \Omega$ of zero measure.

__Theorem .1__    Let $B: J\mathcal{D}^k(C) \to \Lambda^o_{n-1} J\mathcal{D}^k(C)$ be a physical entity.  Then, B is

a quantum conserved entity for a scattering process (A',c';A",c") if the

mean value of the charge $Q.s = \int_A d(s^* B)$   is zero for any n-dimensional compact

region.

__Proof.__  In fact, the function $Q: \Omega \to \mathbb{R}$, given by $Q(s) \equiv Q.s = \int_A d(s^* B)$ should be

zero in a subset of $\Omega$ of zero measure for any n-dimensional compact region

$A \subset M$.

__Corollary .1__  Let  $B: J\mathcal{D}^k(C) \to \Lambda^o_{n-1} J\mathcal{D}^k(C)$  be a conserved entity for a

dynamic configuration c, $(d((D^k c)^* B) = 0)$.  Let $\Omega_o \equiv \{ s \in \Omega \mid d(s^* B) \neq 0 \}$ , where

$\Omega_o$ is the quantum fluctuation of $D^k c$.  If the measure of $\Omega_o$ is different

to zero, the classic conserved entity $c^* B$  has a __quantum anomaly__, that is

$\ll \int_A d(s^* B) \gg \neq 0$.

__Examples.__    The usual instantons that interpolate vacuum configurations to

which correspond tunneling phenomena, are examples that fit our geometric

framework.  Furthermore, in our geometric formulation can be directly uti-

lized the pwerful tools of global analysis .  (See e.g. ref. $\begin{bmatrix}5:i\end{bmatrix}$ where the

Atiyah-Singer index theorem  is related to number of parameters of

instantons solutions and anomalies of Noether current. See also refs. $\begin{bmatrix}8\end{bmatrix}$).


Note that in some interesting physical circustances we can restrict the

statistical set $\Omega$  by adding some additional requirements.  More pre-

cisely,let us give the following.

__Definition .7__   A continuum system $E_k \subset J\mathcal{D}^k(C)$ has a __quantum-variational__

__constraint__  if the configuration bundle $\pi_C : C \to M$ is such that one can

recognize a submanifold $H^r(C)$ of the manifold of $C^k$ sections $(0 \leq k \leq \infty)$

of $\pi_C$ such that: (a)  the canonical map $H^r(C) \to C^k(C)$ is continuous

for $r > n/2+k$, n=dimM, (__Sobolev lemma__); (b) The space $\Gamma(TH^r(C))$ of

tangent vectors to $H^r(C)$ admits a Hilbert structure; (c) on $H^r(C)$ there

is a Riemannian structure $\gamma$ ;   (d) a numerical function $f: J\mathcal{D}^k(C) \to \mathbb{R}$

on $J\mathcal{D}^k(C)$ is assigned.  This identifies a functional $I_f : H^r(C) \to \mathbb{R}$ given

by  $I_f : c \mapsto f \circ D^k c$.

<u>Example</u>. If M is compact and $H^r_k(C)$ is the set of sections of $\pi_C$ of classe $C^k$ such that in any chart on M the derivations of order $\leq r$ are integrable. Then, for $n/2 < r < \infty$, $H^r(C)$ is a submanifold of the Banach-modelled $\infty$-dimensional manifold $C^k(C)$ on which can be recognized a Riemannian structure.

<u>Definition</u> .8 Let $E_k \subset J\mathcal{D}^k(C)$ have a quantum variational constraint $(H^r(C),\gamma,f)$. Then, we call <u>Scroedinger equation</u> of $E_k$ the equation for the lines in $H^r(C)$ with most rapid decreasing for the functional $I_f$, that is

(S) $$\partial t.\psi = \zeta_f.\psi$$

where: (a) $\zeta_f$ is the Euler-Lagrange field corresponding to the functional $I_f$; (b) $\psi:\mathbb{R}\times H^r(C) \to H^r(C)$ is a one-parameter semi-group of deformations of sections $H^r(C)$.

Then, the <u>quantum cobordism problem</u> in a scattering process characterized by a statistical set $\Omega$ is to find the solutions $\psi$ of the Schroedinger equation (S) which satisfies the internal constraint $\Omega$ , namely $\psi$ must be a map $\psi:\mathbb{R}\times\Omega' \to \Omega'$ where $\Omega' \equiv \Omega \cap H^r(C)$.

<u>Remark.</u> The construction of the field $\zeta_f$ may be made also if $r < n/2$, where $H^r(C)$ is not a proper manifold by means of a vector field $H^o(TC)$, in the following way: $\langle \zeta_f.\psi,u \rangle = \nabla_u(I_f \circ D^k c)$. Furthermore, if $\pi_C:C \to M$ is a vector fiber bundle one has that $J\mathcal{D}^k(C)$ is also a vector fiber bundle and $H^r(C)$ is an Hilbert manifold. In these cases we can consider the variational densities $f:J\mathcal{D}^k(C) \to \mathbb{R}$ that are quadratic on each fiber. The corresponding Euler-Lagrange field $\zeta_f$ defines on each point $c \in H^r(C)$ a discontinuous linear operator.

<u>Examples.</u> 1. If $C \equiv T^*M$ and $f:J\mathcal{D}^k(C) \to \mathbb{R}$, $f\circ D^k c = \frac{1}{2}\left[ (dc)^2+(\delta c)^2\right]$, being M a Riemannian manifold, then $I_f:H^1(C) \to \mathbb{R}$ is the Dirichlet integral for differential forms. By giving to $H^1(C)$ the scalar product of $H^o(C)$ we get $\zeta_f.\psi = \Delta\psi$. So, the Scroedinger equation is the heat equation $\partial t.\psi = \Delta\psi$.

2. Let $K:J\mathcal{D}^k(C) \to \mathbb{K}$ be a linear differential operator such that the fiber bundles C and $\mathbb{K}$ are Riemannian on M. In this case we can canonically associate to K a functional $f_K \circ D^k c = \frac{1}{2}\left| K\circ D^k c\right|^2$, the corresponding Euler-Lagrange field $\zeta_f$ is given by $\zeta_f.\psi = K^*K.\psi$, where $K^*$ is the adjoint of K. So, the Schroedinger equation is $\partial t.\psi = K^*K.\psi$.

## 10. - GEOMETRIC THEORY OF QUANTIZED PDE

We shall , now , study in some details the theory of measure on a statistical

set $\Omega$ of a scattering process of a PDE $E_k \subset J\mathcal{D}^k(W)$.

We shall first recall some useful definitions and notations.

We call indicator function on a subset $E$  the function $\chi_E : \Omega \to \mathbb{R}$  given by:

(a)  $\chi_E|E=1$;  (b)  $\chi_E|(\Omega-E)=0$.  One has  $\chi_E \leq \chi_F$  iff $E \subset F \subset \Omega$.

Suppose $A \equiv E \cup F$, $B \equiv E \cap F$, $C \equiv E \triangle F \equiv (E-F) \cup (F-E)$, (<u>symmetric difference</u>), then one

has  $\chi_B = \chi_E \cdot \chi_F$, $\chi_A = \chi_E + \chi_F - \chi_B$, $\chi_C = |\chi_E - \chi_F|$.

For a sequence $E_1, E_2, \ldots$ of sets we put  $\lim \sup E_i = \bigcap_{n=1}^{\infty} (\bigcup_{i=n}^{\infty} E_i)$, $\lim \inf . E_i =$

$\bigcup_{n=1}^{\infty} (\bigcap_{i=n}^{\infty} E_i)$  and if $\{ E_i \}$ is such that $\lim \sup E_i = \lim \inf E_i = E$  we say that

the sequence <u>converges</u>  to the set E.  A sequence  $\{E_i\}$   is said to be

<u>increasing</u> if, for each positive integer n, $E_n \subset E_{n+1}$; it is said to be <u>decreasing</u>

if for each positive integer n, $E_n \supset E_{n+1}$ . A <u>monotone</u> sequence of sets is one

which is either increasing or decreasing.

Note that any monotone sequence converges to a limit.  If  $\chi_n$ is the indicator

function at $E_n \subset \Omega$  (n=1,2,...) and $A=\lim \sup E_n$, $B=\lim \inf E_n$, one has that for

all $x \in X$,  $\chi_A(x) = \lim_{n \to \infty} \sup \chi_n(x)$, $\chi_B(x) = \lim_{n \to \infty} \inf \chi_n(x)$.

<u>Definitions</u>    1. A <u>quantum semi-ring</u> for $E_k$ is a class $S$ of subsets of a statistical set $\Omega$ of $E_k$ such that $S$ is a semi-ring that is :(i) $\emptyset \in S$ ; (ii) $A,B \in S \Rightarrow A \cap B \in S$ ; (iii) $A,B \in S \Rightarrow A-B = \bigcup_{i=1}^{n} E_i$ , whenever the $E_i$ are disjoint sets in $S$.

2.    A <u>quantum ring</u> for $E_k$ is a non-empty class $R$ of subsets of a statistical set $\Omega$ of $E_k$ that is a ring , namely: $A,B \in R \Rightarrow A \cap B \in R$ and $A \triangle B \in R$. (Since $\emptyset = A \triangle A$, $A \cup B = (A \triangle B) \triangle (A \cap B)$, and $A-B = A \triangle (A \cap B)$ we see that a quantum ring of $E_k$ is a class of sets of $\Omega$ closed under the operations of union, intersection and difference and $\emptyset \in R$. Thus, a quantum ring is a quantum-semi-ring also . Furthermore, if we define operations $\odot$ =multiplication and $\oplus$=addition by $E \odot F = E \cap F$, $E \oplus F = E \triangle F$, $R$ becomes a ring in the algebraic sense).

3. A <u>quantum field</u> (or <u>quantum algebra</u>) for $E_k$ is any class of subsets of a statistical set $\Omega$ of $E_k$ which is a ring and contains $\Omega$ .  (Thus a quantum ring is a quantum field iff it is closed under the operations of taking the complement).

4.  A <u>quantum $\sigma$-ring</u> for $E_k$ is a quantum ring such that it is closed under countable unions, namely if $A_i \in R$  (i=1,2,...)  $= \bigcup_{i=1}^{\infty} A_i \in R$. (If $R$ is a quantum $\sigma$-ring for $E_k$ and $\{A_n\}$ is a sequence of sets from $R$ then $\lim \sup_n A_n$ and $\lim \inf_n A_n$ both belong to $R$ ).

5.  A <u>quantum $\sigma$-field</u> (<u>quantum Borel field</u> , <u>quantum $\sigma$-algebra</u>) is any class $S$ of sets of a statistical set $\Omega$ of $E_k$ which contains the whole space $\Omega$ and is a quantum $\sigma$-ring (namely a quantum $\sigma$-field is a quantum field which is closed under countable unions).

<u>Remark.</u>  If $\Omega$ is endowed with a topological structure the quantum $\sigma$-field $B$ generated by the open sets of $\Omega$ is called a <u>class of quantum Borel sets of</u> $E_k$. Furthermore, the quantum $\sigma$-field $K$ generated by the compact sets is called a class of quantum Borelian sets of $E_k$. (Recall that in $\mathbb{R}^n$ Borel sets and Borelian sets are the same).

<u>Definition</u> .6  A <u>quantum monotone class</u> of $E_k$ is any class $M$ of subsets of $\Omega$ such that, for any monotone sequence $\{E_n\}$ of sets in $M$ we have $\lim E_n \in M$ .

<u>Proposition</u> .1  A quantum $\sigma$-ring of $E_k$ is a quantum monotone class of $E_k$, and any quantum monotone class of $E_k$ which is a quantum ring of $E_k$ is also a quantum $\sigma$-ring of $E_k$.

**Theorem .1** Given any class $C$ of subsets of $\Omega$ (=statistical set of $E_k$), there is a unique z-class (that is any one of the types 2,3,4,5,6 of Definition 5.5) $S$ containing $C$ such that, if $T$ is any other z-class containing $C$ we must have $T \supset S$ . $S$ is called the <u>quantum z-class of $E_k$</u> generated by $C$ . It is the smallest z-class of subsets of $\Omega$ which contains $C$ .

**Theorem .2** 1. The quantum ring $R(S)$ generated by a quantum semi-ring $S$ of $E_k$ consists precisely of the sets which can be expressed in the form $E = \bigcup_{k=1}^{n} A_k$ of a finite disjoint union of sets of $S$ .

2. If $R$ is any quantum ring of $E_k$, the quantum monotone class $M(R)$ generated by $R$ is the same as the quantum $\sigma$-ring $S(R)$ generated by $R$.

3. Any quantum monotone class $M$ of $E_k$ which contains a quantum ring $R$ contains the quantum $\sigma$-ring $S(R)$ generated by $R$.

4. Suppose $C$ is a z-class of subsets of Y, $f:\Omega \to Y$ is any mapping and $f^{-1}(C)$ denotes the class of subsets of $\Omega$ of the form $f^{-1}(E)$, $E \in C$ . Then, $f^{-1}(C)$ is a quantum z-class of $E_k$ .

**Definition .7** Let $C$ be a non-empty class of sets containing the empty set $\emptyset$. A set function $\mu: C \to \bar{\mathbb{R}}$ (=compactified of $\mathbb{R}$: $\bar{\mathbb{R}} = \mathbb{R} \cup \{-\infty, +\infty\}$) is said to be (finitely) additive if: (i) $\mu(\emptyset) = 0$; (ii) for every finite collection $E_1, \ldots, E_n$ of disjoint sets of $C$ such that $\bigcup_{i=1}^{n} E_i \in C$ we have $\mu(\bigcup_{i=1}^{n} E_i) = \sum_{i=1}^{n} \mu(E_i)$.

**Theorem .3** Suppose $\tau: C \to \mathbb{R}$ is an additive set function defined on a quantum ring $C$ of $E_k$ and $E, F \in C$. Then, (i) if $E \supset F$ and $\tau(F)$ is finite $\tau(E-F) = \tau(E) - \tau(F)$; (ii) if $E \supset F$ and $\tau(F)$ is infinite $\tau(E) = \tau(F)$; (iii) if $\tau(E) = +\infty$ , then $\tau(F) \neq -\infty$ .

**Definition .8** A set function $\mu: C \to \bar{\mathbb{R}}$ is said to be <u>$\sigma$-additive</u> (or <u>completely additive or countably additive</u>) if: (i) $\mu(\emptyset) = 0$; (ii) for any disjoint sequence $E_1, E_2, \ldots$ of sets of $C$ such that $E = \bigcup_{i=1}^{\infty} E_i \in C$ , $\mu(E) = \sum_{i=1}^{\infty} \mu(E_i)$.

**Definition .9** A <u>quantum measure</u> on $E_k$ is a measure on a class $C$ of subsets of a statistical set $\Omega$ of $E_k$, that is any non-negative set function $\mu: C \to \bar{\mathbb{R}}^+ \equiv \{x \in \bar{\mathbb{R}}; x \geq 0\}$ , such that it is $\sigma$-additive. A <u>quantum measure space</u> of $E_k$ is a quantum measure $\mu$ on a quantum $\sigma$-ring $(\Omega, C)$. We write: $(\Omega, C, \mu)$.

**Definition .10** Suppose $R$ is a (quantum) ring and $\mu: R \to \bar{\mathbb{R}}$ is additive with $\mu(E) > -\infty$ for all $E \in R$. Then, for any $E \in R$ we say that:

(i) $\mu$ is <u>continuous from below</u> at E if ($\bullet$) $\lim_{n\to\infty} \mu(E_n)=\mu(E)$ for every monotone increasing sequence $\{E_n\}$ of sets in $R$ which converges to E;

(ii) $\mu$ is <u>continuous from above</u> at E if ($\bullet$) is satisfied for any monotone decreasing sequence $\{E_n\}$ in $R$ with limit $E$ which is such that $\mu(E_n) < \infty$ for some n;

(iii) $\mu$ is <u>continuous at</u> E if it is continuous at E from below and from above, (when $E=\emptyset$ the first requirement is trivially sutisfied).

<u>Theorem</u> .4 Suppose hat $R$ is a (quantum) ring and $\mu:R \longrightarrow \bar{R}$ is additive with $\mu(E) > -\infty$ for all $E \in R$.

(i) If $\mu$ is $\sigma$-additive, then $\mu$ is continuous at E for all $E \in R$ ;

(ii) If $\mu$ is continuous from below at every set $E \in R$ , then $\mu$ is $\sigma$-additive;

(iii) If $\mu$ is finite and continuous from above at $\emptyset$, then $\mu$ is $\sigma$-additive.

<u>Definition</u> .11 A <u>quantum outer measure on</u> $E_k$ is a fonction $\mu:C \longrightarrow \bar{R}$, where $C$ is the class of all subsets of a statistical set $\Omega$ of $E_k$, such that:

(i) $\mu(\emptyset)=0$; (ii) $\mu$ is <u>monotone</u> in the sense that $E \subset F \Rightarrow \mu(E) \leq \mu(F)$;

(iii) $\mu$ is <u>countably and additive</u> in the sense that for any sequence $\{E_i\}$ of sets $E \subset \bigcup_{i=1}^{\infty} E_i \Rightarrow \mu(E) \leq \sum_{i=1}^{\infty} \mu(E_i)$.

<u>Note.</u> Every quantum measure on the class of all subsets of a statistical set $\Omega$ of $E_k$ is a quantum outer measure of $E_k$. The viceversa is not true.

<u>Theorem</u> .5 1. Given a countable additive function $\tau:F \longrightarrow \bar{R}$ defined on a (quantum) $\sigma$-field $F$ , there are measure $\tau_+$ and $\tau_-$ defined on $F$ and subsets P, N in $F$ such that $P \cup N =\Omega$ , $P \cap N = \emptyset$ and for each $E \in F$ $\tau_+(E)=\tau(E \cap P) \geq 0$, $\tau_-(E)=-\tau(E \cap N) \geq 0$, $\tau(E)= \tau_+(E)- \tau_-(E)$, so that $\tau$ is the difference of two measure $\tau_+, \tau_-$ on $F$. At least one of $\tau_+, \tau_-$ is finite, and if $\tau$ is finite or $\sigma$-finite so are both $\tau_+, \tau_-$.

2. If $\mu:C \longrightarrow R^+$ is a non-negative additive set function defined on a (quantum) semi-ring $C$. There is a unique additive set function defined on the generated ring $R=R(C)$ such that $\upsilon$ is an extension of $\mu$. $\upsilon$ is non-negative on $R$ and is called the extension of $\mu$ from $C$ to $R(C)$. More precisely, for any $A \in R$ we put $\upsilon(A)=\upsilon(\bigcup_{k=1}^{n} E_k)=\sum_{k=1}^{n}\mu(E_k)$ , where the sets $E_k$ are disjoint and $E_k \in C$ .

3. If $\mu:C \longrightarrow R^+$ is a (quantum) measure defined on a (quantum) semi-ring $C$, the (unique) additive extension of $\mu$ to the generated ring $R(C)$ is also a (quantum) measure.

372

4. Suppose $\mu: R \longrightarrow \mathbb{R}^+$ be non-negative and additive on a (quantum) ring $R$ . Then:

(i) if $E \subset R$ , and $\{E_i\}$ is a sequence of disjoint sets of $R$ such that $E \supset \bigcup_{i=1}^{\infty} E_i$ , $\mu(E) \geq \sum_{i=1}^{\infty} \mu(E_i)$;

(ii) $\mu$ is a measure iff for any sequence $\{E_i\}$ of sets in $R$ such that $\bigcup_{i=1}^{\infty} E_i \supset E \in R$, $\mu(E) \leq \sum_{i=1}^{\infty} \mu(E_i)$.

<u>Definition</u> .12 Let $(\Omega, F, \mu)$ be a quantum measure space for $E_k \subset \mathcal{D}^k(W)$. A function $f: \Omega \longrightarrow \bar{\mathbb{R}}$ is <u>F-measurable</u> iff $f^{-1}(B) \in F$ for every $B \in \bar{\mathcal{B}} \equiv$ $\sigma$-field of Borel sets of $\bar{\mathbb{R}}$: $\bar{\mathcal{B}} \equiv \{`B \subset \bar{\mathbb{R}}, B=B' \cup A, A \in \bar{\mathbb{R}} -\mathbb{R} \equiv \{-\infty, +\infty\}, B' \equiv$ Borel set
of $\mathbb{R}\}$

<u>Theorem</u> .6 1. In order that $f: \Omega \longrightarrow \bar{\mathbb{R}}$ be F-measurable each of the following conditions is necessary and sufficient:

(i) $\{x: f(x) \leq c\} \in F$ , $\forall c \in \mathbb{R}$ ; (ii) $\{x: f(x) > c\} \in F$ , $\forall c \in \mathbb{R}$;

(iii) $\{x: f(x) \geq c\} \in F$ , $\forall c \in \mathbb{R}$ ; (iv) $\{x: f(x) < c\} \in F, \forall c \in \mathbb{R}$.

2. Any <u>F-simple function</u> , that is $f: \Omega \longrightarrow \mathbb{R}$, $f(x) = \sum_{i=1}^{n} c_i \chi_{E_i}(x), \Omega = \bigcup_{i=1}^{n} E_i$, $E_i$=disjoint, $c_i \in \mathbb{R}$, is F-measurable.

3. Any non-negative measurable function $f: \Omega \longrightarrow \mathbb{R}^+$ is the limit of a monotone increasing sequence of non-negative simple functions.

4. If $f$ and $g$ are measurable functions $\Omega \longrightarrow \bar{\mathbb{R}}$ and $k \in \mathbb{R}$, then each of the functions: $f+k$, $kf$, $f+g$, $f^2$, $f.g$, $1/f$ , $\max(f,g)$, $\min(f,g)$, $f_+$, $f_-$, $|f|$ , is measurable.

5. Suppose $\{f_n\}$ , n=1,2,..., is a sequence of measurable functions $\Omega \longrightarrow \bar{\mathbb{R}}$; then: (i) the functions $\sup_n f_n$ and $\inf_n f_n$ are measurable; (ii) the functions $\lim_{n \to \infty} \sup f_n$, $\lim_{n \to \infty} \inf f_n$ are measurable; (iii) if the sequence $\{f_n\}$ converges and in particular if it is monotone, $\lim_{n \to \infty} f_n$ is measurable.

<u>Remark.</u> If $\Omega$ is endowed with a topological structure and $\mathcal{B}$ is the $\sigma$-field of Borel sets in $\Omega$ , a function $f: \Omega \longrightarrow \bar{\mathbb{R}}$ that is $\mathcal{B}$-measurable is called <u>Borel measurable function on</u> $\Omega$ . In particualr any continuous function $f: \Omega \longrightarrow \mathbb{R}$ on the topological space $\Omega$ is Borel measurable.

<u>Definition</u> .13 A <u>quantum complete measure space</u> on $E_k$ is a quantum measure space $(\Omega, F, \mu)$ such that $\mu: F \longrightarrow \mathbb{R}^+$ and the class $F$ is <u>complete</u> with respect to $\mu$ , namely if $E \subset F$, $F \in F, \mu(F)=0 \Rightarrow E \in F$. In these cases $\mu$ is called a <u>quantum complete measure on</u> $E_k$.

<u>Theorem .7</u>  1.  All measures  $\mu$  which are obtained by restricing a quantum

outer measure  $\mu^*$  to the class  $M$  of sets which are measurabie  $(\mu^*)$  are

quantum complete measures on  $E_k \subset JD^k(W)$.

E.  Given a quantum measure  $\mu$  on a quantum  $\sigma$-ring  $I$  of  $E_k$, let  $\bar{I}$  be

the class of all sets of the form  $E\Delta N$  where  $E \in I$ and      $N \subset F$, with

$\mu(F)=0$.  Then,  $\bar{I}$  is a quantum  $\sigma$-ring of  $E_k$  and if we put  $\bar{\mu}(E\Delta N)=\mu(E)$

then  $\bar{\mu}:\bar{I} \longrightarrow IR^+$  is a (uniquely) defined extension of  $\mu$  from  $I$  to  $\bar{I}$  and

$\mu$  is a quantum complete measure on   $\bar{I}$ .

<u>Definition  .14</u>   <u>(Definition of integral on a quantum measure space of $E_k$)</u>

Let  $(\Omega,F,\mu)$  be a quantum measure space of  $E_k \subset JD^k(W)$.

1. Let   $f:\Omega \longrightarrow IR^+$  be a non-negative simple function $f=\sum_{i=1}^{p}c_i \chi_{E_i}$ ,  $c_i \geq 0$

$(i=1,2,\ldots)$, we define   $\int_\Omega f \, d\mu = \sum_{i=1}^{n} c_i \mu(E_i)$.

2.  Let  $f:\Omega \longrightarrow IR^+$  be a non-negative measurable function, we define

$\int_\Omega f \, d\mu = \lim_{n \to \infty} \int f_n \, d\mu$,  where  $\{f_n\}$  is a monotone increasing sequence of

simple functions such that $f_n \longrightarrow f$  ; (the limit is independent of the parti-

cular sequence used).

3.  Let   $f:\Omega \longrightarrow \bar{R}$  be a measurable function, we define   $\int_\Omega f d\mu = \int_\Omega f_+ \, d\mu - \int_\Omega f_- \, d\mu$

where $f=f_+ - f_-$  being $f_+$  and $f_-$  non-negative measurable functions.

4.  Let A be a set belonging to  $F$ . Put    $\int_A f \, d\mu = \int_\Omega f \chi_A d\mu$ .

<u>Theorem .8</u>  1.  Let  $\Omega$   be a finite set,  $\mu(E)$ the number of points on

E. All functions on  $\Omega$  are simple functions and the theory of integration

reduces to the theory of finite sums.

2.  Let  $(\Omega,F,\mu)$  be a quantum measure space for $E_k$;  Let A,B be disjoint

sets in  $F$   and $f:\Omega \longrightarrow \bar{R}$, $g:\Omega \longrightarrow \bar{R}$  two functions integrable (over $\Omega$ )

with respect to  $\mu$ .  Then,  f is integrable over A, f+g and  $|f|$  are

integrable (over $\Omega$) and one has the following results:

(i)  $\int_{A \cup B} f \, d\mu = \int_A f \, d\mu + \int_B f \, d\mu$ ; (ii) f is finite almost every where;

(iii)  $\int(f+g) \, d\mu = \int f \, d\mu + \int g \, d\mu$   ; (iv) $|\int f \, d\mu| \leq \int |f| \, d\mu$    ;

(v)  for any  $c \in IR$, cf is integrable and   $\int cf \, d\mu = c \int f \, d\mu$      ;

(vi)  $f \geq 0 \Rightarrow \int f \, d\mu \geq 0$; $f \geq g \Rightarrow \int f \, d\mu \geq \int g \, d\mu$  ;

(vii) if $f \geq 0$ and   $\int f \, d\mu = 0$, then f=0 a.ë. ; (viii) f=g a.e. $\Rightarrow$

$\int f \, d\mu = \int g \, d\mu$   ; (ix) if  $h:\Omega \longrightarrow \bar{R}$ is  $F$-measurable and $|h| \leq f$

then  h is integrable.

3.  Any function $f:\Omega \longrightarrow \bar{R}$ which is bounded, $F$-measurable, and zero outside

374

a set E in $F$ of finite $\mu$-measure is integrable with respect to $\mu$.

<u>Definition 15.</u>   Let $(\Omega,F,\mu)$ be a measure space.

1. A set function $\nu: F \to \bar{\mathbb{R}}$ is <u>absolutely continuous</u> if $\nu(E)=0$ for every $E \in F$ with $\mu(E)=0$ and we write $\nu \ll \mu$ .

2. A set function $\nu: F \to \bar{\mathbb{R}}$ is <u>singular</u> if there is a set $E_0 \in F$ for which $\mu(E_0)=0$ and $\nu(E)=\nu(E \cap E_0)$, all $E \in F$.

<u>Theorem 9.</u>   Let $(\Omega,F,\mu)$ be a measure space.

1. If $f:\Omega \to \bar{\mathbb{R}}$ is $\mu$-integrable, the set function $\nu(E)= \int_E fd\mu$ , $E \in F$ is finite valued and absolutely continuous.

2. Let be given a $\sigma$-additive $\sigma$-finite set function $\nu: F \to \bar{\mathbb{R}}$ . Then, there exists a unique decomposition (<u>Lebesgue decomposition</u>) $\nu = \nu_1 + \nu_2$ into $\sigma$-additive set functions $\nu_i$ which are $\sigma$-finite and such that $\nu_1$ is singular and $\nu_2 \ll \mu$. Further, there is a a.e. unique finite valued measurable function $f:\Omega \to \mathbb{R}$ (<u>Radon-Nikodym derivative</u> $f \equiv d\nu/d\mu$ ) such that $\nu_2(E)= \int_E fd\mu$ , all $E \in F$.

3. Let $(X,F,\mu)$ and $(Y,G,\nu)$ be two measure spaces and $f:X \to Y$ a measure transformation (that is $f^{-1}(E) \in F$ for every $E \in G$). Then, for any $G$-measurable function $g:Y \to \bar{\mathbb{R}}$ the function $g \cdot f:X \to \bar{\mathbb{R}}$ is $F$-measurable and one has $\int_Y gd(\mu f^{-1})= \int_X g \circ fd\mu = \int_Y g \cdot \phi d\nu$ where $\phi \equiv d\bar{\nu}/d\nu$ is the Radon-Nikodym derivative of $\bar{\nu}$ with respect to $\nu$ , being $\bar{\nu} \equiv \mu f^{-1}$ the measure defined by $\bar{\nu}(E)=\mu(f^{-1}(E))$ , for $E \in G$.

<u>Definition 16.</u>   1. A <u>quantum probability space</u> for $E_k \subset J\mathcal{S}^k(W)$ is a triple $(\Omega,F,P)$ where $\Omega$ is a statistical set, $F$ is a $\sigma$-field of subsets of $\Omega$ and $P$ a complete probability measure on $(\Omega,F)$ , that is a measure satisfying $P(\Omega)=1$.

2. A <u>quantum random variable</u> for $E_k \subset J\mathcal{S}^k(W)$ is a measurable function $f:\Omega \to \bar{\mathbb{R}}$ with respect to a quantum probability space $(\Omega,F,P)$ for $E_k \in J\mathcal{S}^k(W)$.

3. For a quantum random variable $f:\Omega \to \bar{\mathbb{R}}$ we define:

(a) <u>Expectation:</u> $E(f)= \int_\Omega fdP$.   E is a $\mathbb{R}$-linear map on the space of random variables of $(\Omega,F,P)$.

(b) <u>Moments:</u>   $E(f^n)$, $n=1,2,\dots$

(c) <u>Moments generating function:</u> $E(e^{tf})$, $(t \in \mathbb{R})$.

(d) <u>Variance:</u>   $\sigma^2(f) \equiv var(f)=E((f-E(f))^2)= \int_\Omega (f-E(f))^2 dP$.

__Theorem__  .10  (__Tchebychev's inequality__) Let  f  be a quantum random variable

for which  $\mu = E(f)$  and  $\sigma^2 = var(f)$  are finite . Then, for any  $\lambda \geq 1$,

$P\{|f-\mu| \geq \lambda \sigma\} \leq \lambda^{-2}$.

__Definition__  .17  Let $f:(\Omega, F) \rightarrow \bar{R}$ be a quantum random variable for $E_k$.

1. The distribution of f is the measure P' on $(\bar{R}, F')$ given by $P'(B) = P\{f^{-1}(B)\}$

where  $F'$  is the completion of  $B$  with respect to P'.

2. If the quantum random variable  f   is finite with probability one,

then the non-decreasing and right-continuous function $F: \mathbb{R} \rightarrow [0,1]$  defined

by $F(x) = P(f^{-1}(x))$  is called the __distribution function of__ f.

__Theorem__  .11  1. Let F be the distribution function of f.  Then, one has:

(a) $\lim_{x \rightarrow -\infty} F(x) = 0$; (b) $\lim_{x \rightarrow +\infty} F(x) = 1$; (c) For any Borel set $B \subset \mathbb{R}$, $P(f^{-1}(B)) = \int_B dF(x)$.

2. If the quantum random variable f is finite with probability one, and

if $g: \mathbb{R} \rightarrow \bar{R}$  is Borel measurable, then $E\{g(f)\} = \int_{\mathbb{R}} g(x) dF(x)$, where F

is the distribution function of f.

__Definition__  .18  The __characteristic function__ of a (finite, real) quantum

random variable f is   $\phi: \mathbb{R} \rightarrow \mathbb{R}$ , $\phi(t) = E(e^{itf})$.

__Theorem__  .12   Let  $\phi$  be the characteristic function of a quantum random

variable  f  of $E_k \subset \mathcal{D}^k(W)$.

1. One has  $|\phi(t)| \leq 1$ , $\phi(0) = 1$.  If for some  $\tau \neq 0$, $\phi(\tau) = 1$, then  f is

a discrite random variable, taking only values which are integral multiple

of $2\pi/\tau$ .  $\phi$ is uniformly continuous function of t.

2. If a quantum random variable f has distribution F, then its characteristic

function is   $\phi(t) = \int_{\mathbb{R}} e^{itx} dF(x)$, that is the Fourier-Stieltjes transform of F.

3. If  F and G have the same characteristic function, then F=G.

__Definition__  .19 Let $(\Omega, F, P)$ be a quantum probability space for $E_k$ on which

is defined a collection $f = (f_1, f_2, \ldots, f_n)$ of quantum random variables  of

$E_k$.  We shall assume for simplicity that each $f_r$ is almost finite, or that

$f_r$ is a measurable function from $(\Omega, F)$  to $(\mathbb{R}, B)$. Therefore, the function

$f: \Omega \rightarrow \mathbb{R}^n$  is a measurable function  $(\Omega, F) \longrightarrow (\mathbb{R}^n, B)$.  The measure P induces

a measure P' on $(\mathbb{R}^n, B^n)$ by $P'(B) = P\{f^{-1}(B)\}$ , $B \in B^n$.

376

If $F'$ is the completion of $B^n$ with respect to P', then $(\mathbb{R}^n, F', P')$ is a probability space. The measure P' is called the joint probability distribution of the collection f.

__Theorem .13__  If the collection $f \equiv (f_1, \ldots, f_n)$ has joint distribution P' and if $g: \mathbb{R}^n \longrightarrow \mathbb{R}$ is Borel measurable, then $E\{g(f_1, \ldots, f_n)\} = \int_{\mathbb{R}^n} g(x_1, \ldots, x_n) dP'$.

__Remark.__  Note that any charge on a differential equation identifies a quantum random variable.  So, the quantum description of physical entities is made by means of quantum random variables.


Another  important concept that we shall introduce is the relation between the measure theory on a statistical set of PDE  and the spectral theory.

__Definition .20__  A __quantum spectral measure__ for $E_k \subset JD^k(W)$  is a triplet $(\Omega, \Sigma; E)$ where  $\Omega$ is a statistical set  of $E_k$, $(\Omega, \Sigma)$ is a measurable space and E is a mapping $E: \Sigma \longrightarrow L(H) \equiv$ space of linear operators on an Hilbert space H such that:

(i)  $E(\emptyset) = O$, $E(\Omega) = id_H$, $E(A)^* = E(A)$;

(ii)  $E(A \cap B) = E(A) \cdot E(B)$  for any  $A, B \in \Sigma$ ;

(iii)  E is countably additive for the topology simple-strong on L(H), that is for any disjoint sequence $(A_n)$  in  $\Sigma$  any  $x \in H$  one has $E(\bigvee_n A_n)(x) = \sum_n E(A_n)(x)$  in the sense of the topology of H.

The map  $A \longmapsto (E(A)(x)|y) \in \mathbb{R}$, $A \in \Sigma$ ,is a measure denoted by $E_{x,y}$ , for fixed x,y  H;  $(|)$ denotes the scalar product in H.  We write also $E_x \equiv E_{x,x}$.

__Theorem .14__  Let  $BM(\Omega, \Sigma)$  be the algebra of bounded measurable functions $\Omega \longrightarrow \mathbb{C}$  with respect to a quantum spectral measure for $E_k \subset JD^k(W)$: $(\Omega, \Sigma; E)$.  Then, $E: \Sigma \longrightarrow L(H)$ identifies a $*$-morphism of unitary $C^*$-algebras: $\Phi: BM(\Omega, \Sigma) \longrightarrow L(H)$  such that:

(i)  $(\Phi(f)(x)|y) = \int_\Omega f \, dE_{x,y}$ ;  (ii)  $\|\Phi(f)(x)\|^2 = \int_\Omega |f|^2 dE_{x,y}$ ;

(iii)  $\|\Phi(f)\| \leq \|f\|$  .

This theorem admits an extension to non-bounded measurable functions  $\Omega \longrightarrow \mathbb{C}$ on $(\Omega, \Sigma)$.

__Theorem__ .15  Let $M(\Omega,\Sigma)$ be the algebra of measurable functions $\Omega \rightarrow \mathbb{C}$ with respect to a quantum spectral measure $(\Omega,\Sigma;E)$ of $E_k \subset JD^k(W)$. Then, the quantum spectral measure $E: \Sigma \rightarrow L(H)$ identifies a map $\Phi: M(\Omega,\Sigma) \rightarrow L(H)$ such that: (i) the operator $\Phi(f)$ is defined on the dense domain $D(\Phi(f)) \equiv \{x \in H \mid \int |f|^2 dE_x < +\infty\} \equiv D(f)$ ;

(ii)  $(\Phi(f)(x)|y) = \int f\, dE_{x,y}$    $\forall x \in D(f)$    , $\forall y \in H$;

(iii)  $\|\Phi(f)(x)\|^2 = \int |f|^2 dE_x$    , $\forall x \in D(f)$.

Furthermore, $\Phi(f)$ is a normal operator and if $f$ is (positive) real $\Phi(f)$ is (positive) self-adjoint. We call the operator $\Phi(f)$, $f \in M(\Omega,\Sigma)$ , a __quantum random operator__ on $E_k$.

__Corollary__ .1   1.  Under the hypotheses of above theorem let $L^\infty(E) \equiv BM(\Omega,\Sigma)/N$ , where $N$ is the sheaf of functions $E$-neglectable endowed with the quotient norm $\| \|_\infty$ . Then,    $\Phi(f)$    is bounded on $H$ iff $f \in L^\infty(E)$    and one has  $\|\Phi(f)\| = \|f\|_\infty$ .

2.    $\Phi(f)$    is reversable and with bounded inverse on $H$ iff $1/f \in L^\infty(E)$.

3.  $Sp(\Phi(f)) = \bigcap_{\substack{A \in \Sigma \\ E(A) = id_H}} \overline{f(A)}$    , where $S_p(\Phi(f))$ is the spectrum of  $\Phi(f)$.

__Theorem__ .16   (Quantum spectral resolution of PDE)

Let $E_k \subset JD^k(W)$ be a PDE . Then, any quantum random operator $\Phi(f) \in L(H)$ gives to $H$ a splitted representation $H = \oplus_n H_n$ , where $H_n$ are closed subspaces such that:

(i)  $H_n \subseteq D(\Phi(f))$;   (ii)  $H_n$ are invariant for  $\Phi(f)$  and  $\Phi(f)^*$ ;

(iii)  $\Phi(f)_n \equiv \Phi(f)|H_n$ is a bounded normal operator.  $D(\Phi(f))$  is the set of vectors $x = \sum_n x_n \in H$ such that $\sum_n \|\Phi(f)_n(x_n)\|^2 < +\infty$  and one has $\Phi(f)(x) = \sum_n \Phi(f)_n(x_n)$,   $\Phi(f)^*(x) = \sum_n \Phi(f)^*_n(x_n)$, $\forall x \in D(\Phi(f))$.

Furthermore , $\Phi(f)(\pi_n(x)) = \pi_n(\Phi(f)(x))$, $\forall x \in D(\Phi(f))$, being $\pi_n: H \rightarrow H_n$ the orthogonal projection;

(iv)  $S_p(\Phi(f)) = \bigcup_n S_p(\Phi(f)_n)$ ;

(v)  $S_p(\Phi(f))_p = \bigcup_n S_p(\Phi(f)_n)_p$, where  "p" denotes __punctual__.

(vi)  $S_p(\Phi(f))_r = \emptyset$ , where  "r" denotes __residus__.

$\Phi(f)$ is self-adjoint iff $Sp(\Phi(f)) \subset \mathbb{R}$. In this case $\Phi(f)$ is expressed by means of a unique spectral measure on $\mathbb{R}$, such that $\Phi(f)=\int f dE(t)$. Furthermore, the support of $E$ is the spectrum $Sp(\Phi(f))$ of $\Phi(f)$.

<u>Remark.</u> Let $Q=\int_A (K.c)^* \beta$ be a charge of a continuum system $E_k \subset J\mathcal{D}^k(C)$ and let $\hat{Q}=\Phi(f) \in L(H)$ be the corresponding linear operator associated to a spectral measure on $\Omega$, where $f:\Omega \to \mathbb{R}$ is the function on the statistical set $\Omega$ of a scattering process of $E_k$, given by $f(D^k c)=\int_A (K \bullet D^k c)^* \beta$. Being $\Phi(f)$ normal we can suppose that the spectral measure $E$ on $\Omega$ should be defined on $\mathbb{R}$ by substituting the image measure $f(E)$, that is spectral also, such that $f(E)(B)=E(f^{-1}(B))$, for any borel set $B$ of $\mathbb{R}$. So that the operator $\hat{Q}$ is necessarily self-adjoint and can be written as $\hat{Q}=\int_{Sp(\hat{Q})} \lambda dE(\lambda)$ , (<u>spectral resolution of</u> $\hat{Q}$) with domain $D(\hat{Q})=\{\psi \in H \mid \int |\lambda|^2 dE_\psi < +\infty\}$ . Furthermore, one has: (a) $(\hat{Q}(\psi)|\psi)=\int_{Sp(\hat{Q})} \lambda dE(\lambda), \forall \psi \in D(\hat{Q})$; (b) $||Q(\psi)||^2 = \int_{Sp(\hat{Q})} |\lambda|^2 dE_\psi(\lambda)$. A <u>quantum state</u> (of the quantized system) is a positif Hermitian nuclear operator $T \in L_1(H)$ such that $tr(T)=1$. If $T=\psi \boxtimes \psi$, with $\psi \in H$ and $\|\psi\|=1$, $T$ is called a <u>pure quantum state</u>. The space $E$ of quantum states can be identified with a convex part of the unit bubble into $K(H)'$ , (see Appendix). One can see that if $T \in E$ the following propositions are equivalent: (a) $T$ is a pure state; (b) $T$ is a projection $(T^2=T)$; (c) $tr(T^2)=1$; (d) $T$ is an extremal point of the convex $E \subset K(H)'$.

A <u>quantum value</u> of $Q$ is a point $\lambda$ of the spectrum of $\hat{Q}$: $\lambda \in Sp(\hat{Q})$. Furthermore, the <u>probability</u> to find a quantum value of $Q$ in the borel set $\omega \subset \mathbb{R}$, if the system is in the quantum state $T$ is given by the following formula:

$$(20) \qquad p(Q;T;\omega) = tr(E(\omega)\,T)=tr(T\,E(\omega)).$$

The corresponding <u>mean value</u> is

$$(21) \quad \ll Q \gg_T = \int_{Sp(\hat{Q})} \lambda dp(Q;T;\lambda) = tr \int \lambda dE(\lambda)T = tr(T\,\hat{Q}) = tr(\hat{Q}\,T) = \Sigma \alpha_n (\hat{Q}(e_n)|e_n)$$

The corresponding <u>variance</u> is

$$(22) \qquad \sigma^2(Q) = \int_{Sp(\hat{Q})} (\lambda - \ll Q \gg_T)^2 \, dp(Q;T;\lambda) \quad = tr(\int \lambda^2 dE(\lambda)T) - \ll Q \gg_T^2$$

$$= tr(\hat{Q}^2 T) - |tr(\hat{Q}\,T)|^2 \ .$$

In particular, if $T=\psi\otimes\psi$ is a pure quantum state, we have:

(23) (<u>probability</u>)  $p(Q;T;\omega) = \int_\omega dE_\psi(\lambda)$ ;

(24) (<u>mean value</u>)  $\langle\!\langle Q \rangle\!\rangle_T = \int \lambda\, dE_\psi(\lambda)$ ;

(25) (<u>variance</u>)  $\sigma^2(Q)_T = \int (\lambda - \langle\!\langle Q \rangle\!\rangle_\psi)^2 dE_\psi(\lambda) = \|\hat{Q}(\psi)\|^2 - (\hat{Q}(\psi)|\psi)^2$ .

Now, let $\hat{Q}$, $\hat{Q}'$ be two elements of $H(H)$ corresponding to two charges $Q,Q'$ of a continuum system $E_k \subset J\mathcal{D}^k(C)$. If $[\hat{Q},\hat{Q}'] = 0$, we say that $Q$ and $Q'$ are <u>simultaneously observable</u>. Then, the probability to find the quantum value $\lambda$ of $Q$ in the borel set $\omega \subset IR$ if the system is in the pure quantum state $\psi$ is given by

(26)  $p(Q,Q';\psi;\omega,\omega') = \iint_{\omega\omega'} dG_\psi(\lambda,\lambda')$ ,

where $G_\psi(\bullet,\bullet) \equiv E_\psi(\bullet)\otimes F_\psi(\bullet)$ is the spectral measure on $IR^2$ obtained by product of the spectral measure $E$ and $F$ of $Q$ and $Q'$ respectively.

Furthermore, one has the Heisenberg uncertainty relation:

(27)  $\sigma^2(Q)_\psi \cdot \sigma^2(Q')_\psi \geqslant \frac{1}{4} |([\hat{Q},\hat{Q}'](\psi)|\psi)|^2$ .


## 11.- QUANTUM COBORDISM AND DIRAC'S APPROACH TO QUANTIZATION

In this section the quantum cobordism is related to the Dirac's approach to quantization, where the fields are seen as linear operators on Hilbert spaces. As the space of wave functions is obtained as a space of sections of a vector bundle, let us first recall some fundamental results on the space of sections of vector bundles on a (non-necessarily compact) n-dimensional manifold. (See also e.g. refs. [7]) .

Let $\pi:E \to X$ be a complex vector fiber bundle on a paracompact and locally compact $C^\infty$ manifold $X$ of dimension n. Let $E$ have a finite rank r. The space $C^\infty(E)$ of smooth sections of $E$ is a vector space with a natural topological structure of Fréchet and Montel space. Let $C^\infty(E)'$ be the topological dual of $C^\infty(E)$. Let $C^\infty_0(E)$ be the space of sections with support compact of $E$. $C^\infty_0(E)$ has a natural structure of $LF$ and Montel topological vector space. Let $C^\infty_0(E)'$ be the topological dual of $C^\infty_0(E)$. Let $C^\infty(\Lambda^o_n(X)) \equiv \Omega_n(X)$ (resp. $\Omega^+_n(X) \equiv C^\infty(\Lambda^o_n(X)^+)$ be the space of $C^\infty$ sections of the vector fiber bundle $\Lambda^o_n(X)$ of n-forms on X (resp. the space of $C^\infty$ sections of the subbundle $\Lambda^o_n(X)^+$ of $\Lambda^o_n(X)$, characterized by the fibre $\mathbb{R}^+$). The regular Radon measures (resp. positif regular Radon measures (+)) on X are in correspondence one-to-one with the vectors of $\Omega_n(X)$ (resp. $\Omega_n^+(X)$).

Set $E' \equiv E^* \otimes \Lambda_n^o(X)$ be the geometric adjoint of E. One has the canonical isomorphism $E'' \cong E'$. One has the following exact sequences of morphisms of topological vector spaces:

(a) $0 \longrightarrow C^\infty(E') \overset{\gamma}{\longrightarrow} C_o^\infty(E)'$ , where $\gamma$ is given by $\gamma: \sigma \mapsto \gamma(\sigma): f \mapsto \int_X \langle \gamma(\sigma), f \rangle \in \mathbb{C}$;

(b) $0 \longrightarrow C^\infty(E) \overset{\omega}{\longrightarrow} C_o^\infty(E')\iota$, where $\omega$ is given by: $\omega : f \mapsto \omega(f): \sigma \mapsto \int_X \langle \sigma, f \rangle \in \mathbb{C}$.

We call $C_o^\infty(E)'$ (resp. $C_o^\infty(E')'$) the space of distribution -sections of E (resp. E').

If E has a hermitian structure h one has the canonical isomorphisms:
$E \cong E^*$ ; $C_o^\infty(E) \cong C_o^\infty(E')$ ; $C^\infty(E) \cong C^\infty(E')$ ; $C_o^\infty(E)' \cong C_o^\infty(E')'$.

Example. If $E = X \times \mathbb{C}^r$ , with X oriented by a volume form $\eta$ and h is the hermitian structure induced by the scalar product on $\mathbb{C}^r$. Then, in this case one has the following isomorphism $C_o^\infty(E') \cong C_o^\infty(E)'$.

Let $\pi: F \longrightarrow Y$ be another vector fiber bundle of rank s on a m-dimensional manifold Y. The exterior tensor product of E by F is the following vector fiber bundle $E \boxtimes F$ over $X \times Y$ of rank r.s defined by $E \boxtimes F = \bigcup_{(x,y)} E_x \otimes F_y$. The space $C_o^\infty(E \boxtimes F')$ is called the space of kernels. One has the following isomorphisms:

(a) $\Lambda_{n+m}^o(X \times Y) \cong \Lambda_n^o(X) \boxtimes \Lambda_m^o(Y)$; (b) $E' \boxtimes F' \cong (E \boxtimes F)'$.

One has a map $\phi : C_o^\infty(E \boxtimes F) \times C_o^\infty(F)' \longrightarrow C_o^\infty(E)$ given by $\phi(f,T) = \langle f, T \rangle$ ; the partial map $\phi_T : C_o^\infty(E \boxtimes F) \longrightarrow C_o^\infty(E)$ is linear and continuous.

The tensor product $S \otimes T$ of two distribution-sections $S \in C_o^\infty(E)'$, $T \in C_o^\infty(F)'$ is an element of $C_o^\infty(E \boxtimes F)'$ given by $\langle f, S \otimes T \rangle = \langle \langle f, T \rangle, S \rangle$ .

The sub-space $C_o^\infty(E)' \otimes C_o^\infty(F)'$ of $C_o^\infty(E \boxtimes F)'$ generated by the elements of type $S \otimes T$ is called the tensor product of two spaces of distributions $C_o^\infty(E)'$ and $C_o^\infty(F)'$. One has the following properties:

-----

(+) A Radon measure on a topological space X is a continuous linear form on $C^o(X)$.

A Radon measure $\mu$ on a manifold X is called regular (resp. positif regular) if for any chart $(U, \psi)$ of X the induced measure on $\psi(U) \subset \mathbb{R}^n$ is like $\rho\lambda$ where $\rho$ is a $C^\infty$ function (resp. $C^\infty$ positif function) and $\lambda$ is the Lebesgue measure on $\mathbb{R}^n$.

I) <u>Theorems of duality.</u>

(a)  Any element $f \in C_o^\infty(E \boxtimes F)$ is limit of a sequence $\{f_i\}$ of elements of $C_o^\infty(E) \otimes C_o^\infty(F)$.

(b)  $C_o^\infty(E)$ is dense into $C_o^\infty(E')'$, $(C_o^\infty(E'))'$ being a Montel space it is reflexif and its dual can be identified with $C_o^\infty(E'))$.

(c)  $C_o^\infty(E)' \otimes C_o^\infty(F)'$ is dense into $C_o^\infty(E \boxtimes F)'$.

II) <u>Theorems of nuclearity.</u>

(a)  $C^\infty(E)$, $C_o^\infty(E)$ and $C_o^\infty(E)'$ are nuclear spaces.

(b)  One has the topological vector spaces isomorphism: $L(C_o^\infty(F), C_o^\infty(E)') \cong C_o^\infty(E)' \otimes C_o^\infty(F)'$ if the first space is endowed with the strong topology.

III)  <u>Theorems of kernels.</u>

One has the following maps:

(a)  $\Psi_1 : C_o^\infty(E \boxtimes F)' \times C_o^\infty(F) \to C_o^\infty(E)'$  ,  $\langle \Psi_1(K,f), 1 \rangle = \langle K, 1 \otimes f \rangle$ ;

(b)  $\Psi_2 : C_o^\infty(E \boxtimes F)' \times C_o^\infty(E) \to C_o^\infty(F)'$  ,  $\langle \Psi_2(K,1), f \rangle = \langle K, 1 \otimes f \rangle$ .

(c)  One has the following isomorphism of topological vector spaces:

$C_o^\infty(E \boxtimes F)' \cong L(C_o^\infty(F), C_o^\infty(E)')$.   So, any kernel $K$ can be identified with a continous linear map $C_o^\infty(F) \to C_o^\infty(E)'$.

(d)  If $B : C_o^\infty(E) \times C_o^\infty(F) \to \mathbb{C}$ is a bilinear form, continuously separated, there exists a unique kernel $K \in C_o^\infty(E \boxtimes F)'$ such that $B(f,g) = \langle K, f \otimes g \rangle$, $\forall f \in C_o^\infty(E)$, $\forall g \in C_o^\infty(F)$.

In order to satisfy the general requirement of full covariance we shall introduce the following fundamental

__Definition. 1__   Let $F$ be the category whose objects are open subbundles of fiber bundles and whose morphisms are the local fiber bundle diffeomorphisms between those objects. Let $F_H$ be the category whose objects are open subbundles of Hermitian vector fiber bundles and whose morphisms are the local vector fiber bundle diffeomorphisms between these objects preserving the hermitian structures.

Let $Q$ be a covariant functor $Q: F \longrightarrow F_H$ such that:

(a) if $W|U, U \subset M$, $Ob(F)$ $\Rightarrow$ $Q(W|U) = \pi_H^{-1}(U) \in Ob(F_H)$ being $\pi_H: H \to M$ an hermitian vector bundle such that $H = Q(W)$;

(b) if $f \equiv (f_W, f_M) \in Hom_F(W|U, \bar{W}|\bar{U})$, $U \subset M$, $\bar{U} \subset \bar{M}$, then
$Q(f) \in Hom_{F_H}((H \equiv Q(W))|U, (H \equiv Q(W))|U)$ and satisfies:

(c) $\pi_{\bar{H}} \bullet Q(f) = f_M \bullet \pi_H$ ; (d) if $W|U \in Ob(F)$, $U' \subset U$ $\Rightarrow$ $Q(f)|\pi_H^{-1}(U') = Q(f|W|U')$.

We call such a functor $Q$ a __quantum functor__ and the triplet $(W, H; Q)$ such that $Q$ is a quantum functor, $W \in Ob(F)$, $H \in Ob(F_H)$ and $H = Q(W)$ a __quantum super-bundle of geometric objects__ . (Really it is a particular case of super-bundle of geometric objects $[12, 13]$ ).

__Definition .2__   Let $E_k \subset J\mathcal{D}^k(W)$ be a formally integrable differential equation on $\pi: W \to M$. A (h-order) __Dirac-quantization__ of $E_k$ is a triplet $(H, Q, Q)$ where:

(1) $Q$ is a fiber bundle morphism $Q: J\mathcal{D}^{k+q}(W) \longrightarrow J\mathcal{D}^h(H) \ast H$ , $q \geq 0)$, over $M$, where $\bar{\pi}: H \to M$ is a complex Hilbert bundle over $M$ such that the following diagram

$$
\begin{array}{ccc}
E_{k+q} & \xrightarrow{\;Q|E_{k+q}\;} & H^h \\
\downarrow & & \downarrow \\
J\mathcal{D}^{k+q}(W) & \xrightarrow{\quad Q \quad} & J\mathcal{D}^h(H) \ast H
\end{array}
$$

is commutative , being $\pi': H^h \to M$ the subbundle of $J\mathcal{D}^h(H) \ast H$ characterized by means of the sections of $J\mathcal{D}^h(H) \ast H$ that identify (anti-)hermitian operators with respect to the Hilbert structure induced on the $\mathbb{C}$-vector space $C_o^\infty(H)$ of $C^\infty$-sections of $\bar{\pi}$ with support compact. We call Q a __Dirac-quantization operator__.

(2)  Furthermore, we require that $(W,H;Q)$  should be a quantum super-bundle of geometric objects for some quantum functor $Q$ .

If W is a vector bundle and Q is a morphism of vector fiber bundles we say that Q is a <u>linear quantization operator</u>.

A Dirac-quantization operator $Q:J\mathcal{D}^{k+q}(W) \to J\mathcal{D}^{h}(H)^{*}\boxtimes H$ of a PDE $E_k \subset J\mathcal{D}^{k}(W)$, identifies:

(a) a Hilbert space  $H$ (<u>space of wave functions</u>) given by the $C^{\infty}$-sections of  $\bar{\pi}$ with support compact and with inner product given by $(\psi|\psi')=\int_M h(\psi,\psi')$ where h is the product induced by the hermitian structure on H;

(b) a map  $\wedge:C^{\infty}(W) \to L(H)$,  $\wedge:s \mapsto \hat{s}$, such that  $\wedge|S(E_k):S(E_k) \to H(H)=$ space of (anti-)hermitian operators on  $H$ .    $(S(E_k)$=set of solutions of $E_k)$.

If the fiber bundle W has on the fibers a product syructure denoted by  $\{,\}$ we say that a Dirac-quantization operator is <u>normal</u> if the following cor-respondence  $\{s_1,s_2\}^{\wedge}=[\hat{s}_1,\hat{s}_2]\equiv\hat{s}_1\bullet\hat{s}_2-\hat{s}_2\bullet\hat{s}_1$ , $\forall s_1,s_2 \in C^{\infty}(W)$ , is satisfied. We call  $\{,\}$  <u>classical bracket</u>  and   $[,]$   <u>quantum bracket</u> .


We have the following important theorem that relates the quantization of $E_k$ with the spaces of kernels of H:

<u>Theorem</u>  .1    Any Dirac-quantization $Q:J\mathcal{D}^{k}(C) \to J\mathcal{D}^{h}(H)^{*}\boxtimes H$ of a differential equation $E_k \subset J\mathcal{D}^{k}(C)$, identifies a map (the kernel map of Q)

$$K:C^{\infty}(C) \to C_o^{\infty}(H\boxtimes H)'.$$

Furthermore, if Q is a linear quantization, then one has the following exact sequence of morphisms of topological vector spaces:  $0 \to C^{\infty}(C) \xrightarrow{K} C_o^{\infty}(H\boxtimes H)'$.

<u>Proof.</u>   In fact,  $K$ is given by means of the following composition  of maps:

$$C^{\infty}(C) \xrightarrow{Q} L(C_o^{\infty}(H),C_o^{\infty}(H)) \cong L(C_o^{\infty}(H),C_o^{\infty}(H')) \cong L(C_o^{\infty}(H),C_o^{\infty}(H)') = \hat{C}_o^{\infty}(H\boxtimes H)'. \qquad \square$$

<u>Definition .3</u> The category of differential equations $\mathcal{DE}$ is the category having as objects differential equations on fiber bundles and whose morphisms are morphisms of fiber bundles $f_{E_k}:E_k \to \overline{E}_{\overline{k}}$ over differentiable maps $f_M:M \to \overline{M}$ such that $f_{E_k}=\tilde{f}|_{E_k}$ where $\tilde{f}$ is a differentiable map $J\mathcal{D}^k(W) \to J\mathcal{D}^{\overline{k}}(\overline{W})$ over $f_W:W \to \overline{W}$ (and over $f_M:M \to \overline{M}$), i.e. the following diagram

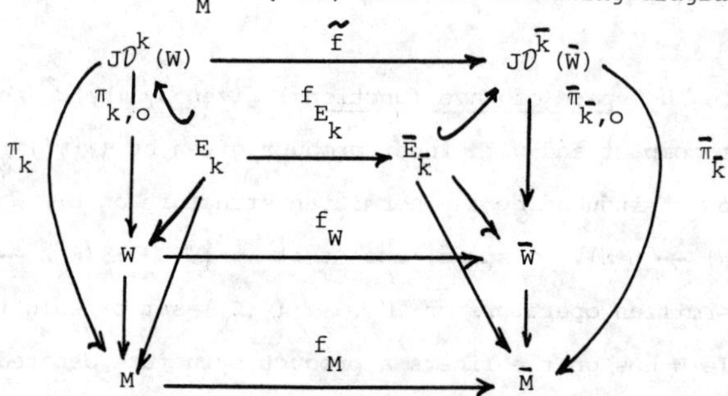

<u>Definition .4</u>  We say that two differential equations $E_k \subset J\mathcal{D}^k(W)$, $\overline{E}_{\overline{k}} \subset J\mathcal{D}^{\overline{k}}(\overline{W})$ belonging to $\mathrm{Ob}(\mathcal{DE})$ are equivalent if there exists a fiber bundle diffeomorphism $(f_W, f_M):W \to \overline{W}$ over $f_M:M \to \overline{M}$ such that it  induces a one to one map between the set of solutions of $E_k$ and $\overline{E}_{\overline{k}}$ respectively:
$$f_*:S(E_k) \to S(\overline{E}_{\overline{k}}) \qquad \text{given by} \qquad s \mapsto s'\equiv f_W \circ s \circ f_M^{-1}.$$
We denote by $[\mathcal{DE}]$    the set of equivalence classes in $\mathcal{DE}$   and $[E_k]$ denotes the equivalence classe in $[\mathcal{DE}]$ identified by $E_k$.

<u>Examples.</u>    Let $E_k \subset J\mathcal{D}^k(W)$  be a formally integrable differential equation of order k on   $\pi:W \to M$. Then, $E_{k+r} \subset J\mathcal{D}^{k+r}(W)$  is a differential equation of order k+r on W equivalent to $E_k$. Of course $W=\overline{W}$, $M=\overline{M}$, $f_M=\mathrm{id}_M$, $f_W=\mathrm{id}_W$. Note, that  there exists a canonical morphism belonging to $\mathrm{Hom}_{\mathcal{DE}}(E_{k+r}, E_k)$, but none belonging to $\mathrm{Hom}_{\mathcal{DE}}(E_k, E_{k+r})$.  More precisely $\pi_{k+r,k}:E_{k+r} \to E_k$ is the canonical morphism induced by the canonical surjection $J\mathcal{D}^{k+r}(W) \to J\mathcal{D}^k(W)$.

2.  Let $E_k \subset J\mathcal{D}^k(W)$  be a differential equation of order k on  $\pi:W \to M$. Let  $f_W:W \to \overline{W}$  be a fiber bundle diffeomorphism over a diffeomorphism $f:M \to \overline{M}$ being  $\overline{\pi}:\overline{W} \to \overline{M}$ a fiber bundle over $\overline{M}$. Then, $\overline{E}_k \subset J\mathcal{D}^k(f_W)(E_k) \subset J\mathcal{D}^k(\overline{W})$  is a differential equation equivalent to $E_k$.   In this case there is a canonical morphism belonging to $\mathrm{Hom}_{\mathcal{DE}}(E_k, \overline{E}_k)$ and its inverse belongs to $\mathrm{Hom}_{\mathcal{DE}}(\overline{E}_k, E_k)$,  this is the map $J\mathcal{D}^k(f_W)$.

The following theorem allows us to relate two equivalent equations by means of prolongations and diffeomorphisms.

<u>Theorem .2</u>  Let $\bar{E}_{\bar{k}} \in [E_k]$ .  We can distinguish three cases:

(1)  $\bar{k}=k$:  then there exists a fibered diffeomorphism $f_W : W \longrightarrow \bar{W}$ such that $\bar{E}_{\bar{k}} = J\mathcal{D}^k(f_W)(E_k)$.

(2)  $\bar{k} > k$, $(\bar{k}=k+q)$:  then there exists a fibered diffeomorphism $f_W : W \longrightarrow \bar{W}$ such that $\bar{E}_{\bar{k}} = J\mathcal{D}^{\bar{k}}(f_W)(E_{k+q})$.

(3)  $\bar{k} < k$, $(k=\bar{k}+q)$:  then there exists a fibered diffeomorphism $f_W : \bar{W} \longrightarrow W$ such that $E_k = J\mathcal{D}^k(f_{\bar{W}})(\bar{E}_{\bar{k}+q})$.

<u>Definition .5</u>  Let $E_k \subset J\mathcal{D}^k(W)$, $\bar{E}_{\bar{k}} \subset J\mathcal{D}^{\bar{k}}(W)$ be two differential equations. We say that the quantum situs $\Omega(E_k)$ is <u>equivalent</u> to $\Omega(\bar{E}_{\bar{k}})$ if there exists a fiber bundle diffeomorphism $f=(f_W, f_M) : W \longrightarrow \bar{W}$ over a diffeomorphism $f_M : M \longrightarrow \bar{M}$ such that it induces a one to one map $\Omega(f):\Omega(E_k) \longrightarrow \Omega(\bar{E}_{\bar{k}})$ given by: $\sigma \equiv D^k s(M) \longmapsto \Omega(f)(\sigma)=D^{\bar{k}}(f_W \circ s \circ f_M^{-1})(f_M(M)) \equiv f^* \sigma$.

<u>Theorem . 3</u>  $\Omega(E_k)$ is equivalent to $\Omega(\bar{E}_{\bar{k}})$ iff $E_k$ is equivalent to $\bar{E}_{\bar{k}}$.

<u>Proof.</u>  First let us assume that $E_k$ is equivalent to $E_k$ by means of a diffeomorphism $f \equiv (f_W, f_M) : W \longrightarrow \bar{W}$. If $\sigma \equiv D^k s(M) \subset \Omega(E_k)$ cobords two Cauchy data $(c' \equiv D^k s'(N'), c'' \equiv D^k s''(N'')) \in C(E_k) \times C(E_k)$, we have that $(\bar{c}' \equiv D^{\bar{k}}(f_W \circ s' \circ f_M^{-1})(f_M(N')), \bar{c}'' \equiv D^{\bar{k}}(f_W \circ s'' \circ f_M^{-1})(f_M(N''))) \in C(\bar{E}_{\bar{k}}) \times C(\bar{E}_{\bar{k}})$. Furthermore, $\bar{\sigma} \equiv D^{\bar{k}}(f_W \circ s \circ f_M^{-1})(f_M(M))$ is a Vinogradov submanifold of $J\mathcal{D}^{\bar{k}}(\bar{W})$ that necessarily cobords $\bar{c}'$ and $\bar{c}''$.

Vice versa, let $\Omega(E_k)$ be equivalent to $\Omega(\bar{E}_{\bar{k}})$ by means of the fiber bundle diffeomorphism $(f_W, f_M) : W \longrightarrow \bar{W}$ over $f_M : M \longrightarrow \bar{M}$. Then, by restriction to the dynamic cobords we obtain the equivalence between $E_k$ and $\bar{E}_{\bar{k}}$.

Then, we have the following functorial property for $\Omega(E_k)$.

<u>Theorem .4</u>  Let  $[E_k]$  be the category whose objects are differential equations equivalent to $E_k$  and whose morphisms are the fiber bundle diffeomorphisms between the corresponding fiber bundles.  Then,  $\Omega(E_k)$ identifies a covariant functor  $\Omega:[E_k] \longrightarrow S=$category of sets, such that $\Omega: E_k \longmapsto \Omega(E_k)$  and if  $f \in \text{Hom}_{[E_k]}(E_k, \bar{E}_{\bar{k}})$ ,  $\Omega(f):\sigma \longmapsto f^* \sigma$.

$\underline{\text{Theorem}}$ .5  Let $E_k$, $\bar{E}_{\bar{k}}$ be equivalent differential equations.  Then, any Dirac-quantization for $E_k$ canonically identifies a Dirac-quantization for $\bar{E}_{\bar{k}}$.

$\underline{\text{Proof}}$.  Let us assume be fixed a Dirac-quantization $(H=Q(W),Q,Q)$ for $E_k$. According to Theorem 5.19 we should distinguish three cases:

(1)  $k=\bar{k}$.  There exists the following canonical Dirac-quantization for $\bar{E}_{\bar{k}}$:
$(\bar{H}=Q(\bar{W}),Q,\bar{Q}=JD^h(Q(f_W^{-1}))^* \boxtimes Q(f_W) \circ Q \circ JD^k(f_W^{-1}))$; h=order of the quantization.

(2)  $\bar{k}>k$ , $(\bar{k}=k+q)$.  There exists the following canonical Dirac-quantization for $\bar{E}_{\bar{k}}$:  $(\bar{H}=Q(\bar{W}),Q,\bar{Q}=JD^h(Q(f_W^{-1})\boxtimes Q(f_W) \circ Q \circ_{\bar{k},k} JD^{\bar{k}}(f_W^{-1}))$.

(3)  $\bar{k}<k$, $(k=\bar{k}+q)$.  There exits the following canonical Dirac-quantization for $\bar{E}_{\bar{k}}$:  $(\bar{H}=Q(W),Q,\bar{Q}=JD^h(Q(f_W)^* \boxtimes Q(f_W) \circ Q \circ JD^{\bar{k}}(f_{\bar{W}}))$.  $\square$

$\underline{\text{Theorem}}$ .6  Let $E_k$, $\bar{E}_{\bar{k}}$ be equivalent equations.  Let $(H=Q(W),Q,Q)$ be a Dirac-quantization for $E_k$ and let $(\bar{H}=Q(\bar{W}),Q,\bar{Q})$ be the corresponding Dirac-quantization for $\bar{E}_{\bar{k}}$ (see Theorem .5) .  Let $E:\Omega(E_k) \longrightarrow L(H)$, $\bar{E}:\Omega(\bar{E}_{\bar{k}}) \longrightarrow L(\bar{H})$ be the maps induced by $Q$ and $\bar{Q}$ respectively.  Then, $E$ and $\bar{E}$ are realated each other by means of the following commutative diagram:

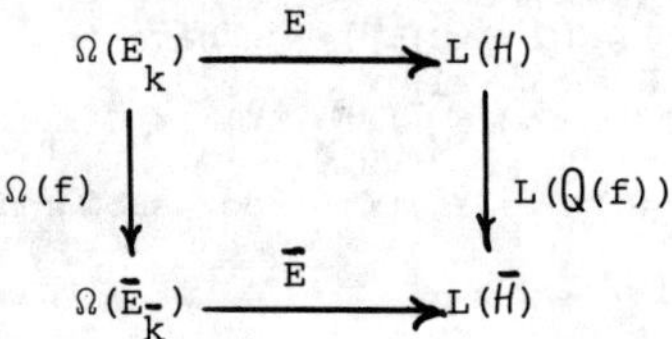


where  $L(Q(f))$ is the functorial extension of  $Q(f)$ by means of the functor L.


$\underline{\text{Definitions}}$  .6   A $\underline{\text{full covariant quantum spectral measure}}$  on a PDE $E_k \subset JD^k(W)$ is any quantum spectral measure $E:(\Omega,\Sigma) \longrightarrow L(H)$, where the Hilbert space is obtained by means of the sections with support compact of Hermitian vector bundle  $\pi:H \longrightarrow M$  such that $(W,H;Q)$ is a quantum super bundle of geometric objects for some quantum functor  $Q:F \longrightarrow F_H$.

$\underline{\text{Theorem}}$  .7  Let $E_k \subset JD^k(W)$,  $\bar{E}_{\bar{k}} \subset JD^{\bar{k}}(\bar{W})$  be equivalent equations by means of the fiber bundle diffeomorphirsm $f:W \longrightarrow \bar{W}$.  Let  $E:(\Omega,\Sigma) \longrightarrow L(H)$ be a full covariant spectral measure of $E_k$.  Then, there is a canonical full covariant quantum spectral measure $\bar{E}:(\bar{\Omega},\bar{\Sigma}) \longrightarrow L(\vec{H})$ on $\bar{E}_k$ with $\bar{\Omega}=\Omega(f)(\Omega)$ and $\bar{\Sigma}=\Omega(f)(\Sigma)$.

<u>Proof.</u> In fact E is identified by means of the following commutative diagram:

$$
\begin{array}{ccc}
(\Omega,\Sigma) & \xrightarrow{\;\;E\;\;} & L(H) \\
{\scriptstyle \Omega(f)} \big\downarrow & & \big\downarrow {\scriptstyle L(\;(f)_{\ast})} \\
(\bar{\Omega},\bar{\Sigma}) & \xrightarrow{\;\;\bar{E}\;\;} & L(H)
\end{array}
$$

where $\phi=Q(f)_{\ast}$ is the natural extension of $Q(f)$.

<u>Corollary 1.</u> For any function $\ell:\bar{\Omega}\to\mathbf{C}$ one has

$$(\Phi(\ell)(\bar{x})\,|\,\bar{y}) = \int_{\bar{\Omega}} \ell.\ d\bar{E}_{\bar{x},\bar{y}} = \int_{\Omega} (\ell\circ\Omega(f))\,dE_{\phi^{-1}(\bar{x}),\phi^{-1}(\bar{y})}$$

$$= (\Phi(\ell\circ\Omega(f))(\phi^{-1}(\bar{x}))\,|\,\phi^{-1}(\bar{y})).$$

<u>REMARK.</u> Above corollary translates the QUANTUM COVARIANCE in a form suitable to be applied in all practical cases.

<u>REMARK.</u> We have seen that any Dirac-quantization $Q:C^{\infty}(W) \longrightarrow L(H)$ of a PDE $E_k \subset J\mathcal{D}^k(W)$ identifies a map $E:\Omega \longrightarrow L(H)$ that can be considered as a particular spectral measure $E:(\Omega,\Sigma)\longrightarrow L(H)$ on any statistical set $\Omega$ of a scattering process, where the family $\Sigma$ of subsets of $\Omega$ is just $\Omega$. Then, to E corresponds a morphism $\Phi:BM(\Omega,\Sigma)\longrightarrow L(H)$ that associates to any (bounded) numerical function f on $\Omega$ an operator $\Phi(f)$ on $H$ such that:

($\bullet$) $(\Phi(f)(\psi)\,|\,\psi') = \int_{\Omega} f\,dE_{\psi,\psi'}$ , where $E_{\psi,\psi'}$ is the measure on $(\Omega,\Sigma)$ associated to E for $\psi,\psi'\in H$. Really, $E_{\psi,\psi'}$ is a good measure if $\Omega$ is a discrete set, otherwise the integral in ($\bullet$) has sense if $dE_{\psi,\psi'}=E_{\psi,\psi'}d\mu$, where $\mu$ is any measure on $\Omega$. In this case $E_{\psi,\psi'}$ can be interpreted as a Radon-Nikodym derivative on $\Omega$.

Therefore, we have the following important theorem that connects Dirac-quantization and quantum cobordism of PDEs.

<u>Theorem .8</u>   Let $E_k \subseteq J\mathcal{D}^k(W)$ be a PDE on the fiber bundle $\pi:W \to M$. Let $Q:J\mathcal{D}^k(W) \to J\mathcal{D}^h(H)^* \otimes H$ be a full covariant Dirac-quantization of $E_k$. Let $\Omega$ be a statistical set of $E_k$ endowed with a measure $\mu$.
Then, there exists a spectral measure on $\Omega$.

<u>Proof.</u>   In fact to the map $E:\Omega \to L(H)$, identified by $Q$, corresponds the following spectral measure:
$\check{E}:(\Omega,\Sigma) \to L(H)$ , $\check{E}(A)= \int_A E d\mu$ , $\forall A \in \Sigma$, where $\Sigma$ is the family of subsets of $\Omega$ associated to $\mu$ and $\check{E}(A)$ is the <u>Bochner integral</u> of $E$, that is a linear operator on $H$ such that $(\check{E}(\psi)|\psi')=\int_A (E(\psi)|\psi')d\mu$ , $\forall \psi, \psi' \in H$ , $\forall A \in \Sigma$.   $\square$

<u>Corollary 2.</u>   Under the hypotheses of theorem .8 we have that to any function $f:\Omega \to \mathbb{C}$ on $\Omega$ (quantum random variable), corresponds an operator $\Phi(f) \in L(H)$ given by $\Phi(f)= \int_\Omega f E d\mu$ such that $(\Phi(f)(\psi)|\psi')=\int_\Omega f(E(\psi)|\psi')d\mu$ , $\forall \psi, \psi' \in H$ .

So, in all cases where we have a canonical criterion to recognize a subset $\Omega_d$ of $\Omega$ that is whether discrete or endowed with a canonical measure, a Dirac-quantization gives a canonical quantum spectral measure on $E_k$.

We have two situations where these circumstances can be done.

<u>Theorem 9.</u>   Let $E_k \subseteq J\mathcal{S}^k(W)$ be a PDE endowed with a quantum variational constraint $(H^r(W), \gamma, f)$. Then, any solution of the corresponding quantum cobordism problem defines a 1-dimensional subset $\Omega_t$ of a statistical set that has a canonical measure .

<u>Theorem 10.</u>   Let $E_k \subseteq J\mathcal{S}^k(W)$ be a PDE endowed with a quantum-variational constraint $(H^r(W), \gamma, f)$. Let $I_f:H^r(W) \to \mathbb{R}$ satisfy the following conditions:
(C) (Palais-Smale) If $S$ is a subset of $H^r(W)$ on which $|I_f|$ is bounded but on which $\|\mathrm{grad}I_f\|$ is not bounded away from zero, then there is a critical point of $I_f$ in the closure of $S$.
(D)   All critical points of $I_f$ are non-degenerate.
Then for any statistical set $\Omega$ of $E_k$ we can recognize a discrete subset $\Omega_d$.
<u>Proof.</u> In fact for any real numbers $a < b$ there are only finitely many critical points of $I_f$ in $I_f^{-1}[a,b]$ .   $\square$

# 12 - APPLICATIONS

In this section, we shall apply former considerations to some fundamental differential equations of the field theory.

## 12.1 - KLEIN-GORDON EQUATION

Let $(M,g)$ be a 4-dimensional hyperbolic manifold. Consider the trivial vector bundle $\pi: E \equiv M \times \mathbb{R} \to M$. Klein-Gordon equation (KG) is the kernel of the following epimorphism of costant rank 5: $K_G \equiv (\Box + k): J\mathcal{D}^2(E) \to M \times \mathbb{R}$ being $\Box$ the Laplacian for scalar fields: $\Box \equiv \delta \bullet d$, and $k$ is a numerical function on $M$, $k: M \to \mathbb{R}$, given by $k \equiv \xi R + m^2$, $\xi \in \mathbb{R}$, $m \in \mathbb{R}^+$ (mass) and R is the Ricci scalar curvature. So, (KG) is a vector subbundle of $J\mathcal{D}^2(E)$ locally characterized by the following local numerical function on $J\mathcal{D}^2(E)$:

$F \equiv g^{\alpha\beta} y_{\alpha\beta} - \Gamma^{\alpha}_{\beta\gamma} g^{\gamma\beta} y_{\alpha} + k y$ , with respect to local coordinates: $\{x^{\alpha}, y, y_{\alpha}, y_{\alpha\beta}\}$ on $J\mathcal{D}^2(E)$. The symbol $g_2$ of (KG) has dim $g_2 = 9$. Furthermore, by considering the first prolongation $(KG)_{+1} \subset J\mathcal{D}^3(E)$ of (KG):

$$(KG)_{+1} \begin{cases} g^{\alpha\beta} y_{\alpha\beta} - \Gamma^{\alpha}_{\beta\gamma} g^{\gamma\beta} y_{\alpha} + k y = 0 \\[2mm] g^{\alpha\beta} y_{\alpha\beta\gamma} - \Gamma^{\alpha}_{\beta\gamma} g^{\gamma\beta} y_{\alpha\gamma} + (\partial x_{\gamma} \cdot g^{\alpha\beta}) y_{\alpha\beta} + k\, y_{\gamma} - \left[\partial x_{\gamma} \cdot (\Gamma^{\alpha}_{\beta\delta} g^{\delta\beta})\right] y_{\alpha} \\[2mm] \qquad\qquad\qquad\qquad\qquad\qquad + (\partial x_{\gamma} \cdot k)\, y = 0 \end{cases}$$

we see that $(KG)_{+1}$ is the kernel of an epimorphism of constant rank 9, so $(KG)_{+1}$ is a vector subbundle of $J\mathcal{D}^3(E)$ and dim $g_3 = 16$. Then, by using Proposition 3.2 we can prove that (KG) has an involutive symbol.
Furthermore, note that as $g_3$ is the kernel of an epimorphism of constant rank 9, $g_3$ is a vector bundle over (KG). So, (KG) is formally integrable iff the canonical map $(KG)_{+1} \to (KG)$ is an epimorphism. On the other hand as $\dim F_1 = 0$, from Theorem 3.4 we gat that the map $\pi_{3,2}: (KG)_{+1} \to (KG)$ is surjective, (and $\dim(KG)_{+1} - \dim(KG) = 34 - 18 = 16 = \dim g_3$). So, we conclude that (KG) is formally integrable.

Let us calculate, now, the distribution $\blacksquare R_2$ on (KG) defining the infinitesimal symmetries of (KG).

A vector field $X:E \to TE$ on a vector field $\tilde{X}:M \to TM$ is an infinitesimal symmetry for (KG) iff, with respect to a local coordinate system $\{x^\alpha, y\}$ on E, one has $\overset{(2)}{X} . F = \lambda\, F$ , where $\lambda : J\mathcal{D}^2(E) \to \mathbb{R}$. As $\overset{(2)}{X}$ is given by:

$$\overset{(2)}{X} = X^\alpha\, \partial x_\alpha + Y\partial y + \left[ -y_\sigma(\partial x_\alpha x^\sigma) + y_\alpha(\partial y.Y) + (\partial x_\alpha.Y) \right] \partial y^\alpha +$$
$$+ \left[ -y_{j_1\sigma}(\partial x_{j_2}.x^\sigma) - y_{\sigma j_2}(\partial x_{j_1}.x^\sigma) - (\partial x_{j_1}\partial x_{j_2}.x^\sigma)y_\sigma + y_{j_1 j_2}(\partial y.Y) + \right.$$
$$\left. +(\partial x_{j_1}\partial y.Y)y_{j_2} + (\partial x_{j_1}\partial x_{j_2}.Y) + (\partial x_{j_2}\partial y.Y)y_{j_1} + (\partial y\partial y.Y)y_{j_1}y_{j_2} \right] \partial y^{j_1 j_2}$$

we have:

$$X^\alpha(\partial x_\alpha . g^{\omega\beta})y_{\omega\beta} - X^\alpha(\partial x_\alpha . \Gamma^\omega_{\beta\gamma})g^{\gamma\beta}\, y_\omega - X^\alpha \Gamma^\omega_{\beta\gamma}(\partial x_\alpha . g^{\gamma\beta})y_\omega + Yk +$$
$$\left[ -y_\sigma(\partial x_\alpha.x^\sigma) + y_\alpha(\partial y.Y) + (\partial x_\alpha.Y) \right] g^{\gamma\beta}\Gamma^\alpha_{\beta\gamma} + \left[ -y_{j_1\sigma}(\partial x_{j_2}.x^\sigma) - y_{\sigma j_2}(\partial x_{j_1}.x^\sigma) \right.$$
$$(\uparrow) \qquad - (\partial x_{j_1}\partial x_{j_2}.x^\sigma)y_\sigma + y_{j_1 j_2}(\partial y.Y) + (\partial x_{j_1}\partial y.Y)y_{j_2} + (\partial x_{j_1}\partial x_{j_2}.Y) + (\partial x_{j_2}\partial y.Y)y_{j_1} +$$
$$\left. +(\partial y\partial y.Y)y_{j_1}y_{j_2} \right] g^{j_1 j_2}$$
$$= \lambda\, g^{\omega\beta}y_{\omega\beta} - \lambda \Gamma^\omega_{\beta\gamma}g^{\gamma\beta}\, y_\omega + \lambda ky.$$

Now, as the second derivative $y_{\omega\beta}$ occur only linearly in the first terms of equation $(\uparrow)$ $\lambda$ can only depend on $(x^\alpha, y, y_\alpha)$. Then, the coefficient of $y_{\omega\beta}$ is:

$$X^\alpha(\partial x_\alpha . g^{\omega\beta}) - g^{\omega j_2}(\partial x_{j_2}.x^\beta) - g^{\beta j_1}(\partial x_{j_1}.x^\omega) + g^{\omega\beta}(\partial y.Y) = \lambda\, g^{\omega\beta} \ .$$

So, we must have $\ (\downarrow)\ $ $(\ L_{\tilde{X}}g)^{\omega\beta} = g^{\omega\beta}\left[ \lambda - (\partial y.Y) \right]$ . From $(\downarrow)$ we get: $\lambda - (\partial y.Y) = f(x^\alpha)$. So $\lambda$ can only depend on $(x^\alpha, y)$.

The coefficients of $y_\alpha y_\beta$ and $y_\alpha$ are respectively:

(a$\downarrow$) $(\partial y\partial y.Y)y_{j_1}y_{j_2}\, g^{j_1 j_2} = 0$ $\Rightarrow$ $Y = \phi(x^\alpha)y + c$ , $\phi:M \to \mathbb{R}$ , $c \in \mathbb{R}$; $\lambda = f + \phi$;

(b$\downarrow$) $-(\partial x_{j_1}\partial x_{j_2}.x^\omega)g^{j_1 j_2} - (\partial x_\alpha.x^\omega)A^\alpha - X^\alpha(\partial x_\alpha.A^\omega) + 2\phi A^\omega + 2(\partial x_\sigma.\phi)g^{\omega\sigma} + fA^\omega = 0.$

From the coefficient of y we get: $\square\phi - kf = 0$ . Furthermore, we have also $c = 0$. Assembling the above equations we get that X is given by $X = X^\alpha\,\partial x_\alpha + \phi y\, \partial y$, where $X^\alpha, \phi : M \to \mathbb{R}$ are real functions on M solutions of the following linear differential equation:

$$\blacksquare(KG) \left\{ \begin{array}{l} \square\phi - \dfrac{k}{tr(g)}\, tr(L_{\tilde{X}}g) = 0 \\[2em] -(\partial x_\alpha.\partial x_\beta.x^\omega)g^{\alpha\beta} - (\partial x_\alpha.x^\omega)A^\alpha - X^\alpha(\partial x_\alpha.A^\omega) + 2\phi A^\omega + 2(\partial x_\sigma.\phi)g^{\omega\sigma} + \dfrac{A^\omega k}{tr(g)}\, tr(L_{\tilde{X}}g) = 0 \end{array} \right.$$

where $A^\omega \equiv \Gamma^\omega_{\beta\gamma}g^{\gamma\beta} : M \to \mathbb{R}$.

Let us, now, consider the Dirac-quantization of (KG). We shall consider the following natural geometric objects canonically associated to (KG):

1) **Hilbert bundle over M:** $H = \mathrm{Clif}^C(M) = $ complex Clifford bundle over M. The hermitian structure of $H$ is the natural extension on $\mathrm{Clif}^C(M)$ of the hermitian structure $\underline{h}$ on $T^C M \equiv \mathbf{C} \otimes TM$ defined for real vector fields by: $\underline{h}(i\xi,\nu) = ig(\xi,\nu)$, $\underline{h}(\xi,i\nu) = -ig(\xi,\nu)$ and extended by linearity to complex vector fields.

2) $H$ = space of borelian sections square integrable of $H$, endowed with the inner product : $(\psi \mid \psi') = \int_M h(\psi,\psi').\eta$ , where $\eta = \star 1$ is the canonical volume form on M.

3) **First-order linear Dirac-quantization operator:**

$$(\spadesuit) \qquad Q: \mathcal{JD}(E). \longrightarrow \mathcal{JD}(H)^{\star} \otimes H \quad , \quad Q \cdot Df \equiv Q.f = \overset{\lor}{\nabla}_{\mathrm{grad} f} + \frac{1}{2} kf \ ,$$

where $k$ is as in above equation (KG) and $\overset{\lor}{\nabla}$ is the absolute differential on $\mathrm{Clif}^C(M)$ canonically associated to the Levi-Civita connection corresponding to g. More precisely $\overset{\lor}{\nabla}$ is the extension on $\mathrm{Clif}^C(M)$ of the complexificated absolute differntial $\nabla^C$ on $T^C M$ given by: $\nabla^C_\xi \nu = \nabla_\xi \nu$ , $\nabla^C_{(i\xi)}\nu = \nabla^C_\xi(i\nu) = i\nabla_\xi \nu$ for real vector fields $\xi, \nu : M \to TM$ and extended by linearity to arbitrary complex vector fields on M. One has:

(a) $\quad \overset{\lor}{\nabla}_{f\xi}\psi = f\nabla_\xi \psi \qquad , \qquad \overset{\lor}{\nabla}_\xi(f\psi) = f\overset{\lor}{\nabla}_\xi\psi + (\xi.f)\psi \qquad , \forall f: M \to \mathbf{C} \ , \ \forall \psi \in C^\infty(H)$

(b) $\quad \zeta.h(\psi,\psi') = h(\overset{\lor}{\nabla}_\zeta \psi,\psi') + h(\psi,\overset{\lor}{\nabla}_\zeta \psi') \quad , \ \forall \zeta \in C^\infty(T^C M), \forall \psi,\psi' \in C^\infty(H)$.

(c) The following diagram is commutative:

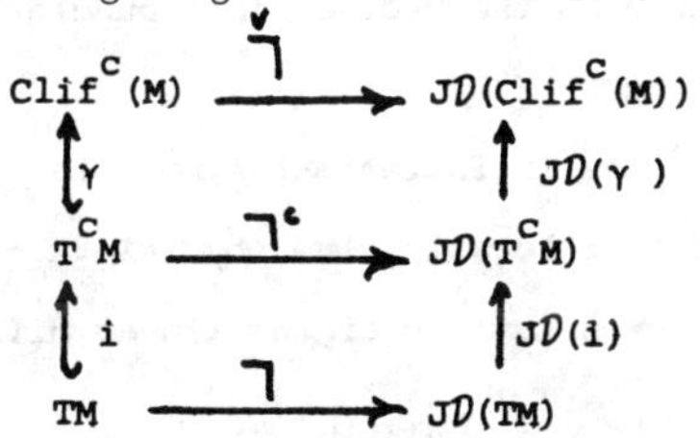

$$
\begin{array}{ccc}
\mathrm{Clif}^C(M) & \overset{\overset{\lor}{\urcorner}}{\longrightarrow} & \mathcal{JD}(\mathrm{Clif}^C(M)) \\
\Big\uparrow{\scriptstyle \gamma} & & \Big\uparrow{\scriptstyle \mathcal{JD}(\gamma)} \\
T^C M & \overset{\urcorner^c}{\longrightarrow} & \mathcal{JD}(T^C M) \\
\Big\uparrow{\scriptstyle i} & & \Big\uparrow{\scriptstyle \mathcal{JD}(i)} \\
TM & \overset{\urcorner}{\longrightarrow} & \mathcal{JD}(TM)
\end{array}
$$

where $\gamma$ and i are the canonical embeddings , $\urcorner$ is the Levi-Civita connection, $\urcorner^c$ is the complexificated connection induced by $\urcorner$ on $T^C M$ and $\overset{\lor}{\urcorner}$ is the corresponding connection induced on $\mathrm{Clif}^C(M)$.

Now, let us impose the condition of anti-hermitianicity for Q.f with respect to the product of $H$ . From above equation (b) we can write

$$h(Q.f(\psi),\psi') + h(\psi,Q.f(\psi')) = \langle dh(\psi,\psi') \mid \mathrm{grad} f \rangle + h(f\tfrac{k}{2}\psi,\psi') + h(\psi,\tfrac{k}{2}f\psi')$$

$$= \langle dh(\psi,\psi') \mid \mathrm{grad} f \rangle + kf\, h(\psi,\psi').$$

Then, by integrating on M and taking into account that

$$\int_M \langle dh(\psi,\psi') \,|\, \mathrm{grad}f \rangle \cdot \eta = \int_M dh(\psi,\psi') \wedge *df = \langle dh(\psi,\psi'), df \rangle = \langle h(\psi,\psi'), \delta df \rangle$$

$$= \langle h(\psi,\psi'), \Box f \rangle$$

where $\langle , \rangle$ is the usual scalar product in $C_o^\infty(T^* M)$, we get that Q.f is anti-hermitian iff f is a solution of the equation (KG).  So, the following diagram:

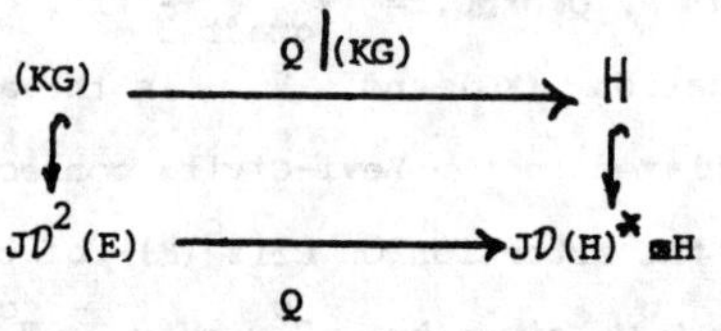

$$
\begin{array}{ccc}
(KG) & \xrightarrow{\;\;Q\,|(KG)\;\;} & H \\[2pt]
\Big\downarrow & & \Big\downarrow \\[2pt]
J\mathcal{D}^2(E) & \xrightarrow[\;\;Q\;\;]{} & J\mathcal{D}(H)^* \boxempty H
\end{array}
$$

is commutative, where we have denoted by Q also the natural lifting of Q to $J\mathcal{D}^2(E)$.

Therefore, we have proven the following

<u>Theorem  .1</u>   Klein-Gordon equation (KG)  is an involutive formally integrable linear second-order differential equation on the vector fiber bundle $\pi: E \equiv M \times \mathbb{R} \to M$. (KG)  is also completely integrable.

The distribution $\boxempty R_2$ on (KG) defining the infinitesimal symmetries of (KG) is generated by vector fields $X: E \to TE$ on E locally given by $X = X^\alpha \partial x_\alpha + y\phi \partial y$, where $X^\alpha, \phi: M \to \mathbb{R}$ are real functions on M solutions of the linear differential equation given in $\boxempty$(KG).

(KG) admits a canonical Dirac-quantization: $Q(KG) \equiv (H \equiv \mathrm{Clif}^C(M), \mathbb{Q} \equiv \mathrm{Clif}^C, Q)$, where : (a) $\mathrm{Clif}^C$ is the complexificated Clifford functor; (b) Q is given in $(\boxempty)$.

## 12.2 - MAXWELL EQUATION

Let M be a four dimensional manifold, with a Lorentz metric $g:M \to \overset{o}{S}_2 M$.

Let $E = \overset{*}{T} M$ and let $C \equiv \Gamma(E) \subset J\mathcal{D}(E)$ be the Levi-Civita connection.

Maxwell-equation is the kernel $(M) \subset J\mathcal{D}^2(E)$ of the following second order differential operator $M:J\mathcal{D}^2(E) \to E$ given by $M \bullet D^2 A = \delta dA$. Let $\{x^\alpha, \overset{.}{x}_\alpha, \overset{.}{x}_{\alpha\beta}, \overset{.}{x}_{\alpha\beta\gamma}\}$ be a fibered coordinate system on $J\mathcal{D}^2(E)$, then one has the following local expression for (M):

$$(M) \qquad F_\alpha \equiv g^{\beta\gamma}(\overset{.}{x}_{[\alpha\beta]\gamma} + \Gamma^\delta_{\beta\gamma} \overset{.}{x}_{[\alpha\delta]} + \Gamma^\delta_{\beta\alpha} \overset{.}{x}_{[\gamma\delta]}) = 0,$$

being $\Gamma^\delta_{\beta\gamma} :M \to \mathbb{R}$ the connection coefficients of C. $M$ is an epimorphism of constant rank 8, so (M) is a vector subbundle of $J\mathcal{D}^2(E)$ of dimension $\dim(M)=60$, with symbol $g_2$ of dimension: $\dim g_2 = 36$. Furthermore, by considering the first prolongation $(M)_{+1} \subset J\mathcal{D}^3(E)$ of (M):

$$(M)_{+1} \begin{cases} F_\alpha = 0 \\ (\partial x_\omega \cdot g^{\beta\gamma})(\overset{.}{x}_{[\alpha\beta]\gamma} + \Gamma^\delta_{\beta\gamma} \overset{.}{x}_{[\alpha\delta]} + \Gamma^\delta_{\beta\alpha} \overset{.}{x}_{[\gamma\delta]}) + g^{\beta\gamma}(\partial x_\omega \cdot \Gamma^\delta_{\beta\gamma})\overset{.}{x}_{[\alpha\delta]} + g^{\beta\gamma}(\partial x \Gamma^\delta_{\beta\alpha})\overset{.}{x}_{[\gamma\delta]} \\ \qquad + g^{\beta\gamma}(\Gamma^\delta_{\beta\gamma} \overset{.}{x}_{[\alpha\delta]\omega} + \Gamma^\delta_{\beta\alpha} \overset{.}{x}_{[\gamma\delta]\omega}) + g^{\beta\gamma}\overset{.}{x}_{[\alpha\beta]\gamma\omega} = 0 \end{cases}$$

we see that $(M)_{+1}$ is the kernel of an epimorphism of constant rank 24. So, $(M)_{+1}$ is a vector subbundle of $J\mathcal{D}^3(E)$ and $\dim(M)_{+1}=124$. Furthermore, $\dim g_3 = 64$. Note that as $g_3$ is the kernel of an epimorphism of constant rank, $g_3$ is a vector bundle over (M). Let us calculate the dimension of $F_1$:

$$\dim F_1 = 4 \times 4 - 4\frac{(4+2)!}{3! \, 3!} + 64 = 0.$$

So, we can conclude that the map $(M)_{+1} \to (M)$ is surjective. Finally, we can prove that (M) is involutive. In fact:

$$(\dim g_3 = 64) = (\dim g_2 = 36) + (\dim g_2^{(1)} = 20) + (\dim g_2^{(2)} = 8) + (\dim g_2^{(3)} = 0) = 64.$$

Therefore, (M) is a formally integrable differential equation and, as we are in the analytical case, it is also completely integrable.

Now, let us consider the Dirac-quantization of (M). We have the following natural geometric objects canonically associated to (M):

1)  Hilbert bundle over M:   $H = Clif^C(M)$;

2)  Hilbert space $H$  as in section 6.1;

3)  First-order linear Dirac-quantization operator:

$(\clubsuit)$    $Q : J\mathcal{D}(E) \longrightarrow J\mathcal{D}(H)^{\bowtie} \boxtimes H$,   $Q \bullet DA \equiv Q.A = \overset{\lor}{\nabla}_{X(A)} + \frac{1}{2} \langle \nabla A, {}_A\bar{F} \rangle$

being $\overset{\lor}{\nabla}$  the absolute differential canonically associated to the Levi Civita connection,  $X(A) : M \rightarrow TM$ is the vector field canonically associated to the differential 1-form A given by  $X(A) = (A_\alpha F^{\alpha\beta}) \partial x_\beta$  ,  ${}_A F \equiv dA = F_{\alpha\beta} dx^\alpha \wedge dx^\beta = (\partial x_\alpha . A_\beta - \partial x_\beta . A_\alpha) dx^\alpha \wedge dx^\beta$  and ${}_A\bar{F}$  is the controvariant expression of ${}_A F$. Then, by imposing the condition of anti-hermitianicity to Q.A, and taking into account that  $h(\overset{\lor}{\nabla}_{X(A)} \psi, \psi') + h(\psi, \overset{\lor}{\nabla}_{X(A)} \psi') = \langle dh(\psi, \psi') | X(A) \rangle$   we get

$$0 = \int_M \langle dh(\psi, \psi') | X(A) \rangle . \eta + \int_M h(\psi, \psi') . \langle \nabla A, {}_A\bar{F} \rangle . \eta$$

$$\|$$

$$\int_M \{ h(\psi, \psi') . [ \delta \underline{X}(A) + \langle \nabla A, {}_A\bar{F} \rangle ] \} . \eta \quad ,$$

being $\underline{X}(A)$ the covariant expression of X(A).   So, Q.A  is anti-hermitian iff $\delta X(A) + \langle \nabla A, {}_A\bar{F} \rangle = 0$.  Taking into account that $\delta \underline{X}(A) = -\nabla_\beta (A_\alpha F^{\alpha\beta}) = -(\nabla_\beta A_\alpha) F^{\alpha\beta} + -A_\alpha (\nabla_\beta F^{\alpha\beta})$, we get that  Q.A  is anti hermitian iff  $A_\alpha (\nabla_\beta F^{\alpha\beta}) = 0$. This condition is certainly verified if A is a solution of (M).

So, we can conlcude with the following

<u>Theorem</u> .2    Maxwell equation (M)  is an involutive formally integrable linear second-order differential equation on the vector fiber bundle $\pi : E \equiv \overset{x}{T} M \rightarrow M$. (M)  is also completely integrable.  (M)  admits a canonical Dirac-quantization: $Q(M) \equiv (H = Clif^C(M), \hat{Q} \equiv Clif^C, Q)$, where  Q is given in  $(\clubsuit)$.

<u>Remark</u> .1    If one requires that the e.m. potential A satisfies the Lorentz equation (L):

(L)                $\delta A \equiv g^{\alpha\beta} [ (\partial x_\alpha . A_\beta) - \Gamma^\delta_{\alpha\beta} A_\delta ] = 0$;

then the Dirac quantization operator $(\clubsuit)$  is  substituted by the following one

$(\clubsuit L)$            $Q.A = \overset{\lor}{\nabla}_{X(A) + \bar{A}} + \frac{1}{2} \langle \nabla A, {}_A\bar{F} \rangle$ ,

where $\bar{A}$  is the controvariant expression of A.  In fact, in this case Q.A is anti-hermitian iff  $-A_\alpha (\nabla_\beta F^{\alpha\beta}) - \delta A = 0$.  So, on the combined equations (L)+(M) the operator $(\clubsuit L)$ is anti-hermitian.

## 12.3 - DIRAC EQUATION

Let M be a 4 dimensional hyperbolic manifold with metric field g and volume form $\eta = *1$. Set $E = \mathrm{Clif}^{\mathbb{C}}(M)$.

As pointed out first in ref.[12] , Dirac equation (D) can be described in a full covariant way as the kernel of the following differential operator:
$\mathcal{D}_{ir} : J\mathcal{D}(E) \rightarrow \mathrm{Cli}^{\mathbb{C}}(M)$, $\mathcal{D}_{ir}.\psi = \overset{\vee}{\pi} \cdot \overset{\vee}{\nabla}\psi + m.\psi$ , where $\overset{\vee}{\nabla}$ is the absolute differential on $\mathrm{Clif}^{\mathbb{C}}(M)$ associated to the Levi-Civita connection on $(M,g)$ and $\overset{\vee}{\pi}$ is the canonical map $\overset{\vee}{\pi} : T^* M \otimes \mathrm{Cli}^{\mathbb{C}}(M) \rightarrow \mathrm{Clif}^{\mathbb{C}}(M)$ given by composition
$T^* M \otimes \mathrm{Clif}^{\mathbb{C}}(M) \rightarrow T^{\mathbb{C}} M^* \otimes \mathrm{Clf}^{\mathbb{C}}(M) \cong T^{\mathbb{C}} M \otimes \mathrm{Clif}^{\mathbb{C}}(M) \overset{a}{\rightarrow} \mathrm{Clif}^{\mathbb{C}}(M)$ being a the map imduced by Clifford multiplication. Furthermore, $m \in \mathbb{R}$ is the __mass__. The local expression of (D) with respect to local coordinates $\{ x^{\alpha}, z^{i_1 \ldots i_s}, z^{i_1 \ldots i_s}_{\alpha} \}$
( $0 \leq i_1 < \ldots < i_s \leq 3$ , $0 \leq \alpha \leq 3$ ), on $J\mathcal{D}(E)$ is the following:

$$(D) \quad \sum_{\substack{0 \leq i_1 < \ldots < i_s \leq 3 \\ 0 \leq \alpha \leq 3}} (A^{j_1 \ldots j_s \alpha}_{i_1 \ldots i_s} z^{i_1 \ldots i_s}_{\alpha} + B^{j_1 \ldots j_s}_{i_1 \ldots i_s} z^{i_1 \ldots i_s}) = 0 \ , \quad 0 \leq j_1 < \ldots < j_s \leq 3 \ , \quad 1 \leq s \leq 4$$

being
$$A^{j_1 \ldots j_s \alpha}_{i_1 \ldots i_s} \equiv \sum_{0 \leq \beta \leq 3} \gamma^{j_1 \ldots j_s}_{\beta \, i_1 \ldots i_s} g^{\alpha \beta} : M \rightarrow \mathbb{C};$$

$$B^{j_1 \ldots j_s \alpha}_{i_1 \ldots i_s} \equiv \sum_{\substack{0 \leq 1_1 < \ldots < 1_s \leq 3 \\ 0 \leq \alpha, \beta \leq 3}} (\gamma^{j_1 \ldots j_s}_{\beta 1_1 \ldots 1_s} g^{\alpha \beta} \Gamma^{1_1 \ldots 1_s}_{\alpha i_1 \ldots i_s}) + m \, \delta^{j_1 \ldots j_s}_{i_1 \ldots i_s} : M \rightarrow \mathbb{C},$$

where $\Gamma^{1_1 \ldots 1_s}_{\alpha i_1 \ldots i_s}$ are the connection coefficients of the connection on $\mathrm{Clif}^{\mathbb{C}}(M)$ and $\gamma^{j_1 \ldots j_s}_{\beta \, 1_1 \ldots 1_s}$ are the components of $a(\partial x_{\beta} \otimes \partial x_{p_1} \ldots \partial x_{p_s})$.

Note that we are working with mixed real-complex coordinates. Of course, it is better to consider only real coordinates. So, put:
$x^{i_1 \ldots i_s} \equiv \mathrm{Re}\, z^{i_1 \ldots i_s}, \ y^{i_1 \ldots i_s} \equiv \mathrm{Im}\, z^{i_1 \ldots i_s}, \ x^{i_1 \ldots i_s}_{\alpha} \equiv \mathrm{Re}\, z^{i_1 \ldots i_s}_{\alpha}, \ y^{i_1 \ldots i_s}_{\alpha} \equiv$
and $\mathrm{Im}\, z^{i_1 \ldots i_s}_{\alpha}$

$\overline{A}^{j_1 \ldots j_s \alpha}_{i_1 \ldots i_s} \equiv \mathrm{Re}\, A^{j_1 \ldots j_s \alpha}_{i_1 \ldots i_s}, \ \tilde{A}^{j_1 \ldots j_s \alpha}_{i_1 \ldots i_s} \equiv \mathrm{Im}\, A^{j_1 \ldots j_s \alpha}_{i_1 \ldots i_s}, \ \overline{B}^{j_1 \ldots j_s}_{i_1 \ldots i_s} \equiv \mathrm{Re}\, B^{j_1 \ldots j_s}_{i_1 \ldots i_s}$

$\tilde{B}^{j_1 \ldots j_s}_{i_1 \ldots i_s} \equiv \mathrm{Im}\, B^{j_1 \ldots j_s}_{i_1 \ldots i_s}$

Then, equation (D) becomes in real coordinates:

$$(D)\begin{cases}
\bar{A}^{j_1\cdots j_s\,\alpha}_{i_1\cdots i_s}\,x_\alpha^{i_1\cdots i_s} - \overset{*}{A}^{j_1\cdots j_s\,\alpha}_{i_1\cdots i_s}\,y_\alpha^{i_1\cdots i_s} + \bar{B}^{j_1\cdots j_s}_{i_1\cdots i_s}\,x^{i_1\cdots i_s} - \overset{*}{B}^{j_1\cdots j_s}_{i_1\cdots i_s}\,y^{i_1\cdots i_s} = 0 \\[2ex]
\bar{A}^{j_1\cdots j_s\,\alpha}_{i_1\cdots i_s}\,y_\alpha^{i_1\cdots i_s} + \overset{*}{A}^{j_1\cdots j_s\,\alpha}_{i_1\cdots i_s}\,x_\alpha^{i_1\cdots i_s} + \bar{B}^{j_1\cdots j_s}_{i_1\cdots i_s}\,y^{i_1\cdots i_s} + \overset{*}{B}^{j_1\cdots j_s}_{i_1\cdots i_s}\,x^{i_1\cdots i_s} = 0.
\end{cases}$$

$\mathcal{D}_{ir}$ is an epimorphism of constant rank 36. So, (D) is a vector subbundle of $J\mathcal{D}(E)$ of dimension $\dim(D)=132$ with symbol $g_1$ of dimension $\dim g_1=96$.

Furthermore, by considering the first prolongation $(D)_{+1}\subset J\mathcal{D}^2(E)$ of (D) we see that $(D)_{+1}$ is the kernel of an epimorphism of constant rank, so $(D)_{+1}$ is a vector subbundle of $J\mathcal{D}^2(E)$ with dimension $\dim(D)_{+1}=324$. Moreover, the first prolongation $g_2$ of the symbol $g_1$ is a vector bundle over (D) of fiber dimension $\dim g_2=192$. Furthermore, the dimension of $F_1$ is $\dim F_1=4\times32-32\frac{(4+1)!}{2!3!}+192=0$. So, we can conclude that the canonical map $(D)_{+1}\longrightarrow(D)$ is surjective.

Finally, we can prove that (D) is involutive. In fact:

$(\dim g_2=192)=(\dim g_1=96)+(\dim g_1^{(1)}=64)+(\dim g_1^{(2)}=32)+(\dim g_1^{(3)}=0).$

Therefore, (D) is a formally integrable differential equation and, as we are in the analytic case, it is also completely integrable.

Now, consider the Dirac-quantization of (D). We have the following natural geometric objects canonically associated to (D):

1) Hilbert bundle over M: $H\equiv\mathrm{Clif}^C(M)$;

2) Hilbert space $H$ as in section 6.1;

3) First-order linear Dirac-quantization operator:

($\bullet$)  $Q:J\mathcal{D}(E)\longrightarrow J\mathcal{D}(H)\boxtimes H$ , $Q\bullet D\phi\equiv Q.\phi=\overset{v}{\nabla}_{X(\phi)} + \left[-m.h(\phi,\phi)+f(\phi)\right]$ ,

being $\overset{v}{\nabla}$ the canonical absolute derivative associated to the Levi Civita connection, and $X(\phi):M\to TM$ is a vector field, canonically associated to the field $\phi:M\to\mathrm{Clif}^C(M)$, given by : $X(\phi)=Y(\phi)+\bar{Y}(\phi)$, $\bar{Y}(\phi)\equiv$complex conjugate of $Y(\phi)$, $Y(\phi)\equiv(id_{TM}\boxtimes h_\phi)\circ(id_{TM}\boxtimes a)\circ(g\boxtimes\phi):M\to TM\boxtimes TM\boxtimes E\to TM\boxtimes E\to TM\boxtimes C\cong \mathfrak{f}_M$, where $h_\phi$ is the map induced by the hermitian product on H and $\phi$ .

The local expression of $Y(\phi)$ is the following: $Y(\phi)=g^{\nu\mu}\,\phi^A\,\gamma^B_{\nu A}\,\bar{\phi}_B\,\partial x_\mu$ .

Moreover, $f(\phi)$ is the following complex numerical function on M:

$$f(\phi) \equiv \phi^A \nabla_\mu (g^{\mu\nu} \gamma^B_{\nu A} \bar{\phi}_B) : M \to \mathbb{C} \, .$$  Then, by imposing the condition of anti-hermitianicity for $Q.\phi$ , a calculus similar to previous one gives that $Q.\phi$ is anti-hermitian iff: $\left[ (\nabla_\mu \phi^A) g^{\mu\nu} \gamma^B_{\nu A} + m\delta^B_A \phi^A \right] \bar{\phi}_B + c.c. = 0$. On the other hand on equation (D) above condition is verified.

So, we have proven the following:

<u>Theorem  .3</u>  Dirac equation (D) is an involutive formally integrable linear first-order differential equation on the vector fiber bundle $\pi : E \equiv \mathrm{Clif}^C(M) \to M$. (D) is also completely integrable.  (D) admits a canonical Dirac-quantization: $Q(D) \equiv (H \equiv \mathrm{Clif}^C(M)$, $\hat{Q} \equiv \mathrm{Clif}^C, Q)$, where $Q$ is given in (●).

<u>Remark  .2</u>  Some authors (e.g. Lichnerowicz) have proposed to consider for spinor fields a second order linear differential equation that we can generalize as the kernel of the following differential operator:
$$L : JD^2(E) \to E, \quad L \circ D^2 \phi = \hat{g} \bullet (\nabla^2 \phi) + (m^2 - \tfrac{R}{4}) \phi \, ,$$
where $\nabla^2 \phi$ is the second-order absolute differential  induced on $E$ by the Levi-Civita connection; $\hat{g}$ is the map $T^*M \otimes T^*M \otimes E \to {\rm I\!R} \otimes E \cong E$ canonically induced by the metric and R is the Ricci scalar curvature . The local expression of $(L) \equiv \ker L \subset JD^2(E)$ is the following:

$$(L) \qquad g^{\alpha\beta} y_{\alpha\beta} + \Gamma^A_{\alpha\beta\, B}\, y^B_\nu + \Gamma^A_{\alpha\beta}\, y^A + (m^2 - \tfrac{R}{4}) y^A = 0,$$

where  $\{x^\alpha, y^\alpha, y^A_\alpha, y^A_{\alpha\beta}\}$ is a local coordinate system on $JD^2(E)$  and $\Gamma^{\cdot\cdot}_{\cdot\cdot}$ are the connection coefficients; A and B are multindices.

The Dirac-quantization of the combined equations $(L)+(D)$ is obtained by substituting the operator $Q$ in (●) by the following one:

$$(●L) \qquad Q. \equiv \overset{\vee}{\nabla}_{X(\phi)+Z(\phi)} + \left[ m.h(\phi,\phi) + f(\phi) + \ell(\phi) + (m^2 - \tfrac{R}{4}) h(\phi,\phi) \right] , \quad Z(\phi) = U(\phi) + \bar{U}(\phi),$$

being  $X(\phi)$  and $f(\phi)$ as in Theorem 6.3 . Moreover, $U(\phi)$ is the vector field associated to  $\phi$ given by composition: $U(\phi) \equiv h_\phi \circ ('g \otimes id_E) \circ \nabla \phi : M \to T^*M \otimes E \cong TM \otimes E \to TM$
Locally  $U(\phi)$ is given by $U(\phi) = g^{\mu\nu} (\nabla_\mu \phi^A) \bar{\phi}_A \partial x_\nu$ ;  $\ell(\phi)$ is the complex numerical function on M given by $\ell(\phi) = \nabla_\nu (g^{\mu\nu} \bar{\phi}_A) ( \nabla_\mu \phi^A) : M \to \mathbb{C}$.

In fact, by imposing the condition of anti-hermitianicity for $Q.\phi$ , a calculus similar to previous ones  gives that $Q.\phi$ in (●L) is anti-hermitian iff: $\left\{ \left[ (\nabla_\mu \phi^A) g^{\mu\nu} \gamma^B_{\nu A} + m. \delta^B_A \phi^A \right] + \left[ g^{\mu\nu} (\nabla_\nu \nabla_\mu \bar{\phi}^B) + (m^2 - \tfrac{R}{4}) \delta^B_A \phi^A \right] \right\} \bar{\phi}_B + c.c. = 0$.
Above condition is certainly sutisfied on the combained equations $(L)+(D)$.

## 12.4 - EINSTEIN EQUATION

Let M be a compact four dimensional manifold .  Before we  consider Einstein equation, we shall introduce some fundamental machinary on the linear connections on a vector bundle.

<u>Proposition</u> .1  1.  A linear differential connection $\bar{\Gamma}:TE \to vTE$ on a vector bundle  $\pi:E \to M$  identifies a fiber bundle morphism $\daleth D \equiv r \cdot \bar{\Gamma}:TE \to E$ over  $\pi$  called the <u>connection map</u> of $\bar{\Gamma}$ , being  $r:vTE \to E$  the canonical surgection.

2.  Let  $\{x^\alpha, y^j\}$  be a coordinate system on E and let

($\phi$)  TE: $\{ \breve{x}^\alpha, \breve{y}^j, \dot{x}^\alpha, \dot{y}^j \}$  ,  vTE: $\{ \breve{x}^\alpha, \breve{y}^j, \dot{y}^j \}$  be the corresponding coordinate systems on TE and vTE respectively.  Then, the local expression of $\bar{\Gamma}$ and $\daleth D$ are the following:

$$\bar{\Gamma} : \begin{cases} \breve{x}^\alpha \cdot \bar{\Gamma} = \breve{x}^\alpha \\ \breve{y}^j \cdot \bar{\Gamma} = \breve{y}^j \\ \dot{y}^j \cdot \bar{\Gamma} = \Gamma^j_\beta \dot{x}^\beta + \dot{y}^j \end{cases} \quad ; \quad \daleth D : \begin{cases} \breve{x}^\alpha \cdot \daleth D = \breve{x}^\alpha \\ y^j \cdot \daleth D = \Gamma^j_\beta \dot{x}^\beta + \dot{y}^j \end{cases}$$

where  $\Gamma^j_\beta :E \to \mathbb{R}$  are numerical functions on E called the connection coefficients of $\bar{\Gamma}$ and that are defined by  $\bar{\Gamma} \cdot \partial x_\beta = \Gamma^j_\beta \partial y_j$ .

<u>Proposition</u>  .2  Let  $C_1 \equiv \daleth(E) \subset J\mathcal{D}(E)$  be a linear connection on a vector fiber bundle  $\pi:E \to M$. $C_1$  can be identified with a section $\gamma$ of the fiber bundle  $E^* \otimes T^*M \otimes TE \to M$  given by

( x )  $\gamma(p)(X_p) = \daleth(X_p) \in T_p M \otimes T_{X_p} E$ , for any  $X_p \in E$.

The local expression of  $\gamma$  , taking fibered coordinates  $\{x^\alpha, y^j\}$  on E and $\{x^\alpha, \breve{y}^j, \dot{x}^\alpha, \dot{y}^j\}$  on TE respectively,  is the following:

$$\gamma = \delta^\alpha_\beta \cdot e_\beta \, \theta^\beta \otimes dx^\gamma \otimes \partial x_\alpha \, - \Gamma^\alpha_{\beta\gamma} \theta^\beta \otimes dx^\gamma \otimes \partial y_\alpha \quad ,$$

where $\Gamma^\alpha_{\beta\gamma} \equiv \Gamma^\alpha_\gamma \cdot e_\beta$ , being $\Gamma^\alpha_\gamma :E \to \mathbb{R}$ the connection coefficients of $\bar{\Gamma}$, $\{e_\beta\}$  is a local basis for  $C^\infty(E)$, and $\{\theta^\beta\}$  is its dual, such that the following diagram is commutative:

$$\begin{array}{ccc} E & \longleftarrow & vTE \\ e_\beta \uparrow & & \uparrow \partial y_\beta \\ M & \xrightarrow{\;\;o\;\;} & E \end{array}$$

<u>Corollary  .1</u>  In particular for E=TM  we have  that a linear connection on E identifies a map   $\gamma$: M $\longrightarrow$ T*M⊗T*M⊗TM.

<u>Remark  .3</u>  As $C_1$  can be equivalently defined by means of a morphism $\Gamma$:J$\mathcal{D}$(E)$\longrightarrow$ T*M⊗E or by a connection map $_\daleth$D:TE$\longrightarrow$ E, we have that the rela- tion with $\daleth$:E$\longrightarrow$ J$\mathcal{D}$(E)  is given by the following commutative diagram:

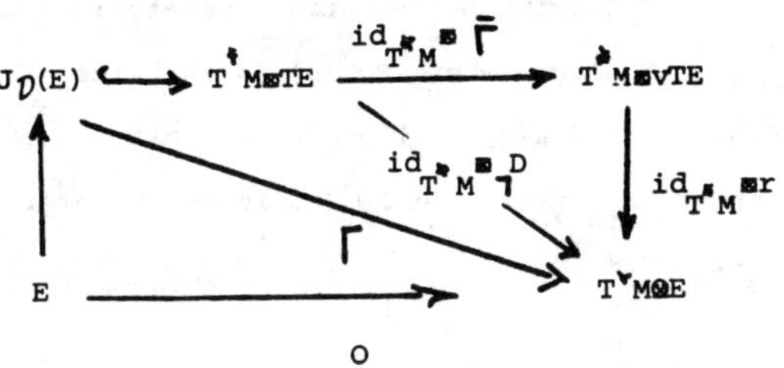

where  O  is the zero section.  In particular one has  $\Gamma$·$\daleth$ =O, $(id_{T^*M}⊗D)$·$\daleth$ =O. From this last equation we get also the following commutative diagram

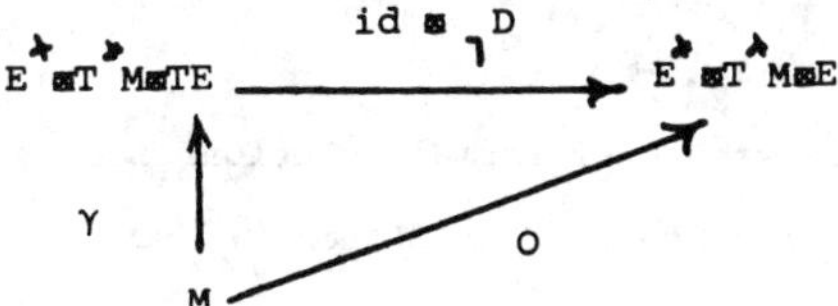

Then, we have the following local relation: $\Gamma^j_{\ \beta}·\gamma\ (\delta^\beta_{\ \gamma}·\partial x_k)=\Gamma^j_{\ \gamma k}$.

<u>Remark   .4</u>  Let  $C_1\equiv\daleth(TM)\subset J\mathcal{D}(TM)$  be a linear differential connection on TM. The curvature  of $C_1$  gives commutative the following diagram:

$$J\mathcal{D}^2(TM) \xrightarrow{\ \nabla^2\ } T^*M⊗T^*M⊗TM \cong \Lambda^o_2M⊗TM⊕S^o_2M⊗TM$$

$$D^2X \uparrow \qquad\qquad \downarrow pr_1$$

$$TM \xrightarrow[\ _\daleth R\ ]{} \Lambda^o_2M⊗TM$$

for any $X\in C^\infty(TM)$, where  $\nabla^2$  is the second absolute differential (see ref. [10] ) canonically associated to  $\Gamma$ .  More precisely, if  $\Gamma^\alpha_{\ \beta\gamma}$  are the con- nection coefficients of  $\daleth$ , and  $\widetilde{\Gamma}^{\ j\ i'}_{\lambda i\ j'} =\Gamma^{\ j}_{\lambda i}\ \delta^{i'}_{\ j'}+\Gamma^{\ i'}_{\lambda j'}\ \delta^{j}_{\ i'},\ \Gamma^{\ i'}_{\lambda j'}=-\Gamma^{\ i'}_{\lambda j'}$, are the connection coefficients of the connection  $\widetilde{\daleth}$  on T*M⊗TM canonically induced by  $\daleth$ , one has for any  $X\in C^\infty(TM)$:

$$\nabla^2 X\equiv\widetilde{\nabla}(\nabla X)=\left[\widetilde{\Gamma}^{\ i'j}_{\lambda i\ j'}\ (\nabla X)^{j'}_{\ j}+(\partial x_\lambda\cdot(\nabla X)^{i'}_{\ i})\right]dx^\lambda⊗\partial \dot{x}^i⊗\dot{x}_{i'}\cdot\nabla X$$

$$=\{-\Gamma^j_{\lambda i}\Gamma^{i'}_{jk}X^k+\Gamma^j_{\lambda i}(\partial x_j\cdot x^{i'})+\Gamma^{i'}_{\lambda j'}\Gamma^{j'}_{ik}X^k-\Gamma^{i'}_{\lambda j'}(\partial x_i\cdot x^{j'})-\Gamma^{i'}_{ik}(\partial x_\lambda\cdot x^k)$$

$$-(\partial x_\lambda\cdot\Gamma^{i'}_{ik})X^k+(\partial x_\lambda\partial x_i\cdot x^i) )\}\ dx^\lambda⊗dx^i⊗\partial x_{i'}\ .$$

Then, by imposing the antisymmetrization condition we get:

$$\gamma R = (\Lambda \blacksquare id_{T^*M}) \bullet \nabla^2 X = R^{i'}_{k\lambda i} X^k \, dx^\lambda \wedge dx^i \blacksquare \partial x_{i'}$$

$$= \left[ \Gamma^{i'}_{\lambda j'} \cdot \Gamma^{j'}_{ik} - \Gamma^{j}_{\lambda i} \Gamma^{i'}_{jk} - (\partial x_\lambda \cdot \Gamma^{i'}_{ik}) + (\partial x_i \cdot \Gamma^{i'}_{\lambda k}) \right] X^k \, dx^\lambda \wedge dx^i \blacksquare \partial x_{i'} .$$

Let $E \subset S_2^{\circ} M$ be the open subbundle of non-degenerate metrics on M.

Einstein equation is the kernel of the following differential operator:

$E_{in} : J\mathcal{D}^2(E) \longrightarrow S_2^{\circ}M$, $\qquad E_{in} \bullet D^2 g = R_{\bullet\bullet}(g) - \lambda R(g) \cdot g$, where $R_{\bullet\bullet}(g)$ is the Ricci

tensor associated to the metric tensor g and R(g) is the Riccic scalar curvature ; $\lambda \in \mathbb{R}$. The local expression of $(E_{in}) = \ker E_{in} \subset J\mathcal{D}^2(E)$ is as follows:

$$(E'_{in}) \qquad G_{\alpha\beta} \equiv R_{\alpha\beta} - \lambda R_{\gamma\delta} g^{\gamma\delta} g_{\alpha\beta} = T_{\alpha\beta} ,$$

where:

$$R_{\alpha\beta} = R^{\lambda}_{\alpha\lambda\beta} = (\partial x_\beta \cdot \Gamma^{\lambda}_{\lambda\alpha}) - (\partial x_\lambda \cdot \Gamma^{\lambda}_{\beta\alpha}) + \Gamma^{\lambda}_{\lambda\omega} \Gamma^{\omega}_{\beta\alpha} - \Gamma^{\omega}_{\lambda\beta} \Gamma^{\lambda}_{\omega\alpha}$$

$$\Gamma^{\lambda}_{\nu\omega} = \tfrac{1}{2} g^{\lambda\alpha} \left[ (\partial x_\nu \cdot g^{\omega\alpha}) + (\partial x_\omega \cdot g_{\alpha\nu}) - (\partial x_\alpha \cdot g_{\nu\omega}) \right] ,$$

$T \equiv T_{\alpha\beta} \, dx^\alpha \blacksquare dx^\beta$ is an assigned section of $S_2^{\circ}M \to M$.

One can see , by using the criterion of existence of formal solutions for

PDE of section 3, that $(E_{in})$ plus the divergence free condition $\nabla_\alpha T^\alpha_\beta = O$,

is involutive formally integrable.

Now, consider the Dirac-quantization of $(E_{in})$. We have the following natural

geometric objects canonically associated to $(E_{in})$:

1)  Hilbert bundle over M:  $H = Clif^c(M)$;

2)  Hilbert space $\mathcal{H}$ :  for any $g \in C^\infty(E)$ we have a canonical Hilbert

space builded as in section 6.1.

3)  First-order Dirac-quantization operator:

$(\bullet) \qquad Q : J\mathcal{D}(E) \longrightarrow J\mathcal{D}(H)^* \blacksquare H$ , $\qquad Q \bullet Dg \equiv Q.g = \check{\nabla}_{X(g)} - f(g)$ ,

where:

(a)  $X(g) : M \longrightarrow TM$ is a vector field on M canonically associated to $g \in C^\infty(E)$,

as the dual of the differential 1-form $\underline{X}(g) : M \longrightarrow T^*M$ such that

$R(g) = \delta \underline{X}(g) + \phi(g)$, where $\phi(g) : M \longrightarrow \mathbb{R}$ is an harmonic function on M.

Note that X(g) is unique, up to divergence free vector fields, by means of

Hodge's decomposition theorem.

(b)  $\check{\nabla}$ is the connection on $Clif^c(M)$ canonically associated to the Levi-

Civita connection generated by g;

(c)  $f(g):M \to \mathbb{R}$  is a real numerical function on M, canonically associated
to g given by:  $f(g)=-\lambda R(g)g_{\alpha\beta}g^{\alpha\beta}+T_{\alpha\beta}g\alpha\beta\,'-\phi(g)=-\lambda R(g) + \langle T,g \rangle - \phi(g)$.
In fact, by using the condition of anti-hermitianicity for Q.g, a calculus
similar to previous ones gives that  Q.g in (●) is anti-hermitian iff:
$g^{\alpha\beta}(G_{\alpha\beta}-T_{\alpha\beta})=0$.  This condition is certainlv sutisfied on equation $(E_{in})$.
So, we can conclude by the following

<u>Theorem</u>  .4   Einstein equation $(E_{in})$ is an involutive formally integrable
non-linear second-order differential equation on the fiber bundle E $\subset S_2^o M$
of non-degenerate metrics on M.  $(E_{in})$ is also completely integrable.
For any compact manifold M, $(E_{in})$ admits a canonical Dirac-quantization:
$Q(E_{in})\cong(H=Clif^c(M),\hat{Q}=S_2^o$ , Q) , where Q is given in (●).


## 12.5 - YANG-MILLS EQUATION


Let (M,g) be a compact four dimensional oriented (pseudo)-Riemannian mani
fold with canonical volume form $\eta = *1$.  Let G be a semisemple Lie group and
denote by A(G) the Lie algebra of G.  Consider a principal G fiber bundle
$(P,M,\pi;G)$ over M.  A principal connection $C_1=\daleth(P)\subset J\rho(P)$ on P is characterized
by a A(G)-valued differential form $_\daleth\omega:P \to T^*P\otimes A(G)$ such that $\phi_a^*{}_\daleth\omega=ad\ a^{-1}\circ{}_\daleth\omega$ ,
$\forall a\in G$, where  $\phi:G\times P \to P$  is the action map of G on P.  $_\daleth\omega$ is called the
<u>Heresmann connection map</u> of $C_1$.  Furthermore, a principal connection can be
also identified with a section A of an affine fiber bundle C(P) over M which
is an open subbundle of the vector fiber bundle $T^*M\otimes E(P) \to M$, being E(P)
the vector fiber bundle of G-invariant vector fields on P.  C(P)  has as
associated vector bundle $T^*M\otimes A(P)$, being $A(P)\equiv P\times A(G)/G$, where G acts on A(G)
by means of the adjoint representation.  So, A(P) is a vector bundle on M,
associated to P, with tipical fiber A(G).  Set E=C(P).
Yang-Mills equation is a second order equation on the fiber bundle E given
as the kernel (YM) of the following differential operator:

$I : J\mathcal{D}^2(E) \longrightarrow T^*M \otimes A(P)$, $I\, D^2_A A = {}_A\delta\,({}_A dA)$, where ${}_A dA \equiv F$; is the curvature of A, being ${}_A d$ the covariant exterior differential by the connection A and ${}_A\delta$ is the corresponding codifferential induced by the metric g.

Let $\{x^\alpha, y^H_\alpha\}$, $0 \leq \alpha \leq \beta$, $1 \leq H \leq \dim A(G)$, be a coordinate system on E and let $\{x^\alpha, y^H_\alpha, y^H_{\alpha\beta}, y^H_{\alpha\beta\gamma}\} \equiv \{x^\alpha, \xi^H_I\}$ be the corresponding coordinate system on $J\mathcal{D}^2(E)$. Note that $0 \leq \alpha, \beta, \gamma \leq \beta$, $1 \leq H \leq \dim A(G) \equiv m$. Then, the local expression of (YM) is the following:

$$(\text{YM})\ \begin{cases} F^{\alpha\beta\,H} \equiv g^{\alpha\delta} g^{\beta\gamma} F^H_{\delta\gamma} \equiv g^{\alpha\delta} g^{\beta\gamma}\left[(\partial x_\delta \cdot A^H_\gamma) - (\partial x_\gamma \cdot A^H_\delta) + C^H_{BC}\, A^B_\delta A^C_\gamma\right] \\[2ex] A^H_\omega \equiv g_{\alpha\omega}\left[(\partial x_\beta \cdot F^{\alpha\beta H}) + C^H_{BC}\, A^C_\beta F^{\alpha\beta B}\right] = 0. \qquad (C^H_{BC} = \text{structure constants of } G). \end{cases}$$

Putting $B^{\alpha\delta\beta\gamma} \equiv g^{\alpha\delta} g^{\beta\gamma}$, $C^{\alpha\beta\delta\gamma\,H}_{BC} \equiv B^{\alpha\delta\beta\gamma} C^H_{BC}$, we can write (YM) in the following more explicit form:

$$(\text{YM})\ \begin{aligned} A^H_\omega \equiv{}& g_{\omega\alpha} B^{\alpha\delta\beta\gamma} y^H_{[\gamma\delta]\beta} + g_{\alpha\omega} C^{\alpha\delta\beta\gamma\,H}_{BC}\left(y^C_\beta y^B_{[\gamma\delta]} + y^B_{\delta\beta} y^C_\gamma + y^C_{\gamma\beta} y^B_\delta\right) + \\[1ex] & g_{\alpha\omega}(\partial x_\beta \cdot B^{\alpha\delta\beta\gamma}) y^H_{[\gamma\delta]} + g_{\alpha\omega}(\partial x_\beta \cdot C^{\alpha\delta\beta\gamma\,H}_{BC}) y^B_\delta y^C_\gamma + \\[1ex] & + g_{\alpha\omega} C^{\alpha\delta\beta\gamma\,H}_{BC} C^H_{DE}\, y^C_\beta y^D_\delta y^E_\gamma = 0 \end{aligned}$$

Let us now study the formal integrability of (YM).

First note that $I$ is an epimorphism of constant rank (4+4xm). So, (YM) is a submanifold of $J\mathcal{D}^2(E)$ of dimension : dim(YM)=(4+56xm). The symbol $(g_2)_q \subset (S^0_2 M \otimes vTE)_q$, $\forall q \in$ (YM), has dimension : dim $g_2$=36xm.

Consider the first prolongation $(\text{YM})_{+1}$ of (YM):

$$(\text{YM})_{+1}\ \begin{cases} (\partial x_\alpha \cdot A^H_\omega) + (\partial \xi^I_K \cdot A^H_\omega)\, \xi^K_{I\alpha} = 0 \\[1ex] A^H_\omega = 0. \end{cases}$$

$(\text{YM})_{+1}$ is the kernel of $I^{(1)} : J\mathcal{D}^3(E) \longrightarrow J\mathcal{D}(T^*M \otimes A(P))$ that is an epimorphism of constant rank 4+20xm. So, $(\text{YM})_{+1}$ is a submanifold of $J\mathcal{D}^3(E)$ with dimension 4+120xm. Furthermore, the symbol $g_3$ of $(\text{YM})_{+1}$ is a vector bundle over (YM) with fiber dimension: dim $g_3$=64xm. Note that: dim$(\text{YM})_{+1}$−dim(YM)=64xm=dim $g_3$. Let us calculate the dimension of $F_1$: $\dim F_1 = 4 \times 4xm - 4xm\,\dfrac{(4+2)!}{3!\,3!} + 64xm = 0.$

So, we have proven that the canonical map $(YM)_{+1} \to (YM)$ is surjective.

Finally, let us prove that (YM) is involutive. In fact, we have:

$$(\dim g_3 = 64xm) = (\dim g_2 = 36xm) + (\dim g_2^{(1)} = 20xm) + (\dim g_2^{(2)} = 8xm) + (\dim g_2^{(3)} = 0).$$

So, we can use Theorem 3.3 to conclude that (YM) is an involutive formally integrable PDE.

Now, consider the Dirac quantization of (YM). We have the following natural geometric objects canonically associated to (YM):

1)  Hilbert bundle over M:  $H = Clif^c(M)$;

2)  Hilbert space $H$ associated to H as in section 6.1;

3)  First-order Dirac-quantization operator:

($\bullet$)  $Q: J\mathcal{D}(E) \to J\mathcal{D}(H)^* \boxtimes H$ ,  $Q \bullet DA \equiv Q.A = \overset{\smile}{\nabla}_{X(A)} - f(A)$, where:

(a)  $X(A): M \to TM$ is a vector field on M canonically associated to $A \in C^\infty(E)$, as the dual of the differential 1-form $\underline{X}(A): M \to T^*M$ such that $\langle A, {}_A\delta_A F \rangle = \delta \underline{X}(A) + \phi(A)$, being $\phi(A): M \to \mathbb{R}$ an harmonic function. Note that the pairing $\langle , \rangle$ is taken with respect to the metric isomorphism $T^*M \cong TM$ and the isomorphism $A(P)^{\wedge} \cong A(P)$ induced by means of the Killing matric on G.

(b)  $\overset{\smile}{\nabla}$ is the absolute differential associated to the connection on $Clif^c(M)$;

(c)  $f(A) \equiv \frac{1}{2}\phi(A): M \to \mathbb{R}$.

In fact, by using the condition of anti-hermitianicity for Q.A, one has that Q.A is anti-hermitian iff: $\langle A, {}_A\delta_A F \rangle = 0$. This condition is certainly satisfied on equation (YM). So, we can conclude by the following

<u>Theorem</u> .5 Yang-Mills equation (YM) is an involutive formally integrable, non-linear second-order differential equation on the fiber bundle $\hat{\pi}: E \equiv C(P) \to M$ of the principal connections on a principal fiber bundle $(P, M, \pi; G)$ , being G a semisemple Lie group. (YM) is also completely integrable.

For any compact manifold M, (YM) admits a canonical Dirac-quantization: $Q(YM) = (H = Clif^c(M), \hat{Q} = \mathbb{C}, Q)$, where $\mathbb{C}$ is the functor given in ref. [12] characterizing the fiber bundle C(P) as a super-bundle of geometric objects on P and Q is given in formula ($\bullet$).

## 13. - CONCLUSIONS

The geometric framework developed allows us to recognize dynamic conservation laws whether on macroscopic level or microscopic one.

The worth of the paper is to show that the classical conservation laws and quantum conservation laws of a continuum system are directly related to the geometric formal properties of the dynamic equation and to develop an intrinsic proceeding able to calculate these conserved entities.

In particular, from the quantum point of view, the emphasis is placed on a new concept of cobordism: the quantum-cobordism , to which is related another new geometric concept: the quantum situs. These geometric structures appear to play a central role in the understanding of the quantum fluctuations and tunneling  phenomena that one detects on microscopic scale.  Parentetically, we belive that it will be worthile also any work directed to attain further interesting informations on these structures.

Moreover, a new fully covariant equation is proposed that, added to the classic dynamic equation of a continnum system, allows a selection of possible quantum configurations.  We call this new equation Schroedinger equation whether as formally it looks like the homonymous famous equation or its physical meaning is the exact translation in our formalism of Schroedinger's equation.

Finally, a new concept of Dirac-quantization (=representation of fields by means of linear operators on suitable Hilbert spaces) is introduced that is directly related to the quantum-cobordism.

## 14.- APPENDIX

## A.1    -    TOPOLOGICAL VECTOR SPACES, C ALGEBRAS AND SPECTRAL THEORY

Let E be a topological vector space on $\mathbb{K}$ ($\equiv \mathbb{R}$ or $\mathbb{C}$). We denote by $E^*$ the <u>algebraic dual</u> and $E' \subseteq E^*$ the <u>topological dual</u>. If $\dim E = n < \infty$ then $\bar{E}^* = E'$. If E and F are normed topological vector spaces we denote by $E \,\bar{\otimes}\, F$ the <u>topological tensor product</u> of E by F that is the vector space $E \otimes F$ normed by the following norm:

$$\|u\| = \inf_{u = \sum x_j \otimes y_j} \lim \{ \textstyle\sum_j \|x_j\| \|y_j\| \}$$ and completed with respect to this norm.

If H is a separated Hilbert space, a continuous operator $U : H \to H$ is called of <u>Hilbert-Schmidt</u> if for any Hilbert basis $\{a_n\}$ of H the addition $\sum_n \|U(a_n)\|^2$ is finite. This implies that U is compact (+); if U is self-adjoint this is equivalent to say that the set of eigenvalues $\{\lambda_n\}$ (all real numbers) is such that $\sum_n \lambda_n^2 < +\infty$ .

A <u>nuclear operator</u> T on a Hilbert space H is given as the composition of two Hilbert-Schmidt operators : $T = U \bullet V$. In these cases we can define a <u>trace</u>: $Tr(T) = (V|U^*) = \sum (T(e_n)|e_n)$ , for any Hilbert basis $\{e_n\}$ . (++)

The <u>self-adjoint nuclear operators</u> are that for which $\sum_n |\lambda_n| < +\infty$ for the set $\{\lambda_n\}$ of their eigenvalues.

A <u>normed algebra</u> is an algebra A on C endowed with a structure of vector space such that: (i) $\|u.v\| \leq \|u\| . \|v\|$ ; (ii) $\|e\| = 1$, e=unit of A.

A <u>Banach algebra</u> is a normed algebra such that its topology is that of a complete space.

<u>Example.</u> Let E be a Banach space over $\mathbb{C}$, then $L_c(E)$=space of continuous linear operators on E, (•) is a Banach algebra with respect to the product given by composition and the norm defined by $\| T \| = \sup_{\|x\| \leq 1} \| T(x) \|$ . In fact, $\| T \bullet U \| \leq \|T\| . \|U\|$ and

---

(+) Compact means that U transforms any limited set into a set with compact aderence. This is equivalent to say that U transforms any weakly convergent succession $\{a_n\}$ , $(\lim_{n \to \infty} \| a_n - a \| = 0)$, into a strongly convergent succession $\{U(a_n)\}$ , $(\lim_{n \to \infty} (U(a_n), g) = (h, g)$ , $\forall\, g \in H)$.

---

(++) $(V|U^*)$ denotes the scalar product of the operators U and V in the space $L_2(H)$ of the Hilbert-Schmidt operators on H.

---

(•) We shall simply write L(E) instead of $L_c(E)$ if no confusion arises.

$\| \mathrm{id}_E \| = 1$.

Let $G(A)$ be the group of invertible elements of a Banach algebra with unit e.
$G(A)$ is an open in A and the map $u \mapsto u^{-1}$ is continuous in $G(A)$.

The _spectrum_ of an element $u \in A$ is the set $Sp(u) = \mathbb{C} - \rho(u)$, where
$\rho(u) \equiv \{ \lambda \in \mathbb{C} \mid (u - \lambda e) \in G(A) \} \subset \mathbb{C}$. The spectrum of $u \in A$ is always compact
and non-empty. Further, $Sp(u) \subseteq \{ \lambda \in C \mid |\lambda| \leq \|u\| \}$. The map $R(u):Sp(u) \to G(A)$
given by $R(u)(\lambda) = (u - \lambda e)^{-1}$ is called _resolution map_ of $u \in A$. Further, $R(u)(\lambda) = \sum_0^\infty \lambda^{-(n+1)} u^n$. In particolar, $R(u)(\lambda) \to 0$, $\lambda \to \infty$. $R(u)$ is a holomorphic map.
One has: $r(u) = \lim_{n \to \infty} \|u^n\|^{1/n}$.
The _spectral radius_ is the number $r(u) = Sup\{ |\lambda|, \lambda \in Sp(u) \}$. If B is a sub-
algebra (with unit) of A one has $Sp(b)_B \supset Sp(b)_A$, $b \in B$ and $\partial Sp(b)_B \subset \partial Sp(b)_A$,
where $Sp(b)_B$, (resp. $Sp(b)_A$) denotes the spectrum of b with respect to B (resp.
A). One has $Sp(b)_A = Sp(b)_B$ if $Sp(b)_B$ is rare or $Sp(b)_A$ is connected, or B is a
full sub-algebra of A (that is $G(B) = G(A) \cap B$; for example the commutative
part $M^c$ of A is a full sub-algebra of A).

A _character_ of a Banach algebra A is a unitary, moltiplicatif linear function
$\chi : A \to \mathbb{C}$. One has: $\chi(a) \in Sp(a)$, $\forall a \in A$. Further, the set $Ch(A)$ of characters
of A is a subset, weakly compact, of the unitary bubble of A!. $Ch(A)$ is called
the _spectrum of the algebra A_. We also write $Sp(A) = Ch(A)$.

If A is a unitary commutative Banach algebra $Ch(A)$ coincides with the set of
maximal ideals of A and it is not empty. Further, $Ch(A)$ is a compact Hausdorff
space under Gel'fand topology, that is the weakly topology on $Ch(A)$ under which
all the functions $\hat{a}:Ch(A) \to \mathbb{C}$, $\hat{a}(\chi) = \chi(a)$ are continuous.

The _Gel'fand's transform_ is the map $\wedge : A \to C^0(Ch(A))$ given by $\hat{} : a \to \hat{a} : \chi \mapsto \hat{a}(\chi) = \chi(a)$.
The Gel'fand's transform is a morphism of unitary algebra, with kernel the
almost-nil-potent elements of A:

(a) $\|\hat{a}\| = r(a) \leq \|a\|$ ;

(b) $Sp(a) = \hat{a}(Ch(A)) = \{ \chi(a); \chi \in Ch(A) \}$ ;

(c) $a \in G(A)$ iff $\chi(a) \neq 0$, $\forall \chi \in Ch(A)$, i.e. the function $\hat{a}$ is not zero on $Ch(A)$.

Let A be a unitary Banach algebra. For two elements $a, b$ A that commute, one
has:

(a) $Sp(a+b) \subset Sp(a) \oplus Sp(b)$; $Sp(a.b) \subset Sp(a).Sp(b)$;

(b) $r(a+b) \leq r(a) + r(b)$ ; $r(a.b) \leq r(a).r(b)$, $r(\mathbb{1}) = 1$; $r(\lambda a) = |\lambda|.r(a)$.

The spectral radius r is a semi-norm of unitary algebra on any unitary commutative
sub-algebra of A.

Let M be a non-empty subset of A.  Then, we call PL(M) the litlest full sub-algebra of A which contains M (and it exists). If M is commutative then PL(M) is so.

Further, one has that Sp(a) is homomorphic to Ch(PL(a)):  $Sp(a) \cong Ch(PL(a))$

Let A be a commutative Banach algebra with unit e, $\|e\| = 1$ . Then,

$$\|\hat{a}\| = \sup\{ \ |\chi(a)| \ : \chi \in Ch(A)\} \ = \lim_{n \to \infty} \| \ a^n \|^{1/n} \ .$$

<u>Definition A1</u>  Let A be an algebra over C.  An <u>involution</u> on A is a bijection $*:A \to A$ such that: (i) $(x^*)^* = x$ ; (ii) $(x+y)^* = x^* + y^*$ ; (iii) $(x.y)^* = y^*.x^*$ ; (iv) $(\lambda x)^* = \bar{\lambda} \ x^*$ , $\lambda \in \mathbb{C}$.

One has $e^* = e$ , and $(x^{-1})^* = (x^*)^{-1}$ if x is invertible.

An element $x \in A$ such that $x^* = x$ is called <u>hermitian</u> ; if $xx^* = x^*x$, x is called <u>normal</u>, and if $x \ x^* = x^* x = e$, x is called <u>unitary</u>.

An <u>involutive algebra</u> is an algebra endowed with an involution.

An <u>involutive Banach algebra</u> is a Banach algebra endowed with an involution such that $\|x^*\| = \|x\|$.

A <u>C*-algebra</u> is an involutive Banach algebra such that $\|x\|^2 = \|x^*.x\|$ .

<u>Examples</u>. 1. Let H be a Hilbert space then $L_c(H)$ is a C*-algebra where the involution is the adjoint: if $T \in L_c(H)$ , $(T^*(x)|y) = (x|T(y))$.

2.  If H is of infinite dimension, the Hilbert-Schmidt operators form a sub-algebra $L_2(H)$ of $L_c(H)$ stable with respect to the involution $T \mapsto T^*$ , but endowed with another norm $\|T\|_2 = \sqrt{\sum_n \|T(a_n)\|^2}$ for a Hilbert basis $\{a_n\}$ of H.  $L_2(H)$ is an involutive Banach algebra but not a C*-algebra.

<u>Definition A2</u>  A (unitary) <u>representation</u> of an involutive algebra A is a homomorphism of algebras $U:A \to L_c(H)$, where H is a Hilbert space, such that: (i) $U(s^*) = U(s)^*$ ; (ii) $U(e) = id_H$.

<u>Theorem A1</u>  (C*-algebras and spectrum)

1.  If A is a commutative C*-algebra $\Rightarrow$ $r(x) = \|x\|$ , $\forall x \in A$.

Further, there are not almost nil-potent elements $\neq 0$.

2.  If A is a unitary C*-algebra:

(a) for any hermitian element $h \in A$ the spectrum $Sp(h) \subset \mathbb{R}$;

(b) for any unitary element $u \in A$ the spectrum $Sp(u) \subset \mathbb{T} = S^1$;

(c) any C*-sub-algebra B is full in A;

(d) for any non-empty part M the C*-algebra B generated by M coincides with $PL(M \cup M^*)$;

(e)  for any normal element a  A , r(a)=$\|$a$\|$ . So, any normal element, almost nil-potent is zero.

3.  If A is a unitary commutatif C$^*$-algebra:

(a)  any character $\chi$ of A is hermitian, that is $\chi(x^*)=\overline{\chi(x)}$ , $\forall x \in A$.

(b)  (<u>Theorem of Gel'fand-Naimark</u>) The Gel'fand transform a $\mapsto \hat{a}$ is an involutif isometry , surjectif of A onto the C$^*$-algebra C$^o$(Ch(A)).

<u>Examples</u>.  1. Let T be a completely regular space.  One has the isometric isomorphism: C$^\infty$(T) $\cong$ C$^o$(Ch(C$^\infty$(T))).  Ch(C$^\infty$(T)) results the compactified of Stone-Čech of T.

2. Let A$\equiv$BM($\Omega,\Sigma$) be the algebra of bounded and measurable functions $\Omega \longrightarrow \mathbb{C}$ with respect to the family $\Sigma$ of $\Omega$ .  The compactified of A (compactification of a measurable space) is just K$\equiv$Ch(A).  So, we can identify BM($\Omega,\Sigma$)$'$ $\equiv$ba($\Sigma$) with a space of Radon measures M(K) $\equiv$C$^o$(K)$'$.

<u>Theorem A2</u> (C$^*$-algebras and spectrum , continued)

1. (Functional calculus). Let A be a unitary C$^*$-algebra.  Let M$\equiv\{a_i\}_{i\in I} \equiv a_i$ be a family of normal commutatif elements of A: that is $a_i a_j = a_j a_i$ and $a_i a_j^* = a_j^* a_i$.  Then, the map $\hat{a}$:Ch(B) $\longrightarrow \mathbb{C}^I$, B$\equiv$PL(M$\cup$M$^*$), $\hat{a}(\chi)=(\chi(a_i))_{i\in I} \in \mathbb{C}^I$ , is continuous.  We call <u>simultaneous spectrum</u> of M the image $\hat{a}$(Ch(B))$\equiv$Sp(a)$\subset \mathbb{C}^I$.  Then, we can associate a $*$-morphism of unitary algebras, that is isometric: $\phi$:C$^o$(Sp(a)) $\longrightarrow$A, that is unique, $\phi(e_i)=a_i$, with $e_i$:z$\rightarrow z_i$ .  Im$\phi$=PL(M$\cup$M$^*$) with M=$(a_i)_{i\in I}$, that is the C$^*$- sub-algebra generated by the $a_i \in$A. Set  $\phi$(f)=f(a).

2. (<u>Spectral theorem</u>). Let a=$(a_i)_{i\in I}$ be as above.  Then, for any f$\in$ C$^\infty$(Sp(a)), one has   Sp(f(a)) =f(Sp(a)).

3. (<u>Simultaneous spectral theorem</u>). Let a=$(a_i)$ be given under the hypotheses of Theorem A2 and f=$(f_j) \subset$ C$^o$(Sp(a)). Set f(a)=$(f_j(a))$. One has Sp(f(a))=f(Sp(a)).

4. Let a=$(a_i)$ be as above and f=$(f_j) \subset$ C$^o$(Sp(a)). Set b=f(a)=$(f_j(a))$. Then, for any g $\in$ C$^o$(Sp(b)) one has g(b)=g(f(a))=(g$\circ$f)(a).

5. Let H be a complex Hilbert space of finite dimension. Let $(T_i)_{i\in I}$=T be a family of normal operators that commute.  The simultaneous spectrum Sp(T)$\subset \mathbb{C}^I$ is finite and there exists a orthonormal basis of H, propre for each $T_i$.

Let us, now, recall some fundamental properties of operators on Hilbert spaces.

An <u>operator</u> on H is a couple (T,D(T)), where D(T) is a sub-vector space of H (that we shall assume dense in H), and such that T$\in$ L$_c$(D(T),H).

An <u>extension</u> of (S,D(S)) is an operator (T,D(T)) such that D(S) $\subseteq$ D(T) and T(x)=S(x), $\forall$ x$\in$D(S). We write  S $\subseteq$ T.

The <u>sum</u> of two operators $(T,D(T))$, $(S,D(S))$ is the operator $(T+S,D(T)\cap D(S))$.

The <u>graph</u> $G(T)$ of an operator $(T,D(T))$ is given by $G(T)=\{\,(x,T(x))\in H\times H,\ x\in D(T)\}$.

$T$ is <u>closed</u> if $G(T)$ is closed in $H\times H$, so for $\{x_n\}\in D(T)$, $x_n\to x$ and $T(x_n)\to y$ we get $x\in D(T)$ and $y=T(x)$. If $T$ is closed then $\ker(T)$ is closed in $H$. If $S$ is a bounded operator on $H$, then $T$ is closed iff $S+T$ is closed. If $T$ is injectif, then $T$ is closed iff $T^{-1}$ is closed.

Two of following properties imply the other: (a) $T$ is bounded in $D(T)$; (b) $T$ is closed; (c) $D(T)$ is closed int $H$.

$T$ is <u>closeable</u> if there is a closed extension. In these cases $T$ admits the litlest closed extension, $\overline{T}$, called <u>closure</u> of $T$.

The <u>adjoint</u> $(T^*,D(T^*))$ of $(T,D(T))$ is given by $D(T^*)=\{\,x\in H;\ y\mapsto (T(y)\,|\,x)\ \text{is } C^0\ \text{in } D(T)\}$ further, one has $\langle y\,|\,T^*(x)\rangle=(T(y)\,|\,x)$. If $T\subseteq S\Rightarrow D(S^*)\subseteq D(T^*)$ : $S^*\subseteq T^*$. But $D(T^*)$ cannot be dense in $H$. $T$ is <u>self-adjoint</u> if $T=T^*$ and $D(T^*)=D(T)$.

$T$ is <u>normal</u> if $T$ is closed and $T\cdot T^*=T^*\cdot T$.

Let $J:H\times H\to H\times H$ be the $C^0$ operator given by $J(x,y)=(y,-x)$. Then, $T^*$ is closed, and one has $G(T^*)=J(\overline{G(T)}^{\perp})=J(\overline{G(T)})^{\perp}=J(G(T))^{\perp}$.

If $T^{-1}$ exists and if $\mathrm{Im}(T)$ is dense in $H$, then $T^*$ is invertible and $(T^*)^{-1}=(T^{-1})^*$.

If $S$ is a bounded operator on $H$ then: (a) $(S+T)^*=S^*+T^*$; (b) $(S\cdot T)^*=T^*\cdot S^*$; (c) $S^*\cdot T^*\subseteq (T\cdot S)^*$; (d) $\ker(T^*)=(\mathrm{Im}(T))^{\perp}$.

An operator $T$ is closeable iff $D(T^*)$ is dense in $H$. Under these hypotheses $(T^*)^*=T^{**}$ is the <u>closure</u> $\overline{T}$ of $T$. In particular, $T=T^{**}$ if $T$ is closed. In general $(T^{**})^*=T^*$.

<u>Theorem A3</u> (<u>von Neumann</u>) 1. Let $T$ be a closed operator. Then $(id_H+T^*\cdot T)$ is a bijection of $D(T^*\cdot T)$ on $H$; the operator $B=(id_H+T^*\cdot T)^{-1}$ is continuous on $H$, hermitian and such that $Sp(B)\subseteq[0,1]$.

2. If $T$ is a closed operator, then the domain of $T^*\cdot T$ is dense into $H$ and one has $(T^*\cdot T)^*=T^*\cdot T$.

3. If $T$ is a closed operator, it is normal iff $D(T)=D(T^*)$ and $\|T(x)\|=\|T^*(x)\|$, $\forall x\in D(T)$.

<u>Definition A3</u> 1. Let $T\in L(H)$. We define the following sets:

(a) <u>Point spectrum of $T$</u>: $Sp(T)_p\equiv\{\,\lambda\in\mathbb{C}\ |\ (\lambda id_H-T)\ \text{is not injectif}\}$ ; $\lambda\in Sp(T)_p$ is called an <u>eigenvalue</u> of $T$.

(b) <u>Continuous spectrum of $T$</u>: $Sp(T)_c\equiv\{\lambda\in\mathbb{C}\ |\ \ker(\lambda id_H-T)=0\ \text{and }\mathrm{Im}(\lambda id_H-T)\ \text{is dense}\}$ and distinguished in $H$.

(c) <u>Residual spectrumm of $T$</u>: $Sp(T)_r\equiv\{\lambda\in\mathbb{C}\ |\ \ker(\lambda id_H-T)=0\ \text{and }\overline{\mathrm{Im}(\lambda id_H-T)}\neq H\}$ .

(d) <u>resolvent of T</u>: $\rho(T) \equiv \{ \lambda \in \mathbb{C} \mid (\lambda \, \mathrm{id}_H - T)^{-1}$ exists and is bounded on all $H\}$.

(e) <u>spectrum of T</u>: $Sp(T) \equiv \mathbb{C} - \rho(T)$.

One has $Sp(T) \supseteq Sp(T)_p \cup Sp(T)_c \cup Sp(T)_r$  $(\bullet)$. The equal holds if $T$ is bounded or $T$ is closed.

The <u>resolvent function</u>: $R(\bullet, T): \rho(T) \to L(H)$, $R(\lambda, T) = (\lambda \, \mathrm{id}_H - T)^{-1}$. One has:

(a) $\rho(T)$ is an open of $\mathbb{C}$; (b) $R(\bullet, T)$ is analytic on $\rho(T)$; (c) The resolvent equation $R(\lambda, T) - R(\mu, T) = (\mu - \lambda) R(\lambda, T) R(\mu, T)$ $\forall \lambda, \mu \in \rho(T)$ is satisfied.

<u>Theorem A4</u>  1. For bounded operators $T$ one has : (a) $Sp(T^*) = \overline{Sp(T)}$  but the homomorphism $(\bullet)$ is not more true; (b) $\lambda \in Sp(T)_p \Rightarrow \bar{\lambda} \in Sp(T^*) \cup Sp(T^*)_r$ ; (c) $\lambda \in Sp(T)_r \Rightarrow \bar{\lambda} \in Sp(T^*)$.

2.  For any operator $T$ : (a) $\lambda \in \rho(T) \Rightarrow \bar{\lambda} \in Sp(T^*)$; (b) $R(\bar{\lambda}, T^*) = R(\lambda, T)^*$ , $\forall \lambda \in Sp(T)$; (c) $\lambda \in Sp(T)_p \Rightarrow \bar{\lambda} \in Sp(T^*) \cup Sp(T^*)_r$; (d) $\lambda \in Sp(T)_r \Rightarrow \bar{\lambda} \in Sp(T^*)_p$.

3.  If $T$ is closed $\lambda \in \rho(T) \iff \bar{\lambda} \in \rho(T^*)$.

4.  For a bounded operator $T$, $Sp(T)$ is a compact non empty set of $\mathbb{C}$.

For a non-bounded operator $T$, $Sp(T)$ is a closed set of $\mathbb{C}$.

<u>Definition A4</u>  Let $H$ be a Hilbert space.  We call <u>spectral measure</u> on $H$ a measurable space $(\Omega, \Sigma)$ with a map $E: \Sigma \to L(H)$ such that : (a) $E(\emptyset) = 0$, $E(\Omega) = \mathrm{id}_H$, $E(A)^* = E(A)$; (b) $E(A \cap B) = E(A) \circ E(B)$ , for any $A, B \in \Sigma$; (c) $E(\bigcup_n A_n)(x) = \sum_n E(A_n)(x)$, for any disjoint sequence $\{A_n\}$ in $\Sigma$ , $\forall x \in H$.

From (b) we try that $E(A)$ is a hermitian projection and if $A \cap B = \emptyset$ one has that $E(A)$ and $E(B)$ are orthogonal.  So, from the condition (c) one has $\|E(A)(x)\|^2 = \sum_n \|E(A_n)(x)\|^2$ with $A = \bigcup_n A_n$. The map $E_{x,y}: \Sigma \to \mathbb{R}$ given by $E_{x,y}(A) \equiv (E(A)(x) \mid y)$ is a measure on $(\Omega, \Sigma)$. $(E_x \equiv E_{x,x})$.

<u>Theorem A5</u> (Spectral theorems)

1.  Let $M(\Omega, \Sigma) \equiv$ algebra of measurable functions $f: \Omega \to \mathbb{C}$, with respect to the family $\Sigma$ of subsets of $\Omega$ . For any $f \in M(\Omega, \Sigma)$ and spectral measure $E$ on $H$ set $\Delta(f) = \{ x \in H \mid \int |f|^2 dE_x < +\infty \} \subset H$ , it is a dense sub-space of $H$. Further, the set $\Delta^\infty(f) = \bigcap_p \Delta(f^p)$ is a dense subspace of $H$.

2.  Let $x \in \Delta(f)$; for any $y \in H$, a function $f \in M(\Omega, \Sigma)$ is integrable with respect to the variation $|E_{x,y}|$ and one has $\int |f| \, d|E_{x,y}| \leq ( \int |f|^2 dE_x )^{1/2} . \|y\|$

3.  In particular, let $BM(\Omega, \Sigma) =$ algebra of bounded measurable functions $f: \Omega \to \mathbb{C}$ with respect to the family $\Sigma$ of subsets of $\Omega$ . Then, to any spectral measure $E$ of $(\Omega, \Sigma)$ there corresponds a map $\phi: BM(\Omega, \Sigma) \to L(H)$ such that $\phi(f)$ is a bounded operator given by $(\phi(f)(x) \mid y) = \int_\Omega f dE_{x,y}$.

Furthermore, $\phi$ is a $*$-morphism from the unitary $\overset{*}{C}$-algebra $BM(\Omega,\Sigma)$ into the unitary $\overset{*}{C}$ algebra $L(H)$ such that : (a) $\| \phi(f)\| \leq \|f\|$ ; (b) $\| \phi(f)(x)\|^2 = \int |f|^2 dE_x$.

4. In general one has: (a) $\phi$ is a multiplicatif operator ; (b) $\phi(\bar{f})=\phi(f)^*$ and $D[\phi(f)^*]= D[\phi(f)]$ ; (c) Each $\phi(f)$ is normal in $L(H)$; (d) $\phi(f)$ is a self-adjoint operator, $\underset{\text{if f is a real function}}{}$ (e) if f is real positif , $\phi(f)$ is a positif self-adjoint operator; (g) $Sp(\phi(f))\subset \overline{f(\Omega)}$; (h) $D[\phi(f)\circ\phi(g)] = D[\phi(g)]\cap D[\phi(f,g)] = \Delta(fg)$ ; (i) $\phi(f)\circ\phi(g) \subseteq \phi(fg)$ if $g \in L^\infty(E)$ then the equal holds.

5. To any function $f \in M(\Omega,\Sigma)$ there corresponds an operator $\phi(f)$ defined on a dense domain $D(\phi(f))=\Delta(f)$ and such that: (a) $(\phi(f)(x)|y)=\int fdE_{x,y}$ ,$\forall x \in \Delta(f)$, $\forall y \in H$; (b) $\| \phi(f)(x)\|^2 = \int |f|^2 dE_x$ , $\forall x \in \Delta(f)$.

6. Let $h:(\Omega,\Sigma) \to (\Omega',\Sigma')$ be a measurable map. Then, for any spectral measure $E:\Sigma \to L(H)$ we have the <u>image spectral measure</u> $h(E)=E':\Sigma' \to L(H)$, given by $E'(B)=E(h^{-1}(B))$,$\forall B \in \Sigma'$. So, we can define the integral $\int f'dE'=\int f'\circ hdE$ , $\forall f' \in BM(\Omega',\Sigma')$.

7. $\forall f \in BM(\Omega,\Sigma)$, $\phi(f)$ being normal, we can suppose E defined on $\mathbb{C}$ by substituting the image measure $f(E)$, that is spectral also such that $f(E)(B)=E(f^{-1}(B))$, for any borel set B of $\mathbb{C}$. One has: (a) $D(T)\equiv\{ x| \int |\lambda|^2 dE_x < +\infty \}$ ; (b) $T=\int_{\mathbb{C}}\lambda dE(\lambda)$; (c) $(T(x)|x) = \int_{\mathbb{C}} \lambda dE_x(\lambda)$ ,$\forall x \in D(T)$; (d) $\| T(x)\|^2 = \int |\lambda|^2 dE_x(\lambda)$.

8. $\phi(f)$ is bounded on H iff $f \in L^\infty(E)$ and one has $\| \phi(f)\| = \|f\|$

9. Let $ba(S)$ be the Baire set of a topological space S. Let $T \subset L(H)$ be a normal operator on the Hilbert space H. We call <u>spectral resolution of T</u> the spectral measure on H associated to $(Sp(T),ba(Sp(T)))$, $E:ba(Sp(T)) \to L(H)$. More precisely E is the <u>unique</u> spectral measure on H such that $(T(x)|y)=\int_{Sp(T)} \lambda dE_{x,y}$ with $E_{x,y}:ba(Sp(T)) \to \mathbb{C}$ , $\lambda:Sp(T) \to \mathbb{C}$ is the identity map on $\mathbb{C}$. The integral being an ordinary Riemann-Stieltjes integral.

10. Let T be a normal operator on H. Then, the spectral resolution E of the adjoint $T^*$ is the spectral measure of the image of the spectral resolution E of T by means of the conjugation $\lambda \to \bar{\lambda}$ ; $Sp(T) \to Sp(T^*) = \overline{Sp(T)}$.

11. If T is a normal operator and E is its spectral measure, considered as a spectral measure on $\mathbb{C}$, one has:

(a) $Supp(E)=Sp(T) \Rightarrow E(\omega)=0 \Leftrightarrow \omega \cap Sp(T)=\emptyset$ , $\forall \omega$ open in $\mathbb{C}$.

(b) $Sp(T)_p = \{ \lambda \in Sp(T) \Leftrightarrow E(\{\lambda\})\neq 0 \}$. Further, one has $E(\{\lambda\})(H)=ker(\lambda id_H -T)$ $\lambda \in Sp(T)_p$

(c) $Sp(T)_r = \emptyset$.

(d) If $Sp(T)$ is numerable, then $Sp(T)_p \neq \emptyset$ and there exists in H a proper orthonormal basis for T.  Further, if $Sp(T)_c \neq \emptyset$, one has that any point of $Sp(T)_c$ is accomulation point for eigenvalues, i.e. $Sp(T)_c = \overline{Sp(T)_p}$.

12.  If T is a normal compact operator, then $Sp(T)_p$ is at most numerabile and each eigenvalue $\lambda \neq 0$ has a finite multiplicity, i.e. $\ker(\lambda id_H - T)$ has finite dimension $\alpha(\lambda)$.  Further, one has the following cases:

(a)  H is of finite dimension and $Sp(T)_p = Sp(T)$ is finite;

(b)  H is of infinite dimension, $Sp(T)_p$ is finite and $Sp(T)_c = \emptyset$ and $\ker(T) \neq 0$, i.e. $0 \in Sp(T)_p$;

(c)  H is of infinite dimension, $Sp(T)_p$ is infinite and it is ordoned in a sequence $\{\lambda_n\}$ such that $|\lambda_n| \downarrow 0$ .  If $\ker(T) \neq 0$, then $0 \in Sp(T)_p$ and $Sp(T)_c = \emptyset$. If T is injectif then $Sp(T)_c = \{0\}$ and H is separated.

13.   Let T be a compact operator on the Hilbert space H.  There exists a sequence $\{\alpha_n\}$, $\alpha_n > 0$, weakly decreasing, finite or numerable $(\alpha_n \downarrow 0)$ and two orthonormal sequences $(e_n)$ and $(\varepsilon_n)$ in H, such that: (a) $T(x) = \Sigma_n \alpha_n (x|e_n)\varepsilon_n$; (b) $T^*(x) = \Sigma_n \alpha_n (x|\varepsilon_n)e_n$ ; (c) $(T^* \bullet T)(x) = \Sigma_n \alpha_n^2 (x|e_n)e_n$ ; (d) $(T \bullet T^*)(x) = \Sigma_n \alpha_n^2 (x|\varepsilon_n)\varepsilon_n$, $\forall x \in H$.   Further, $\|T\| = \|T^*\| = \sup_n \alpha_n$ and the operator T is of Hilbert-Schmidt iff $\Sigma \alpha_n^2 < +\infty$ ; in this case $\|T\|^2 = \Sigma \alpha_n^2$ .
(Remark that the numbers $\alpha_n$ are not the eigenvalues of T, but they are eigenvalues of $[T] \equiv (T^* \bullet T)^{1/2}$ or $[T^*] \equiv (T \bullet T^*)^{1/2}$.   $\alpha_n$ are called __singular values__ of T.

__Definition A5__     $\lambda \in \mathbb{C}$ belongs to the __essential spectrum__ of T, $\lambda \in Sp(T)_{ess}$, iff for any open neighborhood $\omega$ of $\lambda$ , the projection $E(\omega)$ is of infinite dimension.  We call __discrete spectrum__ $Sp(T)_d \equiv Sp(T) - Sp(T)_{ess}$.

__Theorem A6__ (on the essential spectrum)

1.  $\lambda \in \mathbb{C}$ belongs to $Sp(T)_{ess}$ iff $\lambda \quad Sp(T)_c$ or $\lambda \quad Sp(T)_p$, $\alpha(\lambda) = \infty$ , or $\lambda \in Sp(T)_p$, $\lambda$ is not separated in $Sp(T)$.

2.  $Sp(R)_{ess}$ is closed.

3. (H.Weyl)  The following propositions are equivalent:

(a)  $\lambda \in Sp(T)_{ess}$;  (b)  There exists a sequence $\{x_n\}$ of orthonormal vectors of H such that $(T(x_n) - \lambda x_n) \to 0$.

(c)  There exists a sequence $\{x_n\}$ , $\|x_n\| = 1$, weakly convergent $\to 0$ such that $(T(x_n) - \lambda x_n) \to 0$.

4.   Let S and T two normal operators such that R = S-T is a compact operator (not necessarily normal) .  Then, $Sp(S)_{ess} = Sp(T)_{ess}$ .

<u>Definition A6</u>  1. An operator $T \in L(H)$ is called <u>simple</u>  (or <u>without multiplicity</u>) if is unitary equivalent to the operator $Z:L^2(\mu) \to L^2(\mu)$, $Z:f \mapsto \lambda.f$, where $L^2(\mu)$=space of functions $f:\mathbb{C} \to \mathbb{C}$ , square integrable, with respect a positif measure on $\mathbb{C}$, with compact support.

2.  A vector $x \in H$  is <u>cyclic</u> for the normal operator T if the set of vectors $T^m \cdot T^n(x)$, $m,n \geqslant 0$,is total in H.

<u>Theorem A7</u>  1.  The normal operator T on H is simple iff T has a cyclic vector x.   In this case T is unitary equivalent to $Z:L^2(\mu) \to L^2(\mu)$, with  $\mu=E_x$. Further,  $Sp(T)=SuppE_x$.

2.  Let  $\mu$ and $\nu$  two positif measures , with compact support on $\mathbb{C}$.  Let S and T be the operator Z on $L^2(\mu)$ and $L^2(\nu)$ respectively. S is unitary equivalent to T iff $\mu$ is equivalent to $\nu$ in the sense of Radon-Nikodym.

3.  Let x and y two cyclic vectors for T in H.  Then, the measures $E_x$ and $E_y$ are Radon-Nikodym equivalent.

4.  Let T be a normal operator on H.  Then, there exists a Hilbert decomposition $H=\oplus_{i \in I} H_i$ , I=non-necessarily finite, or numerable, into sub-spaces $H_i$, two by two orthogonal, closed under T and $T^*$ such that the restrictions $T_i = T|H_i$ are simple operators.  If  $\mu_i$ is a measure on $\mathbb{C}$ associated to $T_i$ (identified minus of equivalences) then $Sp(T) = \bigcup_i supp\mu_i$.

<u>Theorem A8</u>  <u>(Spectral resolution and functional calculus for non-bounded normal operators)</u>

1.  Let $(H_n)$ be a sequence of closed sub-spaces of H such that H is the Hilbert direct sum of $H_n$ : $H = \oplus H_n$. Let $T_n$ be bounded operator on $H_n$.  Then, there exists a unique closed operator T such that: (a) $H_n \subseteq D(T)$;  (b) $T|H_n = T_n$;  (c) $T(\pi_n(x)) = \pi_n(T(x))$, $\forall x \in D(T)$, where $\pi_n:H \to H_n$ is the orthogonal projection; (d) D(T) is the set of points $x = \Sigma_n x_n \in H$ such that $\Sigma \| T_n(x_n) \|^2 < +\infty$   and one has $T(x) = \Sigma_n T_n(x)$.

2.  Under the above hypotheses there exists a unique normal operator T satisfying conditions (a),(b),(c) and such that $T^*(x) = \Sigma_n T_n^*(x_n)$, $\forall x \in D(T)$.

3.  Let T be a normal operator. Then, H is Hilbert direct sum of closed sub-spaces $H_n$ such that: (a) $H_n \subseteq D(T)$;  (b) $H_n$ is stable for T and $T^*$ for any n; (c) The restriction $T_n$ of T to the sub-space $H_n$ is a bounded normal operator.

4.  Let T be a normal operator; one has: (a) $Sp(T) = \overline{\bigcup_n Sp(T_n)}$;  (b) $Sp(T)_p = \bigcup_n Sp(T_n)_p$; (c) $Sp(T)_r = \emptyset$;  (d) T is self-adjoint iff $Sp(T) \subset \mathbb{R}$;  (e) T is self-adjoint positif, that is $(T(x)|x) \geqslant 0$, $\forall x \in D(T))$ iff $Sp(T) \subseteq \mathbb{R}^+$.

5. Let $T$ be a normal operator. To any Borel function $f:\mathbb{C}\to\mathbb{C}$, we can associate a normal operator $f(T)$ such that $f(T)^{*}=\bar{f}(T)$. Further, if $e$ is the identity function on $\mathbb{C}$, one has $e(T)=T$.

So, to any normal operator $T$ we can associate the spectral resolution $E:ba(\mathbb{C})\to L(H)$ given by $E(A)=1_A(T)$. In particular one has: $f(T)=\int_{\mathbb{C}}f(\lambda)dE(\lambda)$.

The decomposition of $H$ into subspaces $H_n$ is obtained by means of the partition of $\mathbb{C}$ into sets $A_n=\{\,n\leq|f|<n+1\,\}$ One has: $(f(T)(x)|y)=\sum_n\int_n f_n(\lambda)dE_{x,y}=\sum_n\int_n (1_{A_n}f)(\lambda)dE_{x,y}=\int f(\lambda)dE_{x,y}=(\phi(f)(x)|y)$. In particular $T=\int\lambda dE(\lambda)$.

6. Let $f$ and $g$ two borel-functions $\mathbb{C}\to\mathbb{C}$ one has: (a) $f(T)+g(T)\subseteq(f+g)(T)$; (b) $f(T)g(T)\subseteq(fg)(T)$. If the function $g$ is bounded then $f(T)g(T)=(fg)(T)$.

<u>Definition A7</u> $T$ is symmetric if $(T(x)|y)=(x|T(y))$, $\forall x,y\in D(T)$.

<u>Proposition A1</u> A self-adjoint operator is symmetric but the inverse is not true. In fact one has $D(T)\subseteq D(T^{*})$ and $\bar{T}=T^{**}$ if $T$ is symmetric.

<u>Theorem A9</u> 1. Let $T$ be a closed symmetric operator. One has $Sp(T)_p\cup Sp(T)_{,c}\subseteq\mathbb{R}$ and $Im(\lambda id_H-T)$ is closed for any $\lambda\in\mathbb{C}-\mathbb{R}$.

Further, one has one of the following four cases: (a) $Sp(T)_r=\emptyset$, $Sp(T)\subseteq\mathbb{R}$ and $T$ is self-adjoint; (b) $Sp(T)_r\supset\pi_+$ and $Sp(T)=\bar{\pi}_+\equiv\{\,\lambda\,,Im\,\lambda\geq 0\,\}$; (c) $Sp(T)_r\supset\pi_-$ and $Sp(T)=\bar{\pi}_-\equiv\{\lambda\,,\,Im\,\lambda\leq 0\,\}$; (d) $Sp(T)_r\supset\pi_+\cup\pi_-$ and $Sp(T)=\mathbb{C}$.

2. If $T$ is a closed symmetric operator the following propositions are equivalent: (a) $T$ is self-adjoint; (b) $Sp(T)_r=\emptyset$; (c) $Sp(T)\subseteq\mathbb{R}$; (d) $\forall\lambda\in\mathbb{C}-\mathbb{R}\Rightarrow Im(\lambda id_H-T)=H$; (e) $Im(\lambda id_H-T)=H$ for a point of $\pi_+$ and a point of $\pi_-$; (f) $Im(T\pm iid_H)=H$; (g) For any $\lambda\in\mathbb{C}-\mathbb{R}$ one has $ker(\lambda id_H-T^{*})=(0)$; (h) $ker(\lambda id_H-T^{*})=(0)$ for any point of $\pi_+$ and a point of $\pi_-$; (i) $ker(T^{*}+iid_H)=(0)$.

3. If $\rho(T)$ contains a real point then $T$ is self-adjoint.

4. If $T$ is a symmetric operator the following conditions are equivalent: (a) $T$ is essentially self-adjoint; (b) $ker(T^{*}+iid_H)=0$; (c) The spaces $Im(T+iid_H)$ and $Im(T-iid_H)$ are dense into $H$.

<u>Theorem A10</u> (<u>Spectral theorems for self-adjoints operators</u>)

Any self-adjoint operator $T$ on $H$ can be related to a spectral measure $E$ on $\mathbb{R}$: $T=\int tdE(t)$. Further, the spectral measure $E(.)$ is unique and its support is the spectrum $Sp(T)$.

<u>Remark</u>. (On the nuclear operators). Let us conclude this appendix with a spectral theorem for nuclear operators.

The space $L_1(H)$ of nuclear operators on H can be normed by means of the following norm: $\|T\|_1 = \mathrm{tr}(\,[T]\,)$, where $[T] = (\,T^* \circ T\,)^{1/2}$ . If $K(H)$ denotes the space of compact operators on H, ($K(H)$ is the aderence in $L(H)$ of the space of operators with finite rank), we get the inclusions: $L_1(H) \subset L_2(H) \subset K(H) \subset L(H)$. From Theorem A5/13 we see that any $T \in L_1(H)$ has a decomposition like $T = \sum_n \alpha_n \, \varepsilon_n \boxtimes e_n$ where $\{\varepsilon_n\}$ and $\{e_n\}$ are two orthonormal sequences and $\{\alpha_n\}$ is a decresing sequence of eigenvalues of $T = \sum_n \alpha_n \, e_n \boxtimes e_n$ . Then, $\|T\|_1 = \sum_n \alpha_n < +\infty$ and $\mathrm{tr}(T) = \sum_n \alpha_n (\varepsilon_n \,|\, e_n)$ . Then, one can see that $L_1(H)$ is a Banach space with respect to this norm. Further, for any $A \in L(H)$, the operators $A \circ T$ and $T \circ A$ belong to $L_1(H)$ and $\mathrm{tr}(A \circ T) = \mathrm{tr}(T \circ A) = \sum_n \alpha_n (A(\varepsilon_n) \,|\, e_n)$ . Therefore , we can obtain the following inequalities:

(a) $\displaystyle \|T\|_1 = \sup_{\substack{\|A\| \leq 1 \\ A \in K(H)}} |\,\mathrm{tr}(A \circ T)\,| = \sup_{\substack{\|A\| \leq 1 \\ A \in L(H)}} |\,\mathrm{tr}(A \circ T)\,|$ ; (b) $\displaystyle \|A\| = \sup_{\|T\|_1 \leq 1} |\,\mathrm{tr}(A \circ T)\,|$ .

From these relations and by using the canonical bilinear form $\langle A, T \rangle = \mathrm{tr}(A \circ T) = \mathrm{tr}(T \circ A)$, we try also the following natural identifications:

(c) $K(H)' \cong L_1(H)$ ; (d) $L_1(H)' \cong L(H)$ ; (e) $K(H)'' \cong L(H)$.

<u>Theorem A 11 ( Spectral theorem for nuclear operators)</u>.

1. If $T = \sum_i \alpha_i \, x_i \boxtimes x_i$ , with $\alpha_i \geq 0$ , $\sum \alpha_i = 1$, $\|x_i\| = 1$ one has that: (a) $T \in L_1(H)$; (b) T is hermitian positif; (c) $\mathrm{tr}(T) = \sum_i \alpha_i \, \|x_i\|^2 = \sum_i \alpha_i = 1$.

2. Conversely, any nuclear operator can be written in the following spectral decomposition $T = \sum_n \alpha_n \, e_n \boxtimes e_n$ , where $\{e_n\}$ is a orthonormal system and such that $\mathrm{tr}(T) = \sum_n \alpha_n$.

3. Further, if we assume that T is positif and $\mathrm{tr}(T) = 1$ , we get $\alpha_n \geq 0$ and $\sum_n \alpha_n = 1$.

## A.2 - LOCAL CHARACTERIZATION OF SOME GEOMETRIC STRUCTURES RELATED TO PDE

Let $E_k \subset \mathcal{JD}^k(W)$ be a k-order PDE locally characterized by equations $A^\alpha = 0$, $\alpha = 1$, and let $\{\bar{x}^i, y^j_{i_1 \dots i_\alpha}\}_{0 \leq \alpha \leq n}$ be a fibered coordinate system on $\mathcal{JD}^k(W)$.

● <u>Symbol</u> $g_k$ of $E_k$. $g_k$ is the set of vectors $v_u = X^j{}_{\gamma_1 \dots \gamma_k} \partial Y_j^{\gamma_1 \dots \gamma_k}(u)$ such that $(\partial y_j^{\gamma_1 \dots \gamma_k} . F^\alpha)(u) X^j_{\gamma_1 \dots \gamma_{k-}}(u) = 0$, $F^\alpha(u) = 0$, $u \in \mathcal{JD}^k(W)$.

In particular if $E_k$ is a linear PDE locally given by

$$F^\alpha \equiv \sum_{0 \leq \omega \leq k} A^{\alpha \, \alpha_1 \cdots \alpha_\omega}_j (\partial x_{\alpha_1} \dots \partial x_{\alpha_\omega} . f^j) = 0,$$ then $g_k$ is locally identified by the following equations:

$$A^{\alpha \, \alpha_1 \cdots \alpha_k}_j (p) X^j_{\alpha_1 \cdots \alpha_k}(u) = 0, \quad p = \pi_k(u).$$

● <u>1-prolongation of $E_k$</u>:

$$E_{k+1} \subset \mathcal{D}^{k+1}(W) : \sum_{1 \leq j \leq q} (\partial \xi_j . F^\alpha) \bullet D^k f \, (\partial x_i . (D^k f)^j) = 0,$$

where $(D^k f)^j \equiv \xi^j \bullet D^k f$.

● <u>Linearization of a Lie equation</u>: Let $R_k \subset \Pi^k(X)$ be a Lie equation locally characterized by the equations $\phi^{\bar{\omega}} = 0$, then $R_k \subset \mathcal{D}^k(TX)$ is locally given by:

$$\sum_{\substack{0 \leq \alpha \leq n \\ 1 \leq i_1, \dots, i_\alpha \leq n \\ 1 \leq j \leq n}} (\partial Y_j^{i_1 \dots i_\alpha} . \phi^\omega) \xi^j_{i_1 \dots i_\alpha} = 0, \quad n = \dim X.$$

● <u>k-Prolongation of $\pi$-related vector fields</u> on a fiber bundle $\pi: W \to M$.

Let $X = X^\alpha \partial x_\alpha + Y^j \partial y_j$ be the local expression of a vector field on $W$ $\pi$-related with a vector field $X = X^\alpha \partial x_\alpha$ on $M$. Then, the k-prolongation $\overset{(k)}{X}$ is locally given by: $\overset{(k)}{X} = X^\alpha \partial x_\alpha + \sum_{0 \leq \alpha \leq n} X^j_{i_1 \dots i_\alpha} \partial y_j^{i_1 \dots i_\alpha}$, where $X^j_{i_1 \dots i_\alpha}$ are numerical functions on $\mathcal{J}^k(W)$ given by:

$$X^j_{i_1 \dots i_\alpha} = \square_{i_1} \dots \square_{i_\alpha} . x^j - \sum_{\substack{\sigma \in B_\alpha \\ 1 \leq P \leq \alpha}} (\partial x_{\sigma(i_p)} \dots \partial x_{\sigma(i_1)} . x^\omega) y^j_{\sigma(i_{p+1}) \dots \sigma(i_\alpha)}$$

with $\square_i . f \equiv (\partial x_i . f) + \sum_{\substack{0 \leq s \leq q \\ 1 \leq i_1 \leq \dots \leq i_s \leq n}} (\partial y_j^{i_1 \dots i_s} . f) y^j_{i i_1 \dots i_s}$ for any function $f: \mathcal{D}^k(W) \to \mathbb{R}$, and $B_\alpha \equiv$ set of cyclic permutations of $\alpha$ objects. Set $X^j_{i_1 \dots i_p} = x^j$ for $p = 0$.

● <u>Dimension of the k-jet-derivative space $\mathcal{D}^k(W)$ over $\pi: W \to M$</u>

Let $\dim M = n$, $\dim$ fiber $W = m$. $\dim \mathcal{D}^k(W) = n + m \dfrac{(k+n)!}{k! \, n!}$.

● <u>Dimension of p-simmetric tensor product of a vector space of dimension n.</u>

$S^p(V) \equiv V \odot \underbrace{\dots}_{\text{p-times}} \odot V$. $\dim S^p(V) = \dfrac{(p+n-1)!}{p! \, (n-1)!}$.

# REFERENCES

[1] Here are reported some fundamental references on intrinsic variational calculus, conservation laws and related arguments.


(a) BAUDERON,M., Ann.Inst.Henri Poincarè, 36(2)(1982),159-179.

(b) BENN,I.M., Ann.Inst.Henri Poincarè, 37(1)(1982),67-91.

(c) DEDECKER,P., Lect.Notes Math., 570(1977),395-496.

(d) FISCHER,A., Gen.Rel.Grav.,14(7)(1982),683-689.

(e) FOMENKO,A.T., Russian Math. Surv.,36(6)(1981),127-165.

(f) GARCIA,P.L., Symposia Math.,14(1974),219-246.

(g) GOLDSCHMIDT,H. and STERNBERG,S., Ann.Inst.Fourier, Grenoble,27(1)(1973), 203-267.

(h) JADCZYK,A., Ann.Inst.Henri Poincarè, 38(2)(1983),99-111.

(i) KUPERSHMIDT,B.A., Lect.Notes Math., 775(1980),168-218.

(l) MASQUÉ ,J.M., Geometrodynamics Proceedings (1983), 49-55, Pitagora Editrice, Bologna 1984.

(m) PALAIS,R.S., *Foundations of Global Non-Linear Analysis*, W.A.Benjamin, N.Y.,1968.

(n) PRASTARO,A., Boll.Un.Mat.Ital.,(5)18-A(1981),411-416.

(o) TAKENS,F., J.Differential Geom.,14(1979),543-562.

(p) TULCZYJEW,W.M., Lect.Notes Math., 836(1980),22-48.

(q) VINOGRADOV,A.M., Soviet Math.Dokl.,18(1977),1200-1204.

(r) VINOGRADOV,A.M., J.Math.Analys.Appl.,100(1)(1984),1-129.

(s) WESTENHOLZ,C., Ann.Inst.Henri Poincarè, 30(4)(1979),353-367.


[2] Here are reported some fundamental references on the formal theory of partial differential equations.


(a) GOLDSCHMIDT,H., J.Differential Geom., 1(1967),269-307.

(b) GOLDSCHMIDT,H. and SPENCER,D.G., Acta Math., 1(36)(1976),103-239.

(c) GUILLEMIN,V., Trans.Amer.Math.Soc., 116(1965),544-560.

(d) GUILLEMIN,V. and STERNBERG,S., J.Differential Geom., 1(1967),127-131.

(e) KUMPERA,A.K. and SPENCER,D.G., *Lie equations:General theory*, Annals of Math. Studies, No. 93, Princeton Univ.Press, 1972.

(f) LEWY,H., Ann. Math., 65(1)(1957),155-158.

(g) LEHMANN,D., Ann. Inst. Fourier, Grenoble, 21(3)(1971),83-94.

(h) LIBERMANN,P., *Connexions d'ordre supérieure et tenseurs de structure*, Atti Convegno Inter. Geom. Differenziale, Bologna 1967.

(i) OLVER,P.J., J.Differential Geom., 14(1979),497-542.

(l) OVSIANNIKOV,L.V., *Group Analysis of Partial Differential Equations*, Nauka, Moscow 1978.

418

(m)  POLLACK,A.S., J.Differential Geom., 2(1974),355-390.

(n)  POMMARET,J.F., *Systems of Partial Differential Equations and Lie Pseudo-groups*, Gordon and Breach, N.Y., 1978.

(o)  POMMARET,J.F., *Differential Galois Theory*, Gordon and Breach, N.Y. 1983.

(p)  REINHART,B.L., *Differential Geometry of Foliations*, Springer-Verlag, Berlin 1983.

(q)  RODRIGUES,A.M., Ann. Inst. Fourier, Grenoble, 31(3)(1981),245-274.

(r)  SINGER,I.M. and STERNBERG,S., J.Analys. Math., 15(1965),1-114.

(s)  SPENCER,D.G., Bull. Am. Math. Soc., 75(1965),1-114.

(t)  VINOGRADOV,A.M., J.Soviet.Math., 17(1)(1981),1624-1649.

[3]  For the Dirac's approach to the quantization see the following references:

(a)  BIRREL,N.D. and DAVIES,P.C., *Quantum Fields in Curved Spaces*, Cambridge Univ.Press, Cambridge 1982.

(b)  COOK,J.M., Trans.Amer.Math.Soc., 74(1953),222-245.

(c)  DE WITT,B.S., I.,Phys.Rev.,160(5)(1967),1113-1148; II.,III.,Phys.Rev.,162(5)(1967),1195-1255.

(d)  DIMOCK,J., Comm.Math.Phys., 77(1980),219-228.

(e)  DIRAC,P.A.M., *The Principles of Quantum Mechanics*, Oxford Clarendon Press, Oxford 1958.

(f)  KAY,B.S., I.,Comm.Math.Phys., 62(1978),55-70; II.,Comm.Math.Phys.,71(1980),29-46.

(g)  KASTLER,D., *Introduction a l'Electrodynamique Quantique*, Dunod, Paris 1961.

(h)  WEINLESS,M., J.Functional Analy.,4(1969),350-379.

[4]  For symplectic geometric approaches to the quantization see the following references:

(a)  ALDAYA,V. and DE AZCARRAGA,J.A., J.Math.Phys.,23(7)(1982),1297-1305.

(b)  BAYEN,F.,FLATO,L.,FRONSDAL,C.,LICHNEROWICZ,A. and STERNHEIMER,D., I.,II., Ann.Phys.,110(1978),61-151.

(c)  CZYZ,J., Rep.Math.Phys., 15(1979),57-97.

(d)  KOSTANT,B., Lect.Notes Math., 170(1970),87-213.

(e)  KOSTANT,B., Lect.Notes Math., 570(1975),177-306.

(f)  LICHNEROWICZ,A., Ann.Inst.Fourier, Grenoble, 32(1)(1982),157-209.

(g)  SOURIAU,J.M., *Structure des Systemes Dynamiques*, Dunod, Paris 1970.

(h)  VAISMAN,I., Ann.Inst.Henri Poincarè, 31(1)(1979),15-24.

(i)  VAISMAN,I., Rend.Sem.Mat.,Torino, 39(3)(1981),140-152.

[5] For other geometric point of views on the quantization see the
following references:

(a) ALVAREZ-GAUME',L., <u>J.Phys.,A:Math.Gen.</u>,<u>16</u>(1983),4177-4182.

(b) CRUMEYROLLE,A., <u>Ann.Inst.Henri Poincarè</u>, <u>A29</u>(2)(1978),217-231.

(c) EASTWOOD,M.G.,PENROSE,R. and WELLS,JR.,R.D., <u>Comm.Math.Phys.</u>,<u>78</u>(1981),
305-351.

(d) HAWKING,S.W., <u>Nucl.Phys.</u>,B 144(1978),349-362.

(e) MIEKE,E., <u>Gen.Rel.Grav.</u>,<u>8</u>(3)(1977),175-196.

(f) PENROSE,R., <u>Int.J.Theoret.Phys.</u>,<u>1</u>(1968),61-99.

(g) PENROSE,R., <u>Reports Math.Phys.</u>,<u>12</u>(1977),65-76.

(h) PENROSE,R. and MAcCALLUM,A.H., <u>Phys.Rep.</u>,<u>6</u>(1972),241-316.

(i) ROMER,H., <u>Lect.Notes Phys.</u>,<u>139</u>(1981),167-211.

(l) WELLS,JR.R.O., <u>Bull.Am.Math.Soc.</u>,<u>1</u>(2)(1979),296-336.

(m) WESTENHOLZ,C., <u>Ann.Inst.Henri Poincarè</u>,<u>29</u>(3)(1978),285-303

(n) WESTENHOLZ,C. *Differential Forms in Mathematical Physics*, North-Holland,
Amsterdam 1981.

[6] Here are reported some useful references on homological algebra and
algebraic topology.

(a) HILTON,P.J. and STAMMBACH,U., *A Course in Homological Algebra*,
Springer-Verlag, Berlin 1970.

(b) HIRSCH,M.W., *Differential Topology*, Springer-Verlag, Berlin 1976.

(c) HIRZEBRUCH,F., *Topological Methods in Algebraic Geometry*, Springer-
Verlag, Berlin 1966.

(d) MILNOR,J.W., *Topology from the Differentiable View Point*, The University
Press of Virginia, Charlottesville 1965.

(e) SWITZER,R.M., *Algebraic Topology-Homotopy and Homology*, Springer-
Verlag, Berlin 1975.

(f) VAISMAN,I., *Cohomology and Differential Forms*, Marcel Dekker, N.Y. 1973.

(g) WALL,C.J., *A Geometric Introduction to Topology*, Addison-Wesley, Mass.,1972.

(h) WARNER,F.W., *Foundations of Differentiable Manifolds and Lie Groups*,
Glenview, Ill.,1971.

(i) WHITEHEAD,G.W., *Elements of Homotopy Theory*, Springer-Verlag, Berlin 1978.

[7] Here are reported some useful references on topological vector spaces
and manifolds of maps.

(a) BOURBAKI,N., *Espaces Vectoriels Topologiques*, Hermann, Paris ,
tome 1 (1963), tome 2 (1955).

(b) GARSOUX,J., *Espaces Vectoriels Topologiques et Distributions*, Dunod,
Paris 1963.

(c)    BOURBAKI,N., *Varietés Différentielles et Analytiques*, Hermann, Paris 1971.

(d)    EELLS,J. and ELWORTHY,K.D., Symp.Pure Math.,Am.Math.Soc., (    ),41-44.

(e)    EELLS,J. and SAMPSON,J.H., Ann.Inst.Fourier, Grenoble, 14(1)(1964),61-70.

(f)    ELIASSON,H.I., J.Differential Geom., 1(1967),169-194.

(g)    PALAIS,R.S. and SMALE,S., Bull.Amer.Math. Soc.,70 (1964), 165-172.

(h)    SEELEY,R.T., Trans.Am.Math.Soc.,117(1965),167-204.

[8]    Here are reported some fundamental references relating manifolds of maps to some physical models.

(a)    ATIYAH,M.,BOTT,R.and PATODI,V.K., Inventiones Math.,19(1973),279-330.

(b)    ATIYAH,M.and WARD,R.S., Comm.Math.Phys.,55(1977),117-124.

(c)    ATIYAH,M. and JONES,J.D.S., Comm.Math.Phys.,61(1978),97-118.

(d)    BABELON,O. and VIALLET,C.M., Comm.Math.Phys., 81(1981),515-525.

(e)    MIATTER,P.K. and VIALLET,C.M., Comm.Math.Phys.,79(1981),457-472.

(f)    NARASIMHAN,M.S. and RAMADAS,T.R., Comm.Math.Phys.,67(1979),121-136.

[9 ]    PRASTARO,A., Stochastica, 3(2)(1979),15-31.

[10]    PRASTARO,A., I.,Boll.Un.Mat.Ital.,(5)17-B(1980),704-726;
II.,Boll.Un.Mat.Ital.,(5)S.-FM(1981),69-106;
III.,Boll.Un.Mat.Ital.,(5)S.-FM(1981),107-129.

[11]    PRASTARO,A., Boll.Un.Mat.Ital.,(6)1-B(1982),1015-1028.

[12]    PRASTARO,A., Riv.Nuovo Cimento, 5(4)(1982),1-122.

[13]    PRASTARO,A., *Geometrodynamics of Non-Relativistic Continous Media;*
I.,Rend.Sem.Mat.,Torino, 40(2)(1982),89-117;
II.,Rend.Sem.Mat.,Torino, (to appear).

[14]    PRASTARO,A., Geometrodynamics Proceedings (1983), 65-90.
Pitagora Editrice, Bologna 1984.

GEOMETRODYNAMICS PROCEEDINGS (1985), pp. 421-442
edited by A. Pràstaro
© 1985 by World Scientific Publishing Co.

# THE DOULBEAULT-KOSTANT COMPLEX AND GEOMETRIC QUANTIZATION

Mircea Puta

Seminarul de Geometrie si Topologie,
University of Timisoara
1900 Timisoara, Romania

During the last twenty years the use of differential geometric
methods in physics undergone a steady increase.  Perhaps the most extensive
applications have been to the symplectic formulation of classical mechanics
and quantization.  Today, symplectic geometry represents a privileged area
in which both pure and applied mathematicians can enjoy fruitful cooperation.

The main goal of my lecture is to present some aspects of the
Doulbeault-Kostant complex theory and also to point out some of its applications in the Kostant cohomological correction of geometric quantization.

The material is divided up in five sections as follows.  In the first
section some elements of the geometric quantization are reviewed and the
necessity of the Kostant cohomological correction of geometric quantization
is pointed out.  The theory of the Doulbeault-Kostant complex is developed
in the following three sections.  Finally in the last section the Kostant
cohomological corrections of geometric quantization is discussed and some
of its applications are sketched.

## 1.    Some elements on Geometric Quantization

The theory of geometric quantization was founded independently by
Bert Kostant [3] and Jean-Marie-Souriau [13] around 1964.  It is essentially
a globalization of classical quantization schemes in which all objects are
expressed in geometrical terms.

The need for a geometrical approach to quantization arises from the
mathematical difficulties inherent in the correspondence procedure of
standard canonical quantization.  These are:

422

i. Dependence upon arbitrary canonical coordinates and lack of invariance under general canonical transformation.

ii. In general, any operator representing a quantum observable will depend upon the ordering of the canonical coordinates in its corresponding classical counterpart.

Problems such as these become increasingly difficult to overcome when quantization is applied to general constrained classical systems, or to systems without symmetries. By placing the traditional quantization procedure in a geometrical setting, the Kostant-Souriau theory clarifies the mathematical ambiguities which may arise during passage from the classical to the quantum domain of more complicated physical systems.

We begin by reviewing the fundamental notions of geometric quantization.

Let $(M, \omega)$ be a <u>symplectic manifold</u>, i.e. $M$ is a real $C^\infty$-manifold and $\omega$ is a closed non-degenerate 2-form on $M$. The 2-form $\omega$ sets up a one-one correspondence between vector fields $X$ on $M$ and 1-forms $\alpha$ on $M$ via the formula

$$\alpha(Y) = \omega(X, Y) \quad , \tag{1.1}$$

for arbitrary vector fields $Y$. In particular, every $f \in C^\infty(M, \mathbb{R})$ gives rise to a vector field $X_f$ known as the <u>Hamiltonian vector field generated by $f$</u>, namely the vector field corresponding to the 1-form $df$.

If $f, g \in C^\infty(M, \mathbb{R})$, we define:

$$\{f, g\} = X_f(g) \quad , \tag{1.2}$$

called the <u>Poisson bracket</u> of $f$ and $g$.

Under $\{\,,\,\}$, $C^\infty(M, \mathbb{R})$ becomes a Lie algebra over $\mathbb{R}$ called the <u>Poisson algebra on $M$</u> and moreover $f \mapsto X_f$ is a Lie algebra homomorphism.

An example of this set up is $M = \mathbb{R}^{2n}$ with coordinates labeled $q^1, \ldots, q^n, p_1, \ldots, p_n$ and with

$$\omega = \sum_{j=1}^{n} dp_j \wedge dq^j \quad .$$

In this example

$$X_f = \sum_{j=1}^{n} \left( \frac{\partial f}{\partial q^j} \frac{\partial}{\partial p_j} - \frac{\partial f}{\partial p_j} \frac{\partial}{\partial q^j} \right)$$

so that

$$\{f, g\} = \sum_{j=1}^{n} \left( \frac{\partial f}{\partial q^j} \frac{\partial g}{\partial p_j} - \frac{\partial f}{\partial p_j} \frac{\partial g}{\partial q^j} \right) \quad .$$

The theorem of Darboux-Weinstein says that every symplectic manifold is locally isomorphic to this example for some $n$, [14].

Geometric quantization is a procedure which allows us to obtain a representation of a subalgebra of the Poisson algebra $C^\infty(M, \mathbb{R})$ by operators on a Hilbert space associated with the symplectic manifold $M$. Restriction to a subalgebra of $M$ is necessary because of the so called "no-go" theorems of Van Hove, which state that the representation for the entire Poisson algebra does not exist if physically reasonable assumptions are imposed on the quantization procedure.

The first step in geometric quantization is the construction of a complex line bundle over $M$. Such a line bundle does not always exist. In fact, we say that $M$ is _quantizable_ if there exists a Hermitian line bundle $L^\omega$ over $M$, called _pre-quantum line bundle_, with a compatible connection $\nabla^\omega$ such that $(1/\hbar)\omega$ ($\hbar$ is the Planck constant divided by $2\pi$) is the curvature of $\nabla^\omega$.

The existence of such a line bundle $L^\omega$ imposes a topological condition on $\omega$. The De Rham cohomology class of the 2-form $(1/\hbar)\omega$ must be an integral valued element of $H^2(M, \mathbb{R})$. In other words, the integral of $\omega$ over each closed surface in $M$ must be an integral multiple of $\hbar$. This is a quantization condition on the phase space $M$, and moreover can be proved that $M$ is quantizable if and only if this condition is satisfied. It is indirectly related to the old Bohr-Sommerfeld quantization rule.

When $M$ is quantizable and simply-connected there is up to an isomorphism exactly one Hermitian line bundle $L^\omega$ over $M$ with a compatible connection $\nabla^\omega$, having $(1/\hbar)\omega$ as curvature form. For a non simply connected quantizable space, the set of distinct Hermitian line bundle $L^\omega$ with compatible connections having $(1/\hbar)\omega$ as curvature form is parametrized by the group $H^1(M, S^1)$, where $S^1$ is the circle group.

424

When $\omega$ is an exact form, as for example in the case when $M$ arises from a configuration space, i.e. $M = T^*Q$ and $\omega = d\theta$, the quantization condition is trivially satisfied since when $\omega$ represents the zero De Rham cohomology class. In this case $M$ is quantizable and its prequantum bundle is a trivial one, $L^\omega = M \times \mathbb{C}$.

The Hilbert space of the prequantization $\mathcal{H}$ is the completion space of the space of all sections $s$ of $L^\omega$ with compact support with respect to the inner product

$$< t_1, \, t_2 > = (\frac{1}{2\pi\hbar})^n \int_M (t_1, \, t_2)\omega^n \qquad . \tag{1.3}$$

If $f \in C^\infty(M, \, \mathbb{R})$ is a classical observable then its representation as an operator on the Hilbert space $\mathcal{H}$ is given by:

$$\delta_f = - i\hbar \nabla^\omega_{X_f} + f \qquad . \tag{1.4}$$

An example of this set up is $M = T^*Q$ with the usual symplectic form $\omega = d\theta$. Then $M$ is quantizable and its pre-quantum bundle is a trivial one. $L^\omega = M \times \mathbb{C}$. Then the Hilbert representation space $\mathcal{H}$ can be identified with the Hilbert space of complex $L^2$-functions on $M$, and for each $f \in C^\infty(M, \, \mathbb{R})$ the operator $\delta_f$ is given via (1.4) by:

$$\delta_f = - i\hbar \left[ X_f - \frac{i}{\hbar} X_f \lrcorner \, \theta \right] + f \qquad .$$

However, the Schrödinger prescription for quantum mechanics implies that the pre-quantum Hilbert space should instead be $L^2(Q)$. Also the uncertainty principle is violated for functions in $L^2(T^*Q)$. In this sense the prequantum Hilbert space is too large. Thus we need to reduce the size of our Hilbert space. This is accomplished by introducing the notion of polarization. A <u>polarization</u> of $(M, \, \omega)$ is a maximally isotropic involutive complex subtangent bundle $F$ of $T(M)_\mathbb{C}$ ( = complexified tangent bundle of $T(M)$). If $N_F^{\frac{1}{2}}$ denotes the bundle of half-forms normal to $F$ and $L = L^\omega \otimes N_F^{\frac{1}{2}}$, $L$ has an F-connection $\nabla$ and we denote by $S_F(M, \, L)$ the space of polarized sections of $L$, i.e. $S_F(M, \, L) = \{s \in \bar{\Gamma}(L) | \nabla_{X} s = 0,$ $\forall \, X \in \mathcal{X}(M, \, F)\}$, where $\mathcal{X}(M, \, F) = \{X \in \mathcal{X}(M) | X_x \in F_x, \, \forall \, x \in M\}$. Then there exists a natural inner product on $S_F(M, \, L)$, $[15]$, and let $\mathcal{H}_F^{\frac{1}{2}}$ be its completion with respect to this inner product.

Let $C_F^0(M, \mathbb{R})$ be the space of all functions $f \in C^\infty(M, \mathbb{R})$ such that $X_f$ is a section of $F$, and $C_F^1(M, \mathbb{R})$ the space of all functions of $C^\infty(M, \mathbb{R})$ whose Hamiltonian vector fields are infinitesimal automorphisms of $F$, i.e. $C_F^1(M, \mathbb{R}) = \{f \in C^\infty(M, \mathbb{R}) \mid [X_f, X] \in \mathfrak{X}(M, F), \forall X \in \mathfrak{X}(M, F)\}$. Then $C_F^1(M, \mathbb{R})$ is a Lie algebra under the Poisson bracket, $C_F^0(M, \mathbb{R})$ is an abelian ideal in $C_F^1(M, \mathbb{R})$, and if $f$ is in $C_F^1(M, \mathbb{R})$ there is a natural Lie derivative action of $X_f$ in $\Gamma(N_F^{1/2})$. Combining $\delta_f$ with this Lie derivative gives a differential operator $(\delta_f^{1/2})_f$ on $L$ which preserves $\mathcal{H}_F^{1/2}$. This action of $(\delta_F^{1/2})_f$ on $\mathcal{H}_F^{1/2}$ is known as <u>quantization</u> and is defined for functions $f$ in $C_F^1(M, \mathbb{R})$. Functions in $C_F^0(M, \mathbb{R})$ quantize as zeroth order differential operators, that is as multiplication operators, and so can be considered to have been quantized in an already diagonal form.

An example of this set up is $(M, \omega) = (\mathbb{R}^{2n} \sum_{j=1}^{n} dp_j \wedge dq^j)$ which is quantizable and its prequantum bundle is a trivial one $L^\omega = M \times \mathbb{C}$. Let $F$ be the vertical polarization on $M$ spanned by $\{x_{q^1}, \ldots, x_{q^n}\}$. Then it is not hard to see that $N_F^{1/2}$ is globally spanned by $(dq^1 \wedge \ldots \wedge dq^n)^{1/2}$ and moreover $\mathcal{H}_F^{1/2}$ can be identified with $L^2(\mathbb{R}^n)$. Moreover $q^j$, $p_j \in C_F^1(M, \mathbb{R})$, $j = 1, 2, \ldots, n$, and we refined the classical Schrödinger representation

$$
\begin{cases}
(\delta_F^{1/2})_{-q^j} = q^j \quad ; \\[2ex]
(\delta^{1/2})_{-p_j} = -i\, \dfrac{\partial}{\partial q^j} \quad , \qquad j = 1, 2, \ldots, n \quad .
\end{cases}
$$

It often turns out that the bundle $L = L^\omega \otimes N_F^{1/2}$ may not have any smooth global sections which are covariantly constant along vector fields in the polarization. The construction of the Hilbert space via the standard Kostant-Souriau procedure becomes impossible. This problem may arise even with the simplest of classical phase spaces. For example, let $M = \mathbb{R}^2 \setminus \{o\}$ be the punctured plane with Euclidean coordinates $p$ and $q$ and let $\omega = dp \wedge dq$. Suppose that $f$ is a function such that $X_f$ is everywhere tangent to the polarization $F$. When $H = (1/2)(p^2 + q^2)$ (such as the Hamiltonian for the 1-dimensional harmonic oscillator), then $F$ consists of the tangents to the concentric circles $p^2 + q^2 = $ constant, and there are no smooth global sections of $L^\omega \otimes N_F^{1/2}$ covariantly constant along each of these leaves.

In these pathological cases when the Hilbert representation space $\mathcal{H}_F^{1/2}$ is trivial, B. Kostant [4] has suggested using higher cohomology groups of $M$ for the quantization process. Then the Doulbeault-Kostant complex (see section 5) gives a convenient representation of these cohomological groups in terms of L-valued forms defined on $F$. Therefore there are ample motivations for a general study of the Doulbeault-Kostant complex.

## 2.    Complex Foliations and Differential Forms

Let $M$ be an $(n+m)$-dimensional, orientable, paracompact, smooth $(= C^\infty)$ and real manifold. We denote by $T(M)$ [resp. $T(M)_\mathbb{C} = T(M) \otimes \mathbb{C}$] the tangent bundle [resp. the complexified tangent bundle] of $M$ and by $C^\infty(M, \mathbb{R})$ the space of smooth $(= C^\infty)$ real functions on $M$.

Definition 2.1. ([1]) A complex foliation on $M$ is a complex subbundle $F \subset T(M)_\mathbb{C}$ satisfying the following two conditions:

  i.  $F \cap \bar{F}$ is of constant rank;

  ii.  both $F$ and $F + \bar{F}$ are integrable.

Definition 2.2. ([1]) A differential form of type $(o, p)$ on $M$ is a smooth section of the bundle $\wedge^p F^*$. We denote by $A_F^p(M)$ [resp. $\mathcal{A}_F^p(M)$] the space of differential forms [resp. the sheaf of germs of differential forms] of type $(o, p)$ on $M$.

The exterior derivative along $F$, $d_F$ can be defined by

$$d_F : \alpha \in A_F^p(M) \longmapsto d_F\alpha \in A_F^{p+1}(M) \quad ,$$

where:

$$(d_F\alpha)(X_1,\ldots,X_{p+1}) = \sum_{i=1}^{p+1} (-1)^{i+1} X_i(\alpha(X_1,\ldots,\hat{X}_i,\ldots,X_{p+1}))$$

$$+ \sum_{i<j} \alpha([X_i, X_j], X_1,\ldots,\hat{X}_i,\ldots,\hat{X}_j,\ldots,X_{p+1}) \quad ,$$

for any vector fields $X_1,\ldots,X_{p+1} \in \mathcal{X}(M, F)$.

One readily sees that $d_F^2 = 0$, and $d_F f = df_{|F}$, for $f \in C^\infty(M, \mathbb{R})$.

Let $C_F(M)$ be the space of F-holomorphic functions on $M$, i.e. $C_F(M) = \ker d_F \subset A_F^0(M)$, and $\mathcal{C}_F(M)$ the corresponding sheaf of germs of F-holomorphic functions on $M$. Then we can prove.

Theorem 2.1. ([1], [10]). The sequence $0 \longrightarrow \mathcal{C}_F(M) \xrightarrow{i} \mathcal{A}_F^0(M) \xrightarrow{d_F} \mathcal{A}_F^1(M) \xrightarrow{d_F} \cdots \xrightarrow{d_F} \mathcal{A}_F^n(M) \longrightarrow 0$, $n = \mathrm{rank}(F)$, is a fine resolution of the sheaf $\mathcal{C}_F(M)$ and therefore:

$$H^p(M, \mathcal{C}_F(M)) = H^p(M, d_F)$$

for each $p \geq 0$. In particular, $H^p(M, \mathcal{C}_F(M)) = 0$ for $p > n$.

Examples

1. In the case $F = T(M)_{\mathbb{C}}$ this theorem is just the usual De Rham theorem because $\mathcal{C}_F(M)$ is just the constant sheaf $\mathbb{C}$.

2. In the case $T(M)_{\mathbb{C}} = F \oplus \bar{F}$, the above theorem is just the usual Doulbeault theorem.

If $B$ is a complex vector space then the above theorem still holds for the sheaves $\mathcal{A}^p(M, B)$ of B-valued forms.

Theorem 2.2 ([1]). The cohomology of $M$ with coefficients in the sheaf $\mathcal{C}_F(M, B)$ of germs of B-valued, F-holomorphic functions on $M$ is given by

$$H^p(M, \mathcal{C}_F(M, B)) = H^p(M, B, d_F) \quad .$$

for each $p \geq 0$. In particular $H^p(M, \mathcal{C}_F(M, B)) = 0$ for $p > n = \mathrm{rank}(F)$.

We now consider a further generalization of Theorem 2.1. Let $E$ be a complex vector bundle over $M$. We say that $E$ is _F-holomorphic_ if there exists a vector bundle atlas $(U_\alpha)$ on $M$ such that the 1-cocycle $(c_{ij})$ representing $E$ for this atlas satisfies:

$$d_F\, c_{ij} = 0 \quad , \tag{2.1}$$

meaning that each $c_{ij}$ is an F-holomorphic map defined on $U_i \cap U_j$.

Let $s \in \Gamma(E)$ be a section of $E$ represented by the maps $s_i : U_i \to E$ in the given trivializations $E_{|U_i} \simeq U_i \times E$. Then, over $U_i \cap U_j : s_j = c_{ij} s_i$. Using the fact that $d_F$ on (vector-valued) maps simply is the exterior derivative restricted to $F$, we have that

$$d_F s_j = (d_F c_{ij}) s_i + c_{ij} d_F s_i = c_{ij} d_F s_i$$

because of (2.1). Thus, the maps $d_F s_i$ represent a global section of $E$ and will be denoted by $d_F s$.

Definition 2.3. ([1]). An E-valued differential form of type $(o, p)$ on $M$ is a smooth section of the bundle $\wedge^p F^* \otimes E$. We denote by $S_F^p(M, E)$ [resp. $\mathcal{S}_F^p(M, E)$] the space of E-valued differential forms [resp. the sheaf of germs of E-valued differential forms] of type $(o, p)$ on $M$.

The exterior derivative defined above can be extended in a natural way to $S_F^p(M, E)$.

Definition 2.4. ([1]). The complex:

$$S_F^0(M, E) \xrightarrow{d_F} S_F^1(M, E) \xrightarrow{d_F} \dots \xrightarrow{d_F} S_F^n(M, E) \tag{2.2}$$

is called the Doulbeault-Kostant complex.

Let $S_F(M, E) \overset{\text{def}}{=} \ker(d_F : S_F^0(M, E) \to S_F^1(M, E))$, and $\mathcal{S}_F(M, E)$ the corresponding sheaf. Then we can prove:

Theorem 2.3. ([1]). The cohomology of $M$ with coefficients in the sheaf $\mathcal{S}_F(M, E)$ can be identified with the cohomology of the Doulbeault-Kostant complex (2.2). In particular, $H^p(M, \mathcal{S}_F(M, E)) = 0$, for $p \geq n = \text{rank}(F)$ and moreover we have the following identification:

$$H^0(M, \mathcal{S}_F(M, E)) = S_F(M, E) \quad .$$

Now we shall prove that there exists a generalized version of Mayer-Vietoris theorem for the Doulbeault-Kostant complex.

Let $U, V$ be open subsets of $M$ such that $M = U \cup V$. Then there is a sequence of inclusions:

$$U \cap V \underset{i_V}{\overset{i_U}{\rightrightarrows}} U \sqcup V \to M \quad ,$$

where $U \sqcup V$ is the disjoint union of $U$ and $V$ and $i_U$ and $i_V$ are the inclusions of $U \cap V$ in $U$ and $V$ respectively. Then it can be easily verified that we have the following exact sequence:

$$0 \longrightarrow S_F^*(M, E) \longrightarrow S_F^*(U, E) \oplus S_F^*(V, L) \longrightarrow S_F^*(U \cap V, E) \longrightarrow 0 \quad , \quad (2.3)$$

where $S_F^*(\cdot, E) = \overset{n}{\underset{p=0}{\oplus}} S_F^p(\cdot, E)$.

Theorem 2.4. The sequence (2.3) induces a long exact sequence in cohomology:

$$\ldots \longrightarrow H^p(M, E, d_F) \longrightarrow H^p(U, E, d_F) \oplus H^p(V, E, d_F)$$

$$\longrightarrow H^p(U \cap V, E, d_F) \longrightarrow H^{p+1}(M, E, d_F)$$

$$\longrightarrow H^{p+1}(U, E, d_F) \longrightarrow H^{p+1}(V, E, d_F) \longrightarrow H^{p+1}(U \cap V, E, d_F) \longrightarrow \ldots$$

Let $F^0 \subset T^*M_{\mathbb{C}}$ be the subcotangent bundle of covectors with vanish on F. Thus $f \in C_F(U)$ if and only if $df$ is a section of $F^0$ on U.

If there are functions $f_1, \ldots, f_m$ in $C_F(U)$ with $(df_1, \ldots, df_m)$ a frame of $F^0$ at each point of U, then $(f_1, \ldots, f_m)$ is called a $\underline{C_F\text{-}}$ $\underline{\text{coordinate system}}$ and U a $\underline{C_F\text{-coordinate neighborhood}}$. The atlas of M defined by these local coordinates will be called a $\underline{C_F\text{-atlas}}$. We reminded also that an open cover $\{U_i\}$ of M is called a $\underline{\text{good cover}}$ if all finite intersections $U_{i_0} \cap \ldots \cap U_{i_p}$ are diffeomorphic to $\mathbf{R}^{n+m}$, and a manifold which has such a cover is said to be of $\underline{\text{finite type}}$.

In the sequel we suppose that M has a $C_F$-atlas of finite type. Then a generalized version of Künneth formula can be obtained:

Theorem 2.5. Let M, N be smooth manifolds. F and G respectively complex foliations on M and N and $E_1$ and $E_2$ holomorphic vector bundles on M and N respectively. Then for each $p \leq \text{rank}(F)$ we have

$$H^p(M \times N, E_1 \times E_2, d_F \times d_G) = \underset{r+s=p}{\oplus} H^r(M, E_1, d_F) \times H^s(N, E_2, d_G) \quad .$$

Proof. The two natural projections $\pi : M \times N \to M$, $\rho : M \times N \to N$, give rise to a map on forms $\alpha \otimes \beta \longmapsto \pi^*(\alpha) \otimes \rho^*(\beta)$, which induces a map in cohomology:

$$\psi : H^*(M, E_1, d_F) \otimes H^*(N, E_2, d_G) \longrightarrow H^*(M \times N, E_1 \times E_2, d_F \times d_G) \quad .$$

430

We will show that $\psi$ is an isomorphism. If $M = \mathbf{R}^n \times \mathbf{R}^m$ this is simply the Poincaré lemma for $d_F$.

Let $U, V$ be open sets and $n'$ a fixed integer. From the Mayer-Vietoris theorem we get an exact sequence by tensoring with $H^{n'-p}(N, E_2, d_G)$. Summing over $p = o,\ldots,n'$, yields the exact sequence:

$$\ldots \longrightarrow \bigoplus_{p=o}^{n'} H^p(U \ V, E_1, d_F) \otimes H^{n'-p}(N, E_2, d_G)$$

$$\longrightarrow \bigoplus_{p=o}^{n'} H^p(U, E_1, d_F) \otimes H^{n'-p}(N, E_2, d_G)$$

$$\oplus \ H^p(V, E_2, d_G) \otimes H^{n'-p}(N, E_2, d_G)$$

$$\longrightarrow \bigoplus_{p=o}^{n'} H^p(U \ V, E_1, d_F) \otimes H^{n'-p}(N, E_2, d_G) \longrightarrow \ldots$$

The following diagram is commutative (see Fig. 1). By Five lemma if the theorem is true for $U, V$ and $U \cap V$, then it is also true for $U \cup V$. Then the Künneth formula follows by induction on the cardinality of a $C_F$-good cover.

QED

3.  <u>The Doulbeault-Kostant Complex of a Line-bundle with connection</u>

We add the following remarks on complex line bundles over $M$ in view of their geometric interest.

If the complex line-bundle over $M$ is F-holomorphic, then all the preceding considerations, of course, apply.

If the complex line-bundle over $M$ is not F-holomorphic then let $\nabla$ be the linear connection on $L$ and $\Omega$ its curvature.

$\nabla$ may now be used to define a differential operator $\partial^F$, the covariant exterior derivative along $F$, on $S_F^p(M, L)$ by:

$$\partial^F : \alpha \in S_F^p(M, L) \longmapsto \partial^F \alpha \in S_F^{p+1}(L) \quad ,$$

where:

$$(\partial^F \alpha)(X_1,\ldots,X_{p+1}) \overset{\mathrm{def}}{=\!=\!=} \sum_{i=1}^{p+1} (-1)^{i+1} \nabla X_i \alpha(X_1,\ldots,\hat{X}_i,\ldots,X_{p+1})$$

$$+ \sum_{i<j} \alpha([X_i, X_j], X_1,\ldots,\hat{X}_i,\ldots,\hat{X}_j,\ldots,X_{p+1}) \quad ,$$

for each $X_1, \ldots, X_{p+1} \in \mathfrak{X}(M, F)$.

It is easy to verify that in general $(\partial^F)^2 \neq 0$.

$$
\begin{array}{c}
\vdots \\
\downarrow \\
\bigoplus_{p=0}^{n'} H^p(U \cap V, E_1, d_F) \otimes \\
\otimes H^{n'-p}(N, E_2, d_G) \longrightarrow H^p((U \cap V) \times N, E_1 \times E_2, d_F \times d_G) \\
\downarrow \\
\bigoplus_{p=0}^{n'} H^{p+1}(U \cup V, E_1, d_F) \otimes \\
\otimes H^{n'-p}(N, E_2, d_G) \longrightarrow H^{p+1}((U \cup V) \times N, E_1 \times E_2, d_F \times d_G) \\
\downarrow \qquad\qquad \downarrow \\
\bigoplus_{p=0}^{n'} (H^{p+1}(U, E_1, d_F) \otimes \\
H^{n'-p-1}(N, E_2, d_G) \ \oplus \\
H^{p+1}(V, E_1, d_F) \otimes \\
H^{n'-p-1}(N, E_2, d_G) \longrightarrow H^p(U \times N, E_1 \times E_2, d_F \times d_G) \ \otimes \\
\otimes H^p(V \times N, E_1 \times E_2, d_F \times d_G) \\
\vdots \qquad\qquad \vdots
\end{array}
$$

Fig. 1

Definition 3.1. ([10]). We say that $F$ and $\nabla$ are compatible or that $\nabla$ is an F-connection on $L$ if $\Omega$ vanishes on $F$, i.e. if $F$ is totally isotropic for $\Omega$.

Example. The canonical connections $\nabla^\omega$ and $\nabla$ of the complex line-bundles $L^\omega$ and respectively $L^\omega \otimes N_F^{1/2}$ used in geometric quantization, (see sections 1, 4) are compatible with the polarization $F$.

Now it is not hard to see that if $\nabla$ and $F$ are compatible then $(\partial^F)^2 = 0$. Thus we can also obtain:

<u>Theorem 3.1</u>. ([10]) The cohomology of $M$ with coefficients in the sheaf $\mathcal{S}_F(M, L)$ can be identified with those of the complex $\{S_F^p(M, L), \partial^F\}$. In particular, $H^p(M, \mathcal{S}_F(M, L)) = 0$ for $p > n = \mathrm{rank}(F)$ and moreover we have the following identification:

$$H^0(M, \mathcal{S}_F(M, L)) = S_F(M, L) \quad . \tag{3.1}$$

<u>Remark 3.1</u>. The complex $\{S_F^p(M, L), \partial^F\}$ is called frequently the Doulbeault-Kostant complex.

<u>Remark 3.2</u>. The all results of the above section can be extended in a natural way to the complex $\{S_F^p(M, L), \partial^F\}$, [6].

4.  <u>Spectral Properties of the Doulbeault-Kostant Complex</u>

Let $M$ be an $(n+m)$-dimensional manifold, $F$ a complex foliation on $M$, $\mathrm{rank}(F) = n$ and $E$ a complex, F-holomorphic vector-bundle over $M$.

<u>Theorem 4.1</u>. ([1]). The condition $F + \bar{F} = T(M)_{\mathbb{C}}$ is necessary and sufficient for the following to hold:

i.  The Doulbeault-Kostant complex $\{S_F^p(M, E), d_F\}$ is an elliptic one.

ii. For the line bundle $L$ with a compatible linear connection $\nabla$, the Doulbeault-Kostant complex $\{S_F^p(M, L), \partial^F\}$ is an elliptic one.

<u>Remark 4.1</u>. If $M$ is a polarized symplectic manifold, ellipticity holds if and only if $M$ is Kähler.

<u>Remark 4.2</u>. If $F$ is a real elliptic foliation on $M$, then $F = T(M)$ and $d_F = d$.

In the sequel we suppose that $M$ is a compact manifold and $F$ is a complex, elliptic foliation on $M$. Suppose that $\wedge^p F^* \otimes E$ has a fixed Hermitian structure $h$ and a volume form $\lambda$ is chosen on $M$. These data define a global inner product on $S_F^p(M, E)$, for each $p = 0,1,2,\ldots,n$, by

$$\langle \alpha, \beta \rangle = \int_M h(\alpha, \beta) \cdot \lambda \ . \tag{4.1}$$

Then $S_F^p(M, E)$ becomes a prehilbertian space and let $L^2 S_F^p(M, E)$ be its completion.

The inner product (4.1) permitted us to define the formal adjoint $d_F^*$ of $d_F$ and then we obtain the Laplace operator $\Delta_{F, E}^p = d_F d_F^* + d_F^* d_F : L^2 S_F^p(M, E) \to L^2 S_F^p(M, E)$.

We called <u>E-valued F-harmonic p-forms</u> these E-valued p-forms which are annihilated by $\Delta_{F, E}^p$ and we denote by $\mathcal{H}_F^p(M, E)$ the space of these forms.

<u>Theorem 4.2.</u>  For each $p = 0,1,\ldots,n$ we have

   i.  $L^2 S_F^p(M, E) = \ker(\Delta_{F, E}^p) \oplus \mathrm{Ran}(\Delta_{F, E}^p)$;

   ii. $\mathcal{H}_F^p(M, E) \subset C^\infty S_F^p(M, E)$;

   iii. $H_F^p(M, E, d_F) \simeq \mathcal{H}_F^p(M, E)$;

   iv. $\dim H^p(M, E, d_F) < \infty$ .

<u>Proof.</u>  $\Delta_{F, E}^p$ is an elliptic operator, therefore $L^2 S_F^p(M, E) = \mathrm{Coker}\,(\Delta_{F, E}^p) \oplus \mathrm{Ran}(\Delta_{F, E}^p)$, with $\dim \ker(\Delta_{F, E}^p) < \infty$  $\dim \mathrm{Coker}(\Delta_{F, E}^p) < \infty$ ; moreover $\ker(\Delta_{F, E}^p)$ consists of smooth $(= C^\infty)$ forms of type $(o, p)$. This gives (ii). Since $\Delta_{F, E}^p$ is self-adjoint, $\mathrm{Coker}(\Delta_{F, E}^p) = \ker(\Delta_{F, E}^p)$ and we get (i).

Since $\Delta_{F, E}^p = (d_F + d_F^*)(d_F + d_F^*)$, $\Delta_{F, E}^p \gamma = 0$ implies $(d_F + d_F^*)\gamma = 0$. Therefore $0 = \langle (d_F + d_F^*)\gamma, d_F \gamma \rangle = \| d_F \gamma \|^2 + \langle \gamma, d_F d_F^* \gamma \rangle$  and $\langle (d_F + d_F^*)\gamma, d_F^* \gamma \rangle = \| d_F^* \gamma \|^2 + \langle d_F d_F \gamma, \gamma \rangle$ so that $d_F \gamma = 0$ and $d_F^* \gamma = 0$; in particular, E-valued F-harmonic forms are $d_F$-closed. Therefore the map $\phi : \gamma \in \mathcal{H}_F^p(M, E) \longmapsto \phi(\gamma) \overset{\mathrm{def}}{=\!=\!=} \gamma + \mathrm{ran}(d_F) \in H^p(M, S_F(M, E))$ is well defined. To show that $\phi$ is injective, take $\gamma \in \mathrm{Ran}(d_F) \cap \ker(\Delta_{F, E}^p)$. Then $\gamma = d_F \lambda$, for some $\lambda$ with $\Delta_{F, E}^p (d_F \lambda) = 0$; that is $d_F d_F^* d_F \lambda = 0$. Now, $0 = \langle d_F d_F^* d_F \lambda, d_F \lambda \rangle = \| d_F^* d_F \lambda \|^2$, so that $d_F^* d_F \lambda = 0$. Similarly $0 = \langle d_F^* d_F \lambda, \lambda \rangle = \| d_F \lambda \|^2$, so $d_F \lambda = 0$. This gives $\gamma = 0$ and $\phi$ is injective as desired. We now show $\phi$ is surjective. By (i) any $d_F$-closed p-form $\beta$ may be written as $\beta = \gamma + \Delta^p_{F,E}\alpha$, with $\Delta^p_{F,E}\gamma = 0$. Now $d_F\gamma = 0$ and $d_F\beta = 0$, so $0 = d_F\Delta^p_{F,E}\alpha = d_F(d_Fd_F^* + d_F^*d_F)\alpha = d_Fd_F^*d_F\alpha$. Therefore $0 = \langle d_Fd_F^*d_F\alpha, d_F\alpha \rangle = \| d_Fd_F^*\alpha \|^2$, which implies $\Delta^p_{F,E}\alpha = d_F(d_F^*\alpha) \in \mathrm{Ran}(d_F)$. So $\beta = \gamma + \partial^F$ (something) with $\gamma$-harmonic, and $\phi$ is surjective  The proof of (iii) is complete. Part (iv) is an immediate consequence of the finiteness of $\mathcal{H}^p_F(M, E)$ and (iii).

QED

As a consequence we can deduce

<u>Corollary 4.1.</u>  The Euler characteristic of the Doulbeault-Kostant complex

$$\chi_F(M, E) = \sum_{i=0}^{n} (-1)^i \dim H^i(M, S_F(M, E)) \quad ,$$

is finite.

Using some results characteristic classes, Fischer and Williams [1] proved a generalized version of the Riemann-Roch theorem.  More precisely we have:

<u>Theorem 4.3.</u>  ([1]).  Let $M$ be a compact even dimensional oriented manifold and assume $F$ is elliptic.  If the foliation defined by $D = F \cap \bar{F} \cap T(M)$ defines a fiber bundle $M \to M/D$, then $M/D$ is a complex manifold with complex structure $\mathcal{C}_D$ induced by $F$, and for any $F$-holomorphic bundle $E$ over $M$ we have

$$\chi_F(M, E) = ch(E)\, e\, (D)\mathcal{C}(T(M)/D)[M] \quad ,$$

where $\mathcal{C}(\cdot)$ is the Todd class of a complex vector bundle, $ch(\cdot)$ is the exponential Chern character of a complex vector bundle and $e(\cdot)$ is the Euler class of a real vector bundle.

<u>Examples</u>

1.  If $F$ is real and $E = 1$ the trivial line bundle, we have $d_F = d$, $T(M)/D = 0$ and $\chi(M) = 2(T(M))[M]$.

2.  If $D = 0$, i.e. $M$ is a complex manifold, $d_F = \bar{\partial}$, $T(M)/D = T(M)$ and

$$\chi(M, E) = ch(E)\,\mathcal{C}(M)[M] \quad ,$$

is the Hirzebruch-Riemann-Roch theorem.

Also, Fischer and Williams proved a generalized version of the Serre duality theorem, [1], see also [11].

The Laplace operator $\Delta_{F,E}^p$ is an elliptic, self-adjoint, positive operator it therefore has an infinite sequence of eigenvalues (denoted by $\mathrm{Spec}(M, \Delta_{F,E}^p)$):

$$0 = \lambda_{0,F,E}^p < \lambda_{1,F,E}^p \le \cdots \quad + \infty \quad ,$$

each eigenvalue being repeated as many times as its multiplicity indicates.

Now, for each $\lambda \in \mathbf{R}$ we let $V_\lambda^p(M, F, E)$ the eigenspace of $\Delta_{F,E}^p$ on $S_F^p(M, E)$ associated to $\lambda$:

$$V_\lambda^p(M, F, E) = \{\omega \in S_F^p(M, E) \mid \Delta_{F,E}^p \omega = \lambda\omega\} \quad ,$$

and $m_\lambda^p(M, F, E) = \dim V_\lambda^p(M, F, E)$.

From the theory of elliptic operators we deduce the following basic properties:

<u>Theorem 4.4</u>. For all $\lambda \in \mathbf{R}$, $V_\lambda^p(M, F, E)$ is finite dimensional. Further $V_\lambda^p(M, F, E) = 0$ except for a discrete set of non-negative $\lambda$'s and this countable sequence of subspaces gives an orthogonal direct sum decomposition of the Hilbert space $L^2 S_F^p(M, E)$. Thus

$$L^2 S_F^p(M, E) = \bigoplus_\lambda V_\lambda^p(M, F, E) \quad .$$

Finally, relative to this decomposition we have the Hodge formula:

$$V_0^p(M, F, E) = \ker(\Delta_{F,E}^p) \quad ,$$

while for every $\lambda > 0$

$$\Delta_{F,E}^p : V_\lambda^p(M, F, E) \longrightarrow V_\lambda^p(M, F, E)$$

is an isomorphism.

<u>Theorem 4.5</u>. In the situation described above the series

$$3_{F,E}^p(t) = \sum_\lambda e^{-t\lambda} m_\lambda^p(M, F, E) \tag{4.2}$$

436

converges for every $t > 0$. Furthermore, near $t = 0$, $\mathfrak{Z}^p_{F, E}(t)$ has an asymptotic expansion of the form

$$\mathfrak{Z}^p_{F, E}(t) \sim \sum_{k \geq -n-m} t^{k/2} U_k(\Delta^p_{F, E}) \quad , \qquad k \in \mathfrak{Z} \quad ,$$

where each $U_k(\Delta^p_{F, E})$ is given as the integral over $M$ of a certain measure $\mu_k(\Delta^p_{F, E})$ on $M$, canonically fashioned out of the coefficients of $\Delta^p_{F, E}$. Thus

$$U_k(\Delta^p_{F, E}) = \int_M \mu_k(\Delta^p_{F, E}) \quad .$$

Remark. Relative to the decomposition

$$L^2 S^p_F(M, E) = \oplus\ V^p_\lambda(M, F, E) \quad ,$$

the operator $\Delta^p_{F, E}$ is, of course, in diagonal form

$$\Delta^p_{F, E}\Big|_{V^p_\lambda(M, F, E)} = \lambda \quad ,$$

so that $H^{F, E}_t = e^{-t \Delta^p_{F, E}}$ is a well defined family of bounded operators acting on $L^2 S^p_F(M, E)$ and satisfying the equation

$$\frac{d}{dt} H^{F, E}_t + \Delta^p_{F, E} H^{F, E}_t = 0 \quad ,$$

with initial value $H^{F, E}_0 = \text{Id}$. Furthermore at least formally

$$\mathfrak{Z}^p_{F, E}(t) = \text{Tr}\ e^{-t \Delta^p_{F, E}} \quad .$$

Hence (4.2) should be interpreted as proving that $H^{F, E}_t$ is indeed of trace class for $t > 0$ and for each $p \in \mathbb{N}$, $0 \leq p \leq n = \text{rank}(F)$.

The following theorem will show us that there exists a deep connection between the Euler characteristic of the Doulbeault-Kostant complex and its spectral properties. More precisely we have the following generalized version of the Mckean-Singer telescopage formula:

<u>Theorem 4.6.</u>

$$\sum_{p=o}^{n} (-1)^p m_\lambda^p(M, F, E) = \begin{cases} \lambda_F(M, E) & \text{if} \quad \lambda = 0 \quad, \\[2em] 0 & \text{if} \quad \lambda \neq 0 \quad. \end{cases} \tag{4.3}$$

<u>Proof.</u>  The first equality follows from the Theorem 4.2.  If $\lambda > 0$, then it is easy to verify that the following identities hold:

$$V_\lambda^p(M, F, E) \cap (d_F)^{-1}(0) = V_\lambda^p(M, F, E) \cap d_F S_F^{p-1}(M, E)$$

$$V^p(M, F, E) \cap (d_F^*)^{-1}(0) = V_\lambda^p(M, F, E) \cap d_F^* S_F^{p+1}(M, E) \quad.$$

Setting:

$$A_F^p(\lambda, E) = V_\lambda^p(M, F, E) \cap (d_F)^{-1}(0) \quad,$$

$$B_F^p(\lambda, E) = V_\lambda^p(M, F, E) \cap (d_F^*)^{-1}(0) \quad,$$

we obtain in a natural way the orthogonal decomposition of $V_\lambda^p(M, F, E)$:

$$V_\lambda^p(M, F, E) = A_F^p(\lambda, E) \oplus B_F^p(\lambda, E) \quad.$$

On the other hand, it is not hard to see that  $d_F : B_F^p(\lambda, E) \longrightarrow A_F^{p+1}(\lambda, E)$  is an isomorphism and then our result follows easily.

QED

<u>Theorem 4.7.</u>  If  $n = 2k$  and  $\mathrm{Spec}(M, \Delta_{F, E}^p) = \mathrm{Spec}(M, \Delta_{F, E}^{n-p})$  for each  $p = 0, 1, \ldots, n$,  then  $m_\lambda^k(M, F, E)$  is an even number.

<u>Proof.</u>  Using the telescopage formula (4.3) we can write successively:

$$0 = \sum_{p, \text{ even}} m_\lambda^p(M, F, E) - \sum_{p, \text{ odd}} m_\lambda^p(M, F, E)$$

$$= \sum_{\substack{p < k \\ p, \text{ even}}} m_\lambda^p(M, F, E) + \sum_{\substack{p > k \\ p, \text{ even}}} m_\lambda^p(M, F, E) - \sum_{\substack{p < k \\ p, \text{ odd}}} m_\lambda^p(M, F, E)$$

$$- \sum_{\substack{p > k \\ p, \text{ odd}}} m_\lambda^p(M, F, E) + (-1)^k m_\lambda^k(M, F, E) \quad,$$

438

and then in view of our hypothesis we deduce the announced result:

$$m_\lambda^k(M, F, E) = (-1)^{k+1} 2 \left\{ \sum_{\substack{p < k \\ p \geq o \\ p, \text{ even}}} m_\lambda^p(M, F, E) - \sum_{\substack{p < k \\ p \geq o \\ p, \text{ odd}}} m_\lambda^p(M, F, E) \right\}$$

QED

Remark. The all results of this section can be extended to the Doulbeault-Kostant complex $\{S_F^p(M, L), \partial^F\}$ defined in the above section. More details and proofs can be founded in [5], [7], [9].

5.    Geometric Quantization and the Doulbeault-Kostant Complex

Let $(M, \omega)$ be a quantizable manifold, $(L^\omega, \nabla^\omega)$ its prequantum bundle and $F$ a complex (real) polarization on $M$ such that $M$ is a metaplectic manifold with respect to $F$. The tensor product between $L^\omega$ and $N_F^{1/2}$ gives the complex line bundle $L^\omega \otimes N_F^{1/2}$ which is crucial in geometric quantization. Moreover the complex line-bundle $L^\omega \otimes N_F^{1/2}$ has a connection $\nabla$ defined by

$$\nabla = \nabla^\omega \otimes 1 + 1 \otimes \nabla^{1/2}$$

and this connection is compatible with $F$. Using the identification (3.1) it follows that the cohomological group $H^0(M, S_F(M, L^\omega \otimes N_F^{1/2}))$ is the space on which the quantization takes place. When it vanishes, Kostant [4] has suggested using the higher cohomology groups of $S_F(M, L^\omega \otimes N_F^{1/2})$ for the quantization process. Then, the Doulbeault-Kostant complex $\{S_F^p(M, L^\omega \otimes N_F^{1/2}, \partial^F\}$ gives a convenient representation of these cohomological groups in terms of $L^\omega \otimes N_F^{1/2}$ - valued forms defined on $F$. It follows that the basic quantized objects in the Kostant cohomological correction of geometric quantization are the cohomological groups of $S_F(M, L^\omega \otimes N_F^{1/2})$:

$$H^p(M, S_F(M, L^\omega \otimes N_F^{1/2})) \quad , \quad p = 0,1,\ldots,n \quad .$$

Much of the theory is concerned with how to give these groups a Hilbert space structure. For example suppose that the leaves of the foliation $F$ are simply connected. Then the polarized sections of the bundle $L^\omega \otimes N_F^{1/2}$ are 1-densities on $M$ and then the space of polarized sections of the bundle $L^{\omega} \otimes N_F^{1/2}$, with compact support has a natural structure of pre-Hilbert space. In this instance, its Hilbert space completion is the basic quantized object. Moreover it can be proved that $H^p(M, S_F(M, L^{\omega} \otimes N_F^{1/2}))$ $= 0$ for $p > o$.

For the general case when the leaves of $F$ are not simply connected the problem is open.

Since $(\delta_F^{1/2})_f$, for $f \in C^{\infty}(M, F, 1)$ acts on $S_F(M, L^{\omega} \otimes N_F^{1/2})$ it also acts on each $H^p(M, S_F(M, L^{\omega} \otimes N_F^{1/2}))$ and we shall also regard this action as quantization.

It was quickly verified by Blattner, Rawnsley, Simms and Sniatycki that one certainly obtained the correct spectrum for the harmonic oscillator in one dimension on $H^1(M, S_F(M, L^{\omega} \otimes N_F^{1/2}))$. Their result was extended by Rawnsley [10] to the general case of an n-dimensional harmonic oscillator.

Another interesting application of the Kostant cohomological correction of geometric quantization was recently described by Gotay and Isenberg [2], in the quantization of the Robertson-Walker cosmology with massive scalar field (such a quantization cannot be obtained via traditional quantization methods). For such a system there is a natural choice of polarization. An analysis of the quantum dynamics shows that all quantum states eventually collapse-geometric quantization gives unambiguous results which suggest that quantum effects do not prevent a "Big Crunch", at least for simple cosmologies.

For real polarizations, the groups $H^p(M, S_F(M, L^{\omega} \otimes N_F^{1/2}))$ were considered by Śniatycki [12] and by the author [6], [8].

Example. (Simms).

Let $(M, \omega) = (\mathbb{R}^2 \setminus \{0\}, dp \wedge dq)$ be the phase space of a 1-dimensional harmonic oscillator, and $\theta = pdq$ the global symplectic potential of $\omega$. Then $(M, \omega)$ is a quantizable manifold and its prequantum bundle $L^{\omega}$ is a trivial one.

Making the transformations:

$$\begin{cases} q = r \cos t \\ p = r \sin t \end{cases},$$

we obtain:

$$\omega = r^2 dt\, dr \quad ; \quad \theta = -\frac{1}{2} r^2 dt \quad .$$

Let  F  be the polarization on  M,  globally generated by  $\partial/\partial t$. Then  M  is a metaplectic manifold with respect to  F  and the half-forms bundle  $N_F^{1/2}$  is a trivial one.  Therefore we have the identifications:

$$\Gamma(L^\omega) = C^\infty(M, \mathbb{C}) = \Gamma(N_F^{1/2}) \qquad .$$

Via the above identifications,  $\partial^F$  becomes the covariant derivative along  F.  Then we have the following exact sequence:

$$0 \to H^0(M, \, S_F(M, \, L^\omega \otimes N_F^{1/2})) \to C^\infty(M, \mathbb{C}) \to C^\infty(M, \mathbb{C}) \to H^1(M, \, S_F(M, \, L^\omega \otimes N_F^{1/2})) \to 0 \quad .$$

On the other hand the sections of the bundle  $L^\omega \otimes N_F^{1/2}$  being of the form  $e^{it/2} \cdot \psi \cdot s$  we can write:

$$\partial^F(e^{it/2}\psi s) = -\frac{1}{i}\left(\frac{\partial}{\partial t} - \frac{r^2}{2}\right)(e^{it/2}\psi)s$$

$$= -\frac{s}{i}\left(\frac{1}{2}e^{it/2}\psi + e^{it/2}\frac{\partial\psi}{\partial t} - \frac{r^2}{2}e^{it/2}\psi\right)$$

$$= -\frac{s}{i}\left[e^{it/2}\left(\frac{i}{2}\psi + \frac{\partial\psi}{\partial t} - \frac{r^2}{2}\psi\right)\right]$$

$$= -\frac{s}{i}e^{it/2}\left[\frac{\partial\psi}{\partial t} + \left(\frac{i}{2} - \frac{r^2}{2}\right)\psi\right] = -\frac{s}{i}e^{it/2}\left[\frac{\partial\psi}{\partial t} + \frac{i}{2}\left(1 - \frac{r^2}{i}\right)\psi\right] \quad .$$

Therefore the operator  $\partial^F$  on  $\Gamma(L^\omega \otimes N_F^{1/2})$  corresponds with the operator D  on  $C^\infty(M, \mathbb{C})$  given by

$$D\psi = -\frac{s}{i}\left[\frac{\partial\psi}{\partial t} + \frac{i}{2}\left(1 - \frac{r^2}{i}\right)\psi\right] \quad .$$

Since the unique solution of the equation  $D\psi = 0$  is  $\psi = 0$,  it follows that

$$S_F(M, \, L^\omega \otimes N_F^{1/2}) = H^0(M, \, S_F(M, \, L^\omega \otimes N_F^{1/2})) = 0 \quad ,$$

and then  $H^1(M, \, S_F(M, \, L^\omega \otimes N_F^{1/2}))$  can be considered as the Hilbert representation space.

A straightforward calculation show us that  $H^1(M, \, S_F(M, \, L^\omega \otimes N_F^{1/2}))$  can be in a canonically way identified with the vector space of complex sequences.  It will be the Hilbert representation space in the quantization problem of the 1-dimensional harmonic oscillator.

Now to consider the quantization of the Hamiltonian $H = (1/2)(p^2 + q^2) = (1/2)r^2$ from the 1-dimensional harmonic oscillator problem.

It is easy to see that:

$$(\delta_F^{1/2})_H = \frac{\hbar}{2}\left(\frac{i}{2} + \frac{\partial}{\partial t}\right)$$

as an operator defined on $(L^\omega \otimes N_F^{1/2})$.

On the other hand we have:

$$\frac{\hbar}{i}\left(\frac{\partial}{\partial t} + \frac{i}{2}\right)e^{i\lambda t} = \hbar\left(\lambda + \frac{1}{2}\right)e^{i\lambda t} \quad ,$$

and then the spectrum of $(\delta_F^{1/2})_H$ on $H^1(M, S_F(M, L^\omega \otimes N_F^{1/2}))$ is given by

$$\{\hbar(\lambda + 1/2) \,|\, \lambda \in \mathbf{N}\} \quad ,$$

and we refined the classical result of Schrödinger quantization.

I want to finish with the observation that the pairing problem, the problem of B.K.S. method are yet unsolved for the Kostant cohomological correction of geometric quantization. Also the relation between the Kostant cohomological correction and half-forms correction of geometric quantization when $S_F(M, L^\omega \otimes N_F^{1/2})$ does not vanish is open. Only partial results are known and some examples [6], [8].


## References


[1]    H.R. Fischer, F.L. Williams, Complex foliated structures, cohomology of the Doulbeault-Kostant complexes, Trans. Amer. Math. Society, 252 (1979), 163-194.

[2]    M.J. Gotay, J. Isenberg, Kostant-Souriau quantization of Robertson-Walker cosmologies with a scalar field, Lecture Notes in Physics, 94 (1979), 293-295, Springer, Berlin.

[3]    B. Kostant, Quantization and unitary representations, Lecture Notes in Mathematics, 170 (1970), 87-208, Springer, Berlin.

[4]    B. Kostant, On the definition of quantization, Colloq. Sympl. Aix-en-Provence, 1974.

[5]    M. Puta, Some spectral properties of the Kostant Complex, Journ. Math. Physics 23 (1982), 1749-1751.

442


[6]   M. Puta, Some remarks on the cohomology groups of a complex foliation and geometric quantization.  A IV$^a$ Conferinta de Vibratii in Construcţia de Maşini, Timişoara 1982, Vol. 2, 293-296.

[7]   M. Puta, Differential forms on a complex foliated manifold and geometric quantization, (to appear in Rend. Sem. Mat. Torino).

[8]   M. Puta, Some remarks on the cohomology of a real foliated manifold (to appear in Rend. di Mat. Roma).

[9]   M. Puta, Geometric quantization and the Doulbeault-Kostant complex, (to appear).

[10]  J.H. Rawnsley, On the cohomology groups of a polarization and diagonal quantization, Trans. Amer. Math. Society, 230 (1977), 235-255.

[11]  D.J. Simms, Serre duality for polarized symplectic manifolds, Rep. Math. Physics, 12 (1977), 213-217.

[12]  J. Śniatycki, On cohomology groups appearing in geometric quantization, Lecture Notes in Mathematics, 570, (1977), 46-67, Springer, Berlin.

[13]  J.M. Souriau, Structure des systemes dynamiques, Dunod, Paris, 1970.

[14]  A. Weinstein, Lectures in Symplectic Manifolds, C.B.M.S. 29, Amer. Math. Soc. Providence, 1977.

[15]  N. Woodhouse, Geometric Quantization, Oxford University Press (1980).



Received January 18, 1985.

GEOMETRODYNAMICS PROCEEDINGS (1985), pp. 443-458
edited by A. Pràstaro


# SYMPLECTIC ORIGIN OF SOME PROPERTIES OF GENERALLY COVARIANT FIELD THEORIES

Adam Smolski

Institute of Mathematics,
Technical University of Warsaw,
Warsaw, Poland

1.      One of the most popular topics in mathematical physics is the
theory of symmetries and conservation laws in mechanics and classical field
theory.  Observations about the interrelation between symmetries and conservation laws have been present in the literature from the age of the
founders of analytical mechanics through great geometers like Klein and
Cartan up to our times.  Today the efforts go mainly in two directions:
to make the statements most general, including the cases of theories of
higher order, velocity depending transformations etc., or to make the
formalism most elegant, with the use of the most adequate mathematical
language.  Usually it is the language of differential geometry.

I will consider here only the simplest cases of invariance leading
to conservations laws.  My aim will be to show what a simple and natural
description these questions have in the language of symplectic geometry.
The application of symplectic methods in this field is nothing new, of
course, but it is normally limited to the apparatus of Hamiltonian Systems.
However, the theory of Hamiltonian Systems is not applicable in all
situations - in field theory it is less useful than in mechanics.  In the
approach developed recently by W.M. Tulczyjew [1] symplectic geometry enters
into the description on a more general level.  Here, Lagrangian and
Hamiltonian Systems appear as partial cases only.  This approach could be
called the theory of symplectic relations.  The present paper will proceed
within this framework.  I should mention that the remarks presented here
follow from my contacts with W.M. Tulczyjew and derive from his inspiration.

2.      We usually think about the laws of dynamics of a certain physical
system as a rule for distinguishing some states of a system from the set of

all its states.  These distinguished states should be called <u>dynamically</u>
<u>admissible</u>, whereas all states are <u>possible states</u>.  Usually states are
functions or sections of a bundle and the law which distinguishes the
dynamically admissible states can be formulated as a differential equation.
However, there are integral laws, too.  In particular, integral laws can be
the integrated differential laws.

For example, let us take the mechanics of a system whose states are
sections of the bundle  $\pi : P \to \mathbf{R}$.  The fibre  $P_t = \pi^{-1}(\{t\})$  is the space
of configurations and momenta of the system in the moment of time  $t$.  If
the law of dynamics is a first order differential equation, it can be
identified with a subbundle  $D^1$  of the first jet bundle  $\pi^1 : J^1P \to \mathbf{R}$:  a
state is dynamically admissible if and only if at every moment of time  $t$
its jet belongs to  $D_t^1 = D^1 \cap J_t^1P$.

The dynamical law in the integrated form is given by prescribing a
subset  $D_{[t_1,t_2]} \subset P_{t_1} \times P_{t_2}$  for each pair  $(t_1, t_2)$, $t_1 \neq t_2$.  It is therefore a <u>relation</u> between  $P_{t_1}$  and  $P_{t_2}$.  A state  $t \to p(t)$  is dynamically
admissible if and only if, for each pair  $(t_1, t_2)$, $(p(t_1), p(t_2)) \in D_{[t_1,t_2]}$.
It is quite fruitful to view the differential equation  $D_t^1$  as the limit
of the relations  $D_{[t_1,t_2]}$  when the interval  $[t_1, t_2]$  shrinks to the
point  $\{t\}$.  (Similarly,  $J_t^1P$  is the limit of the cartesian product
$P_{t_1} \times P_{t_2}$.)  This is the reason for calling a differential equation an
<u>infinitesimal relation</u>.

In field theory states are usually sections of the bundle  $\pi : P \to M$
over some n-dimensional manifold  $M$  (space-time for example).  Integral
dynamical law will be connected with a domain  $T \subset M$  with boundary  $\partial T$.
With the boundary  $\partial T$  we associate the space  $P_T$  parametrized by the
values of sections of  $\pi$  at the points of  $\partial T$.  The values of the dynamically admissible sections form a subset  $D_T \subset P_T$.  The subset  $D_T$  can be
seen as a relation between the values at different points of  $\partial T$.  The
infinitesimal limit of such a relation when  $T$  shrinks to the point  $\{t\}$
will be again a differential equation  $D_t^1 \subset J_t^1P$.

The full description of the dynamics may require taking into account,
in addition to configurations and momenta, some other parameters, like
external forces, sources etc.  In such cases the space  $P_T$  associated with
a domain  $T$  contains not only the functions on  $\partial T$  but also the functions

defined in the interior of  T.

In the theories admitting a variational formulation the space  $P_T$  has a canonical structure of the cotangent bundle

$$P_T = T^*Q_T$$

where  $Q_T$  is the <u>configuration space</u> associated with  T.  Therefore  $P_T$  has a canonical symplectic form  $\omega_T$.  It may happen that the subsets  $D_T \subset P_T$  describing the integral laws of dynamics are <u>Lagrangian submanifolds</u>  of  $(P_T, \omega_T)$.  If we interpret  $D_T$  as a relation, then it is a <u>symplectic relation</u>.  We expect also the sets obtained in the limit  $T \to \{t\}$  to be Lagrangian submanifolds of appropriate symplectic manifolds.  In this case they are <u>infinitesimal symplectic relations</u>.  The theory with these pro- perties will be called <u>Lagrangian</u>.  All the theories derived from the Lagrange function belong to this class and the Lagrange function appears as the generating function of appropriate Lagrangian submanifolds, but the Lagrangian theories without a Lagrange function are also imaginable since a Lagrangian submanifold does not always possess a generating function.

In the present paper I apply the concepts of usual differential geometry to the objects of infinite dimensional nature.  This is far from mathematical rigour.  However, we are interested mainly in effects of algebraical, not topological nature.  A rigorous formulation is not difficult in linear field theory [2], but I think we should not limit our- selves to the linear case since it could obscure the geometric picture.

3.    Let us begin a more detailed exposition of the symplectic approach with the example of the mechanics of a system with configurations  $q^A(t)$,  $A = 1,\ldots,N$,  and a Lagrange function  $L(t, q^A, \dot{q}^A)$.  The full state is described by the functions  $q^A(t)$, $p_A(t)$, $f_A(t)$,  where  $p_A$  are coordinates of the momentum and  $f_A$  are coordinates of the external force.  The dynamically admissible states are functions  $q_A(t)$, $p_A(t)$, $f_A(t)$  satisfying the equations

$$p_A = \frac{\partial L}{\partial \dot{q}^A}$$

$$\dot{p}_A + f_A = \frac{\partial L}{\partial \dot{q}^A} \quad .$$

446

The configuration space $Q_{[t_1,t_2]}$ associated with the time interval $[t_1, t_2]$ is parametrized by functions $q^A(t)$, $t \in [t_1, t_2]$, and the space of states $P_{[t_1,t_2]}$ is parametrized by the functions $q^A(t)$, $f_A(t)$ on $[t_1, t_2]$ and the values $p_A(t_1)$, $p_A(t_2)$ of the momentum at the ends of the interval. The identification of $P_{[t_1,t_2]}$ with $T^*Q_{[t_1,t_2]}$ is given by

$$\alpha_{[t_1,t_2]} : P_{[t_1,t_2]} \to T^*Q_{[t_1,t_2]}$$

$$(p) = \int_{t_1}^{t_2} [f_A(t)dq^A(t)]dt + p_A(t_2)dq^A(t_2) - p_A(t_1)dq^A(t_1)$$

for $\quad p = (q^A(t), f_A(t), p_A(t_1), p_A(t_2))$ ,

whence the canonical symplectic form on $P_{[t_1,t_2]}$ reads

$$\omega_{[t_1,t_2]} = \int_{t_1}^{t_2} [df_A \wedge dq^A]dt + dp_A(t_2) \wedge dq^A(t_2) - dp_A(t_1) \wedge dq^A(t_1) \quad .$$

The subset $D_{[t_1,t_2]} \subset P_{[t_1,t_2]}$, which we call in the sequel the "dynamics" for the interval $[t_1, t_2]$, is defined by

$$(p \in D_{[t_1,t_2]}) \Leftrightarrow (\alpha_{[t_1,t_2]}(p) = dW_{[t_1,t_2]})$$

where $W_{[t_1,t_2]} : Q_{[t_1,t_2]} \to \mathbf{R}$ is the function of <u>action</u> for $[t_1, t_2]$:

$$W_{[t_1,t_2]}(q^A(t)) = \int_{t_1}^{t_2} L(t, q^A(t), \dot{q}^A(t))dt \quad .$$

In symplectic terms we should say that $D_{[t_1,t_2]}$ is the Lagrangian sub-manifold of $(P_{[t_1,t_2]}, \omega_{[t_1,t_2]})$ <u>generated</u> by the function $W_{[t_1,t_2]}$. The standard computation gives

$$dW_{[t_1,t_2]} = \int_{t_1}^{t_2} (\frac{\partial L}{\partial \dot{q}^A} - \frac{d}{dt} \frac{\partial L}{\partial \dot{q}^A})dq^A + (\frac{\partial L}{\partial \dot{q}^A} dq^A)(t_2) - (\frac{\partial L}{\partial \dot{q}^A} dq^A)(t_1) \quad ;$$

therefore the subset $D_{[t_1,t_2]}$ has the equations

$$f_A(t) = \frac{\partial L}{\partial q^A} - \frac{d}{dt}\frac{\partial L}{\partial \dot{q}^A} \qquad , \qquad t \in [t_1, t_2]$$

$$p_A(t_i) = \frac{\partial L}{\partial \dot{q}^A}(t_i, q^A(t_i), \dot{q}^A(t_i)) \qquad , \qquad i = 1,2 \quad .$$

We are often interested in a part of the "dynamics" only, namely the part corresponding to the vanishing force: $f_A = 0$. This part is described by the equations

$$\frac{\partial L}{\partial q^A} - \frac{d}{dt}\frac{\partial L}{\partial \dot{q}^A} = 0 \qquad , \qquad t \in [t_1, t_2]$$

$$p_A(t_i) = \frac{\partial L}{\partial \dot{q}^A} \qquad , \qquad i = 1, 2 \quad .$$

Setting $f_A = 0$ in $\omega_{[t_1,t_2]}$, we obtain

$$\tilde{\omega}_{[t_1,t_2]} = dp_A(t_2) \wedge dq^A(t_2) - dp_A(t_1) \wedge dq^A(t_1) \quad ,$$

which is a symplectic form in $P_{t_1} \times P_{t_2}$, ($P_t$ is the space of configurations $q^A$ and momenta $p_A$ at the time $t$). The equations

$$p_A(t_i) = \frac{\partial L}{\partial \dot{q}^A}(t_i, q^A(t_i), \dot{q}^A(t_i)) \qquad , \qquad i = 1,2$$

with the right-hand side evaluated on the configuration $q^A(t)$ satisfying $\frac{\partial L}{\partial q^A} - \frac{d}{dt}\frac{\partial L}{\partial \dot{q}^A} = 0$, describe (under some regularity conditions) a submanifold $\tilde{D}_{[t_1,t_2]}$ in $P_{t_1} \times P_{t_2}$ which is Lagrangian with respect to $\tilde{\omega}_{[t_1,t_2]}$. The passage from $(P_{[t_1,t_2]}, \omega_{[t_1,t_2]})$ and $D_{[t_1,t_2]}$ to $(P_{t_1} \times P_{t_2}, \tilde{\omega}_{[t_1,t_2]})$ and $\tilde{D}_{[t_1,t_2]}$ will be called the <u>elimination of the force.</u> In symplectic terms it can be described by means of a symplectic reduction.

In the limit $[t_1, t_2] \to \{t\}$ the space $P_{t_1} \times P_{t_2}$ goes to $J_t^1 P$ and

$\tilde{\omega}_{[t_1,t_2]}$ goes to

$$\tilde{\omega}_t^1 = d\dot{p}_A \wedge dq^A + dp_A \wedge d\dot{q}^A$$

where $(q^A, p_A, \dot{q}^A, \dot{p}_A)$ are the usual coordinates in $J_t^1 P$.

The "dynamics" $\tilde{D}_{[t_1,t_2]}$ tends in the limit to the submanifold $\tilde{D}_t^1 \subset J_t^1 P$ described by

$$p_A = \frac{\partial L}{\partial \dot{q}^A} \quad , \quad \dot{p}_A = \frac{\partial L}{\partial q^A} \quad ,$$

which are the usual Lagrangian equations of motion with vanishing external forces.

$\tilde{D}_t^1$ is a Lagrangian submanifold of $(J_t^1 P, \tilde{\omega}_t^1)$ which we call the "infinitesimal dynamics" at $t$.

In the same limit $Q_{[t_1,t_2]}$ tends to $J_t^1 Q$ ($Q$ is the bundle whose sections are the configurations), and $\alpha_{[t_1,t_2]}$ passes into the mapping

$$\tilde{\alpha}_t^1 : J_t^1 P \longrightarrow T^*(J_t^1 Q)$$

defined by

$$\tilde{\alpha}_t^1(q^A, p_A, \dot{q}^A, \dot{p}_A) = \dot{p}_A dq^A + p_A d\dot{q}^A \quad .$$

The equations of $\tilde{D}_t^1$ may be put in the form

$$((q^A, p_A, \dot{q}^A, \dot{p}_A) \in \tilde{D}_t^1) \Longleftrightarrow (\tilde{\alpha}_t^1(q^A, p_A, \dot{q}^A, \dot{p}_A) = dL(t, q^A, \dot{q}^A)) \quad .$$

This means that the Lagrangian submanifold $\tilde{D}_t^1$ is generated by $L$.

4.     In field theory the situation is similar. Configurations $q(t) = (q^A(t))$, $A = 1,\ldots,N$, are now sections of a bundle $\xi : Q \to M$, where $M$ is an n-dimensional manifold. The Lagrange function $\mathcal{L}$ is a function on $J_t^1 Q$ with values in scalar densities on $M$. In coordinates we write $\mathcal{L} = L dt^1 \wedge \ldots \wedge dt^n$ where $L$ is a scalar function. The full state of a system is described by $p(t) = (q^A(t), p_A^\lambda(t), f_A(t))$, where $p_A^\lambda$, $\lambda = 1,\ldots,n$, are the coordinates of the momentum and $f_A$ are the coordinates of the

external force.  Dynamically admissible states satisfy the equations

$$p_A^\lambda = \frac{\partial L}{\partial q^A_{,\lambda}}$$

$$p_{A,\lambda}^\lambda + f_A = \frac{\partial L}{\partial q^A} \quad .$$

As an equivalent of a time interval $|t_1, t_2|$ we choose here an n-dimensional non-oriented cell $T$ in $M$.  The configuration space $Q_T$ associated with $T$ is the space of functions $q^A(t)$ on $T$ and the space of states $P_T$ for $T$ is parametrized by the functions $q^A(t)$, $f_A(t)$ on $T$ and $p_A^\perp(t)$ on $\partial T$ — by $\perp$ we have denoted the component transversal to $\partial T$.  The identification of $P_T$ with $T^*Q_T$ is given by

$$\alpha_T(q^A\big|_T, f_A\big|_T, p_A\big|_{\partial T}) = \int_T (f_A dq^A)dt^1 \wedge \ldots \wedge dt^n$$

$$+ \int_{\partial T} (p_A^\lambda dq^A)(\frac{\partial}{\partial t^\lambda} \lrcorner\, dt^1 \wedge \ldots \wedge dt^n)$$

and the canonical symplectic form on $P_T$ is

$$\omega_T = \int_T (df_A \wedge dq^A)dt^1 \wedge \ldots \wedge dt^n + \int_{\partial T} (dp_A^\lambda \wedge dq^A)(\frac{\partial}{\partial t^\lambda} \lrcorner\, dt^1 \wedge \ldots \wedge dt^n) \quad .$$

The "dynamics" $D_T$ is defined by

$$(p \in D_T) \Longleftrightarrow (\alpha_T(p) = dW_T)$$

where

$$W_T(q^A(t)) = \int_T L(t, q^A(t), q^A_{,\lambda}(t))dt^1 \wedge \ldots \wedge dt^n \quad ;$$

it is therefore the Lagrangian submanifold generated by $W_T$.  We have

$$dW_T = \int_T \left[\frac{\partial L}{\partial q^A} - (\frac{\partial L}{\partial q^A_{,\lambda}})_{,\lambda}\right]dt^1 \wedge \ldots \wedge dt^n$$

$$+ \int_{\partial T} (\frac{\partial L}{\partial q^A_{,\lambda}})(\frac{\partial}{\partial t^\lambda} \lrcorner\, dt^1 \wedge \ldots \wedge dt^n)$$

450

and so the equations of $D_T$ are

$$f_A(t) = \frac{\partial L}{\partial q^A} - \left(\frac{\partial L}{\partial q^A_{,\lambda}}\right)_{,\lambda} \qquad , \qquad t \in T$$

$$p^\perp_A(t) = \frac{\partial L}{\partial q^A_{,\perp}} \qquad , \qquad t \in \partial T \quad .$$

The elimination of the force leads to the space $\tilde{P}_T$ parametrized by the functions $q^A(t)$, $p^\perp_A(t)$ on $\partial T$ with the symplectic form

$$\tilde{\omega}_T = \int_{\partial T} (dp^\lambda_A \wedge dq^A)\left(\frac{\partial}{\partial t^\lambda} \ dt^1 \wedge \ldots \wedge dt^n\right)$$

and the "dynamics" $\tilde{D}_T \subset \tilde{P}_T$ described by the equations

$$p^\perp_A(t) = \frac{\partial L}{\partial q^A_{,\perp}} \qquad , \qquad t \in \partial T$$

where the right-hand side is evaluated on the configuration satisfying the Euler-Lagrange equations $\frac{\partial L}{\partial q^A} - \left(\frac{\partial L}{\partial q^A_{,\lambda}}\right)_{,\lambda} = 0.$ Under some regularity conditions $\tilde{D}_T$ will be a Lagrangian submanifold of $(\tilde{P}_T, \tilde{\omega}_T)$.

In the limit $T \to \{t\}$ we obtain the space $\tilde{P}^1_t$ parametrized by $q^A$, $p_A$, $q^A_{,\lambda}$, $p^\lambda_{A,\lambda}$ with the symplectic form

$$\tilde{\omega}^1_t = dp^\lambda_{A,\lambda} \wedge dq^A + dp^\lambda_A \wedge dq^A_{,\lambda} \quad .$$

The limit of $Q_T$ is $J^1_t Q$, and we have an identification

$$\tilde{\alpha}^1_t : \tilde{P}^1_t \longrightarrow T^*(J^1_t Q)$$

$$\tilde{\alpha}^1_t(q^A, \ p_A, \ q^A_{,\lambda}, \ p^\lambda_{A,\lambda}) = p^\lambda_{A,\lambda} \ dq^A + p^\lambda_A dq^A_{,\lambda} \quad .$$

The limit of $\tilde{D}_T$ is $\tilde{D}^1_t$ with the equations

$$p^\lambda_{A,\lambda} = \frac{\partial L}{\partial q^A}$$

$$p^\lambda_A = \frac{\partial L}{\partial q^A_{,\lambda}} \quad ,$$

and so $\tilde{D}_t^1$ is the Lagrangian submanifold generated by the function $L$ on $J_t^1 Q$.

The formulation above admits also the "Lagrangian" theory without the Lagrange function - a theory in which all "dynamics" describing the set of dynamically admissible states are Lagrangian submanifolds of suitable spaces, not necessarily having generating functions. Such a situation really occurs in concrete physical examples.

5.    If we adopt the point of view that the "dynamics" are more fundamental than the Lagrange function for example, it is natural to pose the question of symmetries and conservation laws on the level of the properties of the "dynamics".

I will consider here a very simple type of the invariance of the "dynamics", namely their invariance with respect to the vector fields on the configuration space. It is convenient to introduce this concept in an abstract way. (This is done in [3]; here I report a fragment from there).

Consider the cotangent bundle to the manifold $Q$

$$\pi_Q : T^*Q \to Q \quad .$$

Let $X$ be a vector field on $Q$ and $X^*$ its canonical lift to $T^*Q$. Let $D$ be a Lagrangian submanifold of $T^*Q$ (with respect to the canonical symplectic form $\omega_Q = d\theta_Q$ in $T^*Q$).

We say that the submanifold $D$ is <u>invariant with respect to the vector field $X$</u> if and only if $X^*(D) \subset TD$, i.e. at the points of $D$ the vector field $X^*$ is tangent to $D$. Such a concept is an infinitesimal version of the invariance of $D$ with respect to the (canonical lifts of the) transformations generated by $X$.

Let us define for $C \in R$:

$$N(X, C) = \{p \in T^*Q : <X, p> = C\} \quad .$$

We have

<u>Proposition</u>:  A connected Lagrangian submanifold $D \subset T^*Q$ is invariant with respect to a vector field $X$ on $Q$ if and only if, for a certain $C \in R$, $D \subset N(X, C)$. (See [3] for proof).

If $X$ is nowhere vanishing, $N(X, C)$ is a coisotropic submanifold of $T^*Q$. The case $C = 0$ is particularly interesting - for the invariance with $D \subset N(X, 0)$ I have proposed in [3] the term <u>strict invariance</u>.

If the Lagrangian submanifold $D$ is generated by a function $L$, then $D \subset N(X, C)$ means $< X, dL > = C$. In the case of strict invariance the function $L$ is simply constant on the integral curves of $X$. Such a property of a Lagrange function of a physical system is normally understood as symmetry or invariance.

6.     It turns out that the inclusion $D \subset N(X, C)$, which follows from the invariance, when applied to the "dynamics" of a physical system, can often be interpreted as a conservation law. There are, however, two characteristic situations, which I will illustrate by two very simple examples.

A.   Let us consider a mechanical system with two degrees of freedom, $q^1$, $q^2$, and a Lagrange function $L(t, q^1, q^2, \dot{q}^1, \dot{q}^2)$ which does not depend on $q^2$. Among the equations of motion we have $\dot{p}_2 + f_2 = 0$, and so with the vanishing force we obtain a conservation law $\dot{p}_2 = 0$. The same result can be read out from the form of the "dynamics" for the interval $|t_1, t_2|$. The transformation

$$(q^A(t), f_A(t), p_A(t_i)) \longrightarrow (q^A(t) + \varepsilon^A, f_A(t), p_A(t_i)) \quad ,$$

where $A = 1,2$,  $i = 1,2$,  where $\varepsilon^A = 0$ for $A = 1$,  does not lead out of the "dynamics" $D_{[t_1,t_2]}$. This means that $D_{[t_1,t_2]}$ is invariant with respect to the vector field $X = (\delta q^A(t))$ on $Q_{[t_1,t_2]}$ such that $\delta q^1 = 0$, $\delta q^2 = 1$, $t \in [t, t]$. Similarly the "dynamics" after the elimination of the force $\tilde{D}_{[t_1,t_2]}$ is invariant with respect to the vector field $\tilde{X} = (\delta q^A(t_1), \delta q^A(t_2))$ on $Q_{t_1} \times Q_{t_2}$, where $\delta q^1(t_i) = 0$, $\delta q^2(t_i) = 1$, $i = 1,2$.

In both cases we have strict invariance. The inclusion $\tilde{D}_{[t_1,t_2]} \subset N(\tilde{X}, 0)$ means that, for $p \in \tilde{D}_{[t_1,t_2]}$,

$$0 = < \tilde{X}, p > = p_2(t_2)\delta q^2(t_2) - p_2(t_1)\delta q^2(t_1) = p_2(t_2) - p_2(t_1)$$

hence $p_2(t_2) = p_2(t_1)$. This is the conservation law $\dot{p}_2 = 0$ in the integrated form.

B.  Let us consider a mechanical system with two degrees of freedom, $q^1$, $q^2$, and a Lagrange function $L(t, q^1, q^2, \dot{q}^1, \dot{q}^2)$ which does not depend on $q^2$ and $\dot{q}^2$. Among the equations of motion we have $p_2 = 0$, $\dot{p}_2 + f_2 = 0$ at each $t$, and so the dynamically admissible states satisfy $p_2(t) = 0$, $f_2(t) = 0$ for each $t$. The transformation which changes arbitrarily the component $q^2(t)$ of the element of $P_{[t_1,t_2]}$ does not lead out of the "dynamics" $D_{[t_1,t_2]}$. In the version without forces the invariance will be with respect to two independent vector fields on $Q_{t_1} \times Q_{t_2} : \tilde{X}_i = (\delta q^A(t_1), \delta q^A(t_2))$, $i = 1,2$, where $\delta q^1(t_j) = 0$, $j = 1,2$, $\delta q^2(t_j) = 1$ for $j = i$ and $\delta q^2(t_j) = 0$ for $j \neq i$. The inclusion $\tilde{D}_{[t_1,t_2]} \subset N(\tilde{X}_i, 0)$ means $p_2(t_i) = 0$, $i = 1,2$. Therefore the coisotropic submanifolds $N(\tilde{X}_i, 0)$ are interpreted as <u>constraints</u> in the phase space at $t$.

A comparison of the two examples suggests to us when conservation laws occur as an effect of invariance. Our understanding of what should be called a conservation law is usually the following: it is a relation which involves the value of the state at $t = t_2$ together with the value of the state at $t = t_1$. Therefore, it will follow from the invariance with respect to a transformation, whose action at $t = t_2$ is in some way correlated with its action at $t = t_1$. If there is no such correlation, the equations which follow from the invariance are satisfied at each $t$ separately, and so they are constraints rather than conservation laws. The first, "rigid" type of transformations is characteristic for systems with symmetry (translational, rotational etc.). The second, "soft" type is encountered in gauge theories and in generally covariant theories, which we are going to discuss now.


7.      The pattern for a generally covariant field theory is given by the General Relativity theory. The discussion of such a theory in an abstract way is reasonable, since it is interesting to extract the properties which are due exactly to general covariance and not to more particular features of the theory.

First of all, configurations of the theory must be geometric objects. The fundamental property of a bundle of geometric objects $\xi : Q \to M$ is the existence of a canonical lift to the bundle of each diffeomorphism $\Phi$ of the base manifold $M$. Let us denote the lift of $\Phi$ to $Q$ by $\Phi_Q$. A section $q(t)$ of $\xi$ can be transported by $\Phi$ in the following way

$$q(t) \longrightarrow [\tilde{\Phi}_Q(q)](t) = \Phi_Q(q(\Phi^{-1}(t))) \quad .$$

If configurations are geometric objects, the same is true for momenta and forces and for states in general. Therefore, the transport by $\Phi$ is defined also for states: to the state $p(t)$ corresponds the state $[\tilde{\Phi}_p(p)](t)$.

The theory is <u>generally covariant</u> if the set of dynamically admissible states is invariant under the transport by diffeomorphisms of $M$.

In accordance with our point of view that the "dynamics" are the fundamental elements of the theory the definition of general covariance should be expressed in terms of the "dynamics" associated with the cells $T \subset M$.

The spaces of configurations and states for $T$ are built from the functions $q(t)$, $p(t)$; therefore the mappings of transport $\tilde{\Phi}_Q$, $\tilde{\Phi}_p$ induce the mappings

$$\hat{\Phi}_Q : Q_T \longrightarrow Q_{\Phi(T)}$$

$$\hat{\Phi}_p : P_T \longrightarrow P_{\Phi(T)} \quad .$$

Under certain regularity assumptions one can prove that the theory is generally covariant if and only if for each n-dimensional cell $T \subset M$ and for each diffeomorphism $\Phi$ of $M$, $\hat{\Phi}_p(D_T) = D_{\Phi(T)}$.

Let $Inv(T)$ denote the set of all diffeomorphisms $\Phi$ such that $\Phi(T) = T$. For $\Phi \in Inv(T)$ general covariance means $\hat{\Phi}_p(D_T) = D_T$, i.e., the "dynamics" $D_T$ is invariant under a transformation $\hat{\Phi}_p$.

It follows from the construction of $\hat{\Phi}_Q$ and $\hat{\Phi}_p$ that (taking into account the identification of $P_T$ with $T^*Q_T$) $\hat{\Phi}_p$ is the canonical lift of $\hat{\Phi}_Q$, and so for $\Phi \in Inv(t)$ the invariance of $D_T$ is of the type discussed previously. We need the infinitesimal version of this invariance.

Let $X$ be a vector field on $M$ generating a family of diffeomorphisms $\{\Phi^s\}$, $s \in R$. The condition $\Phi^s \in Inv(T)$, $s \in R$, demands that $X$ should be tangent to $\partial T$ in the points of $\partial T$. For such $X$ we can define the vector fields $\hat{X}_Q$, $\hat{X}_p$ as generators of $\{\hat{\Phi}_Q^s\}$, $\{\hat{\Phi}_p^s\}$. The vector field $\hat{X}_Q = (\delta q(t))$, $t \in T$, is given by

$$\delta q(t) = - \underset{X}{\pounds} q(t) \quad , \qquad t \in T \quad .$$

We conclude that in a generally covariant theory the "dynamics" $D_T$ is invariant with respect to the vector fields of the form $\hat{X}_Q$ (if $X$ is tangent to $\partial T$ on $\partial T$). It turns out that one can prove even strict invariance in this case. For $(q^A{}_{|T}, f_{A|T}, p^\perp_{A|\partial T}) \in D_T$ we therefore have

$$< \hat{X}_Q, \; \alpha_T(q^A{}_{|T}, f_{A|T}, p_{A|\partial T}) > = 0 \quad ,$$

i.e.

$$\int_T [f_A(- \underset{X}{\pounds} q^A)](dt^1 \wedge \ldots \wedge dt^n) + \int_{\partial T} [p^\lambda_A(- \underset{X}{\pounds} q^A)](\frac{\partial}{\partial t^\lambda} \lrcorner\, dt^1 \wedge \ldots \wedge dt^n) = 0 \quad .$$

When $q(t)$ is a tensor field, $\underset{X}{\pounds} q(t)$ depends linearly on some jet of the vector field $X$ at $t$. We assume for simplicity that it is the second jet, as in General Relativity (in the so-called affine formulation, when $q$ is the linear connection).

The integration by parts yields

$$\int_T [f_A(-\underset{X}{\pounds} q^A)](dt^1 \wedge \ldots \wedge dt^n) + \int_{\partial T} [p^\lambda_A(-\underset{X}{\pounds} q^A)](\frac{\partial}{\partial t^\lambda} \lrcorner\, dt^1 \wedge \ldots \wedge dt^n)$$

$$= \int_T \mathcal{B}_\mu X^\mu dt^1 \wedge \ldots \wedge dt^n + \int_{\partial T} (\mathcal{C}^{III}_k X^k + \mathcal{C}^{II}_\mu X^\mu{}_{,\perp} + \mathcal{C}^I_\mu X^\mu{}_{,\perp,\perp})$$

$$\cdot (\frac{\partial}{\partial t^\perp} \; dt^1 \wedge \ldots \wedge dt^n) + \ldots$$

where dots denote the term which are integrals over smaller dimension parts of the boundary $\partial T$; the coefficients $\mathcal{B}_\mu$, $\mathcal{C}^I_\mu$, $\mathcal{C}^{II}_\mu$, $\mathcal{C}^{III}_k$, $\mu = 1,\ldots,n$, $k = 1,\ldots,n$, $k \neq \perp$, are functions of $q^A$, $f_A$, $p^\perp_A$ (one can observe that $\mathcal{B}_\mu$ does not depend on $p^\perp_A$ and $\mathcal{C}^I_\mu$ does not depend on $f_A$).

The vanishing of the above expression with arbitrary $X$ (tangent on $\partial T$ to $\partial T$) implies the vanishing of all the coefficients. The identities $\mathcal{B}_\mu = 0$ on $T$ have been called the <u>generalized Bianchi identities</u>. They are differential relations between the components $q^A$ and $f_A$. In consequence the equations $f_A = 0$ (which we are usually interested in) are not independent.

The identities $\mathcal{C}^I_\mu = 0$, $\mathcal{C}^{II}_\mu = 0$, $\mathcal{C}^{III}_k = 0$ on $T$ we call, respectively, the first, second and third <u>boundary identities</u>. The equations $\mathcal{C}^I_\mu = 0$ involve only $q^A$ and $p_A$ on $T$, they are therefore the <u>constraints</u> in the phase space of functions $q^A$, $p^\perp_A$ on $\partial T$. Similarly, $C^{II}_\mu = 0$, $C^{III}_k = 0$ are constraints after the insertion of $f_A = 0$.

Let me point out that the above analysis does not resort to the Lagrange function, which does not necessarily exist (which is really the case in the description of the free gravitational field — we are still thinking about the affine formulation).

On the other hand, on this level of generality we do not obtain all the equations or constraints which are in fact due to general covariance. For example, the so-called Hamiltonian constraint in General Relativity has a slightly different nature and is not included among our identities.

The interpretation of the boundary identities as constraints of the Hamiltonian type will be clearer in the so-called time-evolution picture. Let us assume $M = \mathbf{R}^{n+1}$ and take

$$T = \{(t^0, t^1,\ldots,t^n) \in M : a \le t^0 \le b\} \quad .$$

$T$ is not a cell but can be regarded as the limit of cells

$$\{(t^0,\ldots,t^n) \in M : a \le t^0 \le b,\ |t^i| \le r,\ i = 1,\ldots,n\}$$

as $r$ tends to infinity. The boundary of $T$ is $\Sigma_a \cup \Sigma_b$ where

$$\Sigma_c = \{(t^0,\ldots,t^n) \in M,\ t^0 = c\} \quad .$$

With the appropriate asymptotics of states in the limit $r \to \infty$ the space of states for $T$ will be the space parametrized by functions $q^A$, $f_A$ on $T$ and $p^0_A$ on $\Sigma_a \cup \Sigma_b$ with the symplectic form

$$\omega_T = \int_T (df_A \wedge dq^A) dt^0 \wedge\ldots\wedge dt^n + \int_{\Sigma_b} (dp^0_A \wedge dq^A) dt^1 \wedge\ldots\wedge dt^n$$

$$- \int_{\Sigma_a} (dp^0_A \wedge dq^A) dt^1 \wedge\ldots\wedge dt^n \quad .$$

We notice the analogy with the space of states for the interval $[a, b]$ in mechanics — the time-evolution picture consists in dealing with field theory as mechanics with an infinite number of degrees of freedom. These degrees of freedom, i.e., configurations at the time $c$, are described here by functions $q^A(t)$ on $\Sigma_c$. The phase space at the time $c$ is the space of functions $q^A(t)$, $p^0_A(t)$  on $\Sigma_c$ with the symplectic form

$$\omega_c = \int (dp^0_A \wedge dq^A)dt^1 \wedge \ldots \wedge dt^n \quad .$$

The boundary identities for  T  after the elimination of the force are constraints in the phase spaces at times  a, b.  The way in which we have found these constraints allows us to expect that they form a coisotropic submanifold of the phase space, i.e., they are the so-called first class constraints.

The approach presented here explains in a simple way the presence of the so-called constraint algorithm in generally covariant field theories [4].  The Dirac-Bergmann constraint algorithm is an observation that in the course of generating of the equations of motion in the Hamiltonian way (through Poisson brackets) some constraint equations result as conditions that other constraints should be preserved in time.  The Hamiltonian equations of motion are the equations of "infinitesimal dynamics" in the sense introduced above in the case of mechanics.  It is the "dynamics" obtained in the limit when the interval  [a, b]  shrinks to the point  {c}.  In the limit (with $f_A = 0$) we obtain the space of states parametrized by functions $q^A$, $p^0_A$, $\dot{q}^A$, $\dot{p}^0_A$  on  $\Sigma_c$ with the symplectic form

$$\tilde{\omega}^1_c = \int_{\Sigma_c} [d\dot{p}^0_A \wedge . dq^A + dp^0_A \wedge d\dot{q}^A]dt^1 \wedge \ldots \wedge dt^n$$

(notice the analogy with  $J^1_t P$  and  $\tilde{\omega}^1_t$  in mechanics).

Passing to the limit  $[a, b] \to \{c\}$  in the equation

$$\int_T \mathcal{B}_k \chi^k dt^0 \wedge \ldots \wedge dt^n + \int_{\Sigma_b} ( C^{III}_k \chi^k + C^{II}_k \dot{\chi}^k + C^I_k \ddot{\chi}^k)dt^1 \wedge \ldots \wedge dt^n$$

$$- \int_{\Sigma_a} ( C^{III}_k \chi^k + C^{II}_k \dot{\chi}^k + C^I_k \ddot{\chi}^k)dt^1 \wedge \ldots \wedge dt^n = 0$$

we obtain

$$\int_{\Sigma_c} [(\mathcal{B}_k + \dot{C}^{III}_k)\chi^k + ( C^{III}_k + \dot{C}^{II}_k)\dot{\chi}^k + ( C^{II}_k + \dot{C}^I_k)\ddot{\chi}^k + C^I_k \dddot{\chi}^k]$$

$$\cdot dt^1 \wedge \ldots \wedge dt^n = 0 \quad ,$$

458

therefore the equations which the "infinitesimal dynamics" satisfies in
the result of general covariance are

$$C_k^I = 0 \quad , \quad C_k^{II} + \dot{C}_k^I = 0 \quad , \quad C_k^{III} + \dot{C}_k^{II} = 0 \quad ,$$

$$\mathcal{B}_k + \dot{C}_k^{III} = 0$$

and we see that we have obtained directly only the first boundary identities
whereas the second appear as the integrability conditions of the first, the
third are the integrability conditions of the second and finally the Bianchi
identities are the integrability conditions of the third constraints. This
behaviour was the content of the "algorithm".

The effects which I have described here have been extensively
elaborated in the works of P.G. Bergmann (a series of papers in the Physical
Review in the Fifties, cf. [5]) and also by A. Trautmann (cf. [6]). The
subject of their analysis, however, is always the Lagrange function and the
geometric picture is not very transparent. In the present paper I tried to
show that these problems have a particularly simple description in the
symplectic approach.


## References

[1]  W.M. Tulczyjew, _Symplectic formulations of physical theories_,
     Rendic. Sem. Matem. Fis. Milano, _50_ (1980), 123-133.

[2]  W.M. Tulczyjew, _A symplectic framework of linear field theories_,
     Annali di Matematica pura et applicata (IV), vol. CXXX, pp. 177-195
     (1982).

[3]  A. Smolski, _Invariance of Lagrangian Submanifolds_, Rend. Sem. Matem.
     Univ. Politecn. Torino, _41_ (1983), 111-128.

[4]  A. Smolski, _On the Hierarchy of Constraints in General Relativity_,
     Bulletin de l'Academie Polonaise des Sciences, Serie des sciences
     physiques et astron. - vol. XXVII, No. 3, (1979).

[5]  J.A. Anderson, P.G. Bergmann, _Constraints in Covariant Field Theories_,
     Phys. Rev. _83_ (1951), 1018.

[6]  A. Trautman, _Lectures on General Relativity_, Brandeis Summer School
     1967.

Received March 16, 1985.

GEOMETRODYNAMICS PROCEEDINGS (1985), pp. 459 466
edited by A. Pràstaro


# SYMPLECTIC SCATTERING THEORY

Shlomo Sternberg

*Department of Mathematics, Harvard University*
*Cambridge, Massachusetts 02138 , USA*

Abstract:  In [1] Souriau developed a classical scattering theory in order to derive a Poincaré covariant version of the first law of thermodynamics from the principles of general relativity.  In [2] this was generalized to Yang-Mills theories.  The present note is a summary of the results of [2].  Full proofs and a more thorough discussion can be found in [2].

§1.  <u>Symplectic scattering</u>.  Let $\tau$ be a closed two form of corank one on a differentiable manifold Y.  We suppose that the null foliation of $\tau$ is fibrating, so that there is a differentiable manifold U and a fiber map $\rho:\ Y \longrightarrow U$ whose fibers are the null curves of $\tau$.  Thus there exists a symplectic form $\omega$ on U such that

$$(1.1) \qquad\qquad \tau = \rho^{*}\omega.$$

We also assume that the fibration is oriented, i.e., that the fibers of $\rho$ have been oriented in a consistent fashion. Since U is oriented (by $\omega^{n}$ where dim U = 2n) an orientation of Y will provide such an orientation of $\rho$.  We let $\Omega^{k}(Y)$ denote the space of smooth k-forms on Y which are compactly supported relative to $\rho$, so $\sigma \in \Omega^{k}_{\tau}(Y)$ if the k-form $\sigma$ is such that supp $\sigma \cap \rho^{-1}(K)$ is compact for every compact subset $K \subset U$.  On the other hand, we will assume that the fibration is not compact; we assume that locally Y looks like $U \times \mathbb{R}$. Similarly, we let $Z^k_\tau(Y)$ denote the space of all closed k forms which are properly supported relative to $\tau$.

Let $\tau'$ be a second closed two form of corank one on Y with $\tau - \tau' \in Z^2_\tau(Y)$. Then the null curves of $\tau'$ coincide with the null curves of $\tau$ outside some subset C which is compact (or proper) relative to $\rho$. Since the fibration

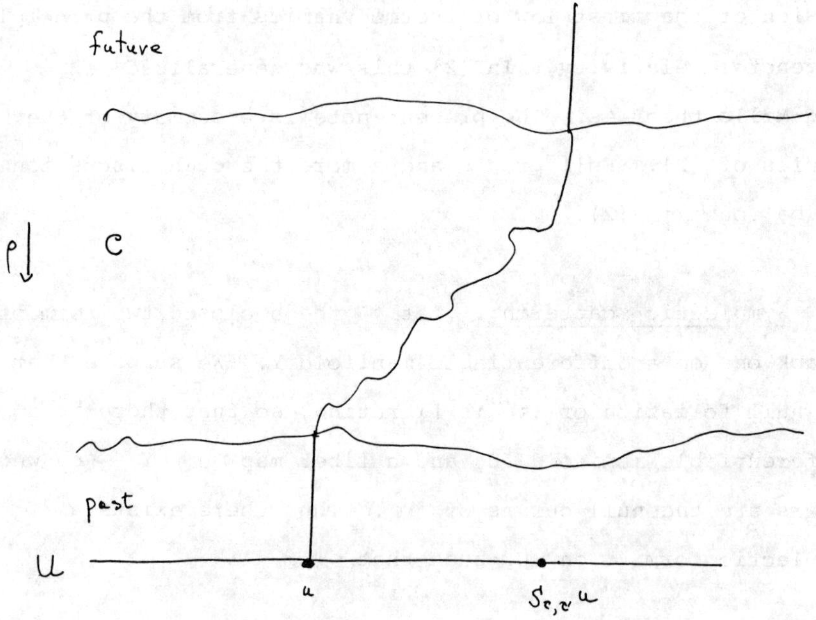


is oriented, each null curve of $\tau'$ connects a null curve of $\tau$ in the "past" (before C) to some other null curve of $\tau$ in the "future" (after C). It therefore determines a transformation $S_{\tau,\tau'}$ of U. The first main result is

(1.2)         the diffeomorphism $S_{\tau,\tau'}$ is symplectic

i.e.,

$$(S_{\tau,\tau'})^* \omega = \omega .$$

Let $\varphi$ be a diffeomorphism of $Y$ such that $\varphi = id$ outside some set $C$ which is compact relative to $\rho$ . Then $\varphi^{*-1}\tau - \tau \in Z_\tau^2(Y)$ . But the null curves of $\varphi^{*-1}\tau$ are just the images under $\varphi$ of the null curves of $\tau$ . Hence

$$(1.3) \qquad S_{\tau,\varphi^{*-1}\tau} = \text{identity.}$$

§2. <u>Infinitesimal scattering.</u> Let $\sigma \in Z_\tau^2(Y)$ . Then (at least locally over $U$), $\tau'_\varepsilon = \tau + \varepsilon\sigma$ will be a closed two form of corank one for sufficiently small $\varepsilon$ , and $\tau'_\varepsilon - \tau \in Z_\tau^2(Y)$ . Hence we get a one parameter family $S_{\tau,\tau'_\varepsilon}$ of symplectic diffeomorphisms on U, which thus determine an infinitesimal symplectic transformation, i.e., a symplectic vector field on U. We shall denote this by $\xi_\sigma$ . Thus

$$D_{\xi_\sigma}\omega = 0 .$$

But $D_{\xi_\sigma}\omega = di(\xi_\sigma)\omega + i(\xi_\sigma)d\omega = di(\xi_\sigma)\omega$ since $d\omega = 0$ . Thus

$$di(\xi_\sigma)\omega = 0.$$

We have thus defined a map, $\sigma \leadsto i(\xi_\sigma)\omega$ of $Z_\tau^2(Y)$ into $Z^1(U)$ . The main computational result of [2] is that this map is just integration along the fiber. (See [3] for the definition and principal properties of integration along the fiber). Thus

$$(2.1) \qquad i(\xi_\sigma)\omega = \rho_*\sigma$$

where $\rho_*$ denotes integration along the fiber. If $H^1(U) = 0$ then we can write $\rho_*\sigma = df_\sigma$ where $f_\sigma$ is determined up to additive constant. If supp $\rho_*\sigma$ has a connected, non-empty complement, we can fix this constant by requiring $f_\sigma$ to vanish there.

For example, let M be a manifold with a Lorentzian structure (and a global sense of past and future). Let Y denote the unit "cosphere" bundle. Let $\tau$ denote the restriction of the canonical symplectic form on T*M to Y. Then the null curves of $\tau$ are (or rather project onto) the unit geodesics (the geodesics whose tangent vectors have length one) on M. We are assuming that this space of geodesics is a smooth manifold, U. Let T be a symmetric tensor of compact support on M. Then adding $\epsilon T$ to the underlying metric gives another Lorentzian metric for sufficiently small $\epsilon$. The scattering operator here carries geodesics into geodesics - one carried into the other by being connected by a geodesic of the modified metric. The corresponding infinitesimal symplectic transformation is given by $\xi_T$ where

$$i(\xi_T)\omega = df_T$$

and

$$(2.2) \qquad f_T(\gamma) = \int_\gamma T(p \otimes p)\, ds$$

and p denotes the cotangent vector of $\gamma$.


§3. <u>Relation with field theory</u>. Let $P \xrightarrow{\pi} M$ be a principal bundle with structure group H, where M is a causal manifold, that is, a Lorentzian manifold with a global notion of future and past. Let Y be an associated bundle of P. Let X be some space of geometric objects associated with P. For example, X might be the space of all connections on P, or sections of

some associated bundle, etc.  We shall assume that the tangent
space to  X  at any point  $\alpha$  is identified with the space,
$\Gamma(V)$, of sections of a fixed vector bundle  $V \longrightarrow M$.  Let us now
assume that we have a rule which assigns, to each  $\alpha \in X$  a closed
two form  $\tau(\alpha)$  on  Y  of corank 1.  We assume that this map is
smooth, local, and equivariant under the action of  Aut $P_G$.  Suppose
that  $\alpha' \in X$  differs from  $\alpha$  only on a compact subset of  M.
Under mild geometric hypotheses, this will imply that  $\tau(\alpha')$
differs from  $\tau(\alpha)$  on a set which is compactly supported relative
to  $\rho = \rho_{\tau(\alpha)}$;  hence we can consider the scattering operator
$S_{\tau(\alpha),\tau(\alpha')}$.  At the infinitesimal level, we then get a linear
map (depending on  $\alpha$) from  $\Gamma_0(V)$  to  $Z^1(U)$, where  $\Gamma_0(V)$  denotes
the space of sections of compact support of V.  If  $H^1(U) = 0$, and
under mild geometrical conditions which allow the fixing of the
arbitrary constants, we get a map

$$F : \quad \Gamma_0(V) \longrightarrow F(U)$$

where  F(U)  denotes the space of smooth functions on  $U_0$.

Let  $G = \text{Aut}_0(P)$  denote the group of automorphisms of
compact support (over M) of P.  For any  $\xi$  in the Lie algebra
of Aut P we let  $D_\xi \alpha \in \Gamma(V)$  denote the corresponding infinitesimal
variation of  $\alpha$.  The  $\xi$ of compact support constitute the Lie
algebra of  G .  We let  $T(G \cdot \alpha) \subset \Gamma_0(V)$  denote the space of all
$D_\xi \alpha$, where  $\xi$  has compact support.  It follows from (1.3) that

$$T(G \cdot \alpha) \quad \subset \quad \ker F .$$

We can also consider the adjoint map $F^*: \text{Dist}_0(U) \longrightarrow \Gamma_0(V)^*$.
Here  $\Gamma_0(V)^*$  is the space of distributional sections of the dual
bundle to V, which is the space of continuous linear functions on
$\Gamma_0(V)$.

In our example of varying the metric, $V = S^2 TM$ is just the vector bundle of symmetric two tensors. Then the computation of section 2 can be regarded as saying that

$$F^*(\delta_\gamma)(T) = \int_\gamma T(p \otimes p)\, ds$$

where $\delta_\gamma$ is the delta function at the geodesic $\gamma$. Thus $F^*(\delta_\gamma)$ is the distributional section along $\gamma$ given by the decomposable tensor $p \otimes p$. Notice that this is precisely the "stress energy tensor" that relativity theory associates with a point of momentum $p$ whose world line is $\gamma$. If $\nu$ were a measure on U, then $F^*(\nu)$ would be a "measure tensor" giving the "stress energy tensor" of a "relativistic dust". Thus we have generalized this idea to the Yang-Mills case.

We can reformulate (3.3) as saying that

$$(3.4) \qquad \mathrm{im}(F^*) \subset T(G \cdot \alpha)^0$$

where $T(G \cdot \alpha)^0$ denotes the space of those $\mu$ which vanish on $T(G \cdot \alpha)$.


§4. <u>Symmetries and conserved quantities</u>. Let $K \subset \mathrm{Aut}\, P$ be a finite dimensional Lie subgroup all of whose elements preserve $\alpha$. Let $\xi$ be an element of the Lie algebra, k, of K. Thus $D_\xi \alpha \equiv 0$. We assumed that M has a global causal structure, in particular, a global notion of the distant past and the distant future. Let w be a function which is $\equiv 0$ in the distant past and $\equiv 1$ in the distant future. Then $D_{w\xi}\alpha \in \Gamma(V)$ vanishes in both the distant past and the distant future. It won't, in general, have compact support, so we cannot evaluate a general element, $\mu$, of $\Gamma_0(V)^*$ on it. But if $\mu$ is "spatially compactly supported" we can evaluate $\mu$ on any section of $V$ which is "compactly supported in the time direction". In particular, $\mu(D_{w\xi}\alpha)$ makes sense. If, in addition, $\mu \in T(G\cdot\alpha)^0$ then it is easy to check that

(4.1) $\qquad \mu(D_{w\xi}\alpha)$ is independent of the choice of w.

This is to be interpreted as a conservation law. To any space like surface we can choose w so as to rapidly go from 0 to 1 from one side of the surface to the other. Then (4.1) says that the value of $\mu(D_{w\xi}\alpha)$ does not depend on the choice of w or on the surface.

We have thus defined a map

$$\Gamma(V)^*_{\text{sp.comp.}} \cap T(G\cdot\alpha)^0 \longrightarrow k^*$$

(4.2) $$\mu \longrightarrow \mu(D_{w\xi}\alpha) \ .$$

Now since $K$ preserves $\alpha$, its action on $Y$ preserves $\tau(\alpha)$ so we get an induced symplectic action of $K$ on $U$. It turns out that this action in Hamiltonian, i.e., has a moment map. Indeed comparing (4.2) with the map $F^*: \text{Dist}_0(U) \longrightarrow \Gamma(V)^*$ gives a map

$$Z: \ \text{Dist}_0(U) \longrightarrow k^* \ .$$

On the other hand, we have a map

$$\delta: \ U \longrightarrow \text{Dist}_0(U)$$

where $\delta(u) = \delta_u$ is the delta function concentrated at $u$.

466

Comparing, we get a map

$$\Phi \;=\; Z \circ \delta \;:\; U \longrightarrow k^*$$

and it turns out that

(4.3)                    $\Phi$  is a moment map for the K action on U.

This factorization of the moment map has far reaching applications
to symplectic geometry and thermodynamics, some of which are
sketched in [2].

## References


[1]  Souriau, J.M.  "Géométrie et Thermodynamique" in Diff.
     Geom. Methods in Math. Phys. Proc. Bonn 1977, Springer
     Lecture Notes in Math. 676 (1978), pp. 370-397.

[2]  Guillemin, V. and Sternberg, S.  "Souriau scattering and
     the Yang-Mills dust", Annals of Physics 1985

[3]  Bott, R. and Tu, L.  Differential forms in algebraic
     topology.  Springer 1983



Received January 18, 1985.